Several Complex Variables and Integral Formulas

Several Complex Variables and Integral Formulas

Kenzo Adachi

Nagasaki University, Japan

World Scientific

NEW JERSEY · LONDON · SINGAPORE · BEIJING · SHANGHAI · HONG KONG · TAIPEI · CHENNAI

Published by

World Scientific Publishing Co. Pte. Ltd.
5 Toh Tuck Link, Singapore 596224
USA office: 27 Warren Street, Suite 401-402, Hackensack, NJ 07601
UK office: 57 Shelton Street, Covent Garden, London WC2H 9HE

British Library Cataloguing-in-Publication Data
A catalogue record for this book is available from the British Library.

SEVERAL COMPLEX VARIABLES AND INTEGRAL FORMULAS

Copyright © 2007 by World Scientific Publishing Co. Pte. Ltd.

All rights reserved. This book, or parts thereof, may not be reproduced in any form or by any means, electronic or mechanical, including photocopying, recording or any information storage and retrieval system now known or to be invented, without written permission from the Publisher.

For photocopying of material in this volume, please pay a copying fee through the Copyright Clearance Center, Inc., 222 Rosewood Drive, Danvers, MA 01923, USA. In this case permission to photocopy is not required from the publisher.

ISBN-13 978-981-270-574-7
ISBN-10 981-270-574-0

Printed in Singapore.

To my family
Machiko,
Hidehiko and Yuko

Preface

The aim of this book is to study some important results obtained in the last 50 years in the function theory of several complex variables that are mainly concerned with the extension of holomorphic functions from submanifolds of pseudoconvex domains and estimates for solutions of the $\bar{\partial}$ problem in pseudoconvex domains.

This book is divided into five chapters.

In Chapter 1 we recall the elementary theory of functions of several complex variables. We prove that every domain of holomorphy is a pseudoconvex open set. Moreover, we give the proof of the Hartogs theorem which means that a separately analytic function is analytic.

In Chapter 2 we deal with L^2 estimates for the $\bar{\partial}$ problem in pseudoconvex domains in $\mathbf{C}^n$ due to Hörmander. As an application, we give the affirmative answer for the Levi problem. Moreover, we prove the Ohsawa-Takegoshi extension theorem by following the method of Jarnicki-Pflug.

In Chapter 3 we construct integral formulas for differential forms on bounded domains in $\mathbf{C}^n$ with smooth boundary, that is, the Bochner-Martinelli formula, the Koppelman formula, the Leray formula and the Koppelman-Leray formula are derived. Using the integral formula, we prove Hölder estimates for the $\bar{\partial}$ problem in strictly pseudoconvex domains with smooth boundary. Moreover, we prove bounded and continuous extensions of holomorphic functions from submanifolds of strictly pseudoconvex domains with smooth boundary which were proved by Henkin in 1972. We also prove H^p and C^k extensions. Finally, we prove Fefferman's mapping theorem by following the method of Range.

In Chapter 4 we discuss the Berndtsson-Andersson formula and the Berndtsson formula. As an application of the Berndtsson-Andersson formula, we give L^p estimates for solutions of the $\bar{\partial}$ problem in strictly pseudoconvex domains in $\mathbf{C}^n$ with smooth boundary. Using the Berndtsson formula, we give counterexamples of L^p ($p > 2$) extensions of holomorphic functions. Finally, we introduce an integral formula which was used by Diederich-Mazzilli to prove bounded extensions of holomorphic functions from affine linear submanifolds of a smooth convex domain of finite type.

Chapter 5 is devoted to the study of classical fundamental theorems in the function theory of several complex variables some of which are used to prove theorems in the previous chapters.

Appendix A is concerned with the compact operator theory in Banach spaces which is used to prove Fefferman's mapping theorem.

In Appendix B we give solutions to the Exercises.

I am grateful to Saburou Saitoh who suggested to me the publication of this book. I am also grateful to Heinrich GW Begehr who suggested that World Scientific might be interested in publishing this book.

I would like to express my sincere gratitude to Joji Kajiwara, Professor Emeritus at the Kyushu University, who introduced me to the function theory of several complex variables when I was a student at the Kyushu University, and to Morisuke Hasumi, Professor Emeritus at the Ibaraki University, who introduced me to the theory of function algebras when I was studying at the Ibaraki University.

Finally, I want to express my thanks to Ms Zhang Ji, Ms Kwong Lai Fun and the staff of World Scientific for their help and cooperation.

Kenzo Adachi

Contents

Preface   vii

1. Pseudoconvexity and Plurisubharmonicity   1
   - 1.1 The Hartogs Theorem 1
   - 1.2 Characterizations of Pseudoconvexity 19

2. The $\bar{\partial}$ Problem in Pseudoconvex Domains   47
   - 2.1 The Weighted L^2 Space 47
   - 2.2 L^2 Estimates in Pseudoconvex Domains 53
   - 2.3 The Ohsawa-Takegoshi Extension Theorem 100

3. Integral Formulas for Strictly Pseudoconvex Domains   117
   - 3.1 The Homotopy Formula 117
   - 3.2 Hölder Estimates for the $\bar{\partial}$ Problem 143
   - 3.3 Bounded and Continuous Extensions 152
   - 3.4 H^p and C^k Extensions 183
   - 3.5 The Bergman Kernel 196
   - 3.6 Fefferman's Mapping Theorem 210

4. Integral Formulas with Weight Factors   245
   - 4.1 The Berndtsson-Andersson Formula 245
   - 4.2 L^p Estimates for the $\bar{\partial}$ Problem 255
   - 4.3 The Berndtsson Formula 264
   - 4.4 Counterexamples for L^p $(p>2)$ Extensions 270
   - 4.5 Bounded Extensions by Means of the Berndtsson Formula . 281

5. The Classical Theory in Several Complex Variables   291

    5.1 The Poincaré Theorem 291
    5.2 The Weierstrass Preparation Theorem 298
    5.3 Oka's Fundamental Theorem 307
    5.4 The Cousin Problem 326

Appendix A   Compact Operators   331

Appendix B   Solutions to the Exercises   343

Bibliography   359

Index   365

Chapter 1

Pseudoconvexity and Plurisubharmonicity

In this chapter we study the properties of holomorphic functions of several complex variables and plurisubharmonic functions. We define the domain of holomorphy and the pseudoconvex open set, and we prove that every domain of holomorphy is pseudoconvex, but the converse (Levi's problem) is left to 2.2.

1.1 The Hartogs Theorem

Definition 1.1 Let $f = u + iv : \Omega \to \mathbf{C}$ be a C^1 function in an open set $\Omega \subset \mathbf{C}^n$. For $z_j = x_j + iy_j$, $j = 1, \cdots, n$, define

$$\frac{\partial f}{\partial z_j} = \frac{1}{2}\left(\frac{\partial f}{\partial x_j} + \frac{1}{i}\frac{\partial f}{\partial y_j}\right)$$
$$= \frac{1}{2}\left(\frac{\partial u}{\partial x_j} + \frac{\partial v}{\partial y_j}\right) + \frac{i}{2}\left(\frac{\partial v}{\partial x_j} - \frac{\partial u}{\partial y_j}\right)$$

and

$$\frac{\partial f}{\partial \bar{z}_j} = \frac{1}{2}\left(\frac{\partial f}{\partial x_j} - \frac{1}{i}\frac{\partial f}{\partial y_j}\right)$$
$$= \frac{1}{2}\left(\frac{\partial u}{\partial x_j} - \frac{\partial v}{\partial y_j}\right) + \frac{i}{2}\left(\frac{\partial v}{\partial x_j} + \frac{\partial u}{\partial y_j}\right).$$

By definition

$$\overline{\frac{\partial f}{\partial z_j}} = \frac{\partial \bar{f}}{\partial \bar{z}_j}, \quad \overline{\frac{\partial f}{\partial \bar{z}_j}} = \frac{\partial \bar{f}}{\partial z_j}.$$

Lemma 1.1 Let $\Omega \subset \mathbf{C}^n$ and $G \subset \mathbf{C}^m$ be open sets and let $f : \Omega \to G$ and $g : G \to \mathbf{C}$ be of class C^k for $k = 0, 1, \cdots, \infty$. Then, $g \circ f : \Omega \to \mathbf{C}$ is of class C^k. Moreover, if we write $f(z) = (f_1(z), \cdots, f_m(z))$, then

$$\frac{\partial}{\partial z_j}(g \circ f)(z) = \sum_{k=1}^{m} \left\{ \frac{\partial g}{\partial w_k}(f(z)) \frac{\partial f_k}{\partial z_j}(z) + \frac{\partial g}{\partial \bar{w}_k}(f(z)) \frac{\partial \bar{f}_k}{\partial z_j}(z) \right\} \quad (1.1)$$

and

$$\frac{\partial}{\partial \bar{z}_j}(g \circ f)(z) = \sum_{k=1}^{m} \left\{ \frac{\partial g}{\partial w_k}(f(z)) \frac{\partial f_k}{\partial \bar{z}_j}(z) + \frac{\partial g}{\partial \bar{w}_k}(f(z)) \frac{\partial \bar{f}_k}{\partial \bar{z}_j}(z) \right\}. \quad (1.2)$$

Proof. We prove (1.1) in case $n = m = k = 1$. Let $f(z) = \alpha(x, y) + i\beta(x, y)$ and $w = u + iv$. Then we have

$$\frac{\partial}{\partial z}(g \circ f)(z) = \frac{1}{2}\left(\frac{\partial}{\partial x} - i\frac{\partial}{\partial y}\right) g(\alpha(x,y), \beta(x,y))$$

$$= \frac{1}{2}\left(\frac{\partial g}{\partial u}\frac{\partial \alpha}{\partial x} + \frac{\partial g}{\partial v}\frac{\partial \beta}{\partial x}\right) + \frac{1}{2i}\left(\frac{\partial g}{\partial u}\frac{\partial \alpha}{\partial y} + \frac{\partial g}{\partial v}\frac{\partial \beta}{\partial y}\right).$$

Then (1.1) follows from the equalities

$$\frac{\partial}{\partial x} = \frac{\partial}{\partial z} + \frac{\partial}{\partial \bar{z}}, \quad \frac{\partial}{\partial y} = i\left(\frac{\partial}{\partial z} - \frac{\partial}{\partial \bar{z}}\right),$$

$$\frac{\partial}{\partial u} = \frac{\partial}{\partial w} + \frac{\partial}{\partial \bar{w}}, \quad \frac{\partial}{\partial v} = i\left(\frac{\partial}{\partial w} - \frac{\partial}{\partial \bar{w}}\right).$$

(1.2) is proved similarly. $\square$

Theorem 1.1 *Let Ω be a bounded open set in $\mathbf{C}$ and let $\partial \Omega$ consist of finite C^1 Jordan curves. For $u \in C^1(\bar{\Omega})$ and $z \in \Omega$, we have*

$$u(z) = \frac{1}{2\pi i}\left\{ \int_{\partial \Omega} \frac{u(\zeta)}{\zeta - z} d\zeta + \iint_{\Omega} \frac{\frac{\partial u}{\partial \bar{z}}(\zeta)}{\zeta - z} d\zeta \wedge d\bar{\zeta} \right\}.$$

Proof. We fix $z \in \Omega$. For $\zeta \in \bar{\Omega} \setminus \{z\}$ we have

$$d_\zeta\left[\frac{u(\zeta) d\zeta}{\zeta - z}\right] = \frac{\bar{\partial} u(\zeta) \wedge d\zeta}{\zeta - z}.$$

For any sufficiently small $\varepsilon > 0$, we set $\Omega_\varepsilon = \{\zeta \in \Omega \mid |\zeta - z| > \varepsilon\}$. It follows from Stokes' theorem that

$$\frac{1}{2\pi i}\int_{|\zeta - z| = \varepsilon} \frac{u(\zeta) d\zeta}{\zeta - z} = \frac{1}{2\pi i}\int_{\zeta \in \partial \Omega} \frac{u(\zeta) d\zeta}{\zeta - z} - \frac{1}{2\pi i}\iint_{\Omega_\varepsilon} \frac{\bar{\partial} u(\zeta) \wedge d\zeta}{\zeta - z}.$$

We have the desired equality by letting $\varepsilon \to 0$. $\square$

Definition 1.2 Let Ω be an open set in $\mathbf{R}^n$. We denote by $\mathcal{D}(\Omega)$ (or $C_c^\infty(\Omega)$) the set of all C^∞ functions f in Ω whose support $\mathrm{supp}(f)$ is a compact subset of Ω.

Theorem 1.2 *Let Ω be a bounded open set in the complex plane and let $K \subset \Omega$ be compact. Then for any open set ω in Ω satisfying $K \subset \omega$, there exist constants C_j, $j = 0, 1, \cdots$, such that*

$$\sup_{z \in K} |f^{(j)}(z)| \leq C_j \|f\|_{L^1(\omega)}$$

for every holomorphic function f in Ω.

Proof. Let K' be a compact set such that $K \subset K'^\circ \subset K' \subset \omega$. Choose a function $\psi \in C_c^\infty(\omega)$ with the properties that $\psi = 1$ in K'. By Theorem 1.1, we have

$$(\psi f)(z) = \frac{1}{2\pi i} \iint_\omega \frac{\partial(\psi f)}{\partial \bar\zeta}(\zeta) \frac{1}{\zeta - z} d\zeta \wedge d\bar\zeta$$

$$= \frac{1}{2\pi i} \iint_\omega \frac{\partial \psi}{\partial \bar\zeta}(\zeta) \frac{f(\zeta)}{\zeta - z} d\zeta \wedge d\bar\zeta.$$

Since $\frac{\partial \psi}{\partial \bar\zeta} = 0$ in K', we have

$$f(z) = \frac{1}{2\pi i} \iint_{\omega \setminus K'} \frac{\partial \psi}{\partial \bar\zeta}(\zeta) \frac{f(\zeta)}{\zeta - z} d\zeta \wedge d\bar\zeta$$

for $z \in K$. By differentiating j times with respect to z, we obtain

$$f^{(j)}(z) = \frac{j!}{2\pi i} \iint_{\omega \setminus K'} \frac{\partial \psi}{\partial \bar\zeta}(\zeta) \frac{f(\zeta)}{(\zeta - z)^{k+1}} d\zeta \wedge d\bar\zeta.$$

If $z \in K, \zeta \in \omega \setminus K'$, then there exists a constant $C > 0$ such that $|z - \zeta| \geq C$. Hence there exists a constant $C_1 > 0$ such that

$$|f^{(j)}(z)| \leq C_1 \iint_{\omega \setminus K'} |f(\zeta)| dx dy \qquad (\zeta = x + iy),$$

which gives the desired inequality. $\square$

Definition 1.3 Let Ω be an open set in $\mathbf{C}$. Then $u : \Omega \to \mathbf{R} \cup \{-\infty\}$ is called subharmonic in Ω if

(1) u is upper semicontinuous in Ω, that is, $\{z \in \Omega \mid u(z) < s\}$ is an open set for any real number s.

(2) For any compact set $K \subset \Omega$ and any continuous function h on K which is harmonic in the interior of K, u satisfies the following properties:
$$u(z) \leq h(z) \quad (z \in \partial K) \implies u(z) \leq h(z) \quad (z \in K).$$

Definition 1.4 Let $u : \Omega \to \mathbf{R}$ be a C^2 function in an open set $\Omega \subset \mathbf{C}$. We say that u is strictly subharmonic in Ω if
$$\frac{\partial^2 u}{\partial z \partial \bar{z}}(z) > 0 \quad (z \in \Omega).$$

Theorem 1.3 *Let Ω be an open set in $\mathbf{C}$. Then a real-valued function $u \in C^2(\Omega)$ is subharmonic in Ω if and only if*
$$\frac{\partial^2 u}{\partial z \partial \bar{z}}(z) \geq 0 \quad (z \in \Omega).$$

Proof. Let $a = \alpha + i\beta \in \Omega$. For r with $0 < r < \text{dist}(a, \partial\Omega)$, define
$$A(r) = \frac{1}{2\pi} \int_0^{2\pi} u(a + re^{i\theta}) d\theta.$$

Then we have
$$\begin{aligned}
\frac{dA(r)}{dr} &= \frac{1}{2\pi} \int_0^{2\pi} \frac{d}{dr} u(a + re^{i\theta}) d\theta \\
&= \frac{1}{2\pi} \int_0^{2\pi} \left\{ \frac{\partial u}{\partial x}(\alpha + r\cos\theta, \beta + r\sin\theta) \cos\theta \right. \\
&\quad \left. + \frac{\partial u}{\partial y}(\alpha + r\cos\theta, \beta + r\sin\theta) \sin\theta \right\} d\theta \\
&= \frac{1}{2\pi r} \int_{|z-a|=r} \left(\frac{\partial u}{\partial x} dy - \frac{\partial u}{\partial y} dx \right) \\
&= \frac{1}{2\pi r} \iint_{|z-a| \leq r} \left(\frac{\partial^2 u}{\partial x^2} + \frac{\partial^2 u}{\partial y^2} \right) dx dy \\
&= \frac{2}{\pi r} \iint_{|z-a| \leq r} \frac{\partial^2 u}{\partial z \partial \bar{z}} dx dy.
\end{aligned}$$

If $\frac{\partial^2 u}{\partial z \partial \bar{z}}(z) \geq 0$, then $\frac{dA(r)}{dr} \geq 0$. Hence $A(r)$ is monotonically increasing. Therefore, we obtain
$$u(a) = A(0) \leq A(r) = \frac{1}{2\pi} \int_0^{2\pi} u(a + re^{i\theta}) d\theta,$$

which means that u is subharmonic. Conversely, suppose u is subharmonic. Suppose there exists a point a satisfying $\frac{\partial^2 u}{\partial z \partial \bar{z}}(a) < 0$. For any sufficiently small $r > 0$, if $|z - a| \leq r$, then $\frac{\partial^2 u}{\partial z \partial \bar{z}}(z) < 0$, which implies that

$$\frac{dA(r)}{dr} = \frac{2}{\pi r} \iint_{|z-a| \leq r} \frac{\partial^2 u}{\partial z \partial \bar{z}} dx dy < 0.$$

Since $A(r)$ is strictly monotonically decreasing, we have

$$u(a) > \frac{1}{2\pi} \int_0^{2\pi} u(a + re^{i\theta}) d\theta,$$

which is a contradiction. Thus, we have $\frac{\partial^2 u}{\partial z \partial \bar{z}}(z) \geq 0$. □

Definition 1.5 For $a \in \mathbf{C}$ and $r > 0$, define

$$B(a, r) := \{z \in \mathbf{C} \mid |z - a| < r\}.$$

The closure of $B(a, r)$ is denoted by $\overline{B}(a, r)$.

Theorem 1.4 *Let u be a continuous real-valued function on $\partial B(0, R)$. For $z = re^{i\theta} \in B(0, R)$, define*

$$U(z) = \frac{1}{2\pi} \int_0^{2\pi} u(Re^{i\varphi}) \frac{R^2 - r^2}{R^2 - 2Rr\cos(\varphi - \theta) + r^2} d\varphi. \quad (1.3)$$

Then U is harmonic in $B(0, R)$. Moreover, if we define $U(z) = u(z)$ for $z \in \partial B(0, R)$, then U is continuous in $\overline{B}(0, R)$. The right side of (1.3) is called the Poisson integral.

Proof. For $|z| < R$, define

$$f(z) = \frac{1}{2\pi} \int_0^{2\pi} u(Re^{i\varphi}) \frac{Re^{i\varphi} + z}{Re^{i\varphi} - z} d\varphi.$$

For $\zeta = Re^{i\varphi}$, we have

$$\frac{Re^{i\varphi} + z}{Re^{i\varphi} - z} = 1 + 2 \sum_{n=1}^{\infty} \left(\frac{z}{\zeta}\right)^n.$$

Since the right side of the above equality converges uniformly on $|\zeta| = R$, we obtain

$$f(z) = \frac{1}{2\pi} \int_0^{2\pi} u(Re^{i\varphi}) d\varphi + \frac{1}{\pi} \sum_{n=1}^{\infty} \left\{ \int_0^{2\pi} \frac{u(Re^{i\varphi})}{(Re^{i\varphi})^n} d\varphi \right\} z^n.$$

Therefore, f is holomorphic in $B(0,R)$. On the other hand we have

$$\begin{aligned}
\operatorname{Re} f(z) &= \frac{1}{2\pi} \int_0^{2\pi} u(Re^{i\varphi}) \operatorname{Re}\left(\frac{Re^{i\varphi}+z}{Re^{i\varphi}-z}\right) d\varphi \\
&= \frac{1}{2\pi} \int_0^{2\pi} u(Re^{i\varphi}) \frac{R^2-r^2}{R^2-2Rr\cos(\varphi-\theta)+r^2} d\varphi \\
&= U(z).
\end{aligned}$$

Hence, U is harmonic in $B(0,R)$. Next we fix a point $\zeta_0 = Re^{i\varphi_0}$. For $\varepsilon > 0$, there exists $\delta > 0$ such that if $\zeta = Re^{\varphi}$, $|\varphi - \varphi_0| < \delta$, then

$$|u(\zeta) - u(\zeta_0)| < \varepsilon.$$

We can choose $\rho > 0$ so small that if $|z| < R$ and $|z - \zeta_0| < \rho$, then there exists a constant $c > 0$ such that

$$R^2 - r^2 < \varepsilon \delta^2, \quad |Re^{i\varphi} - z| > c\delta \ (\varphi_0 + \delta \leq \varphi \leq \varphi_0 - \delta + 2\pi).$$

We set

$$M = \max_{|z|=R} |u(z)|.$$

Then we have

$$\begin{aligned}
|U(z) - U(\zeta_0)| &= |U(z) - u(\zeta_0)| \\
&= \left|\frac{1}{2\pi} \int_0^{2\pi} (u(Re^{i\varphi}) - u(\zeta_0)) \operatorname{Re}\left(\frac{Re^{i\varphi}+z}{Re^{i\varphi}-z}\right) d\varphi\right| \\
&= \left|\frac{1}{2\pi} \int_0^{2\pi} (u(Re^{i\varphi}) - u(\zeta_0)) \frac{R^2-r^2}{|Re^{i\varphi}-z|^2} d\varphi\right| \\
&\leq \left|\frac{1}{2\pi} \int_{\varphi_0-\delta}^{\varphi_0+\delta} (u(Re^{i\varphi}) - u(\zeta_0)) \frac{R^2-r^2}{|Re^{i\varphi}-z|^2} d\varphi\right| \\
&\quad + \left|\frac{1}{2\pi} \int_{\varphi_0+\delta}^{\varphi_0-\delta+2\pi} (u(Re^{i\varphi}) - u(\zeta_0)) \frac{R^2-r^2}{|Re^{i\varphi}-z|^2} d\varphi\right| \\
&\leq \frac{1}{2\pi} \int_{\varphi_0-\delta}^{\varphi_0+\delta} \varepsilon \frac{R^2-r^2}{|Re^{i\varphi}-z|^2} d\varphi + \frac{M}{\pi} \int_{\varphi_0+\delta}^{\varphi_0-\delta+2\pi} \frac{\varepsilon\delta^2}{(c\delta)^2} d\varphi \\
&\leq \varepsilon + \frac{2M\varepsilon}{c^2} = \varepsilon\left(1 + \frac{2M}{c^2}\right).
\end{aligned}$$

Hence, U is continuous in $\overline{B}(0,R)$. □

Lemma 1.2 *Let $\Omega \subset \mathbb{C}$ be an open set and let u be a continuous subharmonic function in Ω, $a \in \Omega$. For r with $0 < r < \text{dist}(a, \partial \Omega)$, define*

$$A(r) = \frac{1}{2\pi} \int_0^{2\pi} u(a + re^{i\theta}) d\theta.$$

Then $A(r)$ satisfies the following:

$$0 < r_1 < r_2 < \text{dist}(a, \partial \Omega) \implies A(r_1) \leq A(r_2).$$

Proof. Let $0 < r < \text{dist}(a, \partial \Omega)$. We denote by φ_r the Poisson integral of u. Then

$$\varphi_r(z) = \frac{1}{2\pi} \int_0^{2\pi} \frac{r^2 - |z-a|^2}{|(z-a) - re^{i\theta}|^2} u(a + re^{i\theta}) d\theta.$$

Moreover, φ_r is harmonic in $B(a, r)$, continuous in $\overline{B}(a, r)$ and $\varphi_r = u$ on $\partial B(a, r)$. Then $u(z) - \varphi_r(z)$ is subharmonic in $B(a, r)$, and equals 0 on $\partial B(a, r)$. By the maximum principle, $u(z) \leq \varphi_r(z)$ for $z \in B(a, r)$. Therefore we have

$$\frac{1}{2\pi} \int_0^{2\pi} u(a + r_1 e^{i\theta}) d\theta \leq \frac{1}{2\pi} \int_0^{2\pi} \varphi_{r_2}(a + r_1 e^{i\theta}) d\theta$$

$$= \frac{1}{2\pi} \int_0^{2\pi} \varphi_{r_2}(a + r_2 e^{i\theta}) d\theta$$

$$= \frac{1}{2\pi} \int_0^{2\pi} u(a + r_2 e^{i\theta}) d\theta.$$

$\square$

Theorem 1.5 *Let $u : \Omega \to \mathbb{R}$ be a continuous real-valued function in an open set $\Omega \subset \mathbb{C}$. Then the following statements are equivalent:*

(a) *u is harmonic in Ω.*
(b) *For any $a \in \Omega$ and any r with $0 < r < \text{dist}(a, \partial \Omega)$, one has*

$$u(a) \leq \frac{1}{2\pi} \int_0^{2\pi} u(a + re^{i\theta}) d\theta.$$

(c) *For any $a \in \Omega$, there exists ε ($0 < \varepsilon < \text{dist}(a, \partial \Omega)$) such that for any r with $0 < r < \varepsilon$ one has*

$$u(a) \leq \frac{1}{2\pi} \int_0^{2\pi} u(a + re^{i\theta}) d\theta.$$

(d) For any $a \in \Omega$ and any r with $0 < r < \text{dist}(a, \partial\Omega)$, if h is continuous in $|\zeta - a| \leq r$, and harmonic in $|\zeta - a| < r$, then u satisfies the following properties:

$$u(\zeta) \leq h(\zeta) \quad \text{for} \quad |\zeta - a| = r \Longrightarrow u(\zeta) \leq h(\zeta) \quad \text{for} \quad |\zeta - a| \leq r.$$

Proof. (b)$\Longrightarrow$(c) and (a)$\Longrightarrow$(d) are trivial. We show that (d)$\Longrightarrow$(b). For $a \in \Omega$, we choose $r > 0$ such that $\overline{B}(a,r) \subset \Omega$. We denote by U the Poisson integral of u for $B(a,r)$. Then U is harmonic in $B(a,r)$, continuous in $\overline{B}(a,r)$, and $U(z) = u(z)$ for $z \in \partial B(a,r)$. Since $u \leq U$ in $B(a,r)$, we have

$$u(a) \leq U(a) = \frac{1}{2\pi}\int_0^{2\pi} U(a + re^{i\theta})d\theta = \frac{1}{2\pi}\int_0^{2\pi} u(a + re^{i\theta})d\theta.$$

This proves (b). Next we show that (c)$\Longrightarrow$(a). Let $K \subset \Omega$ be a compact set. Suppose h is harmonic in the interior of K that is continuous on K, and satisfies $u \leq h$ on ∂K. We set

$$c = \max_{z \in K}(u(z) - h(z)).$$

Suppose $c > 0$. We set

$$K_c = \{z \in K \mid u(z) - h(z) = c\}.$$

Then K_c is compact. We denote by a the nearest point of K_c to ∂K. If we choose $r > 0$ sufficiently small, then we have

$$\frac{1}{2\pi}\int_0^{2\pi}\{u(a + re^{i\theta}) - h(a + re^{i\theta})\}d\theta < \frac{1}{2\pi}\int_0^{2\pi} c\,d\theta = c.$$

On the other hand, if we choose $r > 0$ sufficiently small, then it follows from (c) that

$$c = u(a) - h(a) \leq \frac{1}{2\pi}\int_0^{2\pi}\{u(a + re^{i\theta}) - h(a + re^{i\theta})\}d\theta,$$

which is a contradiction. Thus we have $c = 0$, which implies that $u \leq h$ on K. This proves (a). $\square$

In order to prove the Hartogs theorem we need the following lemma (see Krantz [KR2]).

Lemma 1.3 *Let $\Omega \subset \mathbf{C}$ be an open set and let $f : \Omega \to \mathbf{R} \cup \{-\infty\}$ be upper semicontinuous and bounded above. Then there exists a sequence*

$\{f_j\}$ of real-valued continuous functions in Ω which are bounded above on Ω such that
$$f_1 \geq f_2 \geq \cdots, \quad f_j \to f.$$

Proof. In the case when $f(x) \equiv -\infty$, we may set $f_n(x) = -n$. Thus we may assume that $f(x) \not\equiv -\infty$. For $x \in \Omega$, we set
$$f_j(x) = \sup_{y \in \Omega}\{f(y) - j|x-y|\}.$$
Then $f_1(x) \geq f_2(x) \geq \cdots \geq f(x)$. For $\varepsilon > 0$, if $x_1, x_2 \in \Omega$, $|x_1 - x_2| < \varepsilon/j$, then
$$f(y) - j|x_1 - y| < f(y) - j|x_2 - y| + \varepsilon \quad (y \in \Omega).$$
Therefore we have $f_j(x_1) \leq f_j(x_2) + \varepsilon$. By interchanging x_1 and x_2 we have $|f_j(x_1) - f_j(x_2)| \leq \varepsilon$. Thus each f_j is continuous in Ω. We set
$$\sup_{x \in \Omega} f(x) = M, \quad f(x) = \alpha \quad (\alpha \neq -\infty).$$
For $\varepsilon > 0$, there exists $\delta > 0$ such that if $|x - y| < \delta$, then $f(y) < \alpha - \varepsilon$ since f is upper semicontinuous. If $|y - x| > \delta$, $j > M/\delta$, then $f(y) - j|x-y| \leq M - M = 0$. Thus we have
$$\alpha = f(x) \leq f_j(x) = \sup_{|y-x| \leq \delta}\{f(y) - j|x-y|\} < \alpha + \varepsilon \quad \left(j > \frac{M}{\delta}\right),$$
which shows that $f_j(x) \to f(x)$. Suppose $\alpha = -\infty$. For $N > 0$, there exists $\delta_1 > 0$ such that $f(y) < -N$ whenever $|x - y| < \delta_1$. Hence we have
$$f_j(x) = \max\left[\sup_{|x-y|<\delta_1}\{f(y) - j|x-y|\}, \sup_{|x-y|\geq\delta_1}\{f(y) - j|x-y|\}\right]$$
$$\leq \max\{-N, M - j\delta_1\}.$$
If we choose j sufficiently large, then $j\delta_1 > N + M$. Thus $f_j(x) \to -\infty = f(x)$. □

Corollary 1.1 *For an upper semicontinuous function u in an open set $\Omega \subset \mathbf{C}$, Theorem 1.5 also holds.*

Proof. We show that (d)$\Longrightarrow$(b). Let $a \in \Omega$ and $0 < r < r' < \text{dist}(a, \partial\Omega)$. By Lemma 1.3, there exists a sequence $\{u_j\}$ of continuous functions in $B(a, r')$ such that
$$u_1 \geq u_2 \geq \cdots, \quad u_j \to u.$$

We denote by U_j the Poisson integral of u_j for $B(a,r)$. Then U_j is harmonic in $B(a,r)$, continuous in $\overline{B}(a,r)$, and satisfies $U_j = u_j$ on $\partial B(a,r)$. Thus we have $u(z) \leq U_j(z)$ for $z \in B(a,r)$. Therefore we obtain

$$u(a) \leq U_j(a) = \frac{1}{2\pi}\int_0^{2\pi} U_j(a+re^{i\theta})d\theta = \frac{1}{2\pi}\int_0^{2\pi} u_j(a+re^{i\theta})d\theta.$$

By letting $j \to \infty$, (b) follows from the Fatou lemma. The proof of (c)$\Longrightarrow$(a) is proved in the same way. $\square$

Definition 1.6 For $r_j > 0$ $(j=1,\cdots,n)$, we set $r = (r_1,\cdots,r_n)$. For $a \in \mathbf{C}^n$, define

$$P(a,r) = \{z = (z_1,\cdots,z_n) \mid |z_j - a_j| < r_j, j = 1,\cdots,n\}.$$

If we set

$$P_j = \{z_j \in \mathbf{C} \mid |z_j - a_j| < r_j\},$$

then

$$P(a,r) = P_1 \times \cdots \times P_n.$$

$P(a,r)$ is called a polydisc. When $n=1$, we have $P(a,r) = B(a,r)$.

Definition 1.7 A power series of n variables is denoted by

$$\sum_\nu c_\nu (z-a)^\nu = \sum_{\nu_1=0,\cdots,\nu_n=0}^\infty c_{\nu_1,\cdots,\nu_n}(z_1-a_1)^{\nu_1}\cdots(z_n-a_n)^{\nu_n}. \quad (1.4)$$

The domain of convergence of the power series (1.4) is the interior of the set of points $z \in \mathbf{C}^n$ for which (1.4) converges.

Theorem 1.6 *Every power series converges uniformly on every compact subset of its domain of convergence.*

Proof. Let Ω be the domain of convergence of (1.4). For simplicity, we may assume that $a = (a_1,\cdots,a_n) = 0$. Let $w \in \Omega$. Then we have

$$\sup_\nu |c_\nu w^\nu| = M < \infty.$$

We set $r = (|w_1|,\cdots,|w_n|)$. Let $K \subset P(0,r)$ be a compact set. Then there exists $0 < \lambda < 1$ such that $K \subset P(0,\lambda r)$. If $z \in P(0,\lambda r)$, then we have

$$|c_\nu z^\nu| = |c_\nu z_1^{\nu_1}\cdots z_n^{\nu_n}| \leq |c_\nu|(\lambda r_1)^{\nu_1}\cdots(\lambda r_n)^{\nu_n} = |c_\nu w^\nu|\lambda^{|\nu|} \leq M\lambda^{|\nu|}.$$

On the other hand we have

$$\sum_\nu \lambda^{|\nu|} = \left(\frac{1}{1-\lambda}\right)^n < \infty.$$

Thus $\sum_\nu c_\nu z^\nu$ converges uniformly on K. Let E be any compact subset of Ω. Then there exists $w^1, \cdots, w^k \in \Omega$ such that for each w^j, compact subsets K_j of polydiscs $P(0, \lambda r^j)$ constructed above satisfy

$$E \subset \bigcup_{j=1}^k K_j.$$

Since (1.4) converges uniformly on each K_j, (1.4) converges uniformly on E. □

Definition 1.8 Let $\Omega \subset \mathbf{C}^n$ be an open set. $G \subset\subset \Omega$ means that the closure of G in $\mathbf{C}^n$ is a compact subset of Ω. In this case, G is called relatively compact in Ω.

Definition 1.9 Let $\Omega \subset \mathbf{C}^n$ be an open set. A function $f : \Omega \to \mathbf{C}$ is called holomorphic in Ω if f is continuous in Ω, and for each $a = (a_1, \cdots, a_n) \in \Omega$, if we set $\varphi(z_j) = f(a_1, \cdots, z_j, \cdots, a_n)$, then $\varphi(z_j)$ is holomorphic at a_j. The set of all holomorphic functions in Ω is denoted by $\mathcal{O}(\Omega)$.

Theorem 1.7 For a function $f : \Omega \to \mathbf{C}$ on an open set $\Omega \subset \mathbf{C}^n$, the following statements are equivalent:

(a) f is holomorphic in Ω.
(b) Suppose f is continuous in Ω and $P = P_1 \times \cdots \times P_n \subset\subset \Omega$. Then

$$f(z) = \frac{1}{(2\pi i)^n} \int_{\partial P_1} \cdots \int_{\partial P_n} \frac{f(\zeta) d\zeta_1 \cdots d\zeta_n}{(\zeta_1 - z_1) \cdots (\zeta_n - z_n)} \quad (1.5)$$

for $z \in P$.
(c) For any $\xi \in \Omega$, there exists a neighborhood W of ξ such that

$$f(z) = \sum_{k_1=0,\cdots,k_n=0}^{\infty} a_{k_1,\cdots,k_n} (z_1 - \xi_1)^{k_1} \cdots (z_n - \xi_n)^{k_n} \quad (1.6)$$

for $z \in W$.

Proof. (a)$\Longrightarrow$(b). By iterating the Cauchy integral formula, for $z \in P$

$$f(z) = \frac{1}{2\pi i} \int_{\partial P_1} \frac{f(\zeta_1, z_2, \cdots, z_n)}{\zeta_1 - z_1} d\zeta_1$$

$$= \frac{1}{2\pi i} \int_{\partial P_1} \left\{ \frac{1}{\zeta_1 - z_1} \frac{1}{2\pi i} \int_{\partial P_2} \frac{f(\zeta_1, \zeta_2, z_3 \cdots, z_n)}{\zeta_2 - z_2} d\zeta_2 \right\} d\zeta_1$$

$$= \frac{1}{(2\pi i)^n} \int_{\partial P_1} \int_{\partial P_2} \cdots \int_{\partial P_n} \frac{f(\zeta_1, \cdots, \zeta_n)}{(\zeta_1 - z_1) \cdots (\zeta_n - z_n)} d\zeta_1 d\zeta_2 \cdots d\zeta_n.$$

(b) $\Longrightarrow$ (c). Let $\xi \in \Omega$. We choose $r = (r_1, \cdots, r_n)$ such that

$$P = P(\xi, r) = P_1 \times \cdots \times P_n = \{z \mid |z_j - \xi_j| < r_j \ (j = 1, \cdots, n)\} \subset\subset \Omega.$$

For $\zeta = (\zeta_1, \cdots, \zeta_n) \in \partial P_1 \times \cdots \times \partial P_n$, since

$$\frac{1}{\zeta_1 - z_1} = \frac{1}{(\zeta_1 - \xi_1)(1 - \frac{z_1 - \xi_1}{\zeta_1 - \xi_1})} = \sum_{k=0}^{\infty} \frac{(z_1 - \xi_1)^k}{(\zeta_1 - \xi_1)^{k+1}},$$

we obtain

$$\frac{1}{(\zeta_1 - z_1) \cdots (\zeta_n - z_n)} = \sum_{k_1, \cdots k_n = 0}^{\infty} \frac{(z_1 - \xi_1)^{k_1} \cdots (z_n - \xi_n)^{k_n}}{(\zeta_1 - \xi_1)^{k_1 + 1} \cdots (\zeta_n - \xi_n)^{k_n + 1}}.$$

Since the power series of the right side of the above equality converges uniformly with respect to ζ, substituting into (1.5) and integrating, we obtain

$$f(z) = \sum_{k_1, \cdots, k_n = 0}^{\infty} \frac{1}{(2\pi i)^n} \int_{\partial P_1 \times \cdots \times \partial P_n} \frac{f(\zeta_1, \cdots, \zeta_n)}{(\zeta_1 - \xi_1)^{k_1 + 1} \cdots (\zeta_n - \xi_n)^{k_n + 1}}$$
$$\times d\zeta_1 \cdots d\zeta_n \cdot (z_1 - \xi_1)^{k_1} \cdots (z_n - \xi_n)^{k_n}.$$

We set

$$a_{k_1, \cdots, k_n} = \frac{1}{(2\pi i)^n} \int_{\partial P_1 \times \cdots \times \partial P_n} \frac{f(\zeta_1, \cdots, \zeta_n)}{(\zeta_1 - \xi_1)^{k_1 + 1} \cdots (\zeta_n - \xi_n)^{k_n + 1}} d\zeta_1 \cdots d\zeta_n.$$

Then f is expressed by

$$f(z) = \sum_{k_1, \cdots, k_n = 0}^{\infty} a_{k_1, \cdots, k_n} (z_1 - \xi_1)^{k_1} \cdots (z_n - \xi_n)^{k_n}.$$

This proves (c).

(c)$\Longrightarrow$(a). We choose $r > 0$ such that

$$\{z \mid |z_j - \xi_j| \leq r, j = 1, \cdots, n\} \subset W.$$

By Theorem 1.6, the right side of (1.6) converges uniformly on $P(\xi,r)$. Therefore f is continuous in $P(\xi,r)$. On the other hand, the finite sum

$$\sum_{k_1=0}^{N_1} \cdots \sum_{k_n=0}^{N_n} a_{k_1,\cdots,k_n}(z_1-\xi_1)^{k_1}\cdots(z_n-\xi_n)^{k_n}$$

is holomorphic in each variable z_j and f is the uniform limit of the above finite sum when $N_j \to \infty$. Thus f is holomorphic in $P(\xi,r)$ with respect to each variable z_j. Since $\xi \in \Omega$ is arbitrary, f is holomorphic in Ω. $\square$

Definition 1.10 Let $\Omega \subset \mathbf{C}^n$ be an open set and let f be holomorphic in Ω. For a multi-index $\alpha = (\alpha_1,\cdots,\alpha_n)$, where each α_j is a nonnegative integer, define

$$|\alpha| = \alpha_1 + \cdots + \alpha_n, \quad \alpha! = \alpha_1!\cdots\alpha_n!,$$

$$\partial^\alpha f(z) = \frac{\partial^{|\alpha|}f}{\partial z_1^{\alpha_1}\cdots\partial z_n^{\alpha_n}}(z) \quad (z \in \Omega).$$

Corollary 1.2 *(Cauchy inequality) Let f be a holomorphic function in a polydisc $P(0,r) = \{z \in \mathbf{C}^n \mid |z_j| < r_j, j = 1,\cdots,n\}$. Suppose there exists a constant $M > 0$ such that $|f(z)| \leq M$ for $z \in P(0,r)$. Then*

$$|\partial^\alpha f(0)| \leq \alpha! r^{-\alpha} M$$

for any multi-index $\alpha = (\alpha_1,\cdots,\alpha_n)$, where we define

$$r^\alpha = r_1^{\alpha_1}\cdots r_n^{\alpha_n}.$$

Proof. By Theorem 1.7 f is expressed by $f(z) = \sum_\alpha a_\alpha z^\alpha$. Then it follows from (1.6) that

$$a_\alpha = \frac{\partial^\alpha f(0)}{\alpha!}.$$

On the other hand, by applying the proof of Theorem 1.7, for $0 < s_j < r_j$, $j = 1,\cdots,n$, we have

$$a_\alpha = \frac{1}{(2\pi i)^n} \int_{\{|z_1|=s_1\}\times\cdots\times\{|z_n|=s_n\}} \frac{f(\zeta_1,\cdots,\zeta_n)}{\zeta_1^{\alpha_1+1}\cdots\zeta_n^{\alpha_n+1}} d\zeta_1 \cdots d\zeta_n.$$

Thus we obtain

$$|\partial^\alpha f(0)| \leq \alpha! s^{-\alpha} M.$$

The above inequality holds for any $s > 0$ which satisfies $s_j < r_j$ for $j = 1, \cdots, n$. □

Corollary 1.3 *Let $\Omega \subset \mathbf{C}^n$ be an open set and let $K \subset \Omega$ be a compact set. For any open subset ω of Ω with $K \subset \omega$, there exists a constant C_α such that*

$$\sup_{z \in K} |\partial^\alpha f(z)| \leq C_\alpha \|u\|_{L^1(\omega)} \qquad (f \in \mathcal{O}(\Omega)).$$

Proof. In the case when ω is a polydisc, Corollary 1.3 follows from Theorem 1.2. In the general case, K is contained in the finite union of polydiscs which are contained in ω. Corollary 1.3 follows from Theorem 1.2. □

Lemma 1.4 *(Baire's theorem) Let X be a complete metric space. Then a countable intersection of open dense subsets of X is dense in X.*

Proof. Suppose $\{V_n\}$ is a sequence of open dense subsets of X and W is an open nonempty subset of X. It is sufficient to show that $\cap_{n=1}^\infty V_n \cap W$ is not empty. Let $d(x, y)$ be the metric in X. We set

$$B(x, r) = \{y \in X \mid d(x, y) < r\}.$$

Since $V_1 \cap W \neq \phi$, there exist x_1 and r_1 such that

$$\overline{B}(x_1, r_1) \subset W \cap V_1, \quad 0 < r_1 < 1.$$

Since $V_2 \cap B(x_1, r_1)$ is not empty, there exist x_2 and r_2 such that

$$\overline{B}(x_2, r_2) \subset V_2 \cap B(x_1, r_1), \quad 0 < r_2 < \frac{1}{2}.$$

Repeating this process, there exist x_n and r_n such that

$$\overline{B}(x_n, r_n) \subset V_n \cap B(x_{n-1}, r_{n-1}), \quad 0 < r_n < \frac{1}{n}.$$

Let $i > n$, $j > n$. Then $x_i, x_j \in B(x_n, r_n)$, so that $d(x_i, x_j) < 2r_n < 2/n$, which implies that $\{x_n\}$ is a Cauchy sequence. Since X is complete, $\{x_n\}$ converges. Therefore, there exists $x \in X$ such that $x = \lim_{n \to \infty} x_n$. Since $x_i \in \overline{B}(x_n, r_n)$ $(i > n)$ for each n, $x \in \overline{B}(x_n, r_n) \subset V_n$. Thus we have $x \in \cap_{n=1}^\infty V_n$. On the other hand, $x \in \overline{B}(x_1, r_1) \subset W$, which shows that $\cap_{n=1}^\infty V_n$ is dense in X. □

Lemma 1.5 *Suppose $\{v_k\}$ is a sequence of subharmonic functions in Ω. Assume that v_k, $k = 1, 2, \cdots$, are uniformly bounded from above on every compact subset of Ω and that there exists a constant C such that for any*

$z \in \Omega$, $\overline{\lim}_{k \to \infty} v_k(z) \leq C$. Then for any $\varepsilon > 0$ and any compact subset K of Ω there exists a positive integer N such that for $k > N$

$$v_k(z) \leq C + \varepsilon \qquad (z \in K).$$

Proof. We choose a compact set K_1 with the properties that $K \subset K_1^\circ \subset K_1 \subset \Omega$. Since $\{v_k\}$ is uniformly bounded from above on K_1°, we may assume that $\{v_k\}$ is uniformly bounded from above on Ω. Moreover, when $v_k(z) \leq M$ for $z \in \Omega$ and $M \leq 0$, we can treat $v_k - M$ instead of v_k, so we may assume that $v_k(z) \leq 0$ for $z \in \Omega$. We choose $r > 0$ such that $K \subset \{z \in \Omega \mid \text{dist}(z, \partial\Omega) > 3r\}$. By Corollary 1.1, for $z \in K$, $0 < \rho \leq r$, we have

$$2\pi v_k(z) \leq \int_0^{2\pi} v_k(z + \rho e^{i\theta}) d\theta. \tag{1.7}$$

If we multiply by ρ in (1.7) and integrate with respect to ρ from 0 to r, then we obtain

$$\pi r^2 v_k(z) \leq \iint_{|z-z'|<r} v_k(z') dx' dy' \qquad (z' = x' + iy'). \tag{1.8}$$

It follows from Fatou's lemma that

$$\overline{\lim_{k \to \infty}} \iint_{|z-z'|<r} v_k(z') dx' dy' \leq \iint_{|z-z'|<r} \overline{\lim_{k \to \infty}} v_k(z') dx' dy' \leq \pi C r^2.$$

If we choose k_0 sufficiently large, then

$$\iint_{|z-z'|<r} v_k(z') dx' dy' < \pi \left(C + \frac{\varepsilon}{2}\right) r^2 \qquad (k > k_0).$$

Since $v_k \leq 0$ for $0 < \delta < r$ and $|z - w| < \delta$, it follows from (1.8) that

$$\pi(r+\delta)^2 v_k(w) \leq \iint_{|z'-w|<r+\delta} v_k(z') dx' dy' \leq \iint_{|z-z'|<r} v_k(z') dx' dy'.$$

Hence we have

$$\pi(r+\delta)^2 v_k(w) \leq \pi \left(C + \frac{\varepsilon}{2}\right) r^2.$$

If we choose δ sufficiently small, then

$$v_k(w) < C + \varepsilon \qquad (k > k_0, \ |w - z| < \delta).$$

Since K is compact, $v_k(z) < C + \varepsilon$ for $z \in K$ provided we choose k_0 sufficiently large. $\square$

Lemma 1.6 *Let f be a holomorphic function in $B(0,r)$ such that $|f(z)| \leq M$ for some constant $M > 0$. Then*

$$|f(z_1) - f(z_2)| \leq 2Mr \frac{|z_2 - z_1|}{|r^2 - \bar{z}_1 z_2|} \quad (1.9)$$

for $z_1, z_2 \in B(0,r)$.

Proof. We may assume that f is not constant. Then by the maximum principle (see Exercise 1.3), $|f(z)| < M$. We set $w_1 = f(z_1)$, $w_2 = f(z_2)$. We define $\Phi : B(0,r) \to B(0,1)$ and $\Psi : B(0,M) \to B(0,1)$ by

$$\Phi(z) = \frac{r(z - z_1)}{r^2 - \bar{z}_1 z}, \quad \Psi(w) = \frac{M(w - w_1)}{M^2 - \bar{w}_1 w}.$$

Since $\Psi \circ f \circ \Phi^{-1} : B(0,1) \to B(0,1)$ maps 0 to 0, it follows from Schwarz's lemma (see Exercise 1.7) that

$$|\Psi \circ f \circ \Phi^{-1}(z)| \leq |z|.$$

We set $z = \Phi(z_2)$. Then $|\Psi(w_2)| \leq |\Phi(z_2)|$, which implies (1.9). □

Lemma 1.7 *If a bounded function $f : \Omega \to \mathbf{C}$ on an open set $\Omega \subset \mathbf{C}^n$ is holomorphic with respect to each variable z_j ($j = 1, \cdots, n$), then f is continuous in Ω (hence, f is holomorphic in Ω).*

Proof. Let M be a constant such that $|f(z)| \leq M$. Since the problem is local, we may assume that $\Omega = \{z \in \mathbf{C}^n \mid |z_j| < r_j, j = 1, \cdots, n\}$. By Lemma 1.6 we have

$$|f(z) - f(\zeta)|$$
$$= \left| \sum_{j=1}^{n} \{f(\zeta_1, \cdots, \zeta_{j-1}, z_j, \cdots, z_n) - f(\zeta_1, \cdots, \zeta_j, z_{j+1}, \cdots, z_n)\} \right|$$
$$\leq \sum_{j=1}^{n} 2M \frac{r_j |z_j - \zeta_j|}{|r_j^2 - \bar{\zeta} z|}.$$

Thus $f(z) \to f(\zeta)$ as $z \to \zeta$. Hence f is continuous in Ω. □

Theorem 1.8 *(Hartogs theorem) Let $\Omega \subset \mathbf{C}^n$ be an open set. If f is holomorphic with respect to each variable z_j for $j = 1, \cdots, n$, when the other variables are fixed, then f is holomorphic in Ω.*

Proof. When $n = 1$, Theorem 1.8 is trivial. Assume that Theorem 1.8 has already been proved for $n - 1$ variables.

Under this assumption we prove the following Lemma 1.8 and Lemma 1.9.

Lemma 1.8 *Suppose f is holomorphic in an open set $\Omega \subset \mathbf{C}^n$ with respect to each variable z_j for $j = 1, \cdots, n$. Let $P = P_1 \times \cdots \times P_n$ be a nonempty polydisc such that $\bar{P} \subset \Omega$. Then, there exist discs $P'_j \subset P_j$, $j = 1, \cdots, n$, such that $P_n = P'_n$ and f is bounded on $P' = P'_1 \times \cdots \times P'_n$. Hence f is holomorphic in P'.*

Proof. Define
$$E_M = \{z' \in \bar{P}_1 \times \cdots \times \bar{P}_{n-1} \mid |f(z', z_n)| \leq M \; z_n \in \bar{P}_n\}$$
$$= \cap_{z_n \in \bar{P}_n} \{z' \in \bar{P}_1 \times \cdots \times \bar{P}_{n-1} \mid |f(z', z_n)| \leq M\}.$$

Since Theorem 1.8 is true for functions of $n - 1$ variables, $f(z', z_n)$ is continuous when z_n is fixed. Hence E_M is closed. Further, we have
$$\bigcup_{M=1}^{\infty} E_M = \bar{P}_1 \times \cdots \times \bar{P}_{n-1}.$$

By applying the Baire theorem, if we choose M sufficiently large, then E_M has nonempty interior. If we choose a polydisc P' such that $\bar{P}' \subset E_M \times \bar{P}_n$, $P'_n = P_n$, then f is holomorphic in P'. □

Lemma 1.9 *Let f be defined on a polydisc $P(z^0, R) \subset \mathbf{C}^n$. Suppose that for fixed z_n, f is holomorphic with respect to $z' = (z_1, \cdots, z_{n-1})$ and that there exists $r > 0$ such that f is holomorphic and bounded on $P' = \{z \mid |z_j - z_j^0| < r, j = 1, \cdots, n-1, |z_n - z_n^0| < R\}$. Then f is holomorphic in P.*

Proof. We may assume that $z_0 = 0$. We choose R_1, R_2 such that $0 < R_1 < R_2 < R$. Since $f(z', z_n)$ is holomorphic with respect to z', f is expressed by
$$f(z) = \sum_\alpha a_\alpha(z_n) z'^\alpha \qquad (z \in P), \tag{1.10}$$

where $\alpha = (\alpha_1, \cdots, \alpha_{n-1})$ and each α_j is nonnegative integer. It follows from Theorem 1.7 that
$$a_\alpha(z_n) = \partial^\alpha f(0, z_n)/\alpha!.$$

For polydiscs Q_j, $j = 1, \cdots, n-1$, with centers 0 and sufficiently small radii, by applying Theorem 1.7 we have

$$\partial^\alpha f(0, z_n) = \frac{\alpha!}{(2\pi i)^{n-1}} \int_{\partial Q_1 \times \cdots \times \partial Q_{n-1}} \frac{f(\zeta_1, \cdots, \zeta_{n-1}, z_n)}{\zeta_1^{\alpha_1+1} \cdots \zeta_{n-1}^{\alpha_{n-1}+1}} d\zeta_1 \cdots d\zeta_{n-1}.$$

Thus $\partial a_\alpha(z_n)/\partial \bar{z}_n = 0$. Thus $a_\alpha(z_n)$ is holomorphic with respect to z_n. Since $f(z', z_n)$ is holomorphic with respect to $z' \in \{z' \mid |z_j| < R\}$, by Corollary 1.3 we have

$$|\partial^\alpha f(0, z_n)| \leq \alpha! R'^{-|\alpha|} \sup_{z \in P(0, R')} |f(z)|$$

for $R_2 < R' < R$. Hence we have for fixed z_n

$$|a_\alpha(z_n)| R_2^{|\alpha|} \to 0 \qquad (|\alpha| \to \infty, \ |z_n| < R).$$

On the other hand, if $|f(z)| \leq M$ for $z \in P'$, then by the Cauchy inequality

$$|a_\alpha(z_n)| r^{|\alpha|} \leq M.$$

For two multi-indices $\alpha = (\alpha_1, \cdots, \alpha_n)$ and $\alpha' = (\alpha'_1, \cdots, \alpha'_n)$ we introduce the order such that

$$\alpha < \alpha'$$

if only if there exist i, $1 \leq i \leq n$, such that

$$\alpha_1 = \alpha'_1, \quad \cdots, \quad \alpha_{i-1} = \alpha'_{i-1}, \quad \alpha_i < \alpha'_i.$$

For $\alpha = (\alpha_1, \cdots, \alpha_n)$, we define

$$\varphi_\alpha(z_n) = \frac{1}{|\alpha|} \log |a_\alpha(z_n)|.$$

Then φ_α is subharmonic. Let $\{v_k\}$ be the arrangement of $\{\varphi_\alpha\}$ according to the order of the multi-indices. Thus, $k \to \infty$ is equivalent to $|\alpha| \to \infty$. Since

$$\frac{1}{|\alpha|} \log |a_\alpha(z_n)| \leq -\log r + \frac{1}{|\alpha|} \log M \qquad (|z_n| < R),$$

$\{v_k\}$ is uniformly bounded on $|z_n| < R$. On the other hand, for fixed z_n, if we choose $|\alpha|$ sufficiently large, then we have $|a_\alpha(z_n)| R_2^{|\alpha|} < 1$. Thus for sufficiently large $|\alpha|$ we have

$$\frac{\log |a_\alpha(z_n)|}{|\alpha|} < \log \frac{1}{R_2}.$$

Thus we obtain
$$\varlimsup_{k\to\infty} v_k(z_n) \leq \log \frac{1}{R_2}.$$
It follows from Lemma 1.5 that we have for some sufficiently large k
$$v_k(z_n) \leq \log \frac{1}{R_1} \qquad (|z_n| < R_1),$$
which means that for any sufficiently large $|\alpha|$
$$|a_\alpha(z_n)|R_1^{|\alpha|} \leq 1 \qquad (|z_n| < R_1).$$
Since the above inequality holds for every R_1 satisfying $0 < R_1 < R$, (1.10) converges uniformly on every compact subset of $P(0, R)$. Hence f is continuous in $P(0, R)$. Thus f is holomorphic in $P(0, R)$. $\square$

Proof of Theorem 1.8 Let $\zeta \in \Omega$. We choose $R > 0$ with the properties that a polydisc $\{z \mid |z_j - \zeta_j| \leq 2R\}$ is a subset of Ω. We take $P = P(\zeta, R)$ in Lemma 1.8. Then there exist $r > 0$ and z^0 such that $\max_j |z_j^0 - \zeta_j| < R$, $\zeta_n = z_n^0$, and f is bounded on $P' = \{z \mid |z_j - z_j^0| < r, j = 1, \cdots, n-1, |z_n - z_n^0| < R\} (\subset P(\zeta, R))$. Since f is holomorphic in $P(z_0, R)$ by Lemma 1.9, f is holomorphic at ζ. $\square$

1.2 Characterizations of Pseudoconvexity

We prove that every domain of holomorphy is a pseudoconvex open set. In 2.2 (Corollary 2.4) we prove that an open set in $\mathbf{C}^n$ is a domain of holomorphy if and only if it is pseudoconvex.

Definition 1.11 Let $\Omega \subset \mathbf{C}^n$ be an open set.

(1) A real-valued upper semicontinuous function φ in Ω is called plurisubharmonic if for any $v, w \in \mathbf{C}^n$, $h(\zeta) = \varphi(v + \zeta w)$ is subharmonic in $U = \{\zeta \in \mathbf{C} \mid v + \zeta w \in \Omega\}$. The set of all plurisubharmonic functions in Ω is denoted by $PS(\Omega)$.

(2) A real-valued C^2 function φ in Ω is called strictly plurisubharmonic in Ω if for any $v, w \in \mathbf{C}^n$ ($w \neq 0$), $h(\zeta) = \varphi(v + \zeta w)$ is strictly subharmonic in $U = \{\zeta \in \mathbf{C} \mid v + \zeta w \in \Omega\}$.

Theorem 1.9 *Let $\Omega \subset \mathbf{C}^n$ be an open set.*

(a) If $f \in \mathcal{O}(\Omega)$, then $|f| \in PS(\Omega)$.

(b) If f is a holomorphic function in Ω and ρ is a C^2 subharmonic function in $f(\Omega)$, then $\rho \circ f \in PS(\Omega)$.

(c) Suppose $\{\rho_j\}_{j \in J}$ is a family of plurisubharmonic functions in Ω and $\rho = \sup_{j \in J} \rho_j$. If ρ is finite and upper semicontinuos in Ω, then $\rho \in PS(\Omega)$.

Proof. (a) For $v, w \in C^n$, we set $U = \{\zeta \in C \mid v + \zeta w \in \Omega\}$. For $\zeta \in U$, we set $\varphi(\zeta) = f(v + \zeta w)$. We fix $a \in U$. We choose $r > 0$ such that $r < \text{dist}(a, \partial U)$. Since φ is holomorphic in U, it follows from the Cauchy integral formula that

$$\varphi(a) = \frac{1}{2\pi i} \int_{|\zeta - a| = r} \frac{\varphi(\zeta)}{\zeta - a} d\zeta.$$

Then we have

$$|\varphi(a)| \leq \frac{1}{2\pi} \int_0^{2\pi} |\varphi(a + re^{i\theta})| d\theta.$$

Hence $|\varphi|$ is subharmonic. Thus $|f|$ is plurisubharmonic in Ω.

(b) For $v, w \in C^n$, we set $U = \{\zeta \in C \mid v + \zeta w \in \Omega\}$. For $\zeta \in U$, we set $\varphi(\zeta) = \rho \circ f(v + \zeta w)$. Then we have

$$\frac{\partial^2 \varphi}{\partial \zeta \partial \bar{\zeta}}(\zeta) = \frac{\partial^2 \rho}{\partial w \partial \bar{w}}(f(v + \zeta w)) \left| \frac{\partial}{\partial \zeta}(f(v + \zeta w)) \right|^2 \geq 0,$$

which means that φ is subharmonic in U. Thus $\rho \circ f \in PS(\Omega)$.

(c) For $v, w \in C^n$, we set $U = \{\zeta \in C \mid v + \zeta w \in \Omega\}$. For $\zeta \in U$, we set $\varphi_j(\zeta) = \rho_j(v + \zeta w)$ for $j \in J$. Since φ_j is subharmonic in U, it follows that

$$\varphi_j(a) \leq \frac{1}{2\pi} \int_0^{2\pi} \varphi_j(a + re^{i\theta}) d\theta$$

for $a \in U$ and r with $0 < r < \text{dist}(a, \partial U)$. We set $\varphi(\zeta) = \rho(v + \zeta w)$. Then

$$\sup_{j \in J} \varphi_j(\zeta) = \sup_{j \in J} \rho_j(v + \zeta w) = \rho(v + \zeta w) = \varphi(\zeta).$$

For $\varepsilon > 0$, there exists j_0 such that $\varphi(a) - \varepsilon < \varphi_{j_0}(a)$. Therefore we obtain

$$\varphi(a) < \varphi_{j_0}(a) + \varepsilon \leq \frac{1}{2\pi} \int_0^{2\pi} \varphi_{j_0}(a + re^{i\theta}) d\theta + \varepsilon,$$

$$\leq \frac{1}{2\pi} \int_0^{2\pi} \varphi(a + re^{i\theta}) d\theta + \varepsilon.$$

Since $\varepsilon > 0$ is arbitrary, we obtain
$$\varphi(a) \le \frac{1}{2\pi}\int_0^{2\pi} \varphi(a + re^{i\theta})d\theta.$$
Thus φ is subharmonic in U, which implies that $\rho \in PS(\Omega)$. □

Theorem 1.10 *Let ρ be a real-valued C^2 function in an open set $\Omega \subset \mathbf{C}^n$.*
(a) ρ is plurisubharmonic in Ω if and only if
$$\sum_{j,k=1}^n \frac{\partial^2 \rho}{\partial z_j \partial \bar{z}_k}(z)w_j \bar{w}_k \ge 0$$
for $z \in \Omega$, $w = (w_1, \cdots, w_n) \in \mathbf{C}^n$.
(b) ρ is strictly plurisubharmonic in Ω if and only if
$$\sum_{j,k=1}^n \frac{\partial^2 \rho}{\partial z_j \partial \bar{z}_k}(z)w_j \bar{w}_k > 0$$
for $z \in \Omega$, $0 \ne w = (w_1, \cdots, w_n) \in \mathbf{C}^n$.

Proof. For $v, w \in \mathbf{C}^n$, we set
$$\tilde{\rho}(\zeta) = \rho(v + \zeta w) = \rho(v_1 + \zeta w_1, \cdots, v_n + \zeta w_n).$$
Then Theorem 1.10 follows from the equality
$$\frac{\partial^2 \tilde{\rho}}{\partial \zeta \partial \bar{\zeta}}(\zeta) = \sum_{j,k=1}^n \frac{\partial^2 \tilde{\rho}}{\partial z_j \partial \bar{z}_k}(v + \zeta w) w_j \bar{w}_k.$$
□

Corollary 1.4 *Let Ω be an open set in $\mathbf{C}^n$ and let K be a compact subset of Ω. Suppose ρ is a strictly plurisubharmonic function in Ω. Then there exists a constant $C = C(K) > 0$ such that*
$$\sum_{j,k=1}^n \frac{\partial^2 \rho}{\partial z_j \partial \bar{z}_k}(z)w_j \bar{w}_k \ge C|w|^2$$
for $z \in K$, $w = (w_1, \cdots, w_n) \in \mathbf{C}^n$.

Proof. For $z \in K$, $w \in \mathbf{C}^n$, we set
$$f(z, w) = \sum_{j,k=1}^n \frac{\partial^2 \rho}{\partial z_j \partial \bar{z}_k}(z)w_j \bar{w}_k.$$

Since $f(z,w)$ takes the minimum value $C > 0$ on $K \times \{w \mid |w| = 1|\}$, we have for $z \in K$ and $w \neq 0$

$$\sum_{j,k=1}^{n} \frac{\partial^2 \rho}{\partial z_j \partial \bar{z}_k}(z) \frac{w_j \bar{w}_k}{|w|^2} \geq C.$$

$\square$

Definition 1.12 Let Ω be an open subset of $\mathbf{C}^n$ and let $\Omega \neq \mathbf{C}^n$. For $z \in \mathbf{C}^n$, define

$$\mathrm{dist}(z, \partial \Omega) = \inf\{|z - \zeta| \mid \zeta \in \partial \Omega\},$$

where $\mathrm{dist}(z, \partial\Omega)$ denotes the distance from z to $\partial\Omega$.

Lemma 1.10 Let $\Omega \subset \mathbf{C}^n$ be an open set such that $\Omega \neq \mathbf{C}^n$. For $z \in \Omega$, define $\varphi(z) = \mathrm{dist}(z, \partial \Omega)$. Then φ is continuous in Ω.

Proof. Fix $z_0 \in \Omega$. Then there exists $\xi_0 \in \partial\Omega$ such that $\varphi(z_0) = |z_0 - \xi_0|$. For $z \in \Omega$ there exists $\xi(z) \in \partial\Omega$ such that $\varphi(z) = |z - \xi(z)|$. Then we have

$$\varphi(z) = |z - \xi(z)| \leq |z - \xi_0| \leq |z - z_0| + |z_0 - \xi_0| = |z - z_0| + \varphi(z_0).$$

Thus we have

$$\varphi(z) - \varphi(z_0) \leq |z - z_0|.$$

Similarly, we have

$$\varphi(z_0) = |z_0 - \xi_0| \leq |z_0 - \xi(z)| \leq |z_0 - z| + \varphi(z).$$

Hence we obtain

$$|\varphi(z) - \varphi(z_0)| \leq |z - z_0|,$$

which means that φ is continuous at $z = z_0$. $\square$

Definition 1.13 Let Ω be an open subset in $\mathbf{C}^n$. We say that Ω is a domain of holomorphy if there exists at least one holomorphic function in Ω that cannot be extended as a holomorphic function through any boundary point of Ω.

Remark 1.1 We do not assume that the domain of holomorphy is connected.

Definition 1.14 Suppose $\Omega \subset \mathbf{C}^n$ is an open set and K is a compact subset of Ω. Define

$$\widehat{K}_\Omega^{\mathcal{O}} = \{z \in \Omega \mid |f(z)| \leq \sup_{\zeta \in K} |f(\zeta)|, \ f \in \mathcal{O}(\Omega)\}.$$

$\widehat{K}_\Omega^{\mathcal{O}}$ is called a holomorphically convex hull of K (or $\mathcal{O}$-hull of K). By definition, $K \subset \widehat{K}_\Omega^{\mathcal{O}}$. In case $K = \widehat{K}_\Omega^{\mathcal{O}}$, K is called $\mathcal{O}(\Omega)$-convex.

Definition 1.15 An open set $\Omega \subset \mathbf{C}^n$ is called holomorphically convex if for any compact subset K, $\widehat{K}_\Omega^{\mathcal{O}} \subset\subset \Omega$. (Equivalently, Ω is holomorphically convex if and only if $\widehat{K}_\Omega^{\mathcal{O}}$ is compact for every compact subset K of Ω.)

Lemma 1.11 *Let $\Omega \subset \mathbf{C}^n$ be an open set. Then*

(a) If K and L are compact subsets of Ω with $K \subset L$, then $\widehat{K}_\Omega^{\mathcal{O}} \subset \widehat{L}_\Omega^{\mathcal{O}}$.

(b) We set $N = \widehat{K}_\Omega^{\mathcal{O}}$. If N is compact, then

$$\widehat{N}_\Omega^{\mathcal{O}} = N.$$

Proof. (a) Let $z \in \widehat{K}_\Omega^{\mathcal{O}}$. Then for any $f \in \mathcal{O}(\Omega)$ we have

$$|f(z)| \leq \sup_{\zeta \in K} |f(\zeta)| \leq \sup_{\zeta \in L} |f(\zeta)|.$$

Hence $f \in \widehat{L}_\Omega^{\mathcal{O}}$. This proves (a).

(b) By definition we have $N \subset \widehat{N}_\Omega^{\mathcal{O}}$. If $z \in \widehat{N}_\Omega^{\mathcal{O}}$, then

$$|f(z)| \leq \sup_{\zeta \in \widehat{K}_\Omega^{\mathcal{O}}} |f(\zeta)| \leq \sup_{\zeta \in K} |f(\zeta)|.$$

Hence we have $\widehat{N}_\Omega^{\mathcal{O}} \subset \widehat{K}_\Omega^{\mathcal{O}} = N$. This proves (b). $\square$

Lemma 1.12 *Let $K \subset \mathbf{C}^n$ be compact. We denote by $\widetilde{K}$ the smallest convex set which contains K ($\widetilde{K}$ is called the convex hull of K). Then we have $\widehat{K}_{\mathbf{C}^n}^{\mathcal{O}} \subset \widetilde{K}$.*

Proof. Suppose $w \notin \widetilde{K}$. Then there exists a hyperplane through w $l : \sum_{j=1}^{2n} a_j x_j = b$ which does not intersect $\widetilde{K}$. When $z_j = x_j + ix_{n+j} \in K$, we assume that $\sum_{j=1}^{2n} a_j x_j < b$. If $w_j = u_j + iu_{n+j}$, then $\sum_{j=1}^{2n} a_j u_j = b$. If we set

$$\alpha_j = a_j + ia_{n+j}, \quad f(z) = \exp\left(\sum_{j=1}^{n} \bar{\alpha}_j z_j - b\right),$$

then $f \in \mathcal{O}(\mathbf{C}^n)$. Using the equality

$$\operatorname{Re}\sum_{j=1}^{n}\bar{\alpha}_j z_j = \sum_{j=1}^{2n} a_j x_j,$$

we have

$$|f(z)| = \exp\left(\sum_{j=1}^{2n} a_j x_j - b\right) < 1 \quad (z \in K),$$

$$|f(w)| = \exp\left(\sum_{j=1}^{2n} a_j u_j - b\right) = 1.$$

Thus we have

$$\sup_{z \in K} |f(z)| < 1 = |f(w)|.$$

Hence $w \notin \widetilde{K}_{\mathbf{C}^n}^{\mathcal{O}}$. Thus we obtain $\widehat{K}_{\mathbf{C}^n}^{\mathcal{O}} \subset \widetilde{K}$. $\square$

Lemma 1.13 *Suppose $\Omega \subset \mathbf{C}^n$ is an open set and K is a compact subset of Ω. Then $\widehat{K}_\Omega^{\mathcal{O}}$ is bounded.*

Proof. We denote by $\widetilde{K}$ the convex hull of K. From Lemma 1.12 and the definition of holomorphically convex hull, we have $\widehat{K}_\Omega^{\mathcal{O}} \subset \widehat{K}_{\mathbf{C}^n}^{\mathcal{O}} \subset \widetilde{K}$. Since $\widetilde{K}$ is bounded, $\widehat{K}_\Omega^{\mathcal{O}}$ is bounded. $\square$

Definition 1.16 For $r = (r_1, \cdots, r_n)$, $r_j > 0$, $j = 1, \cdots, n$, $a \in \mathbf{C}^n$ and $\lambda > 0$, define

$$P(a, \lambda r) = \{z \mid |z_j - a_j| < \lambda r_j \, j = 1, \cdots, n\}.$$

For $a \in \Omega \subset \mathbf{C}^n$, we define

$$\delta_\Omega^{(r)}(a) = \sup\{\lambda \mid \lambda > 0, \, P(a, \lambda r) \subset \Omega\}.$$

By definition, $\lambda \leq \delta_\Omega^{(r)}(a)$ if and only if $P(a, \lambda r) \subset \Omega$.

Lemma 1.14 *Let $\Omega \neq \mathbf{C}^n$ be an open set. Then*

$$\operatorname{dist}(a, \partial\Omega) = \inf\{\delta_\Omega^{(r)}(a) \mid |r| = 1\} \quad (a \in \Omega).$$

Proof. We set
$$\delta = \text{dist}(a, \partial\Omega), \quad \eta = \inf\{\delta_\Omega^{(r)}(a) \mid |r| = 1\}.$$
If $|r| = 1$ and $|z_i - a_i| < \lambda r_i$ for $i = 1, \cdots, n$, then $|z - a| < \lambda$, and hence $P(a, \lambda r) \subset B(a, \lambda)$. Thus we have $P(a, \delta r) \subset \Omega$. Therefore, if $|r| = 1$, then $\delta \le \delta_\Omega^{(r)}(a)$, which implies that $\delta \le \eta$. Next we show that $\delta \ge \eta$. For any $\varepsilon > 0$, we choose λ such that $\delta < \lambda < \delta + \varepsilon$. Then $B(a, \lambda) \not\subset \Omega$, which implies that there exists w such that $w \notin \Omega$, $|w - a| < \lambda$. We set
$$|w_i - a_i| = t_i, \quad t = (t_1, \cdots, t_n), \quad r_i = \frac{t_i}{|t|}, \quad r = (r_1, \cdots, r_n).$$
Then $|r| = 1$, $|w_i - a_i| < r_i \lambda$ for $i = 1, \cdots, n$. Hence $w \in P(a, \lambda r)$. Therefore $P(a, \lambda r) \not\subset \Omega$, which means that $\delta_\Omega^{(r)}(a) < \lambda$. Thus we have $\eta \le \delta_\Omega^{(r)}(a) < \lambda < \delta + \varepsilon$. Since $\varepsilon > 0$ is arbitrary, we have $\eta \le \delta$. □

Theorem 1.11 *Suppose $\Omega \subset \mathbf{C}^n$ is an open set and K is a compact subset of Ω. Let $r = (r_1, \cdots, r_n)$ and $r_i > 0$ for $i = 1, \cdots, n$. Assume that $\eta > 0$ satisfies $\delta_\Omega^{(r)}(z) \ge \eta$ for all $z \in K$. Then for any $a \in \widehat{K}_\Omega^\mathcal{O}$ and $f \in \mathcal{O}(\Omega)$, f is holomorphic in $P(a, \eta r)$.*

Proof. We fix $f \in \mathcal{O}(\Omega)$. We choose η' such that $\eta' < \eta$. We set
$$Q = \overline{\underset{a \in K}{\cup} P(a, \eta' r)}.$$
Then Q is a compact subset of Ω. We set $M = \sup_{z \in Q} |f(z)|$. By applying the Cauchy inequality for $P(a, \eta' r)$, we have for $\alpha \in \mathbf{N}^n$
$$\sup_{z \in K} |\partial^\alpha f(z)| \le \frac{\alpha! M}{(\eta' r)^\alpha}.$$
Thus for $a \in \widehat{K}_\Omega^\mathcal{O}$, we obtain
$$|\partial^\alpha f(a)| \le \sup_{z \in K} |\partial^\alpha f(z)| \le \frac{\alpha! M}{(\eta' r)^\alpha}.$$
Hence for $z \in P(a, \eta' r)$, we have
$$\sum_\alpha \left| \frac{\partial^\alpha f(a)}{\alpha!} (z-a)^\alpha \right| \le \sum_k M \left(\frac{|z-a|}{\eta' r} \right)^\alpha < \infty.$$
Thus $\sum_\alpha \frac{\partial^\alpha f(a)}{\alpha!} (z-a)^\alpha$ converges in $P(a, \eta' r)$. If we set
$$\varphi(z) = \sum_\alpha \frac{\partial^\alpha f(a)}{\alpha!} (z-a)^\alpha,$$

then φ is holomorphic in $P(a,\eta'r)$. Since $a \in \Omega$, we have $\varphi = f$ in a neighborhood of a. Therefore f is holomorphic in $P(a,\eta'r)$. Since η is arbitrary so far as $\eta' < \eta$, f is holomorhic in $P(a,\eta r)$. □

Corollary 1.5 *If $\Omega \subset \mathbf{C}^n$ is a domain of holomorphy, then*
$$\operatorname{dist}(K, \partial\Omega) = \operatorname{dist}(\widehat{K}_\Omega^\mathcal{O}, \partial\Omega)$$
for every compact subset K of Ω.

Proof. Since $K \subset \widehat{K}_\Omega^\mathcal{O}$, we have $\operatorname{dist}(K,\partial\Omega) \geq \operatorname{dist}(\widehat{K}_\Omega^\mathcal{O},\partial\Omega)$. We set $\eta = \operatorname{dist}(K,\partial\Omega)$. Let $\eta > \operatorname{dist}(\widehat{K}_\Omega^\mathcal{O},\partial\Omega)$. We choose $a \in \widehat{K}_\Omega^\mathcal{O}$ such that $\operatorname{dist}(a,\partial\Omega) < \eta$. It follows from Lemma 1.14 that there exists r such that $|r| = 1$ and $\delta_\Omega^{(r)}(a) < \eta$. Therefore we have $P(a,\eta r) \not\subset \Omega$. On the other hand, when $|r| = 1$ and $z \in K$, we have $\eta \leq \operatorname{dist}(z,\partial\Omega) \leq \delta_\Omega^{(r)}(z)$. By Theorem 1.11, all $f \in \mathcal{O}(\Omega)$ are holomorphic in $P(a,\eta r)$. Therefore f is holomorphic in a neighborhood of some boundary point of Ω. This contradicts that Ω is a domain of holomorphy. □

Next we state some properties of the infinite product which we will use in the proof of Theorem 1.12.

Definition 1.17 Let $\{z_n\}$ be a sequence of complex numbers. We set
$$P_n = (1+z_1)(1+z_2)\cdots(1+z_n).$$
If $\lim_{n\to\infty} P_n = P$ exist, then we define
$$P = \prod_{n=1}^\infty (1+z_n). \qquad (1.11)$$
The right side of (1.11) is called the infinite product.

Lemma 1.15 *Define*
$$P_N = \prod_{n=1}^N (1+z_n), \quad P_N^* = \prod_{n=1}^N (1+|z_n|).$$
Then

(a) $P_N^* \leq exp(|z_1| + \cdots + |z_N|)$.
(b) $|P_N - 1| \leq P_N^* - 1$.

Proof. Using the inequality
$$1 + |z_i| \leq e^{|z_i|},$$

we have
$$P_N^* = \prod_{n=1}^{N}(1+|z_n|) \le e^{|z_1|+\cdots+|z_N|}.$$

This proves (a).

We prove (b) by induction on N. When $N = 1$, it is trivial. Suppose (b) holds for $N = k$. Since
$$P_{k+1} - 1 = P_k(1+z_{k+1}) - 1 = (P_k-1)(1+z_{k+1}) + z_{k+1},$$
we have
$$\begin{aligned}|P_{k+1}-1| &\le |P_k-1||1+z_{k+1}| + |z_{k+1}| \\ &\le (P_k^*-1)(1+|z_{k+1}|) + |z_{k+1}| \\ &= P_{k+1}^* - 1.\end{aligned}$$

Thus (b) holds for $N = k+1$. □

Lemma 1.16 *Suppose $\{f_k\}$ is a sequence of bounded functions defined on a set $E \subset \mathbf{C}^n$ and $\sum_{j=1}^{\infty}|f_j(z)|$ converges uniformly on E. Then*

(a) $\Pi_{j=1}^{\infty}(1+f_j(z))$ converges uniformly on E.
(b) Let $f = \Pi_{j=1}^{\infty}(1+f_j)$ and $z_0 \in E$. Then $f(z_0) = 0$ if and only if there exists n such that $f_n(z_0) = -1$.
(c) Let $\{k_1, k_2, \cdots\}$ be a permutation of $\{1, 2, \cdots\}$. Then
$$\prod_{j=1}^{\infty}(1+f_j) = \prod_{j=1}^{\infty}(1+f_{k_j}).$$

Proof. Since f_k is bounded on E, there exists a constant c_k such that $|f_k(z)| \le c_k$ for $z \in E$. We set
$$h(z) = \sum_{j=1}^{\infty}|f_j(z)|, \quad h_m(z) = \sum_{j=1}^{m}|f_j(z)|.$$

Since $\{h_m\}$ converges to h uniformly on E, there exists a positive integer n_0 such that for $m \ge n_0$
$$|h(z) - h_m(z)| < 1 \quad (z \in E).$$

Hence for $z \in E$, we have

$$|h(z)| \leq |h_{n_0}(z)| + 1 = \sum_{j=1}^{n_0} |f_j(z)| + 1 \leq \sum_{j=1}^{n_0} c_j + 1 =: \tilde{c}.$$

By Lemma 1.15, if we set

$$P_m(z) = \prod_{j=1}^{m}(1 + f_j(z)), \quad P_m^*(z) = \prod_{j=1}^{m}(1 + |f_j(z)|),$$

then we have

$$|P_m(z)| \leq |P_m^*(z)| \leq \exp(|f_1(z)| + \cdots + |f_m(z)|) = e^{h_m(z)} \leq e^{h(z)} \leq e^{\tilde{c}} =: c.$$

Let $0 < \varepsilon < 1/2$. Then there exists a positive integer t_0 such that

$$|h_{t_0}(z) - h(z)| = \sum_{k=t_0+1}^{\infty} |f_k(z)| < \varepsilon.$$

Let $N \geq t_0$. Since $\{k_1, k_2, \cdots\}$ is a permutation of $\{1, 2, \cdots\}$, we have for some sufficiently large integer M

$$\{1, 2, \cdots, N\} \subset \{k_1, k_2, \cdots, k_M\}.$$

We set

$$q_M(z) = \prod_{j=1}^{M}(1 + f_{k_j}(z)).$$

We set $F = \{k_1, k_2, \cdots, k_M\} - \{1, 2, \cdots, N\}$. Then we have

$$q_M(z) - P_N(z) = \prod_{j=1}^{M}(1 + f_{k_j}(z)) - \prod_{j=1}^{N}(1 + f_j(z))$$

$$= P_N(z) \left\{ \prod_{i \in F}(1 + f_i(z)) - 1 \right\}.$$

By Lemma 1.15 we obtain

$$\left| \prod_{i \in F}(1 + f_i(z)) - 1 \right| \leq \prod_{i \in F}(1 + |f_i(z)|) - 1 \leq \exp\left(\sum_{i \in F} |f_i(z)|\right) - 1.$$

For $z \in E$, we have

$$\begin{aligned}
|q_M(z) - P_N(z)| &\leq |P_N(z)|\left\{\exp\left(\sum_{i=t_0+1}^{\infty} |f_i(z)|\right) - 1\right\} \\
&\leq |P_N(z)|(e^\varepsilon - 1) \\
&= |P_N(z)|(\varepsilon + \frac{\varepsilon^2}{2!} + \cdots) \\
&\leq \varepsilon|P_N(z)|(1 + \frac{1}{2} + \frac{1}{2^2} + \cdots) \\
&= 2\varepsilon|P_N(z)| \leq 2\varepsilon c.
\end{aligned}$$

Thus for N with $N \geq t_0$ and some sufficiently large M we have

$$|q_M(z) - P_N(z)| \leq 2\varepsilon|P_N(z)| \leq 2\varepsilon c \quad (z \in E). \tag{1.12}$$

In particular, when $k_j = j$ we have $q_M = P_M$. It follows from (1.12) that

$$|P_M(z) - P_N(z)| < 2\varepsilon c \quad (z \in E).$$

Thus $\{P_N\}$ converges uniformly on E. This proves (a). Let $k_j = j$. Then for some sufficiently large M it follows from (1.12) that

$$|P_M(z) - P_{t_0}(z)| \leq 2\varepsilon|P_{t_0}(z)|.$$

Thus we obtain

$$|P_{t_0}(z)|(1 - 2\varepsilon) \leq |P_M(z)|.$$

Letting $M \to \infty$ we have

$$|P_{t_0}(z)|(1 - 2\varepsilon) \leq |f(z)|.$$

Since $1 - 2\varepsilon > 0$, $f(z_0) = 0$ implies $P_{t_0}(z_0) = 0$. Thus there exists k such that $f_k(z_0) = -1$. This proves (c). From (a), $\{P_j(z)\}$ converges to f uniformly on E. Therefore, taking N sufficiently large with $N \geq t_0$, if necessary, we have

$$|f(z) - P_N(z)| < \varepsilon \quad (z \in E).$$

For some sufficiently large M, we have

$$|q_M(z) - f(z)| \leq |q_M(z) - P_N(z)| + |P_N(z) - f(z)| < 2\varepsilon c + \varepsilon = \varepsilon(1 + 2c).$$

Thus we have $\lim_{M \to \infty} q_M(z) = f(z)$. $\square$

Theorem 1.12 *For an open set $\Omega \subset \mathbf{C}^n$, the following statements are equivalent:*

(a) Ω is a domain of holomorphy.
(b) For any compact set $K \subset \Omega$, $\widehat{K}_\Omega^\mathcal{O}$ is compact.
(c) For any compact set $K \subset \Omega$, $\operatorname{dist}(K, \partial\Omega) = \operatorname{dist}(\widehat{K}_\Omega^\mathcal{O}, \partial\Omega)$.
(d) If $X \subset \Omega$ is a discrete infinite subset of Ω, then there exists $f \in \mathcal{O}(\Omega)$ such that f is unbounded on X.

Proof. (a) $\Longrightarrow$ (c) follows from Theorem 1.11.
(c)$\Longrightarrow$(b). Since K is compact, we have

$$\operatorname{dist}(\widehat{K}_\Omega^\mathcal{O}, \partial\Omega) \geq \operatorname{dist}(K, \partial\Omega) > 0.$$

Since $\widehat{K}_\Omega^\mathcal{O}$ is closed in Ω, $\widehat{K}_\Omega^\mathcal{O}$ is compact.

(b)$\Longrightarrow$(a). Let $\{K_n\}$ be a sequence of compact sets such that $\Omega = \cup_{n=1}^\infty K_n$ with $K_n \subset K_{n+1}^\circ$, where K_n° denotes the interior of K_n. It follows from Lemma 1.11 that $(\widehat{K}_n)_\Omega^\mathcal{O} \subset (\widehat{K}_{n+1})_\Omega^\mathcal{O}$. If we set $T_n = (\widehat{K}_n)_\Omega^\mathcal{O}$, then by the assumption, T_n is compact and $\Omega = \cup_{n=1}^\infty T_n$. It follows from Lemma 1.11 that $T_n \subset T_{n+1}$, $(\widehat{T}_n)_\Omega^\mathcal{O} = T_n$. We may assume that $T_n \subset T_{n+1}^\circ$. Suppose $X \subset \Omega$ is a countable set and $\overline{X} = \Omega$. Let $X = \{\xi_m\}_{m=1}^\infty$. We denote by B_m the largest open ball with center ξ_m and contained in Ω. Let $\eta_m \in B_m - T_m$. Since $\eta \notin T_m$, there exists $f_m \in \mathcal{D}$ such that

$$|f_m(\eta_m)| > \sup_{\zeta \in T_m} |f_m(\zeta)|.$$

We set

$$g_m(z) = \frac{f_m(z)}{f_m(\eta_m)}.$$

Then $g_m \in \mathcal{O}(\Omega)$ and $g_m(\eta_m) = 1$, $\sup_{\zeta \in T_m} |g_m(\zeta)| < 1$. For some sufficiently large integer k_m

$$\sup_{\zeta \in T_m} |g_m^{k_m}(\zeta)| < \frac{1}{m 2^m}, \quad g_m^{k_m}(\eta_m) = 1.$$

Set $\varphi_m = g_m^{k_m}$. Then $\varphi_m \in \mathcal{O}(\Omega)$, $\varphi_m(\eta_m) = 1$ and $\sup_{\zeta \in T_m} |\varphi_m(\zeta)| < (m 2^m)^{-1}$. We set

$$\varphi(z) = \prod_{j=1}^\infty (1 - \varphi_j(z))^j = (1 - \varphi_1(z))(1 - \varphi_2(z))(1 - \varphi_2(z))(1 - \varphi_3(z)) \cdots.$$

Then for $z \in T_m$

$$m|\varphi_m(z)| + (m+1)|\varphi_{m+1}(z)| + \cdots \leq \frac{1}{2^m} + \frac{1}{2^{m+1}} + \cdots.$$

Therefore, for any positive integer m, $\sum_{j=1}^{\infty} j|\varphi_j(z)|$ converges uniformly on T_m. Thus, $\prod_{j=1}^{\infty}(1-\varphi_j(z))^j$ converges uniformly on every T_m. Thus, φ is holomorphic in Ω. Since $|\varphi_m(z)| < 1$ for $z \in T_1$, $\varphi(z) \neq 0$ for $z \in T_1$. Thus $\varphi(z) \not\equiv 0$. Suppose that there exists a domain V such that $\phi \neq \Omega \cap V \neq V$ and that φ is holomorphic in $\Omega \cup V$. We set $V \cap \Omega = W$. Let $\zeta \in \partial W \cap \partial \Omega \cap V$. Since $X \cap W$ is dense in W, We can choose a subsequence $\{\xi_{m_j}\}$ of X which converges to ζ. If we choose j sufficiently large, then $B_{m_j} \subset W$. Since $\eta_{m_j} \in B_{m_j} - T_{m_j}$, we have $\eta_{m_j} \to \zeta$. In case $k = (k_1, \cdots, k_n)$ with $|k| = k_1 + \cdots + k_n < m_j$, we have

$$\frac{\partial^{|k|} \varphi}{\partial z^k}(z) = \frac{\partial^{|k|}}{\partial z^k}\left\{\left(\prod_{m \neq m_j}(1-\varphi_m(z))^m\right)(1-\varphi_{m_j}(z))^{m_j}\right\}.$$

Hence we have

$$\frac{\partial^k \varphi}{\partial z_1^{k_1} \cdots \partial z_n^{k_n}}(\eta_{m_j}) = 0 \qquad (k < m_j,\ k = k_1 + \cdots + k_n).$$

Since $\zeta \in V$ and φ is holomorphic in V, we obtain

$$0 = \frac{\partial^k \varphi}{\partial z_1^{k_1} \cdots \partial z_n^{k_n}}(\zeta).$$

Thus $\varphi(z) \equiv 0$ in $\Omega \cup V$. This is a contradiction.

(d)$\Longrightarrow$(b). Suppose (b) is not true. There exists a compact set $K \subset \Omega$ such that $\widehat{K}_\Omega^\mathcal{O}$ is not compact. Since $\widehat{K}_\Omega^\mathcal{O}$ is a closed subset of Ω with respect to the relative topology, there exist $\xi_k \in \widehat{K}_\Omega^\mathcal{O}$, $k = 1, 2, \cdots$, such that $\{\xi_k\}$ converges to a boundary point of Ω. Then we have for any $f \in \mathcal{O}(\Omega)$

$$|f(\xi_k)| \leq \sup_{z \in K}|f(z)| < \infty.$$

Since $X = \{\xi_i\}$ is a discrete infinite subset in Ω and f is bounded in X, (d) does not hold.

(b)$\Longrightarrow$(d). We choose a sequence $\{K_n\}$ of compact subsets of Ω such that $\Omega = \cup_{m=1}^{\infty} K_m$, $K_m \subset K_{m+1}$. By the assumption, $T_m = (\widehat{K}_m)_\Omega^\mathcal{O}$ are compact and satisfy $\Omega = \cup_{m=1}^{\infty} T_m$, $T_m \subset T_{m+1}$. We may assume that $T_m \subset (T_{m+1})^\circ$. Suppose $X \subset \Omega$ is a discrete infinite set. Let $X = \{\xi_m\}$. We choose a subsequence $\{T_{m_j}\}$ of $\{T_n\}$ and a subsequence $\{\xi_{\nu_j}\}$ of $\{\xi_m\}$

such that $\xi_{\nu_j} \in T_{m_{j+1}} - T_{m_j}$. For simplicity, we rewrite ξ_{ν_j} by ξ_j and T_{m_j} by T_j. Hence $\xi_j \in T_{j+1} - T_j$. Since $(\widehat{T}_j)_\Omega^{\mathcal{O}} = T_j \not\ni \xi_j$, there exist $f_j \in \mathcal{O}(\Omega)$ such that

$$|f_j(\xi_j)| > \sup_{\zeta \in T_j} |f_j(\zeta)|.$$

Choose α_j such that

$$|f_j(\xi_j)| > \alpha_j > \sup_{\zeta \in T_j} |f_j(\zeta)|.$$

We set $h_j = f_j/\alpha_j$. Then we have $|h_j(\xi_j)| > 1$, $\sup_{\zeta \in T_j} |h_j(\zeta)| < 1$. For any sufficiently large integer k_j, We set $\varphi_j = h_j^{k_j}$. Then we have $\varphi_j \in \mathcal{O}(\Omega)$ and

$$\sup_{\zeta \in T_j} |\varphi_j(\zeta)| < \frac{1}{2^j}, \quad |\varphi_j(\xi_j)| > j + 1 + \sum_{k=1}^{j-1} |\varphi_k(\xi_j)| \quad (j = 1, 2, \cdots).$$

If we set $\varphi = \sum_{k=1}^{\infty} \varphi_k$, then $\varphi \in \mathcal{O}(\Omega)$. Hence we obtain

$$|\varphi(\xi_j)| = \left|\sum_{k=1}^{\infty} \varphi_k(\xi_j)\right| \geq |\varphi_j(\xi_j)| - \left|\sum_{k \neq j} \varphi_k(\xi_j)\right|$$

$$\geq j + 1 + \sum_{k=1}^{j-1} |\varphi_k(\xi_j)| - \sum_{k \neq j} |\varphi_k(\xi_k)|$$

$$= j + 1 - \sum_{k > j} |\varphi_k(\xi_j)|.$$

Since $\xi_j \in T_{j+1}$, we have $\xi_j \in T_k$ for $k \geq j+1$. Thus, we have

$$|\varphi_k(\xi_j)| \leq \sup_{\zeta \in T_k} |\varphi_k(\zeta)| < \frac{1}{2^k},$$

for $k \geq j+1$, which means that

$$\sum_{k > j} |\varphi_k(\xi_j)| \leq \sum_{k > j} \frac{1}{2^k} < 1.$$

Since $|\varphi(\xi_j)| \geq j$, we have $\lim_{j \to \infty} |\varphi(\xi_j)| = \infty$. Thus φ is unbounded on X. $\square$

Definition 1.18 (1) Let $\Omega \subset \mathbf{C}^n$ be an open set such that $\Omega \neq \mathbf{C}^n$. Ω is called a pseudoconvex open set if $-\log \text{dist}(z, \partial\Omega)$ is plurisubharmonic in Ω. In particular, we define $\mathbf{C}^n$ to be pseudoconvex.

(2) Let Ω be a bounded open set in $\mathbf{C}^n$. Ω is called strictly pseudoconvex if there exist a neighborhood W of $\partial\Omega$ and a strictly plurisubharmonic function ρ in W such that $\Omega \cap W = \{z \in W \mid \rho(z) < 0\}$.

Lemma 1.17 *(a) Let $\Omega \subset \mathbf{C}^n$ and $G \subset \mathbf{C}^m$. Suppose Ω and G are pseudoconvex open sets. Then $\Omega \times G$ is a pseudoconvex open set in $\mathbf{C}^{n+m}$.*

(b) Let $\{\Omega_j\}_{j \in J}$ be a family of pseudoconvex open sets in $\mathbf{C}^n$. Then the interior $(\cap_{j \in J} \Omega_j)^\circ$ of $\cap_{j \in J} \Omega_j$ is a pseudoconvex open set.

Proof. (a) We have
$$\partial(\Omega \times G) = (\partial\Omega \times \overline{G}) \cup (\overline{\Omega} \times \partial G).$$
Hence for $(z, w) \in \Omega \times G$, we have
$$\operatorname{dist}((z, w), \partial(\Omega \times G)) = \min\{\operatorname{dist}(z, \partial\Omega), \operatorname{dist}(w, \partial G)\}.$$
Consequently,
$$-\log \operatorname{dist}((z, w), \partial(\Omega \times G)) = -\inf\{\log \operatorname{dist}(z, \partial\Omega),\ \log \operatorname{dist}(w, \partial G)\}.$$
Then $-\log \operatorname{dist}((z, w), \partial(\Omega \times G))$ is plurisubharmonic in $\Omega \times G$, which implies that $\Omega \times G$ is pseudoconvex.

(b) We set $\Omega = (\cap_{j \in J} \Omega_j)^\circ$. For $z \in \Omega$, we have $\operatorname{dist}(z, \partial\Omega) = \inf_{j \in J} \operatorname{dist}(z, \partial\Omega_j)$. Hence we obtain
$$-\log \operatorname{dist}(z, \partial\Omega) = \sup_{j \in J}\{-\log \operatorname{dist}(z, \partial\Omega_j)\}.$$
Then $-\log \operatorname{dist}(z, \partial\Omega)$ is plurisubharmonic in Ω. □

Definition 1.19 Suppose $\Omega \subset \mathbf{C}^n$ is an open set and $K \subset \Omega$ is compact. Define
$$\widehat{K}_\Omega^P = \{z \in \Omega \mid \rho(z) \leq \max_{\zeta \in K} \rho(\zeta),\ \rho \in PS(\Omega)\}.$$
By definition, $\widehat{K}_\Omega^P$ is a closed subset in Ω and $K \subset \widehat{K}_\Omega^P$. In case $K = \widehat{K}_\Omega^P$, we say that K is $PS(\Omega)$-convex.

Lemma 1.18 *Suppose $\Omega \subset \mathbf{C}^n$ is an open set and $K \subset \Omega$ is compact. Then*
$$\widehat{K}_\Omega^P \subset \widehat{K}_\Omega^{\mathcal{O}}.$$

Proof. By Theorem 1.9, if $f \in \mathcal{O}(\Omega)$, then $|f| \in PS(\Omega)$, which completes the proof of Lemma 1.18. □

Theorem 1.13 *Suppose Ω is an open set in $\mathbf{C}^n$. Then the following statements are equivalent:*

(a) Ω is pseudoconvex.
(b) If $K \subset \Omega$ is compact, then $\widehat{K}_\Omega^P$ is compact.
(c) There exists $\rho \in PS(\Omega)$ such that for any real number α, the closure in Ω of the set

$$\Omega_\alpha := \{z \in \Omega \mid \rho(z) < \alpha\}$$

is compact.

Proof. (a) In case $\Omega \neq \mathbf{C}^n$.
(a)$\Longrightarrow$(c). We set

$$\rho(z) = \max\{|z|^2, -\log \operatorname{dist}(z, \partial\Omega)\}.$$

Then $\rho \in PS(\Omega)$. Let $\rho(z) < \alpha$. Then we have

$$|z|^2 < \alpha, \quad -\log \operatorname{dist}(z, \partial\Omega) < \alpha.$$

Thus we have $\operatorname{dist}(z, \partial\Omega) > e^{-\alpha}$, which means that $\Omega_\alpha \subset\subset \Omega$.

(c)$\Longrightarrow$(b). Suppose there exists $\rho_1 \in PS(\Omega)$ such that $\{z \in \Omega \mid \rho_1(z) < \alpha\} \subset\subset \Omega$ for any real α. Let $K \subset \Omega$ be compact. If we choose α such that $\alpha = \sup_{\zeta \in K} \rho_1(\zeta) + 1$, then

$$\widehat{K}_\Omega^P \subset \{z \in \Omega \mid \rho_1(z) \leq \sup_{\zeta \in K} \rho_1(\zeta)\} \subset\subset \Omega,$$

which implies that $\widehat{K}_\Omega^P$ is a compact subset of Ω.

(b)$\Longrightarrow$(a). We set $\varphi(z) = -\log \operatorname{dist}(z, \partial\Omega)$. We show that $\varphi(z)$ is plurisubharmonic in Ω. For $v, w \in \mathbf{C}^n$ and $w \neq 0$, we set

$$U = \{\lambda \in \mathbf{C}^n \mid v + \lambda w \in \Omega\}.$$

We set $g(\lambda) = \varphi(v + \lambda w)$. It is sufficient to show that $g(\lambda)$ is subharmonic in U. For $\lambda_0 \in U$, it is sufficient to show that $g(\lambda)$ is subharmonic in a neighborhood of λ_0. We set $a = v + \lambda_0 w$. Then $a \in \Omega$. Since $a + \lambda w = v + w(\lambda + \lambda_0)$, if we set $\psi(\lambda) = \varphi(a + \lambda w)$, then $\psi(\lambda) = g(\lambda + \lambda_0)$. So it is sufficient to show that $\psi(\lambda)$ is subharmonic in a neighborhood of 0. There exists $r > 0$ such that $\{a + \lambda w \mid |\lambda| \leq r\} \subset \Omega$. Let h be harmonic in a neighborhood of $|\lambda| \leq r$. It is sufficient to show that if $h(\lambda) \geq \psi(\lambda)$ on $|\lambda| = r$, then $h(\lambda) \geq \psi(\lambda)$ on $|\lambda| \leq r$. There exists a harmonic function h^*

in $|\lambda| \leq r$ such that $f = h + ih^*$ is holomorphic in $|\lambda| \leq r$. We have for $|\lambda| = r$

$$e^{h(\lambda)} \geq e^{\psi(\lambda)} = e^{-\log \text{dist}(a+\lambda w, \partial \Omega)} = \frac{1}{\text{dist}(a + \lambda w, \partial \Omega)}.$$

Thus on $|\lambda| = r$ we have

$$\text{dist}(a + \lambda w, \partial \Omega) \geq e^{-h(\lambda)} = |e^{-f(\lambda)}|.$$

Therefore, if $|\lambda| = r$, $|\zeta| < 1$, then $a + \lambda w + \zeta e^{-f(\lambda)} \in \Omega$. We fix ζ with $|\zeta| < 1$. For $0 \leq t \leq 1$, we set

$$\Gamma_t = \{a + \lambda w + t\zeta e^{-f(\lambda)} \mid |\lambda| \leq r\}.$$

Then we have

$$\Gamma_0 = \{a + \lambda w \mid |\lambda| \leq r\} \subset \Omega.$$

We set

$$T = \{t \in [0,1] \mid \Gamma_t \subset \Omega\}.$$

Then $T \subset [0,1]$, $0 \in T$. If T is closed and open in $[0,1]$, then $T = [0,1]$, and hence $1 \in T$. Then for $|\zeta| < 1$ we have

$$\Gamma_1 = \{a + \lambda w + \zeta e^{-f(\lambda)} \mid |\lambda| \leq r\} \subset \Omega.$$

Thus, for $|\lambda| \leq r$, we obtain

$$\text{dist}(a + \lambda w, \partial \Omega) \geq |e^{-f(\lambda)}| = e^{-h(\lambda)},$$

which implies that

$$\psi(\lambda) \leq h(\lambda) \qquad (|\lambda| \leq r).$$

Hence $\psi(\lambda)$ is subharmonic. Finally, we show that T is closed and open in $[0,1]$. Let $t_0 \in T$. Then $\Gamma_{t_0} \subset \Omega$. Since Ω is open, $\Gamma_t \subset \Omega$ for any sufficiently closed point t to t_0. Thus $t \in T$, and hence T is open. Next we show that T is closed. We set

$$K = \{a + \lambda w + t\zeta e^{-f(\lambda)} \mid |\lambda| = r, 0 \leq t \leq 1\}.$$

Then K is compact and $K \subset \Omega$. By the assumption, $\widehat{K}_\Omega^P$ is compact. Let $t \in T$. We set $g(\lambda) = a + \lambda w + t\zeta e^{-f(\lambda)}$. Since $\Gamma_t \subset \Omega$, $g(\lambda) \in \Omega$ for $|\lambda| \leq r$. Hence $g(\lambda)$ is holomorphic in $|\lambda| \leq r$. Let $\rho \in PS(\Omega)$. Then

$\rho \circ g(\lambda)$ is subharmonic in $|\lambda| \leq r$. By applying the maximum principle for subharmonic functions, we have for $|\lambda| \leq r$

$$\rho \circ g(\lambda) \leq \sup_{|\lambda|=r} \rho \circ g(\lambda) = \sup_{|\lambda|=r} \rho(a + \lambda w + \zeta e^{-f(\lambda)}) \leq \sup_{z \in K} \rho(z).$$

Thus we have $g(\lambda) \in \widehat{K}_\Omega^P$. Hence for $t \in T$ we obtain

$$\Gamma_t = \{g(\lambda) \mid |\lambda| \leq r\} \subset \widehat{K}_\Omega^P.$$

Let $t_\nu \in T$ and $t_\nu \to t_0$. Then $\Gamma_{t_\nu} \subset \widehat{K}_\Omega^P$. Since $\widehat{K}_\Omega^P$ is compact, we have $\Gamma_{t_0} \subset \widehat{K}_\Omega^P \subset \Omega$, and hence $t_0 \in T$. Thus T is closed. Hence T is closed and open in $[0,1]$, which shows that (b)$\Longrightarrow$(a).

(b) In case $\Omega = \mathbf{C}^n$. By definition, Ω is pseudoconvex. If $K \subset \mathbf{C}^n$ is compact, then by Lemma 1.13 $\widehat{K}_\Omega^O$ is bounded. Since $\widehat{K}_\Omega^P \subset \widehat{K}_\Omega^O$, $\widehat{K}_\Omega^P$ is compact. We set $\rho(z) = |z|^2$. Then $\rho \in PS(\Omega)$ and $\Omega_\alpha = \{z \in \mathbf{C}^n \mid \rho(z) < \alpha\} \subset\subset \mathbf{C}^n$. $\square$

Corollary 1.6 *Let $\Omega \subset \mathbf{C}^n$ be an open set. If Ω is a domain of holomorphy, then Ω is pseudoconvex.*

Proof. Let $K \subset \Omega$ be compact. It follows from Theorem 1.12 that $\widehat{K}_\Omega^O$ is compact. Since $\widehat{K}_\Omega^P \subset \widehat{K}_\Omega^O$, $\widehat{K}_\Omega^P$ is compact. By Theorem 1.13, Ω is pseudoconvex. $\square$

Lemma 1.19 *(Dini's theorem) Suppose K is a compact subset in $\mathbf{C}^n$ and that $\{f_n\}$ is a sequence of real-valued continuous functions on K that converges to f monotonically on K. Then $\{f_n\}$ converges to f uniformly on K.*

Proof. Suppose

$$f_1(x) \geq f_2(x) \geq \cdots, \quad f_n(x) \to f(x).$$

We set $g_n(x) = f_n(x) - f(x)$. We denote by α_n the maximum of g_n in K. Then α_n is monotonically decreasing. Let $\lim_{n\to\infty} \alpha_n = \alpha$. It is sufficient to show that $\alpha = 0$. Suppose $\alpha > 0$. Let $x_n \in K$ be a point such that $\alpha_n = g_n(x_n)$. We can choose a convergent subsequence $\{x_{k_n}\}$ of $\{x_n\}$. Define $\lim_{n\to\infty} x_{k_n} = x_0$. Then we have $g_{k_n}(x_0) \to 0$ as $n \to \infty$. If we choose N sufficiently large, then we obtain

$$n \geq N \implies g_{k_n}(x_0) < \frac{\alpha}{2}.$$

On the other hand we have for $m \geq n$

$$g_{k_n}(x_{k_m}) \geq g_{k_m}(x_{k_m}) = \alpha_{k_m} \geq \alpha,$$

which implies that $g_{k_n}(x_0) \geq \alpha$. This is a contradiction. $\square$

Theorem 1.14 *Suppose a real-valued function $\lambda \in \mathcal{D}(\mathbf{C}^n)$ satisfies the following properties:*

(1) If $\lambda \geq 0$ and $|z| > 1$, then $\lambda(z) = 0$.
(2) λ depends only on $|z_1|, \cdots, |z_n|$.
(3) $\int_{\mathbf{C}^n} \lambda(z) dV(z) = 1$, where dV denotes the Lebesgue measure on $\mathbf{C}^n$.

Let Ω be an open set in $\mathbf{C}^n$ and let u be a plurisubharmonic function in Ω. For $\varepsilon > 0$, we set

$$\Omega_\varepsilon = \{z \in \Omega \mid dist(z, \partial\Omega) > \varepsilon\}$$

and

$$u_\varepsilon(z) = \int_{|\zeta|<1} u(z - \varepsilon\zeta)\lambda(\zeta) dV(\zeta) \quad (z \in \Omega_\varepsilon).$$

Then u_ε is plurisubharmonic in Ω_ε and $u_\varepsilon \in C^\infty(\Omega_\varepsilon)$. Moreover, we have $u_\varepsilon \downarrow u$ as $\varepsilon \downarrow 0$.

Proof. From the condition (2), we have

$$u_\varepsilon(z) = \int_{|\zeta|<1} \left[\frac{1}{2\pi}\int_0^{2\pi} u(z - e^{it}\varepsilon\zeta) dt\right] \lambda(\zeta) dV(\zeta).$$

We set $h(w) = u(z + w(-\zeta))$. Since h is subharmonic in a neighborhood of 0, it follows from Lemma 1.2 that

$$0 < \varepsilon_1 < \varepsilon_2 \Rightarrow \frac{1}{2\pi}\int_0^{2\pi} h(\varepsilon_1 e^{it}) dt \leq \frac{1}{2\pi}\int_0^{2\pi} h(\varepsilon_2 e^{it}) dt.$$

Thus, $u_\varepsilon \downarrow u$ as $\varepsilon \downarrow 0$. On the other hand, u_ε is expressed by

$$u_\varepsilon(z) = \int u(\zeta) \lambda\left(\frac{z-\zeta}{\varepsilon}\right) \varepsilon^{-2n} dV(\zeta),$$

which implies that $u_\varepsilon \in C^\infty(\Omega_\varepsilon)$. Let $a \in \Omega_\varepsilon$ and $w \in \mathbf{C}^n$. In order that u_ε is plurisubharmonic in Ω_ε, it is sufficient to prove that $h(\eta) = u_\varepsilon(a + \eta w)$ is subharmonic in a neighborhood of 0. Since $u(a - \varepsilon\zeta + \eta w)$ is subharmonic with respect to η in a neighborhood of 0, we have for any sufficiently small $r > 0$

$$\frac{1}{2\pi}\int_0^{2\pi} u_\varepsilon(a + re^{i\theta}w)d\theta$$
$$= \int_{|\zeta|<1}\left[\frac{1}{2\pi}\int_0^{2\pi} u(a + re^{i\theta}w - \varepsilon\zeta)d\theta\right]\lambda(\zeta)dV(\zeta)$$
$$\geq \int_{|\zeta|<1} u(a - \varepsilon\zeta)\lambda(\zeta)dV(\zeta) = u_\varepsilon(a).$$

□

Theorem 1.15 *Let Ω be a pseudoconvex domain in $\mathbf{C}^n$. Then there exists a C^∞ strictly plurisubharmonic function u in Ω such that for any real number C the closure of $\{z \in \Omega \mid u(z) < C\}$ in Ω is compact.*

Proof. We set $\delta(z) = \mathrm{dist}(z, \partial\Omega)$. Then $-\log\delta(z)$ is plurisubharmonic in Ω. Define

$$\Phi(z) = -\log\delta(z) + |z|^2$$

and

$$\Omega_C = \{z \in \Omega \mid \Phi(z) < C\}.$$

Then Ω_C is a relatively compact subset of Ω. For any sufficiently small $\varepsilon > 0$, define

$$\Phi_j(z) = \int_{\Omega_{j+1}} \Phi(\zeta)\lambda\left(\frac{z-\zeta}{\varepsilon}\right)\varepsilon^{-2n}dV(\zeta) + \varepsilon|z|^2,$$

where λ is the function defined in Theorem 1.14. By definition, we have $\Phi_j \in C^\infty(\mathbf{C}^n)$. Let $z \in \overline{\Omega}_j$. We set $(z-\zeta)/\varepsilon = w$. For $|w| \leq 1$ and any sufficiently small ε we have $\zeta = z - \varepsilon w \in \Omega_{j+1}$. Hence Φ_j can be written

$$\Phi_j(z) = \int_{|w|<1} \Phi(z - \varepsilon w)\lambda(w)dV(w) + \varepsilon|z|^2.$$

By Theorem 1.14, if $\varepsilon \downarrow 0$, then $\Phi_j \downarrow \Phi$ in $\overline{\Omega}_j$ and Φ_j is strictly plurisubharmonic in a neighborhood of $\overline{\Omega}_j$. It follows from the Dini theorem (Lemma 1.19) that $\Phi_j < \Phi + 1$ on $\overline{\Omega}_j$. Let $\chi \in C^\infty(\mathbf{R})$ satisfy $\chi(t) = 0$ if $t \leq 0$, $\chi(t) > 0$ if $t > 0$, $\chi'(t) > 0$, $\chi''(t) > 0$ (for example, $\chi(t) = te^{-\frac{1}{t}}$ $(t > 0)$, 0 $(t \leq 0)$). Define

$$\Psi_j = \chi(\Phi_j + 2 - j).$$

Then Ψ_j is strictly plurisubharmonic in a neighborhood of $\overline{\Omega}_j\setminus\Omega_{j-1}$ and $\Psi_j > 0$. Φ_0 is strictly plurisubharmonic and $\Phi_0 \geq \Phi$ in a neighborhood of $\overline{\Omega}_0$. Since Ψ_1 is strictly plurisubharmonic and $\Psi_1 > 0$ in a neighborhood of $\overline{\Omega}_1\setminus\Omega_0$, $\Phi_0 + a_1\Psi_1 > \Phi$ in a neighborhood of $\overline{\Omega}_1\setminus\Omega_0$ if $a_1 > 0$. Further, by Corollary 1.4 there exists a constant $C > 0$ such that

$$\sum_{i,j=1}^n \frac{\partial^2 \Psi_1}{\partial z_i \partial \bar{z}_j}(z) w_i \bar{w}_j \geq C|w|^2.$$

Similarly, there exists a constant $C_1 > 0$ such that

$$\left| \sum_{i,j=1}^n \frac{\partial^2 \Phi_0}{\partial z_i \partial \bar{z}_j}(z) w_i \bar{w}_j \right| \leq C_1 |w|^2.$$

Hence for any sufficiently large $a_1 > 0$, we have

$$\sum_{i,j=1}^n \frac{\partial^2 \Phi_0}{\partial z_i \partial \bar{z}_j}(z) w_i \bar{w}_j + a_1 \sum_{i,j=1}^n \frac{\partial^2 \Psi_1}{\partial z_i \partial \bar{z}_j}(z) w_i \bar{w}_j \geq a_1 C|w|^2 - C_1 |w|^2 > 0.$$

Hence $u_1 = \Phi_0 + a_1\Psi_1$ is strictly plurisubharmonic in a neighborhood of $\overline{\Omega}_1\setminus\Omega_0$. Since Φ_0 is strictly plurisubharmonic, $\Phi_0 \geq \Phi$ in a neighborhood of $\overline{\Omega}_0$ and $\Psi_1 \geq 0$ in a neighborhood of $\overline{\Omega}_1$, u_1 is strictly plurisubharmonic and $u_1 > \Phi$ in a neighborhood of $\overline{\Omega}_1$. Repeating this process, there exist positive numbers $a_1, \cdots, a_m$ such that

$$u_m = \Phi_0 + \sum_{j=1}^m a_j \Psi_j$$

is strictly plurisubharmonic and $u_m > \Phi$ in a neighborhood of $\overline{\Omega}_m$. If $k \geq j+3$, then $\Psi_k = 0$ on Ω_j. Thus there exists $u = \lim_{m\to\infty} u_m$ such that u is strictly plurisubharmonic, $u \in C^\infty(\Omega)$ and $u \geq \Phi$ in Ω. $\square$

Lemma 1.20 *Let f be differentiable at $x = a$ and let $f(a) = 0$. Let h be continuous at $x = a$. Then fh is differentiable at $x = a$. Moreover, we have*

$$\{f(x)h(x)\}'_{x=a} = h(a)f'(a).$$

Proof. By the definition of differentiation, we have

$$\lim_{x\to a} \frac{h(x)f(x) - h(a)f(a)}{x-a} = \lim_{x\to a} \frac{h(x)(f(x) - f(a))}{x-a} = h(a)f'(a).$$

$\square$

Definition 1.20 Let $\Omega \subset \mathbf{R}^n$ be an open set. We say that Ω has a C^k ($k \geq 1$) boundary if there exist a neighborhood U of $\partial\Omega$ and a C^k function ρ in U such that

(1) $\Omega \cap U = \{x \in U \mid \rho(x) < 0\}$.
(2) $d\rho \neq 0$ on $\partial\Omega$, where

$$d\rho(x) = \sum_{j=1}^{n} \frac{\partial \rho}{\partial x_j}(x) dx_j.$$

Lemma 1.21 Let $\Omega = \{x \mid \rho(x) < 0\} \subset \mathbf{R}^n$ be a bounded domain with C^k ($k \geq 1$) boundary and let f be a C^k function in a neighborhood of $\overline{\Omega}$. Assume that $f(x) = 0$ for all $x \in \partial\Omega$. Then for $P \in \partial\Omega$ there exist a neighborhood U of P and a C^{k-1} function h in U such that $f(x) = \rho(x)h(x)$ for $x \in U$.

Proof. Without loss of generality, we may assume that $P = 0$. Since $d\rho \neq 0$ on $\partial\Omega$, we may assume that there exists a neighborhood U of P such that if $x = (x', x_n)$ ($x' = (x_1, \cdots, x_{n-1})$) forms a coordinate system in U. Then we have $\rho(x) = x_n$ for $x \in U$. Since $f(x', 0) = 0$, we have

$$f(x', x_n) = f(x', x_n) - f(x', 0) = \int_0^1 \frac{d}{dt}\{f(x', tx_n)\} dt$$

$$= x_n \int_0^1 \frac{\partial f}{\partial x_n}(x', tx_n) dt.$$

Define

$$h(x', x_n) = \int_0^1 \frac{\partial f}{\partial x_n}(x', tx_n) dt.$$

Then $h(x', x_n)$ is of class C^{k-1} in U and $f(x) = \rho(x)h(x)$ for $x \in U$. □

Theorem 1.16 Let $\Omega \subset \mathbf{C}^n$ be an open set with C^2 boundary. Let $\Omega = \{z \in \widetilde{\Omega} \mid \rho(z) < 0\}$, where ρ is a C^2 function in a neighborhood $\widetilde{\Omega}$ of $\overline{\Omega}$ and satisfies $d\rho \neq 0$ on $\partial\Omega$. Then Ω is pseudoconvex if and only if

$$\sum_{j,k=1}^{n} \frac{\partial^2 \rho}{\partial z_j \partial \bar{z}_k}(z) w_j \bar{w}_k \geq 0 \qquad (1.13)$$

for all z and $w = (w_1, \cdots, w_n)$ satisfying

$$z \in \partial\Omega, \quad \sum_{j=1}^{n} \frac{\partial \rho}{\partial z_j}(z) w_j = 0.$$

Proof. Suppose ρ_1 is a C^2 defining function for Ω. By Lemma 1.21 there exists a C^1 function h in a neighborhood V of $\partial\Omega$ such that $\rho_1 = h\rho$. Since $d\rho_1 = h d\rho$ on $\partial\Omega$, we have $h > 0$ in V. For z and w satisfying $z \in \partial\Omega$ and $\sum_{j=1}^{n} \frac{\partial \rho_1}{\partial z_j}(z) w_j = 0$, we have

$$\sum_{j=1}^{n} \frac{\partial \rho}{\partial z_j}(z) w_j = 0.$$

By Lemma 1.20, we obtain

$$\sum_{j,k=1}^{n} \frac{\partial^2 \rho_1}{\partial z_j \partial \bar{z}_k}(z) w_j \bar{w}_k = h(z) \sum_{j,k=1}^{n} \frac{\partial^2 \rho}{\partial z_j \partial \bar{z}_k}(z) w_j \bar{w}_k \geq 0.$$

Thus the condition (1.13) is independent of the choice of the defining function ρ. Suppose Ω is pseudoconvex. Define

$$\tilde{\rho}(z) = \begin{cases} -\mathrm{dist}(z, \partial\Omega) & (z \in \overline{\Omega}) \\ \mathrm{dist}(z, \partial\Omega) & (z \in \Omega^c). \end{cases}$$

Then $\tilde{\rho}$ is a C^2 function in a neighborhood of $\partial\Omega$ and satisfies $d\tilde{\rho} \neq 0$ on $\partial\Omega$ (see Krantz-Parks [KRP]). If $z \in \Omega$ is sufficiently close to $\partial\Omega$, then for $\delta(z) := \mathrm{dist}(z, \partial\Omega)$, $-\log \delta(z)$ is plurisubharmonic. By Theorem 1.10, we have

$$\sum_{j,k=1}^{n} \frac{\tilde{\partial}^2}{\partial z_j \partial \bar{z}_k}(-\log \delta(z)) w_j \bar{w}_k$$

$$= -\frac{1}{\delta} \sum_{j,k=1}^{n} \frac{\partial^2 \delta}{\partial z_j \partial \bar{z}_k}(z) w_j \bar{w}_k + \frac{1}{\delta(z)^2} \left| \sum_{j=1}^{n} \frac{\partial \delta}{\partial z_j}(z) w_j \right|^2 \geq 0.$$

Thus if $z \in \Omega$ is sufficiently close to $\partial\Omega$, then

$$\sum_{j,k=1}^{n} \frac{\partial^2 \tilde{\rho}}{\partial z_j \partial \bar{z}_k}(z) w_j \bar{w}_k + \frac{1}{\delta(z)} \left| \sum_{j=1}^{n} \frac{\partial \delta}{\partial z_j}(z) w_j \right|^2 \geq 0. \tag{1.14}$$

Suppose that $z \in \partial\Omega$, $\sum_{j=1}^{n} \frac{\partial \tilde{\rho}}{\partial z_j}(z) w_j = 0$. We choose sequences $\{z^{(i)}\}$ and $\{w^{(i)}\}$ satisfying

$$z^{(i)} \in \Omega, \quad z^{(i)} \to z, \quad w^{(i)} \to w, \quad \sum_{j=1}^{n} \frac{\partial \tilde{\rho}}{\partial z_j}(z^{(i)}) w_j^{(i)} = 0.$$

By (1.14) we have

$$\sum_{j,k=1}^{n} \frac{\partial^2 \tilde{\rho}}{\partial z_j \partial \bar{z}_k}(z^{(i)}) w_j^{(i)} \bar{w}_k^{(i)} \geq 0.$$

Letting $i \to \infty$ we have (1.13).

Conversely we assume that (1.13) holds. We set $\varphi(\tau) = \log \delta(z + \tau w)$. It is sufficient to show that $-\varphi(\tau)$ is subharmonic. Assume that

$$\frac{\partial^2 \varphi}{\partial \tau \partial \bar{\tau}}(0) = c > 0.$$

Using Taylor's formula, we obtain

$$\varphi(\tau) = \varphi(0) + 2\mathrm{Re}\left(\frac{\partial \varphi}{\partial \tau}(0)\tau\right) + \mathrm{Re}\left(\frac{\partial^2 \varphi}{\partial \tau^2}(0)\tau^2\right) + \frac{\partial^2 \varphi}{\partial \tau \partial \bar{\tau}}(0)|\tau|^2 + o(|\tau|^2).$$

We set

$$A = 2\frac{\partial \varphi}{\partial \tau}(0), \quad B = \frac{\partial^2 \varphi}{\partial \tau^2}(0).$$

Then

$$\varphi(\tau) = \log \delta(z) + \mathrm{Re}(A\tau + B\tau^2) + c|\tau|^2 + o(|\tau|^2).$$

Suppose $z_0 \in \partial \Omega$ satisfies $\delta(z) = |z - z_0|$. For $0 < s \leq 1$, define

$$\psi_s(\tau) = z + \tau w + s(z_0 - z)e^{A\tau + B\tau^2}.$$

Then

$$\delta(\psi_s(\tau)) = \delta(z + \tau w + s(z_0 - z)e^{A\tau + B\tau^2})$$
$$\geq \delta(z + \tau w) - s|z - z_0||e^{A\tau + B\tau^2}|.$$

On the other hand, we have

$$\delta(z + \tau w) = e^{\varphi(\tau)} = \delta(z)|e^{A\tau + B\tau^2}|e^{c|\tau|^2 + o(|\tau|^2)},$$

which implies that

$$\delta(\psi_s(\tau)) \geq \delta(z)|e^{A\tau + B\tau^2}|e^{c|\tau|^2/2} - s\delta(z)|e^{A\tau + B\tau^2}|$$
$$= \delta(z)|e^{A\tau + B\tau^2}|(e^{c|\tau|^2/2} - s).$$

For s with $0 < s < 1$ we have $\psi_s(0) = z + s(z_0 - z) \in \Omega$. Hence for $0 < s < 1$ and any sufficiently small $|\tau|$, we have $\psi_s(\tau) \in \Omega$, and hence $\psi_1(\tau) \in \overline{\Omega}$. If we set $f(\tau) = \delta(\psi_1(\tau))$, then $f(0) = 0$. So we have for any sufficiently small $|\tau|$

$$f(\tau) \geq \frac{c}{4}\delta(z)|e^{A\tau+B\tau^2}|\|\tau|^2. \tag{1.15}$$

Thus $f(\tau)$ takes a local minimum at $\tau = 0$, and hence $\frac{\partial f}{\partial \tau}(0) = 0$. Further, by (1.15), the case that $\frac{\partial^2 f}{\partial \tau^2}(0) = \frac{\partial^2 f}{\partial \tau \partial \bar\tau}(0) = 0$ does not occur. By Taylor's formula, we have

$$f(\tau) = \mathrm{Re}\left(\frac{\partial^2 f}{\partial \tau^2}(0)\tau^2\right) + \frac{\partial^2 f}{\partial \tau \partial \bar\tau}(0)|\tau|^2 + o(|\tau|^2).$$

In the above equation we set $\tau = e^{it}\lambda$, where t and λ are real numbers. Then in case $\frac{\partial^2 f}{\partial \tau \partial \bar\tau}(0) \leq 0$, $f(\tau)$ is negative for some t, which implies that $\frac{\partial^2 f}{\partial \tau \partial \bar\tau}(0) > 0$. For any sufficiently small $|\tau|$, we have $\psi_1(\tau) \in \overline{\Omega}$, and hence $\tilde\rho(\psi(\tau)) = -\delta(\psi(\tau)) = -f(\tau)$. Thus if we set $\psi_1(\tau) = \lambda(\tau)$, then

$$\sum_{j=1}^{n} \frac{\partial \tilde\rho}{\partial z_j}(z_0)\lambda'_j(0) = 0$$

and

$$\sum_{j,k=1}^{n} \frac{\partial^2 \tilde\rho}{\partial z_j \partial \bar z_k}(z_0)\lambda'_j(0)\overline{\lambda'_j(0)} < 0.$$

This contradicts (1.13). □

Definition 1.21 Let $\Omega \subset\subset \mathbf{C}^n$ be an open set. Ω is called an analytic polyhedron if there exist a neighborhood U of $\overline{\Omega}$ and a finite number of functions $f_1, \cdots, f_k \in \mathcal{O}(U)$ such that

$$\Omega = \{z \in U \mid |f_1(z)| < 1, \cdots, |f_k(z)| < 1\}.$$

The collection of functions $f_1, \cdots, f_k$ is called a frame for Ω.

Theorem 1.17 *Every analytic polyhedron is holomorphically convex.*

Proof. Let $\Omega = \{z \in U \mid |f_1(z)| < 1, \cdots, |f_k(z)| < 1\}$, where U is a neighborhood of $\overline{\Omega}$ and $f_1, \cdots, f_k \in \mathcal{O}(U)$. Let $K \subset \Omega$ be compact. We set $r_j = \sup_K |f_j|$. Then $r_j < 1$. Now we have

$$\widehat{K}_{\Omega}^{\mathcal{O}} \subset \{z \in U \mid |f_1(z)| \leq r_1, \cdots, |f_k(z)| \leq r_k\} \subset\subset \Omega,$$

which implies that Ω is holomorphically convex. □

Theorem 1.18 Let $\Omega \subset \mathbf{C}^n$ be an open set and let K be a compact subset of Ω. Suppose K is $\mathcal{O}(\Omega)$-convex. Then K has a neighborhood basis consisting of analytic polyhedra defined by frames of functions holomorphic in Ω.

Proof. Let $U \subset\subset \Omega$ be a neighborhood of K. Since $K = \widehat{K}_\Omega^\mathcal{O}$, $\widehat{K}_\Omega^\mathcal{O} \cap \partial U$ is empty. Let $a \in \partial U$. Then there exists $f_a \in \mathcal{O}(\Omega)$ such that $|f_a(a)| > \sup_{z \in K} |f_a(z)|$. We choose r such that $|f_a(a)| > r > \sup_{z \in K} |f_a(z)|$, and set $g_a = f_a/r$, then we have $|g_a(a)| > 1$, $\sup_{z \in K} |g_a(z)| < 1$. Thus there exists a neighborhood W_a of a such that for $z \in W_a$, $|g_a(z)| > 1$. By the argument of compactness, there exist open sets $W_1, \cdots, W_k$ and functions $g_1, \cdots, g_k \in \mathcal{O}(\Omega)$ such that

$$\partial U \subset \bigcup_{j=1}^{k} W_j, \quad |g_j(z)| > 1 \ (z \in W_j).$$

We set $\tilde{\Omega} = \{z \in U \mid |f_j(z)| < 1, j = 1, \cdots, k\}$. Then $K \subset \tilde{\Omega} \subset\subset U$, which completes the proof of Theorem 1.18. $\square$

Exercises

1.1 Let f be a C^1 function defined on a domain in $\mathbf{C}$. Show that the following equalities hold.

$$\overline{\frac{\partial f}{\partial z}} = \frac{\partial \bar{f}}{\partial \bar{z}}, \quad \overline{\frac{\partial f}{\partial \bar{z}}} = \frac{\partial \bar{f}}{\partial z}.$$

1.2 Let $\Omega \subset \mathbf{C}^n$ be an open set. Show that a real-valued function u on Ω is upper semicontinuous if and only if

$$\limsup_{\Omega \ni z \to a} u(z) \leq u(a) \quad (a \in \Omega),$$

where we define

$$\limsup_{\Omega \ni z \to a} u(z) = \lim_{\delta \to 0+} \left\{ \sup_{z \in \Omega \cap B(a,\delta)} u(z) \right\}.$$

1.3 (Maximum principle) Let $\Omega \subset \mathbf{C}^n$ be a domain and let f be a holomorphic function in Ω. Suppose there exists a point $\xi \in \Omega$ such that $|f(z)| \leq |f(\xi)|$ for all $z \in \Omega$. Show that f is constant.

1.4 Let $\Omega \subset \mathbf{C}^n$ be an open set and let $\{f_j\}$ be a sequence of holomorphic functions in Ω. Suppose $\{f_j\}$ converges uniformly to f on every compact subset of Ω. Show that f is holomorphic in Ω.

1.5 Let f be a holomorphic function in a domain $\Omega \subset \mathbf{C}^n$. Suppose there exists a point $\xi \in \Omega$ such that for all multi-indices $\alpha = (\alpha_1, \cdots, \alpha_n)$,

$$(\partial^\alpha f)(\xi) := \frac{\partial^{|\alpha|} f}{\partial z_1^{\alpha_1} \cdots \partial z_n^{\alpha_n}}(\xi) = 0,$$

where each α_j is a nonnegative integer and $|\alpha| = \alpha_1 + \cdots + \alpha_n$. Show that $f = 0$.

1.6 Construct the function λ in Theorem 1.14.

1.7 (**Schwarz lemma**) Let f be a holomorphic function in the unit disc $B(0,1) \subset \mathbf{C}$. Assume that $f(0) = 0$ and $|f(z)| \leq 1$ for $z \in B(0,1)$. Prove that

$$|f(z)| \leq |z|, \quad |f'(0)| \leq 1.$$

If either $|f(z)| = |z|$ for some $z \neq 0$ or if $|f'(0)| = 1$, prove that f is expressed by $f(z) = \alpha z$ for some complex constant α of unit modulus.

1.8 (**Schwarz-Pick lemma**) Let $f : B(0,1) \to B(0,1)$ be a holomorphic function in the unit disc $B(0,1) \subset \mathbf{C}$. Assume that $f(z_1) = w_1$ and $f(z_2) = w_2$ for some $z_1, z_2 \in B(0,1)$. Show that

$$\left| \frac{w_1 - w_2}{1 - w_1 \bar{w}_2} \right| \leq \left| \frac{z_1 - z_2}{1 - z_1 \bar{z}_2} \right|$$

and

$$|f'(z_1)| \leq \frac{1 - |w_1|^2}{1 - |z_1|^2}.$$

If the equality holds one of the above inequalities, prove that $f : B(0,1) \to B(0,1)$ is a one-to-one onto mapping.

1.9 Let $f : \Omega \to \mathbf{C}$ and $g : \Omega \to \mathbf{C}$ be holomorphic functions in an open set $\Omega \subset \mathbf{C}$ and let $a \in \Omega$. If $f(a) = g(a) = 0$ and $g'(a) \neq 0$, prove that

$$\lim_{z \to a} \frac{f(z)}{g(z)} = \lim_{z \to a} \frac{f'(z)}{g'(z)}.$$

1.10 (**Uniqueness theorem**) Let $f : \Omega \to \mathbf{C}$ be a holomorphic function in a domain $\Omega \subset \mathbf{C}$. If there exist a point $a \in \Omega$ and a sequence $\{z_n\}$ in Ω which converges a such that $z_n \neq a$ and $f(z_n) = 0$ for all n, then $f = 0$.

1.11 (Open mapping theorem) Let $f : \Omega \to \mathbf{C}$ be a non-constant holomorphic function in an open set $\Omega \subset \mathbf{C}$. Prove that $f(\Omega)$ is an open set.

1.12 Let f be a holomorphic function in a simply connected domain $\Omega \subset \mathbf{C}$. Assume that f never vanishes. Prove the following:

(1) For a natural number m, there exists a holomorphic function g in Ω such that $f = g^m$.

(2) There exists a holomorphic function h in Ω such that $f = e^h$.

1.13 Prove the following:
Let $f : \Omega \to \mathbf{C}$ be a holomorphic function in an open set $\Omega \subset \mathbf{C}$. If f is one-to-one, then $f'(z) \neq 0$ for all $z \in \Omega$.

1.14 Prove the following:
Let $f : \Omega \to \mathbf{C}$ be a holomorphic function in a domain $\Omega \subset \mathbf{C}$. If f is one-to-one, then $f^{-1} : f(\Omega) \to \Omega$ is holomorphic. Moreover, $(f^{-1})'(w) = \{f'(f^{-1}(w))\}^{-1}$.

Chapter 2

The $\bar{\partial}$ Problem in Pseudoconvex Domains

In this chapter we give the proof of L^2 estimates for the $\bar{\partial}$ problem in pseudoconvex domains in $\mathbf{C}^n$ due to Hörmander [HR2]. The assertion that Ω pseudoconvex implies Ω is a domain of holomorphy is known as the Levi problem. The Levi problem was first solved affirmatively in $\mathbf{C}^2$ by Oka [OkA1] in 1942, and in $\mathbf{C}^n$ it was solved independently by Oka [OkA3], Bremermann [BRE] and Norguet [NOR] in the early 1950s. In 2.2 we give the proof of the Levi problem by the method of Hörmander [HR2]. In 2.3 we prove L^2 extensions of holomorphic functions from submanifolds of bounded pseudoconvex domains in $\mathbf{C}^n$ which was first proved by Ohsawa and Takegoshi [OHT].

2.1 The Weighted L^2 Space

For the preparation of the next section, we study the weighted L^2 space whose element consists of differential forms. Moreover, we prove Green's theorem which is useful for the proof of the Ohsawa-Takegoshi extension theorem.

Definition 2.1 Let $\Omega \subset\subset \mathbf{R}^n$ be a domain with C^1 boundary and let ρ be a defining funtion for Ω, that is, ρ is a real-valued C^1 function in a neighborhood G of $\overline{\Omega}$ and satisfies

$$\Omega = \{x \in G \mid \rho(x) < 0\}, \quad d\rho(x) := \sum_{j=1}^{n} \frac{\partial \rho}{\partial x_j}(x) dx_j \neq 0 \quad (x \in \partial\Omega).$$

Define the surface element dS by

$$dS = \sum_{j=1}^{n}(-1)^{j-1}\nu_j dx_1 \wedge \cdots \wedge [dx_j] \wedge \cdots \wedge dx_n, \qquad (2.1)$$

where, $[dx_j]$ means that dx_j is omitted, and $\nu = (\nu_1, \cdots, \nu_n)$ is the unit outward normal vector for the boundary $\partial\Omega$.

If we set

$$|d\rho| = \sqrt{\left(\frac{\partial\rho}{\partial x_1}\right)^2 + \cdots + \left(\frac{\partial\rho}{\partial x_n}\right)^2},$$

then ν can be written

$$\nu = \frac{1}{|d\rho|}\left(\frac{\partial\rho}{\partial x_1}, \cdots, \frac{\partial\rho}{\partial x_n}\right).$$

Now we prove Green's theorem.

Theorem 2.1 *(**Green's theorem**) Let u be a C^1 function on $\overline{\Omega}$. Then*

$$\int_{\partial\Omega}\frac{\partial\rho}{\partial x_j}u\frac{dS}{|d\rho|} = \int_{\Omega}\frac{\partial u}{\partial x_j}dV,$$

where dV is the Lebesgue measure in $\mathbf{R}^n$.

Proof. We set

$$d[x]_k = dx_1 \wedge \cdots \wedge [dx_k] \wedge \cdots \wedge dx_n.$$

Then by (2.1) we obtain

$$\int_{\partial\Omega}\frac{\partial\rho}{\partial x_j}u\frac{dS}{|d\rho|} = \int_{\partial\Omega}\frac{\partial\rho}{\partial x_j}u\frac{1}{|d\rho|^2}\sum_{k=1}^{n}(-1)^{k-1}\frac{\partial\rho}{\partial x_k}d[x]_k$$

$$= \int_{\partial\Omega}\frac{\partial\rho}{\partial x_j}u\frac{1}{|d\rho|^2}\sum_{k\neq j}(-1)^{k-1}\frac{\partial\rho}{\partial x_k}d[x]_k$$

$$+ \int_{\partial\Omega}\frac{\partial\rho}{\partial x_j}u\frac{1}{|d\rho|^2}(-1)^{j-1}\frac{\partial\rho}{\partial x_j}d[x]_j.$$

Since $\rho = 0$ on $\partial\Omega$, we have

$$\frac{\partial\rho}{\partial x_j}dx_j = -\sum_{i\neq j}\frac{\partial\rho}{\partial x_i}dx_i.$$

Consequently,

$$\int_{\partial\Omega} \frac{\partial \rho}{\partial x_j} u \frac{dS}{|d\rho|} = \int_{\partial\Omega} u(-1)^{j-1} d[x]_j = \int_{\Omega} \frac{\partial u}{\partial x_j} dV,$$

which completes the proof of Theorem 2.1. □

Definition 2.2 Let $\Omega \subset \mathbf{C}^n$ be an open set and let $\varphi \in C^\infty(\Omega)$ be a real-valued function. We denote by $L^2(\Omega, \varphi)$ the space of L^2 integrable functions with respect to the measure $e^{-\varphi} dV$, where dV is the Lebesgue measure in $\mathbf{C}^n$. Let p and q be integers with $0 \leq p, q \leq n$. For multi-indices $\alpha = (i_1, \cdots, i_p)$ and $\beta = (j_1, \cdots, j_q)$, where $i_1, \cdots, i_p, j_1, \cdots, j_q$ are integers between 1 and n, define $|\alpha| = p$, $|\beta| = q$ and

$$dz^\alpha = dz_{i_1} \wedge \cdots \wedge dz_{i_p}, \quad d\bar{z}^\beta = d\bar{z}_{j_1} \wedge \cdots \wedge d\bar{z}_{j_q}.$$

We also denote by $L^2_{(p,q)}(\Omega, \varphi)$ the space of all (p,q) forms f on Ω whose coefficients $f_{\alpha,\beta}$ belong to $L^2(\Omega, \varphi)$.

Definition 2.3 Let f be a (p,q) form in Ω. Then f is expressed by

$$f = \sum_{\substack{|\alpha|=p \\ |\beta|=q}}{}' f_{\alpha,\beta} dz^\alpha \wedge d\bar{z}^\beta,$$

where $\sum'$ implies that the summation is performed only over strictly increasing multi-indices. Further, we set

$$|f|^2 = \sum_{\alpha,\beta}{}' |f_{\alpha,\beta}|^2.$$

By definition, $f \in L^2_{(p,q)}(\Omega, \varphi)$ means that

$$\|f\|_\varphi^2 := \int_\Omega |f|^2 e^{-\varphi} dV < \infty.$$

We denote by $L^2_{(p,q)}(\Omega, \text{loc})$ the space of all (p,q) forms f on Ω whose coefficients $f_{\alpha,\beta}$ are L^2 functions on every compact subset of Ω. For f, g $\in L^2_{(p,q)}(\Omega, \varphi)$ with

$$f = \sum_{\alpha,\beta}{}' f_{\alpha,\beta} dz^\alpha \wedge d\bar{z}^\beta, \quad g = \sum_{\alpha,\beta}{}' g_{\alpha,\beta} dz^\alpha \wedge d\bar{z}^\beta,$$

we define the inner product of f and g by

$$(f,g) = {\sum_{\alpha,\beta}}' \int_\Omega f_{\alpha,\beta} \overline{g_{\alpha,\beta}} e^{-\varphi} dV.$$

Then $L^2_{(p,q)}(\Omega, \varphi)$ is a Hilbert space.

Definition 2.4 For $g \in C^1(\Omega)$, define

$$\delta_j g = e^\varphi \frac{\partial}{\partial z_j}(g e^{-\varphi}) = \frac{\partial g}{\partial z_j} - g \frac{\partial \varphi}{\partial z_j}.$$

In order to prove Theorem 2.2 we need the following lemma.

Lemma 2.1 Let Ω be a bounded open set in $\mathbf{C}^n$ with C^1 boundary and let ρ be a defining function for Ω. For

$$f = {\sum_{I,J}}' f_{I,J} dz^I d\bar{z}^J \in C^1_{(p,q)}(\overline{\Omega}), \quad u = {\sum_{I,K}}' u_{I,K} dz^I d\bar{z}^K \in C^1_{(p,q-1)}(\overline{\Omega}),$$

we have

$$(\bar\partial u, f) = (-1)^p \int_\Omega {\sum_{I,K}}' \sum_{j=1}^n \frac{\partial u_{I,K}}{\partial \bar{z}_j} \overline{f_{I,jK}} e^{-\varphi} dV$$

$$= (-1)^{p-1} \int_\Omega {\sum_{I,K}}' \sum_{j=1}^n u_{I,K} \overline{\delta_j f_{I,jK}} e^{-\varphi} dV$$

$$+ (-1)^p \int_{\partial\Omega} {\sum_{I,K}}' u_{I,K} \sum_{j=1}^n \overline{f_{I,jK} \frac{\partial \rho}{\partial z_j}} e^{-\varphi} \frac{dS}{|d\rho|}.$$

Proof. We prove Lemma 2.1 in case $p = 0, q = 1$. The other cases will be left to the reader. We set $z_j = x_{2j-1} + ix_{2j}$. Then it follows from Green's theorem that

$$\int_\Omega \frac{\partial u}{\partial x_j} = \int_{\partial\Omega} u \frac{\partial \rho}{\partial x_j} \frac{dS}{|d\rho|}.$$

If w is a C^1 function on $\overline{\Omega}$, then we obtain

$$\int_{\partial\Omega} \frac{\partial \rho}{\partial \bar{z}_j} u \bar{w} e^{-\varphi} \frac{dS}{|d\rho|} = \int_\Omega \frac{\partial}{\partial \bar{z}_j}(u\bar{w} e^{-\varphi}) dV$$

$$= \int_\Omega \frac{\partial u}{\partial \bar{z}_j} \bar{w} e^{-\varphi} dV + \int_\Omega u \overline{\delta_j w} e^{-\varphi} dV.$$

By setting $w = f_j$ and adding with respect to j, we obtain the desired equality. □

Definition 2.5 For $f = \sum'_{I,J} f_{I,J} dz^I \wedge d\bar{z}^J \in C^1_{(p,q)}(\overline{\Omega})$, define

$$\bar{\partial}^* f = (-1)^{p-1} {\sum_{I,K}}' \sum_{j=1}^n \delta_j f_{I,jK} dz^I \wedge d\bar{z}^K.$$

$f \in \text{Def}(\bar{\partial}^*)$ means that

$$\sum_{j=1}^n f_{I,jK} \frac{\partial \rho}{\partial z_j} = 0 \quad \text{on } \partial\Omega.$$

for every multi-index I and K.

For $f \in \text{Def}(\bar{\partial}^*)$, it follows from Lemma 2.1 that

$$(\bar{\partial} u, f) = (u, \bar{\partial}^* f).$$

The following theorem was proved by Hörmander [HR1].

Theorem 2.2 Let $\Omega \subset\subset \mathbf{C}^n$ be an open set with C^2 boundary and let ρ be a defining function for Ω. Let $\alpha = \sum'_{I,J} \alpha_{I,J} dz^I \wedge d\bar{z}^J$ be a $C^2(p,q)$ form on $\overline{\Omega}$ and let $\alpha \in \text{Def}(\bar{\partial}^*)$, φ a C^2 function on $\overline{\Omega}$. Then

$$\|\bar{\partial}^* \alpha\| + \|\bar{\partial}\alpha\|^2 = {\sum_{I,K}}' \sum_{j,k=1}^n \int_\Omega \alpha_{I,jK} \overline{\alpha}_{I,kK} \frac{\partial^2 \varphi}{\partial z_j \partial \bar{z}_k} e^{-\varphi} dV$$

$$+ {\sum_{I,J}}' \sum_{j=1}^n \int_\Omega \left|\frac{\partial \alpha_{I,J}}{\partial \bar{z}_j}\right|^2 e^{-\varphi} dV$$

$$+ {\sum_{I,K}}' \sum_{j,k=1}^n \int_{\partial\Omega} \alpha_{I,jK} \overline{\alpha}_{I,kK} \frac{\partial^2 \rho}{\partial z_j \partial \bar{z}_k} e^{-\varphi} \frac{dS}{|d\rho|}.$$

Proof. We prove Theorem 2.2 in case $p = 0$, $q = 1$ and the other cases will be left to the reader. Let w be a C^2 function on $\overline{\Omega}$. Then we have

$$\left(\delta_k \frac{\partial}{\partial \bar{z}_j} - \frac{\partial}{\partial \bar{z}_j} \delta_k\right) w = w \frac{\partial^2 \varphi}{\partial z_k \partial \bar{z}_j}.$$

Thus for C^2 functions v and w, we have

$$\int_\Omega \delta_j v \overline{\delta_k w} e^{-\varphi} dV - \int_\Omega \frac{\partial v}{\partial \bar{z}_k} \overline{\frac{\partial w}{\partial \bar{z}_j}} e^{-\varphi} dV$$
$$= \int_\Omega v\bar{w} \frac{\partial^2 \varphi}{\partial z_k \partial \bar{z}_j} e^{-\varphi} dV + \int_{\partial\Omega} \frac{\partial \rho}{\partial z_j} v\overline{\delta_k w} e^{-\varphi} \frac{dS}{|d\rho|}$$
$$- \int_{\partial\Omega} \frac{\partial \rho}{\partial \bar{z}_k} v \overline{\frac{\partial w}{\partial \bar{z}_j}} e^{-\varphi} \frac{dS}{|d\rho|}.$$

On the other hand we have

$$\bar{\partial}\alpha = \sum_{j,k=1}^n \frac{\partial \alpha_j}{\partial \bar{z}_k} d\bar{z}_k \wedge d\bar{z}_j = \sum_{j>k} \left(\frac{\partial \alpha_j}{\partial \bar{z}_k} - \frac{\partial \alpha_k}{\partial \bar{z}_j} \right) d\bar{z}_k \wedge d\bar{z}_j,$$

$$\left| \frac{\partial \alpha_j}{\partial \bar{z}_k} - \frac{\partial \alpha_k}{\partial \bar{z}_j} \right|^2 = \left| \frac{\partial \alpha_k}{\partial \bar{z}_j} \right|^2 + \left| \frac{\partial \alpha_j}{\partial \bar{z}_k} \right|^2 - \frac{\partial \alpha_k}{\partial \bar{z}_j} \overline{\frac{\partial \alpha_j}{\partial \bar{z}_k}} - \frac{\partial \alpha_j}{\partial \bar{z}_k} \overline{\frac{\partial \alpha_k}{\partial \bar{z}_j}}.$$

Therefore we have

$$\|\bar{\partial}^*\alpha\|^2 + \|\bar{\partial}\alpha\|^2 = \sum_{j,k=1}^n \int_\Omega \delta_j \alpha_j \overline{\delta_k \alpha_k} e^{-\varphi} dV$$
$$- \sum_{j,k=1}^n \int_\Omega \frac{\partial \alpha_j}{\partial \bar{z}_k} \overline{\frac{\partial \alpha_k}{\partial \bar{z}_j}} e^{-\varphi} dV$$
$$+ \sum_{j,k=1}^n \int_\Omega \left| \frac{\partial \alpha_k}{\partial \bar{z}_j} \right|^2 e^{-\varphi} dV.$$

Taking account of the boundary condition, we have

$$\|\bar{\partial}^*\alpha\|^2 + \|\bar{\partial}\alpha\|^2$$
$$= \sum_{j,k=1}^n \int_\Omega \left| \frac{\partial \alpha_k}{\partial \bar{z}_j} \right|^2 e^{-\varphi} dV + \sum_{j,k=1}^n \int_\Omega \alpha_j \bar{\alpha}_k \frac{\partial^2 \varphi}{\partial z_j \partial \bar{z}_k} e^{-\varphi} dV$$
$$- \sum_{j,k=1}^n \int_{\partial\Omega} \alpha_j \frac{\partial \rho}{\partial \bar{z}_k} \overline{\frac{\partial \alpha_k}{\partial \bar{z}_j}} e^{-\varphi} \frac{dS}{|d\rho|}.$$

By Lemma 1.21, there exists a C^1 function λ such that

$$\sum_{k=1}^n \alpha_k \frac{\partial \rho}{\partial z_k} = \lambda \rho.$$

Hence we have on $\partial\Omega$

$$\sum_{k=1}^{n}\left(\frac{\partial\alpha_k}{\partial\bar{z}_j}\frac{\partial\rho}{\partial z_k} + \alpha_k\frac{\partial^2\rho}{\partial\bar{z}_j\partial z_k}\right) = \lambda\frac{\partial\rho}{\partial\bar{z}_j}.$$

If we multiply by $\bar{\alpha}_j$ and add with respect to j, we obtain on $\partial\Omega$ using the boundary condition

$$\sum_{j,k=1}^{n}\left(\bar{\alpha}_j\frac{\partial\alpha_k}{\partial\bar{z}_j}\frac{\partial\rho}{\partial z_k} + \bar{\alpha}_j\alpha_k\frac{\partial^2\rho}{\partial\bar{z}_j\partial z_k}\right) = 0,$$

which completes the proof of Theorem 2.2. □

2.2 L^2 Estimates in Pseudoconvex Domains

In this section we study L^2 estimates for the $\bar{\partial}$ problem in pseudoconvex domains in $\mathbf{C}^n$ by following Hörmander [HR2]. In Chapter 1 we proved that every domain of holomorphy is a pseudoconvex domain. Here, we prove that every pseudoconvex domain is a domain of holomorphy by applying L^2 estimates for solutions of the $\bar{\partial}$ problem.

Let H^1 and H^2 be Hilbert spaces. We denote the inner product of H^1 by $(x,y)_1$ for $x,y \in H^1$, and the inner product of H^2 by $(x,y)_2$ for $x,y \in H^2$. Let $\mathcal{D} \subset H^1$ be a dense subset of H^1 and let $T : \mathcal{D} \to H^2$ be a linear operator. Then we set $\mathcal{D} = \mathcal{D}_T$, $T(\mathcal{D}) = \mathcal{R}_T$.

Definition 2.6 Let $T : \mathcal{D} \to H^2$ be a linear operator. Define

$$\mathcal{G}_T := \{(x,Tx) \mid x \in \mathcal{D}_T\} \subset H^1 \times H^2.$$

We say that T is a closed operator if its graph $\mathcal{G}_T$ is a closed subspace of $H^1 \times H^2$.

Definition 2.7 Let $y \in H^2$. We say that $y \in \mathcal{D}_{T^*}$ if there exists a constant $c = c(y) > 0$ such that

$$|(Tx,y)_2| \le c\|x\|_1$$

for all $x \in \mathcal{D}_T$. By definition $\mathcal{D}_{T^*}$ is a subspace of H^2.

Lemma 2.2 For $y \in \mathcal{D}_{T^*}$ there exists a unique $z \in H^1$ such that

$$(x,z)_1 = (Tx,y)_2$$

for all $x \in \mathcal{D}_T$. We set $z = T^*y$. Then $T^* : \mathcal{D}_{T^*} \to H^1$ is a linear operator and satisfies

$$(x, T^*y)_1 = (Tx, y)_2 \tag{2.2}$$

for all $x \in \mathcal{D}_T$, $y \in \mathcal{D}_{T^*}$.

Proof. For $y \in \mathcal{D}_{T^*}$, $x \in \mathcal{D}_T$, define

$$\varphi(x) = (Tx, y)_2.$$

Then φ is a linear functional on $\mathcal{D}_T$ and satisfies $|\varphi(x)| \leq c\|x\|_1$ for some constant c. Hence φ is bounded. Since $\mathcal{D}_T$ is dense in H^1, for $x \in H^1$ there exists $x_\nu \in \mathcal{D}_T$ such that $x_\nu \to x$. We have

$$|\varphi(x_\nu) - \varphi(x_\mu)| \leq c\|x_\nu - x_\mu\| \to 0 \quad (\nu, \mu \to \infty).$$

Hence $\{\varphi(x_\nu)\}$ converges. If we define $\varphi(x) := \lim_{\nu \to \infty} \varphi(x_\nu)$, then φ is a bounded linear functional on H^1. By the Riesz representation theorem, there exists a unique $z \in H^1$ such that

$$\varphi(x) = (x, z)_1$$

for all $x \in H^1$. Thus we have

$$(x, z)_1 = (Tx, y)_2 \qquad (x \in \mathcal{D}_T, y \in \mathcal{D}_{T^*}).$$

Next we show that T^* is linear. For $y_1, y_2 \in \mathcal{D}_{T^*}$ and $x \in \mathcal{D}_T$, we have

$$(x, T^*(y_1 + y_2))_1 = (Tx, y_1 + y_2)_2 = (x, T^*y_1 + T^*y_2)_1.$$

Since $\mathcal{D}_T$ is dense in H^1, we have $T^*(y_1 + y_2) = T^*y_1 + T^*y_2$. Similarly, we have $T^*(\alpha y) = \alpha T^* y$ for $\alpha \in \mathbf{C}$, $y \in \mathcal{D}_{T^*}$, which means that T^* is a linear operator. $\square$

Lemma 2.3 $T^* : \mathcal{D}_{T^*} \to H^1$ is a closed operator.

Proof. It is sufficient to show that

$$\mathcal{G}_{T^*} = \{(y, T^*y) \mid y \in \mathcal{D}_{T^*}\} \subset H^2 \times H^1$$

is closed. Suppose

$$(y_n, z_n) \in \mathcal{G}_{T^*}, \quad (y_n, z_n) \to (y_0, z_0).$$

Then $y_n \in \mathcal{D}_{T^*}$ and $z_n = T^*y_n$. Since $\{z_n\}$ is bounded, there exists a constant $M > 0$ such that $\|z_n\| < M$ for all n. For $x \in \mathcal{D}_T$, we have

$$|(Tx, y_n)_2| = |(x, z_n)_1| \leq \|x\| \|z_n\| \leq M\|x\|_1.$$

Letting $n \to \infty$, we have

$$|(Tx, y_0)_2| \leq M\|x\|_1,$$

which means that $y_0 \in \mathcal{D}_{T^*}$. On the other hand, we have

$$|(x, T^*y_n)_1 - (x, T^*y_0)_1| = |(Tx, y_n - y_0)_2| \leq \|Tx\|_2 \|y_n - y_0\|_2 \to 0.$$

Then $(x, z_n)_1 \to (x, T^*y_0)_1$. Hence we have

$$(x, z_0)_1 = (x, T^*y_0)_1 \quad (x \in \mathcal{D}_T).$$

Thus we have $z_0 = T^*y_0$, and hence $(y_0, z_0) \in \mathcal{G}_{T^*}$, which means that $\mathcal{G}_{T^*}$ is closed. $\square$

Lemma 2.4 *Suppose $\mathcal{D}_{T^*}$ is dense in H^2. Then we have*

$$\mathcal{D}_{T^{**}} \supset \mathcal{D}_T, \quad T^{**}|_{\mathcal{D}_T} = T.$$

Proof. For $x \in \mathcal{D}_T$ and $y \in \mathcal{D}_{T^*}$, we have

$$|(x, T^*y)_1| = |(Tx, y)_2| \leq \|Tx\|_2 \|y\|_2,$$

which means that $x \in \mathcal{D}_{T^{**}}$. Thus $\mathcal{D}_T \subset \mathcal{D}_{T^{**}}$. On the other hand we have

$$(Tx, y)_2 = (x, T^*y)_1 = \overline{(T^*y, x)_1} = \overline{(y, T^{**}x)_2} = (T^{**}x, y)_2.$$

Since $\mathcal{D}_{T^*}$ is dense in H^2, we obtain $Tx = T^{**}x$ for $x \in \mathcal{D}_T$, and hence $T^{**}|_{\mathcal{D}_T} = T$. $\square$

Lemma 2.5 *Let $T : \mathcal{D}_T \to H^2$ be a closed operator. Then $\mathcal{D}_{T^*}$ is dense in H^2 and $T^{**} = T$.*

Proof. Define $\mathcal{H} = H^1 \times H^2$. Let $(x, y) \in \mathcal{H}$ and $(u, v) \in \mathcal{H}$. We define the inner product $<,>$ in $\mathcal{H}$ by

$$<(x,y),(u,v)> := (x,u)_1 + (y,v)_2.$$

Then $\mathcal{H}$ is a Hilbert space. Further, define $J : \mathcal{H} \to \mathcal{H}$ by $J(x, y) = (-x, y)$. We define $\mathcal{G}_T$ and $\mathcal{G}_T^*$ by

$$\mathcal{G}_T := \{(x, Tx) \mid x \in \mathcal{D}_T\} \subset \mathcal{H}$$

and

$$\mathcal{G}_T^* := \{(T^*y, y) \mid y \in \mathcal{D}_{T^*}\} \subset \mathcal{H}.$$

Then for $y \in H^1$, $z \in H^2$ we have

$$< (-x, Tx), (y, z) > = 0 \qquad (x \in \mathcal{D}_T)$$
$$\iff (x, y)_1 = (Tx, z)_2 \qquad (x \in \mathcal{D}_T)$$
$$\iff z \in \mathcal{D}_{T^*}, \quad y = T^*z.$$

Hence we obtain

$$(y, z) \perp J\mathcal{G}_T \iff (y, z) \in \mathcal{G}_T^*,$$

which means that $(J\mathcal{G}_T)^\perp = \mathcal{G}_T^*$. Since T is closed, $\mathcal{G}_T$ is closed, and hence $J\mathcal{G}_T$ is closed. Thus we have $J\mathcal{G}_T = (\mathcal{G}_T^*)^\perp$. Similarly, we have $\mathcal{G}_T = (J\mathcal{G}_T^*)^\perp$. Let $u \in H^2$. Suppose $(u, v) = 0$ for all $v \in \mathcal{D}_{T^*}$. Since $< (0, u), (T^*v, v) > = 0$, we have $(0, u) \in (\mathcal{G}_T^*)^\perp$. Hence $(0, u) \in J\mathcal{G}_T$. Thus there exists $x \in \mathcal{D}_T$ such that $(0, u) = (-x, Tx)$, which implies that $u = 0$. Therefore if $(u, v)_2 = 0$ for every $v \in \mathcal{D}_{T^*}$, then $u = 0$. If φ is a bounded linear functional on H^2, then by the Riesz representation theorem, there exists $z \in H^2$ such that

$$\varphi(v) = (v, z)_2 \qquad (v \in H^2).$$

If $\varphi = 0$ on $\mathcal{D}_{T^*}$, then $z = 0$, which means that $\varphi = 0$ on H^2. By applying the Hahn-Banach theorem, $\mathcal{D}_{T^*}$ is dense in H^2. Thus $T^{**} : \mathcal{D}_{T^{**}} \to H^2$ is defined. We set

$$\mathcal{G}_T^{**} = \{(z, T^{**}z) \mid z \in \mathcal{D}_{T^{**}}\} \subset \mathcal{H}.$$

Since T^* is closed, by using the same method as above we have

$$(J\mathcal{G}_T^*)^\perp = \mathcal{G}_T^{**}.$$

On the other hand, taking account of the equality $(J\mathcal{G}_T^*)^\perp = \mathcal{G}_T$, we have $\mathcal{G}_T = \mathcal{G}_T^{**}$, which means that $\mathcal{D}_T = \mathcal{D}_{T^{**}}$. It follows from Lemma 2.4 that $T = T^{**}$. □

Theorem 2.3 *(Banach-Steinhaus theorem) Suppose X is a Banach space, Y is a normed linear space, and $\{T_\alpha\}_{\alpha \in A}$ is a collection of bounded linear operators of X into Y. Then either (1) or (2) holds:*

(1) There exists a constant $M > 0$ such that
$$\|T_\alpha\| \leq M$$
for every $\alpha \in A$.

(2) There exists a dense subset E of X such that
$$\sup_{\alpha \in A} \|T_\alpha(x)\| = \infty$$
for every $x \in E$.

Proof. We set
$$\varphi(x) = \sup_{\alpha \in A} \|T_\alpha(x)\| \qquad (x \in X)$$
and
$$V_n = \{x \mid \varphi(x) > n\}.$$

If we set $f_\alpha(x) = \|T_\alpha(x)\|$, then $f_\alpha(x)$ is continuous, and hence V_n is an open set. Suppose V_N is not dense in X. Then there exist $x_0 \in X$ and $r > 0$ such that if $\|x\| \leq r$, then $x_0 + x \notin V_N$. Thus $\varphi(x_0 + x) \leq N$. Hence we have
$$\|T_\alpha(x_0 + x)\| \leq N \qquad (\alpha \in A, \|x\| \leq r),$$
which means that
$$\|T_\alpha(x)\| \leq \|T_\alpha(x_0 + x)\| + \|T_\alpha(x_0)\| \leq 2N.$$

Then we obtain
$$\|T_\alpha\| = \sup_{\|x\|=1} \|T_\alpha(x)\| = \sup_{\|x\|=1} \frac{1}{r}\|T_\alpha(rx)\| \leq \frac{2N}{r}.$$

In case that all V_n are dense subset of X, by the Baire theorem, $\cap_{n=1}^\infty V_n$ is dense in X, and for $x \in \cap_{n=1}^\infty V_n$ we have $\varphi(x) = \infty$. □

Theorem 2.4 *Suppose $\mathcal{D}$ is a dense subspace in H^1 and $T : \mathcal{D} \to H^2$ is a closed operator. Let F be a closed subspace of H^2 and let $F \supset \mathcal{R}_T$. Then the following statements are equivalent:*

(a) $F = \mathcal{R}_T$.

(b) There exists a constant $c > 0$ such that

$$\|y\|_2 \le c\|T^*y\|_1$$

for all $y \in F \cap \mathcal{D}_{T^*}$.

Proof. $(a) \Longrightarrow (b)$. Suppose $F = \mathcal{R}_T$. Every element $z \in H^2$ is uniquely expressed by

$$z = z_1 + z_2 \qquad (z_1 \in F,\ z_2 \in F^\perp).$$

Since $z_1 = Tx$ for $x \in \mathcal{D}_T$, we have for $y \in F \cap \mathcal{D}_{T^*}$

$$|(y,z)_2| = |(y,z_1)_2| = |(y,Tx)_2| = |(x,T^*y)_1| \le \|x\|_1 \|T^*y\|_1. \qquad (2.3)$$

We set

$$K = \{y \mid T^*y \ne 0,\ y \in \mathcal{D}_{T^*}\} \cap F.$$

Further, we set for $y \in K$

$$\varphi_y(z) = \frac{(y,z)_2}{\|T^*y\|_1}.$$

Then by the Banach-Steinhaus theorem, either there exists a constant $c > 0$ such that

$$\|\varphi_y\| < c \qquad (2.4)$$

for every $y \in K$, or there exists a dense subset E of H^2 such that

$$\infty = \sup_{y \in K} |\varphi_y(z)| \qquad (2.5)$$

for every $z \in E$. By (2.3), (2.5) does not hold. Thus we have

$$\frac{|(y,z)_2|}{\|T^*y\|_1} < c \qquad (2.6)$$

for all $y \in K$. Substituting $z = y/\|y\|_2$ into (2.6), we have

$$\|y\|_2 < c\|T^*y\|_1$$

for all $y \in K$. In case $T^*y = 0$, we have $y = 0$ by (2.3). This proves (b).

$(b) \Longrightarrow (a)$. Fix $z \in F$. Suppose the equality

$$\|y\|_2 \le c\|T^*y\|_1 \qquad (y \in F \cap \mathcal{D}_{T^*}) \qquad (2.7)$$

holds. If $w \in T^*(F \cap \mathcal{D}_{T^*})$, then we have $w = T^*y$ for $y \in F \cap \mathcal{D}_{T^*}$. We define a linear functional φ on $T^*(F \cap \mathcal{D}_{T^*})$ by $\varphi(w) = (y,z)_2$. If $w = T^*y_1 = T^*y_2$, then by (2.7), we have $y_1 = y_2$. Therefore φ is well defined. Since

$$|\varphi(w)| \leq \|y\|_2 \|z\|_2 \leq c\|w\|_1 \|z\|_2,$$

φ is a bounded linear functional on $T^*(F \cap \mathcal{D}_{T^*})$. By the Hahn-Banach theorem, φ can be extended to a bounded linear functional on H^1. By the Riesz representation theorem, there exists $x_0 \in H^1$ such that

$$\varphi(w) = (w, x_0)_1 \qquad (w \in H^1).$$

Hence we have

$$(y, z)_2 = (T^*y, x_0)_1$$

for all $y \in F \cap \mathcal{D}_{T^*}$. If $y \in F^\perp \cap \mathcal{D}_{T^*}$, then $\mathcal{R}_T \subset F$, which implies that

$$(T^*y, x)_1 = (y, Tx)_2 = 0$$

for all $x \in \mathcal{D}_T$. Thus we have $T^*y = 0$. Hence, for $y \in \mathcal{D}_{T^*}$, if we set $y = y_1 + y_2$ ($y_1 \in F \cap \mathcal{D}_{T^*}$, $y_2 \in F^\perp \cap \mathcal{D}_{T^*}$), then

$$(y, z)_2 = (y_1, z)_2 + (y_2, z)_2 = (y_1, z)_2 = (T^*y_1, x_0)_1 = (T^*y, x_0)_1$$

and

$$|(T^*y, x_0)_1| \leq \|y\|_2 \|z\|_2 \qquad (y \in \mathcal{D}_{T^*}).$$

Thus we have $x_0 \in \mathcal{D}_{T^{**}} = \mathcal{D}_T$. Consequently,

$$(y, z)_2 = (T^*y, x_0)_1 = (y, Tx_0)$$

for $y \in \mathcal{D}_{T^*}$. Hence $z = Tx_0 \in \mathcal{R}_T$, which implies that $F \subset \mathcal{R}_T$. □

Lemma 2.6 *Let $\mathcal{D}$ be a dense subspace of H^1 and let $T: \mathcal{D} \to H^2$ be a closed operator. If $\mathcal{R}_T$ is closed, then $\mathcal{R}_{T^*}$ is closed.*

Proof. We set $F = \mathcal{R}_T$ in Theorem 2.4. Then

$$\|f\|_2 \leq c\|T^*f\|_1 \qquad (f \in F \cap \mathcal{D}_{T^*}).$$

Let $f \in \mathcal{D}_{T^*}$. Then f is uniquely expressed by

$$f = f_1 + f_2 \qquad (f_1 \in F \cap \mathcal{D}_{T^*}, \ f_2 \in F^\perp \cap \mathcal{D}_{T^*}).$$

Since $T\varphi \in F$ for $\varphi \in \mathcal{D}_T$, we obtain

$$(\varphi, T^*f_2)_1 = (T\varphi, f_2)_2 = 0.$$

Thus $T^*f_2 = 0$, and hence $T^*f = T^*f_1$, which means that $T^*(\mathcal{D}_{T^*}) = T^*(F \cap \mathcal{D}_{T^*})$. Suppose $T^*(F \cap \mathcal{D}_{T^*}) \ni T^*f_\nu$, $T^*f_\nu \to g$. Then

$$\|f_\nu - f_\mu\|_2 \le c\|T^*(f_\nu - f_\mu)\|_1 \to 0 \quad (\nu, \mu \to \infty).$$

Hence there exists $f_0 \in H^2$ such that $f_\nu \to f_0$, and hence $(T^*f_\nu, f_\nu) \to (g, f_0)$. Since T^* is a closed operator, we have $f_0 \in \mathcal{D}_{T^*}$, $g = T^*f_0$, and hence, $g \in T^*(\mathcal{D}_{T^*})$. Thus $T^*(\mathcal{D}_{T^*}) = \mathcal{R}_{T^*}$ is a closed subset. $\square$

Definition 2.8 Let $T : \mathcal{D}_T \to H^2$ be a linear operator. Define

$$\operatorname{Ker} T := \{x \in \mathcal{D}_T \mid Tx = 0\}.$$

$\operatorname{Ker} T$ is called a kernel (or a null space) of T.

Lemma 2.7 Let $T : \mathcal{D}_T \to H^2$ be a closed operator. Then $\operatorname{Ker} T$ is a closed subspace. Moreover, we have

$$(\mathcal{R}_T)^\perp = \operatorname{Ker} T^*, \quad \overline{\mathcal{R}_{T^*}} = (\operatorname{Ker} T)^\perp.$$

Proof. Let $Tu_\nu = 0$, $u_\nu \to u$. Since $(u_\nu, Tu_\nu) \to (u, 0)$ and T is closed, we have $0 = Tu$, and hence $u \in \operatorname{Ker} T$. Hence $\operatorname{Ker} T$ is a closed subset. Let $y \in (\mathcal{R}_T)^\perp$. For $x \in \mathcal{D}_T$, we have

$$|(Tx, y)_2| = 0 \le \|x\|_1,$$

which implies that $y \in \mathcal{D}_{T^*}$. Since

$$0 = (Tx, y)_2 = (x, T^*y)_1 \quad (x \in \mathcal{D}_T),$$

$T^*y = 0$, and hence $y \in \operatorname{Ker} T^*$. Thus we have $(\mathcal{R}_T)^\perp \subset \operatorname{Ker} T^*$. On the other hand, for $g \in \operatorname{Ker} T^*$, $f \in \mathcal{D}_T$,

$$(Tf, g)_2 = (f, T^*g)_1 = 0.$$

Hence we have $g \in (\mathcal{R}_T)^\perp$ which implies that $\operatorname{Ker} T^* \subset (\mathcal{R}_T)^\perp$. Thus we obtain $(\mathcal{R}_T)^\perp = \operatorname{Ker} T^*$. Taking account of the equality $(\operatorname{Ker} T^*)^\perp = \overline{\mathcal{R}_T}$, replacing T by T^* we have $\overline{\mathcal{R}_{T^*}} = (\operatorname{Ker} T)^\perp$. $\square$

Lemma 2.8 If $f \in L^2_{(p,q)}(\Omega, \varphi)$, then $f \in L^2_{(p,q)}(\Omega, loc)$.

Proof. For any compact subset K of Ω, there exist constants $c_1 > 0$ and $c_2 > 0$ such that
$$c_1 \leq e^{-\varphi(x)} \leq c_2 \quad (x \in K).$$
Then
$$\int_K |f|^2 dV \leq \frac{1}{c_1} \int_K |f|^2 e^{-\varphi(x)} dV < \infty,$$
which implies that $f \in L^2_{(p,q)}(\Omega, \text{loc})$. □

Lemma 2.9 *If $f \in L^2_{(p,q)}(\Omega, \text{loc})$, then there exists $\varphi \in C^\infty(\Omega)$ such that $f \in L^2_{(p,q)}(\Omega, \varphi)$.*

Proof. Suppose $\{K_n\}$ is a sequence of compact subsets of Ω and that
$$K_n \subset\subset \mathring{K}_{n+1} \subset \Omega, \quad \bigcup_{n=1}^\infty K_n = \Omega.$$
We set
$$\int_{K_n} |f|^2 dV = c_n.$$
We choose functions $a_n \in C_c^\infty(\mathbf{C}^n)$ with the properties
$$0 \leq a_n(z) \leq 1 \ (z \in \mathbf{C}^n), \quad a_n(z) = \begin{cases} 1 & (z \in K_n - K_{n-1}) \\ 0 & (z \notin K_{n+1} - K_{n-2}) \end{cases}.$$
For $z \in \Omega$, define
$$\varphi(z) = \sum_{n=1}^\infty (\log n^2(c_n + 1)) a_n(z).$$
Then $\varphi \in C^\infty(\Omega)$, and, $\varphi(z) \geq \log n^2(c_n + 1)$ $(z \in K_n - K_{n-1})$. Hence we have
$$\int_\Omega |f|^2 e^{-\varphi} dV = \int_{K_1} |f|^2 e^{-\varphi} dV + \sum_{n=2}^\infty \int_{K_n \setminus K_{n-1}} |f|^2 e^{-\varphi} dV$$
$$\leq c_1 + \sum_{n=2}^\infty \int_{K_n \setminus K_{n-1}} |f|^2 e^{-\log n^2(c_n+1)} dV$$
$$= c_1 + \sum_{n=2}^\infty \frac{1}{n^2(c_n+1)} c_n$$
$$\leq c_1 + \sum_{n=2}^\infty \frac{1}{n^2} < \infty,$$

which implies that $f \in L^2_{(p,q)}(\Omega, \varphi)$. □

Definition 2.9 For $f \in L^2_{(p,q)}(\Omega, \varphi)$, define

$$\bar{\partial} f = \sum_{\substack{|\alpha|=p \\ |\beta|=q}}{}' \sum_{k=1}^{n} \frac{\partial f_{\alpha,\beta}}{\partial \bar{z}_k} d\bar{z}_k \wedge dz_\alpha \wedge d\bar{z}_\beta.$$

Then by Lemma 2.8, $\bar{\partial} f$ exists in the sense of distributions.

Definition 2.10 We denote by $\mathcal{D}_{(p,q)}(\Omega)$ the set of all C^∞ (p,q) forms in Ω whose supports are compact subsets of Ω. Further, we set $\mathcal{D}(\Omega) = \mathcal{D}_{(0,0)}(\Omega)$.

Theorem 2.5 $\mathcal{D}_{(p,q)}(\Omega)$ is dense in $L^2_{(p,q)}(\Omega, \varphi_1)$. Further, if we set $T = \bar{\partial}$, then

$$T : \mathcal{D}_T \to L^2_{(p,q+1)}(\Omega, \varphi_2)$$

is a closed operator.

Proof. Since $\mathcal{D}_{(p,q)}(\Omega)$ is dense in $L^2_{(p,q)}(\Omega, \varphi_1)$, $\mathcal{D}_T$ is dense in $L^2_{(p,q)}(\Omega, \varphi_1)$. We set $\mathcal{G}_T = \{(f, Tf) \mid f \in \mathcal{D}_T\}$. Suppose $(f_n, \bar{\partial} f_n) \to (f, g)$. We set $g_n = \bar{\partial} f_n$ and

$$f_n = \sum_{\alpha,\beta}{}' f^n_{\alpha,\beta} dz^\alpha \wedge d\bar{z}^\beta, \quad g_n = \sum_{\alpha,\gamma}{}' g^n_{\alpha,\gamma} dz^\alpha \wedge d\bar{z}^\gamma,$$

$$f = \sum_{\alpha,\beta}{}' f_{\alpha,\beta} dz^\alpha \wedge d\bar{z}^\beta, \quad g = \sum_{\alpha,\gamma}{}' g_{\alpha,\gamma} dz^\alpha \wedge d\bar{z}^\gamma.$$

Then we have

$$g^n_{\alpha,\gamma} = (-1)^p \sum_{\{j\} \cup \beta = \gamma}{}' \epsilon^{j\beta}_\gamma \frac{\partial f^n_{\alpha,\beta}}{\partial \bar{z}_j},$$

where $\epsilon^{j\beta}_\gamma$ means that if the permutation ρ which maps $j\beta$ to γ is even, then $\epsilon^{j\beta}_\gamma$ equals 1, and $\epsilon^{j\beta}_\gamma$ equals -1 if ρ is odd. For $\psi \in \mathcal{D}(\Omega)$ we have

$$\int_\Omega g^n_{\alpha,\gamma} \psi dV = (-1)^{p-1} \sum_{\{j\} \cup \beta = \gamma}{}' \epsilon^{j\beta}_\gamma \int_\Omega f^n_{\alpha,\beta} \frac{\partial \psi}{\partial \bar{z}_j} dV. \qquad (2.8)$$

Since $f_n \to f$ and $g_n \to g$, we have

$$\int_\Omega f^n_{\alpha,\beta} \frac{\partial \psi}{\partial \bar{z}_j} dV \to \int_\Omega f_{\alpha,\beta} \frac{\partial \psi}{\partial \bar{z}_j} dV$$

$$\int_\Omega g^n_{\alpha,\gamma}\psi dV \to \int_\Omega g_{\alpha,\gamma}\psi dV.$$

Letting $n \to \infty$ in (2.8) we have

$$\int_\Omega g_{\alpha,\gamma}\psi dV = (-1)^{p-1} \sum_{\{j\}\cup\beta=\gamma}{}' \epsilon^{j\beta}_\gamma \int_\Omega f_{\alpha,\beta}\frac{\partial \psi}{\partial \bar{z}_j}dV,$$

which means in the sense of distributions that

$$g_{\alpha,\gamma} = (-1)^p \sum_{\{j\}\cup\beta=\gamma}{}' \epsilon^{j\beta}_\gamma \frac{\partial f_{\alpha,\beta}}{\partial \bar{z}_j} = (\bar{\partial}f)_{\alpha,\gamma}.$$

Thus we have $g = \bar{\partial}f$, and hence T is a closed operator. □

We set

$$H^1 = L^2_{(p,q)}(\Omega,\varphi_1), \quad H^2 = L^2_{(p,q+1)}(\Omega,\varphi_2), \quad H^3 = L^2_{(p,q+2)}(\Omega,\varphi_3).$$

Further, we set

$$\mathcal{D}_1 = \{f \in H^1 \mid \bar{\partial}f \in H^2\}, \quad \mathcal{D}_2 = \{f \in H^2 \mid \bar{\partial}f \in H^3\},$$

$$\bar{\partial}|_{\mathcal{D}_1} = T, \quad \bar{\partial}|_{\mathcal{D}_2} = S.$$

Then

$$\mathcal{D}_T = \mathcal{D}_1, \quad \mathcal{D}_S = \mathcal{D}_2.$$

Lemma 2.10 $\eta \in \mathcal{D}(\Omega), f \in \mathcal{D}_S \Longrightarrow \eta f \in \mathcal{D}_S$.

Proof. We have

$$\bar{\partial}(\eta f) = \eta \bar{\partial}f + \bar{\partial}\eta \wedge f. \tag{2.9}$$

Since the right side of (2.9) belongs to $\mathcal{D}_S$, we have $\eta f \in \mathcal{D}_S$. □

Lemma 2.11 $f \in \mathcal{D}_{T^*}, \eta \in \mathcal{D}(\Omega) \Longrightarrow \eta f \in \mathcal{D}_{T^*}$.

Proof. Let $u \in \mathcal{D}_T$. Using the equality

$$(\eta f, Tu)_2 = (f, \bar{\eta}Tu)_2 = (f, T(\bar{\eta}u))_2 - (f, \bar{\partial}\bar{\eta} \wedge u)_2$$
$$= (T^*f, \bar{\eta}u)_1 - (f, \bar{\partial}\bar{\eta} \wedge u)_2,$$

we have

$$|(\eta f, Tu)_2| \leq \|\eta T^*f\|_1 \|u\|_1 + \|f\|_2 \|\bar{\partial}\bar{\eta} \wedge u\|_2.$$

On the other hand, since supp(η) is compact, there exists a constant $c > 0$ such that

$$\|\bar{\partial}\bar{\eta} \wedge u\|_2^2 = \int_\Omega |\bar{\partial}\bar{\eta} \wedge u|^2 e^{-\varphi_1} e^{\varphi_1-\varphi_2} dV \leq c \int_\Omega |u|^2 e^{-\varphi_1} dV = c\|u\|_1^2.$$

Hence we have

$$|(\eta f, Tu)_2| \leq (\|\eta T^* f\|_1 + \sqrt{c}\|f\|_2)\|u\|_1.$$

Thus $\eta f \in \mathcal{D}_{T^*}$. □

Lemma 2.12 *Let $\Omega \subset \mathbf{C}^n$ be an open set and let f be a nonnegative function in Ω. Suppose f is bounded on every compact subset of Ω. Then there exists a function $\varphi \in C^\infty(\Omega)$ such that $f(z) \leq \varphi(z)$ for $z \in \Omega$.*

Proof. Let $\mathcal{A} = \{U_\nu \mid \nu = 1, 2, \cdots\}$ and $\mathcal{B} = \{V_\nu \mid \nu = 1, 2, \cdots\}$ be locally finite open covers of Ω such that $U_\nu \subset\subset V_\nu \subset\subset \Omega$. Choose $a_\nu \in \mathcal{D}(\Omega)$ such that $a_\nu = 1$ on $\overline{U}_\nu$, supp(a_ν) $\subset V_\nu$ and $0 \leq a_\nu \leq 1$. We set $\sup_{z \in V_\nu} f(z) = M_\nu$. Define

$$\varphi(z) = \sum_{\nu=1}^\infty M_\nu a_\nu(z).$$

Then φ satisfies the desired properties. □

Lemma 2.13 *Let $\Omega \subset \mathbf{C}^n$ be an open set. Let $\{K_j\}_{j=0}^\infty$ be a sequence of compact subsets of Ω such that*

$$K_{j-1} \subset\subset (K_j)^\circ, \quad \bigcup_{j=0}^\infty K_j = \Omega,$$

where $(K_j)^\circ$ denotes the interior of K_j. Let $\eta_j \in \mathcal{D}(\Omega)$ be functions such that $\eta_j = 1$ on K_{j-1}, supp(η_j) $\subset K_j^\circ$ and $0 \leq \eta_j \leq 1$. Then there exists a function $\psi \in C^\infty(\Omega)$ such that

$$\sum_{k=1}^n \left|\frac{\partial \eta_j}{\partial \bar{z}_k}\right|^2 \leq e^\psi \quad (j = 1, 2, \cdots).$$

Proof. Define

$$f(z) = \begin{cases} \sum_{k=1}^n \left|\frac{\partial \eta_j}{\partial \bar{z}_k}\right|^2 & (z \in K_j - K_{j-1}, j = 1, 2, \cdots) \\ 0 & (z \in K_0) \end{cases}.$$

Then f is bounded on every compact subset of Ω and satisfies

$$f(z) \geq \sum_{k=1}^{n} \left|\frac{\partial \eta_j}{\partial \bar{z}_k}\right|^2 \quad (j = 1, 2, \cdots).$$

By Lemma 2.12, there exists a function $\psi \in C^\infty(\Omega)$ such that $f \leq \psi$ in Ω. Since $e^{\psi(z)} \geq \psi(z)$, ψ satisfies the desired properties. □

Definition 2.11 Let ψ be the function in Lemma 2.13. For $\psi \in C^\infty(\Omega)$, we set

$$\varphi_1 = \varphi - 2\psi, \quad \varphi_2 = \varphi - \psi, \quad \varphi_3 = \varphi.$$

We assume $\varphi \in C^2(\Omega)$. φ will be determined later. Then by Lemma 2.13 we have

$$e^{-\varphi_3}|\bar{\partial}\eta_j|^2 \leq e^{-\varphi_2}, \quad e^{-\varphi_2}|\bar{\partial}\eta_j|^2 \leq e^{-\varphi_1} \quad (j = 1, 2, \cdots).$$

Lemma 2.14 *Let η_j, $j = 1, 2, \cdots$, be functions in Lemma 2.13 and let $f \in \mathcal{D}_S$. Then $S(\eta_j f) - \eta_j S(f) \to 0$ in H^3 as $j \to \infty$.*

Proof. From the Schwarz inequality, there exists a constant $c > 0$ such that

$$|S(\eta_j f) - \eta_j S(f)|^2 e^{-\varphi_3} = |\bar{\partial}\eta_j \wedge f|^2 e^{-\varphi_3} \leq c|\bar{\partial}\eta_j|^2 |f|^2 e^{-\varphi_3} \leq c|f|^2 e^{-\varphi_2}.$$

On the other hand, $|S(\eta_j f) - \eta_j S(f)|^2 e^{-\varphi_3} \to 0$ as $j \to \infty$. Hence by the Lebesgue dominated convergence theorem, $\|S(\eta_j f) - \eta_j S(f)\|_3 \to 0$. □

Lemma 2.15 *Suppose*

$$f = \sideset{}{'}\sum_{\substack{|\alpha|=p \\ |\beta|=q+1}} f_{\alpha,\beta} dz^\alpha \wedge d\bar{z}^\beta \in \mathcal{D}_{T^*}.$$

Then we have

$$T^* f = (-1)^{p-1} \sideset{}{'}\sum_{\substack{|\alpha|=p \\ |\gamma|=q}} \sum_{k=1}^{n} e^{\varphi_1} \left\{\frac{\partial}{\partial z_k}\left(e^{-\varphi_2} f_{\alpha,k\gamma}\right)\right\} dz^\alpha \wedge d\bar{z}^\gamma,$$

where we set for $\gamma = (j_1, \cdots, j_q)$

$$f_{\alpha,k\gamma} = \begin{cases} 0 & (\text{one of } j_1, \cdots, j_q \text{ equals to } k) \\ (-1)^r f_{\alpha,\delta} & (\delta = (j_1, \cdots, j_r, k, j_{r+1}, \cdots, j_q)) \end{cases}.$$

Proof. We set
$$u = {\sum_{\substack{|\alpha|=p \\ |\gamma|=q}}}' u_{\alpha,\gamma} dz^\alpha \wedge d\bar{z}^\gamma \in \mathcal{D}_{(p,q)}(\Omega).$$

Then we have
$$Tu = {\sum_{\substack{|\alpha|=p \\ |\gamma|=q}}}' \sum_{k=1}^n \frac{\partial u_{\alpha,\gamma}}{\partial \bar{z}_k} d\bar{z}_k \wedge dz^\alpha \wedge d\bar{z}^\gamma$$

$$= (-1)^p {\sum_{\substack{|\alpha|=p \\ |L|=q+1}}}' {\sum_{\{k\} \cup K = L}}' \varepsilon_L^{kK} \frac{\partial u_{\alpha,K}}{\partial \bar{z}_k} dz^\alpha \wedge d\bar{z}^L.$$

Since $f_{\alpha,J}$ is defined to be skew-symmetric with respect to J, we have $f_{\alpha,L} = \varepsilon_L^{kK} f_{\alpha,kK}$. Hence we have

$$\int_\Omega {\sum_{\substack{|\alpha|=p \\ |\gamma|=q}}}' (T^*f)_{\alpha,\gamma} \overline{u_{\alpha,\gamma}} e^{-\varphi_1} dV = (T^*f, u)_1 = (f, Tu)_2$$

$$= (-1)^p \int_\Omega {\sum_{\substack{|\alpha|=p \\ |L|=q+1}}}' {\sum_{\{k\} \cup K = L}}' \varepsilon_L^{kK} \frac{\partial \bar{u}_{\alpha,K}}{\partial z_k} f_{\alpha,L} e^{-\varphi_2} dV$$

$$= (-1)^p \int_\Omega {\sum_{\substack{|\alpha|=p \\ |L|=q+1}}}' {\sum_{\{k\} \cup K = L}}' \frac{\partial \bar{u}_{\alpha,K}}{\partial z_k} f_{\alpha,kK} e^{-\varphi_2} dV$$

$$= (-1)^p {\sum_{\substack{|\alpha|=p \\ |K|=q}}}' \sum_{k=1}^n \int_\Omega \frac{\partial \bar{u}_{\alpha,K}}{\partial z_k} f_{\alpha,kK} e^{-\varphi_2} dV$$

$$= (-1)^{p-1} {\sum_{\substack{|\alpha|=p \\ |K|=q}}}' \sum_{k=1}^n \int_\Omega e^{\varphi_1} \frac{\partial}{\partial z_k} \left(f_{\alpha,kK} e^{-\varphi_2} \right) \bar{u}_{\alpha,K} e^{-\varphi_1} dV,$$

which implies that
$$(T^*f)_{\alpha,\gamma} = (-1)^{p-1} \sum_{k=1}^n e^{\varphi_1} \left\{ \frac{\partial}{\partial z_k} \left(e^{-\varphi_2} f_{\alpha,k\gamma} \right) \right\}. \qquad \square$$

Lemma 2.16 *Let η_j, $j = 1, 2, \cdots$, be functions in Lemma 2.13. Let $f \in \mathcal{D}_{T^*}$. Then*
$$\|T^*(\eta_j f) - \eta_j T^* f\|_1 \to 0 \qquad (j \to 0).$$

Proof. Suppose
$$f = {\sum}'_{\substack{|\alpha|=p \\ |\beta|=q+1}} f_{\alpha,\beta} dz^\alpha \wedge d\bar{z}^\beta$$
and
$$T^*(\eta_j f) - \eta_j T^* f = {\sum}'_{\substack{|\alpha|=p \\ |\gamma|=q}} g^j_{\alpha,\gamma} dz^\alpha \wedge d\bar{z}^\gamma.$$

It follows from Lemma 2.15 that
$$g^j_{\alpha,\gamma} = (-1)^{p-1} \sum_{k=1}^n e^{\varphi_1} \left\{ \frac{\partial}{\partial z_k} \left(e^{-\varphi_2} \eta_j f_{\alpha,k\gamma} \right) \right\}$$
$$-(-1)^{p-1} \eta_j \sum_{k=1}^n e^{\varphi_1} \left\{ \frac{\partial}{\partial z_k} \left(e^{-\varphi_2} f_{\alpha,k\gamma} \right) \right\}$$
$$= (-1)^{p-1} \sum_{k=1}^n e^{\varphi_1} \frac{\partial \eta_j}{\partial z_k} e^{-\varphi_2} f_{\alpha,k\gamma}.$$

By Lemma 2.13 and the Schwarz inequality we have
$$|g^j_{\alpha,\gamma}|^2 \leq e^{2(\varphi_1-\varphi_2)} \left(\sum_{k=1}^n \left|\frac{\partial \eta_j}{\partial z_k}\right|^2 \right) \left(\sum_{k=1}^n |f_{\alpha,k\gamma}|^2 \right)$$
$$\leq e^{\varphi_1-\varphi_2} \left(\sum_{k=1}^n |f_{\alpha,k\gamma}|^2 \right).$$

Consequently,
$$|T^*(\eta_j f) - \eta_j T^* f|^2 e^{-\varphi_1} = {\sum}'_{\substack{|\alpha|=p \\ |\gamma|=q}} |g^j_{\alpha,\gamma}|^2 e^{-\varphi_1}$$
$$\leq {\sum}'_{\substack{|\alpha|=p \\ |\gamma|=q}} \sum_{k=1}^n |f_{\alpha,k\gamma}|^2 e^{-\varphi_2}$$
$$= (q+1) {\sum}'_{\substack{|\alpha|=p \\ |\beta|=q+1}} |f_{\alpha,\beta}|^2 e^{-\varphi_2}$$
$$= (q+1)|f|^2 e^{-\varphi_2}.$$

On the other hand, since $|T^*(\eta_j f) - \eta_j T^* f|^2 e^{-\varphi_1}$ converges to 0 almost everywhere, it follows from the Lebesgue dominated convergence theorem that $\|T^*(\eta_j f) - \eta_j T^* f\|_1 \to 0$ as $j \to 0$. $\square$

Definition 2.12 For $f \in \mathcal{D}_{T^*} \cap \mathcal{D}_S$, we define
$$\|f\|_{\mathcal{G}} := \|f\|_2 + \|T^*f\|_1 + \|Sf\|_3.$$

Lemma 2.17 Let η_j, $j = 1, 2, \cdots$, be functions in Lemma 2.13. Then for $f \in \mathcal{D}_{T^*} \cap \mathcal{D}_S$
$$\|\eta_j f - f\|_{\mathcal{G}} \to 0 \qquad (j \to \infty).$$

Proof. Since $|\eta_j f - f| \leq |f|$, it follows from the Lebesgue dominated convergence theorem that $\|\eta_j f - f\|_2 \to 0$. Similarly, we have $\|\eta_j T^*f - T^*f\|_1 \to 0$. It follows from Lemma 2.16 that
$$\|T^*(\eta_j f - f)\|_1 = \|T^*(\eta_j f) - \eta_j T^*(f)\|_1 + \|\eta_j T^*f - T^*f\|_1 \to 0$$
as $j \to \infty$. Similarly, $\|S(\eta_j f - f)\|_3 \to 0$, and hence $\|\eta_j f - f\|_{\mathcal{G}} \to 0$. □

Lemma 2.18 Let $f \in \mathcal{D}_S$ and $\mathrm{supp}(f) \subset\subset \Omega$. Then for $0 < \delta < 1$, there exist $f_\delta \in \mathcal{D}_{(p,q+1)}(\Omega)$ such that
$$\|f_\delta - f\|_2 \to 0, \quad \|S(f_\delta) - S(f)\|_3 \to 0$$
as $\delta \to 0$.

Proof. Choose a function $\Phi \in \mathcal{D}(\mathbf{C}^n)$ such that
$$\int_{\mathbf{C}^n} \Phi dV = 1, \quad \mathrm{supp}(\Phi) \subset\subset B(0,1).$$
We set $\Phi_\delta(z) = \delta^{-2n}\Phi(z/\delta)$. For a differential form
$$f = \sum_{\alpha,\beta}{}' f_{\alpha,\beta} dz^\alpha \wedge d\bar{s}^\beta,$$
define
$$f_\delta = \sum_{\alpha,\beta}{}' (f_{\alpha,\beta} * \Phi_\delta) dz^\alpha \wedge d\bar{z}^\beta,$$
where
$$f_{\alpha,\beta} * \Phi_\delta(z) = \int_{\mathbf{C}^n} f_{\alpha,\beta}(z - \zeta)\Phi_\delta(\zeta) d\zeta.$$

Then we have $f_\delta \in \mathcal{D}_{(p,q+1)}(\Omega)$ and $\|f_\delta - f\|_2 \to 0$. On the other hand we have

$$S(f_\delta) = {\sum_{\alpha,\beta}}' \left(\frac{\partial f_{\alpha,\beta}}{\partial \bar{z}_k} * \Phi_\delta\right) d\bar{z}_k \wedge dz^\alpha \wedge d\bar{z}^\beta$$

$$= (-1)^p {\sum_{\substack{|\alpha|=p \\ |L|=q+2}}}' \left({\sum_{\{k\} \cup K = L}}' \varepsilon_L^{kK} \frac{\partial f_{\alpha,K}}{\partial \bar{z}_k}\right) * \Phi_\delta dz^\alpha \wedge d\bar{z}^L,$$

which implies that $\|S(f_\delta) - S(f)\|_3 \to 0$. $\square$

Lemma 2.19 *Let $f \in \mathcal{D}_{T^*}$ and $\mathrm{supp}(f) \subset\subset \Omega$. Then there exist $f_\delta \in \mathcal{D}_{(p,q+1)}(\Omega)$ such that $\|T^*(f_\delta) - T^*(f)\|_1 \to 0$ as $\delta \to 0$.*

Proof. Since functions

$$\sum_{j=1}^n e^{\varphi_1}\left\{\frac{\partial}{\partial z_j}(e^{-\varphi_2} f_{\alpha,j\gamma})\right\} = e^{\varphi_1 - \varphi_2} \sum_{j=1}^n \left(\frac{\partial f_{\alpha,j\gamma}}{\partial z_j} - \frac{\partial \varphi_2}{\partial z_j} f_{\alpha,j\gamma}\right)$$

are L^2 functions and $\mathrm{supp}(f)$ is a compact subset of Ω,

$$g_{\alpha,\gamma} = \sum_{j=1}^n \left(\frac{\partial f_{\alpha,j\gamma}}{\partial z_j} - \frac{\partial \varphi_2}{\partial z_j} f_{\alpha,j\gamma}\right)$$

are L^2 functions. Thus, $\|g_{\alpha,\gamma} * \Phi_\delta - g_{\alpha,\gamma}\|_{L^2} \to 0$. On the other hand we have

$$g_{\alpha,\gamma} = (-1)^{p-1} e^{\varphi_2 - \varphi_1} (T_* f)_{\alpha,\gamma}.$$

Therefore we obtain

$$(-1)^{p-1} e^{\varphi_2 - \varphi_1} T^*(f_\delta)$$
$$= {\sum_{\alpha,\gamma}}' \sum_{j=1}^n \left\{\frac{\partial f_{\alpha,j\gamma}}{\partial z_j} * \Phi_\delta - \frac{\partial \varphi_2}{\partial z_j}(f_{\alpha,j\gamma} * \Phi_\delta)\right\} dz^\alpha \wedge d\bar{z}^\gamma$$
$$:= {\sum_{\alpha,\gamma}}' \psi_{\alpha,\gamma}^\delta dz^\alpha \wedge d\bar{z}^\gamma.$$

Then

$$\|\psi^\delta_{\alpha,\gamma} - g_{\alpha,\gamma} * \Phi_\delta\|_{L^2(\Omega)}$$

$$= \left\| \sum_{j=1}^n \left(\frac{\partial \varphi_2}{\partial z_j} f_{\alpha,j\gamma} \right) * \Phi_\delta - \sum_{j=1}^n \frac{\partial \varphi_2}{\partial z_j} (f_{\alpha,j\gamma} * \Phi_\delta) \right\|_{L^2(\Omega)}$$

$$\leq \left\| \sum_{j=1}^n \left\{ \left(\frac{\partial \varphi_2}{\partial z_j} f_{\alpha,j\gamma} \right) * \Phi_\delta - \frac{\partial \varphi_2}{\partial z_j} f_{\alpha,j\gamma} \right\} \right\|_{L^2(\Omega)}$$

$$+ \left\| \sum_{j=1}^n \left\{ \frac{\partial \varphi_2}{\partial z_j} (f_{\alpha,j\gamma} - f_{\alpha,j\gamma} * \Phi_\delta) \right\} \right\|_{L^2(\Omega)}.$$

Consequently,

$$\|\psi^\delta_{\alpha,\gamma} - g_{\alpha,\gamma}\|_{L^2(\Omega)} \leq \|\psi^\delta_{\alpha,\gamma} - g_{\alpha,\gamma} * \Phi_\delta\|_{L^2(\Omega)} + \|g_{\alpha,\gamma} * \Phi_\delta - g_{\alpha,\gamma}\|_{L^2(\Omega)} \to 0,$$

which implies that $\|T^*(f_\delta) - T^*(f)\|_1 \to 0$. □

Theorem 2.6 *For $f \in \mathcal{D}_{T^*} \cap \mathcal{D}_S$, there exist $f_j \in \mathcal{D}_{(p,q+1)}(\Omega)$, $j = 1, 2, \cdots$, such that $\|f_j - f\|_\mathcal{G} \to 0$ as $j \to \infty$.*

Proof. For $\varepsilon > 0$, by Lemma 2.17 there exists j_0 such that

$$\|\eta_{j_0} f - f\|_\mathcal{G} < \frac{\varepsilon}{2}.$$

Since $\mathrm{supp}(\eta_{j_0} f) \subset\subset \Omega$, it follows from Lemma 2.18 and Lemma 2.19 that

$$\|(\eta_{j_0} f)_{\delta_0} - \eta_{j_0} f\|_\mathcal{G} < \frac{\varepsilon}{2}$$

for some $\delta_0 > 0$. Therefore we have $\|(\eta_{j_0} f)_{\delta_0} - f\|_\mathcal{G} < \varepsilon$. Since $(\eta_{j_0} f)_{\delta_0} \in \mathcal{D}_{(p,q+1)}(\Omega)$, Theorem 2.6 is proved. □

Lemma 2.20 *Let*

$$f = {\sum_{\substack{|\alpha|=p \\ |\beta|=q+1}}}' f_{\alpha,\beta} dz^\alpha \wedge d\bar{z}^\beta \in \mathcal{D}_{(p,q+1)}(\Omega).$$

Then

$$|\bar{\partial} f|^2 = {\sum_{\alpha,\beta}}' \sum_{j=1}^n \left| \frac{\partial f_{\alpha,\beta}}{\partial \bar{z}_j} \right|^2 - {\sum_{\alpha,\gamma}}' \sum_{j,k=1}^n \left(\frac{\partial f_{\alpha,j\gamma}}{\partial \bar{z}_k} \right) \left(\overline{\frac{\partial f_{\alpha,k\gamma}}{\partial \bar{z}_j}} \right).$$

Proof. Since

$$\bar{\partial} f = (-1)^p \underset{\substack{|\alpha|=p \\ |L|=q+2}}{\sum}{'} \underset{\{k\} \cup K = L}{\sum}{'} \varepsilon_L^{kK} \frac{\partial f_{\alpha,K}}{\partial \bar{z}_k} dz^\alpha \wedge d\bar{z}^L,$$

we have

$$|\bar{\partial} f|^2 = \underset{\substack{|\alpha|=p \\ |L|=q+2}}{\sum}{'} \left(\underset{\{j\} \cup J = L}{\sum}{'} \varepsilon_L^{jJ} \frac{\partial f_{\alpha,J}}{\partial \bar{z}_j} \right) \left(\underset{\{k\} \cup K = L}{\sum}{'} \varepsilon_L^{kK} \frac{\overline{\partial f_{\alpha,K}}}{\partial z_k} \right)$$

$$= \underset{\substack{|\alpha|=p \\ |K|=q+1}}{\sum}{'} \underset{|J|=q+1}{\sum}{'} \sum_{j,k=1}^{n} \frac{\partial f_{\alpha,J}}{\partial \bar{z}_j} \frac{\overline{\partial f_{\alpha,K}}}{\partial z_k}$$

$$= \underset{\alpha,J,K}{\sum}{'} \sum_{j=k} \frac{\partial f_{\alpha,J}}{\partial \bar{z}_j} \frac{\overline{\partial f_{\alpha,K}}}{\partial z_k} + \underset{\alpha,J,K}{\sum}{'} \sum_{j \neq k} \frac{\partial f_{\alpha,J}}{\partial \bar{z}_j} \frac{\overline{\partial f_{\alpha,K}}}{\partial z_k}$$

$$:= A + B.$$

When $j = k$, we have $J = K$ and $j \notin J$. Hence we have

$$A = \underset{\alpha,J}{\sum}{'} \sum_{j \notin J} \left| \frac{\partial f_{\alpha,J}}{\partial \bar{z}_j} \right|^2.$$

When $k \neq j$, we have $k \in J$ and $j \in K$. Thus if we set $J - \{k\} = K - \{j\} = \xi$, then we have $\varepsilon_{kK}^{jJ} = -\varepsilon_{k\xi}^J \varepsilon_K^{j\xi}$, $f_{\alpha,J} = \varepsilon_{k\xi}^J f_{\alpha,k\xi}$, $f_{\alpha,K} = \varepsilon_K^{j\xi} f_{\alpha,j\xi}$. Thus we obtain

$$B = - \underset{\alpha,\xi}{\sum}{'} \sum_{j \neq k} \frac{\partial f_{\alpha,k\xi}}{\partial \bar{z}_j} \frac{\overline{\partial f_{\alpha,j\xi}}}{\partial z_k}.$$

Consequently,

$$\underset{\alpha,J}{\sum}{'} \sum_{j \in J} \left| \frac{\partial f_{\alpha,J}}{\partial \bar{z}_j} \right|^2 = \underset{\alpha,\xi}{\sum}{'} \sum_{j=1}^{n} \left| \frac{\partial f_{\alpha,j\xi}}{\partial \bar{z}_j} \right|^2,$$

which gives the desired equality. □

Lemma 2.21 *Let f be a nonnegative function on $\mathbf{R}$ which is bounded on every bounded interval. Assume that there exists t_0 such that $f(t) = 0$ for $t \leq t_0$. Then there exists a convex increasing function $\chi \in C^\infty(\mathbf{R})$ such that*

$$\chi(t) \geq f(t), \quad \chi'(t) \geq f(t) \qquad (t \in \mathbf{R}).$$

Proof. For an integer n, choose a function $a_n \in C^\infty(\mathbf{R})$ such that $0 \leq a_n(t) \leq 1$ for $t \in \mathbf{R}$, and

$$a_n(t) = \begin{cases} 1 & (t \in [n-2, n]) \\ 0 & (t \notin [n-3, n+1]) \end{cases}.$$

Define

$$\sup_{t \in [n-2,n]} f(t) = M_n$$

and

$$\varphi(t) = \sum_{n=-\infty}^{\infty} M_n a_n(t).$$

Then $\varphi(t) \geq f(t)$ for every $t \in \mathbf{R}$. For $n-1 \leq x \leq n$, we have

$$\int_{-\infty}^{x} \varphi(x) dx \geq \int_{n-2}^{n-1} M_n dx = M_n \geq f(x).$$

We set

$$\chi_1(x) = \int_{-\infty}^{x} \varphi(t) dt.$$

Then $\chi_1 \in C^\infty(\mathbf{R})$, and

$$\chi_1'(x) = \varphi(x) \geq f(x), \quad \chi_1(x) \geq f(x) \quad (x \in \mathbf{R}).$$

Choose a function $\theta \in C^\infty(\mathbf{R})$ such that $\chi_1'' \leq \theta$, $\theta \geq 0$ and $\mathrm{supp}(\theta) \subset [t_0, \infty)$. If we set

$$\chi(x) = \int_{-\infty}^{x} \left\{ \int_{-\infty}^{t} \theta(y) dy \right\} dt,$$

then

$$\chi'(x) = \int_{-\infty}^{x} \theta(y) dy \geq \int_{-\infty}^{x} \chi_1''(y) dy = \chi_1'(x) \geq f(x),$$

$$\chi(x) \geq \int_{-\infty}^{x} \left\{ \int_{-\infty}^{t} \chi_1''(y) dy \right\} dt = \chi_1(x) \geq f(x),$$

$$\chi''(x) = \theta(x) \geq 0.$$

Thus χ is a desired function. □

Theorem 2.7 *Let $\Omega \subset \mathbf{C}^n$ be a pseuoconvex domain and let $\rho \in C^\infty(\Omega)$. Then there exists a function $\varphi \in C^\infty(\Omega)$ such that*

$$\sum_{j,k=1}^{n} \frac{\partial^2 \varphi}{\partial z_j \partial \bar{z}_k} w_j \bar{w}_k \geq 2(|\bar\partial\psi|^2 + e^\psi) \sum_{j=1}^{n} |w_j|^2 \quad (w \in \mathbf{C}^n),$$

$$\varphi(z) \geq \rho(z) \quad (z \in \Omega),$$

where ψ is the function in Lemma 2.13.

Proof. By Theorem 1.15, there exists a plurisubharmonic C^∞ function Φ in Ω such that for any real number t

$$\Omega_t = \{z \in \Omega \mid \Phi(z) < t\} \subset\subset \Omega.$$

Since Φ is strictly plurisubharmonic in Ω, there exists a continuous function $m(z) > 0$ in Ω such that for $z \in \Omega$ and $w \in \mathbf{C}^n$

$$\sum_{j,k=1}^{n} \frac{\partial^2 \Phi}{\partial z_j \partial \bar{z}_k}(z) w_j \bar{w}_k \geq m(z)|w|^2.$$

Define

$$g(t) = \max_{z \in \overline{\Omega}_t} \rho(z)$$

and

$$h(t) = \max_{z \in \overline{\Omega}_t} \left\{ \frac{2(|\bar\partial\psi(z)|^2 + e^{\psi(z)})}{m(z)} \right\}.$$

By Lemma 2.21 there exists a convex increasing function $\chi \in C^\infty(\mathbf{R})$ such that $\chi(t) \geq g(t)$, $\chi'(t) \geq h(t)$. We set $\varphi(z) = \chi \circ \Phi(z)$. Then we have

$$\chi(\Phi(z)) \geq g(\Phi(z)) = \max_{w \in \overline{\Omega}_{\Phi(z)}} \rho(w) \geq \rho(z)$$

and

$$\sum_{j,k=1}^{n} \frac{\partial^2 \varphi}{\partial z_j \partial \bar{z}_k}(z) w_j \bar{w}_k = \chi''(\Phi(z)) \left| \sum_{j=1}^{n} \frac{\partial \Phi}{\partial z_j}(z) w_j \right|^2$$

$$+ \chi'(\Phi(z)) \sum_{j,k=1}^{n} \frac{\partial^2 \Phi}{\partial z_j \partial \bar{z}_k}(z) w_j \bar{w}_k$$

$$\geq h(\Phi(z)) m(z) |w|^2$$

$$\geq 2(|\bar{\partial}\psi(z)|^2 + e^{\psi(z)}) |w|^2,$$

which completes the proof of Theorem 2.7. $\square$

Remark 2.1 *We are going to prove the inequality*

$$\|f\|_2^2 \leq \|T^* f\|_1^2 + \|Sf\|_3^2 \quad (f \in \mathcal{D}_{T^*} \cap \mathcal{D}_S). \tag{2.10}$$

If (2.10) holds, then we have for $f \in F := \operatorname{Ker} S$

$$\|f\|_2 \leq \|T^* f\|_1 \quad (f \in \mathcal{D}_{T^*} \cap \mathcal{D}_S).$$

Since $\mathcal{R}_T \subset F \subset \mathcal{D}_S$, we obtain

$$\|f\|_2 \leq \|T^* f\|_1 \quad (f \in \mathcal{D}_{T^*} \cap F).$$

By Theorem 2.4, we have $F = \mathcal{R}_T$, which implies that if $\bar{\partial} f = 0$, then there exists $u \in L^2_{(p,q)}(\Omega, \varphi_1)$ such that $\bar{\partial} u = f$.

Definition 2.13 For $g \in C^1(\Omega)$, define

$$\delta_j g = e^{\varphi} \frac{\partial}{\partial z_j}(g e^{-\varphi}) = \frac{\partial g}{\partial z_j} - g \frac{\partial \varphi}{\partial z_j}.$$

Then for $f \in C^2(\Omega)$ we obtain

$$\left[\delta_j, \frac{\partial}{\partial \bar{z}_k} \right] f := \delta_j \frac{\partial f}{\partial \bar{z}_k} - \frac{\partial}{\partial \bar{z}_k}(\delta_j f) = f \frac{\partial^2 \varphi}{\partial \bar{z}_k \partial z_j}.$$

Theorem 2.8 *Let $\Omega \subset \mathbf{C}^n$ be a pseudoconvex domain and let $\varphi \in C^\infty(\Omega)$ be the function in Theorem 2.7. If we set $\varphi_1 = \varphi - 2\psi$, $\varphi_2 = \varphi - \psi$ and $\varphi_3 = \varphi$, then*

$$\|f\|_2^2 \leq \|T^* f\|_1^2 + \|Sf\|_3^2$$

for $f \in \mathcal{D}_{(p,q+1)}(\Omega)$.

Proof. For $f \in \mathcal{D}_{(p,q+1)}(\Omega)$, we have

$$T^*f = (-1)^{p-1} \sum_{\substack{|\alpha|=p \\ |\gamma|=q}}{}' \sum_{j=1}^n e^{\varphi-2\psi} \left\{ \frac{\partial}{\partial z_j}(e^{-\varphi+\psi} f_{\alpha,j\gamma}) \right\} dz^\alpha \wedge d\bar{z}^\gamma$$

$$= (-1)^{p-1} e^{-\psi} \sum_{\substack{|\alpha|=p \\ |\gamma|=q}}{}' \sum_{j=1}^n \delta_j f_{\alpha,j\gamma} dz^\alpha \wedge d\bar{z}^\gamma$$

$$+ (-1)^{p-1} e^{-\psi} \sum_{\substack{|\alpha|=p \\ |\gamma|=q}}{}' \sum_{j=1}^n f_{\alpha,j\gamma} \frac{\partial \psi}{\partial z_j} dz^\alpha \wedge d\bar{z}^\gamma$$

$$:= A + B.$$

Then

$$\|A\|_1^2 = \int_\Omega \sum_{\alpha,\gamma}{}' \sum_{j,k=1}^n \delta_j f_{\alpha,j\gamma} \overline{\delta_k f_{\alpha,k\gamma}} e^{-\varphi} dV$$

$$= \int_\Omega \sum_{\alpha,\gamma}{}' \sum_{j,k=1}^n \delta_j f_{\alpha,j\gamma} \frac{\partial}{\partial \bar{z}_k} \left(\overline{f_{\alpha,k\gamma}} e^{-\varphi} \right) dV$$

$$= -\int_\Omega \sum_{\alpha,\gamma}{}' \sum_{j,k=1}^n \frac{\partial}{\partial \bar{z}_k}(\delta_j f_{\alpha,j\gamma}) \overline{f_{\alpha,k\gamma}} e^{-\varphi} dV.$$

Consequently,

$$\sum_{\alpha,\gamma}{}' \left| \sum_{j=1}^n f_{\alpha,j\gamma} \frac{\partial \psi}{\partial z_j} \right|^2 \leq \sum_{\alpha,\gamma}{}' \left(\sum_{j=1}^n |f_{\alpha,j\gamma}|^2 \right) \left(\sum_{j=1}^n \left|\frac{\partial \psi}{\partial z_j}\right|^2 \right)$$

$$= |\partial \psi|^2 \left(\sum_{\alpha,\gamma}{}' \sum_{j=1}^n |f_{\alpha,j\gamma}|^2 \right).$$

Hence we obtain

$$2\|T^*f\|_1^2 \geq \|A\|_1^2 - 2\|B\|_1^2$$

$$= -\int_\Omega \sum_{\alpha,\gamma}{}' \sum_{j,k=1}^n \frac{\partial}{\partial \bar{z}_k}(\delta_j f_{\alpha,j\gamma}) \overline{f_{\alpha,k\gamma}} e^{-\varphi} dV$$

$$-2\int_\Omega |\partial \psi|^2 \left(\sum_{\alpha,\gamma}{}' \sum_{j=1}^n |f_{\alpha,j\gamma}|^2 \right) e^{-\varphi} dV.$$

On the other hand we have

$$\|Sf\|_3^2 = \int_\Omega {\sum_{\alpha,\beta}}' \sum_{j=1}^n \left|\frac{\partial f_{\alpha,\beta}}{\partial \bar{z}_j}\right|^2 e^{-\varphi} dV$$

$$- \int_\Omega {\sum_{\alpha,\gamma}}' \sum_{j,k=1}^n \left(\frac{\partial f_{\alpha,j\gamma}}{\partial \bar{z}_k}\right) \left(\overline{\frac{\partial f_{\alpha,k\gamma}}{\partial z_j}}\right) e^{-\varphi} dV.$$

Using the equalities

$$\int_\Omega \left(\delta_j \frac{\partial}{\partial \bar{z}_k} f_{\alpha,j\gamma}\right) \overline{f_{\alpha,k\gamma}} e^{-\varphi} dV = \int_\Omega \frac{\partial}{\partial z_j}\left(\frac{\partial f_{\alpha,j\gamma}}{\partial \bar{z}_k} e^{-\varphi}\right) \overline{f_{\alpha,k\gamma}} dV$$

$$= -\int_\Omega \frac{\partial f_{\alpha,j\gamma}}{\partial \bar{z}_k} e^{-\varphi} \overline{\frac{\partial f_{\alpha,k\gamma}}{\partial z_j}} dV,$$

we have

$$\|Sf\|_3^2 = \int_\Omega {\sum_{\alpha,\beta}}' \sum_{j=1}^n \left|\frac{\partial f_{\alpha,\beta}}{\partial \bar{z}_j}\right|^2 e^{-\varphi} dV$$

$$+ \int_\Omega {\sum_{\alpha,\gamma}}' \sum_{j,k=1}^n \left(\delta_j \frac{\partial}{\partial \bar{z}_k} f_{\alpha,j\gamma}\right) \overline{f_{\alpha,k\gamma}} e^{-\varphi} dV.$$

Consequently,

$$2\|T^*f\|_1^2 + \|Sf\|_3^2 \geq \int_\Omega {\sum_{\alpha,\gamma}}' \sum_{j,k=1}^n \left\{\left(\delta_j \frac{\partial}{\partial \bar{z}_k} - \frac{\partial}{\partial \bar{z}_k}\delta_j\right) f_{\alpha,j\gamma}\right\} \overline{f_{\alpha,k\gamma}} e^{-\varphi} dV$$

$$-2 \int_\Omega |\partial\psi|^2 \left({\sum_{\alpha,\gamma}}' \sum_{j=1}^n |f_{\alpha,j\gamma}|^2\right) e^{-\varphi} dV$$

$$= \int_\Omega {\sum_{\alpha,\gamma}}' \sum_{j,k=1}^n \left\{\frac{\partial^2 \varphi}{\partial z_j \partial \bar{z}_k}\right\} f_{\alpha,j\gamma} \overline{f_{\alpha,k\gamma}} e^{-\varphi} dV$$

$$-2 \int_\Omega |\partial\psi|^2 \left({\sum_{\alpha,\gamma}}' \sum_{j=1}^n |f_{\alpha,j\gamma}|^2\right) e^{-\varphi} dV.$$

It follows from Theorem 2.7 that

$$2\|T^*f\|_1^2 + \|Sf\|_3^2 \geq \int_\Omega \sum_{\alpha,\gamma}{}' 2(|\partial\psi|^2 + e^\psi) \sum_{j=1}^n |f_{\alpha,j\gamma}|^2 e^{-\varphi} dV$$

$$-2\int_\Omega |\partial\psi|^2 \left(\sum_{\alpha,\gamma}{}' \sum_{j=1}^n |f_{\alpha,j\gamma}|^2 \right) e^{-\varphi} dV$$

$$\geq \int_\Omega \sum_{\alpha,\gamma}{}' \sum_{j=1}^n 2|f_{\alpha,j\gamma}|^2 e^{\psi-\varphi} dV \geq 2\|f\|_2^2.$$

□

Corollary 2.1 *For $f \in \mathcal{D}_{T^*} \cap \mathcal{D}_S$, we have*

$$\|f\|_2^2 \leq \|T^*f\|_1^2 + \|Sf\|_3^2.$$

Proof. For $f \in \mathcal{D}_{T^*} \cap \mathcal{D}_S$, by Theorem 2.6 there exist $f_j \in \mathcal{D}_{(p,q+1)}(\Omega)$ such that $\|f_j - f\|_{\mathcal{G}} \to 0$. On the other hand, by Theorem 2.8,

$$\|f_j\|_2^2 \leq \|T^*(f_j)\|_1^2 + \|S(f_j)\|_3^2.$$

Hence letting $j \to \infty$, we have

$$\|f\|_2^2 \leq \|T^*(f)\|_1^2 + \|S(f)\|_3^2.$$

□

Theorem 2.9 *Suppose $f \in L^2_{(p,q+1)}(\Omega, loc)$ satisfies $\bar\partial f = 0$. Then there exists $u \in L^2_{(p,q)}(\Omega, loc)$ such that $\bar\partial u = f$.*

Proof. Let $\{K_j\}$ be a sequence of compact subsets of Ω with the following properties:

$$K_j \subset\subset (K_{j+1})^\circ, \quad \bigcup_{j=1}^\infty K_j = \Omega. \quad (j = 1, 2, \cdots).$$

We set

$$\int_{K_j} |f|^2 dV = M_j \quad (j = 1, 2, \cdots).$$

Let $K_0 = \phi$. Choose a function $\tilde\varphi \in C^\infty(\Omega)$ such that

$$e^{-\tilde\varphi(z)} < \frac{1}{2^j M_j} \quad (z \in K_j - K_{j-1}).$$

Then

$$\int_\Omega |f|^2 e^{-\tilde{\varphi}} dV = \sum_{j=1}^\infty \int_{K_j-K_{j-1}} |f|^2 e^{-\tilde{\varphi}} dV$$
$$\leq \sum_{j=1}^\infty \int_{K_j-K_{j-1}} |f|^2 \frac{1}{2^j M_j} dV \leq \sum_{j=1}^\infty \frac{1}{2^j} = 1.$$

Hence we have $f \in L^2_{(p,q+1)}(\Omega, \tilde{\varphi})$. Suppose φ satisfies the condition of Theorem 2.7 for $\rho = \psi + \tilde{\varphi}$. Since

$$\int_\Omega |f|^2 e^{-\varphi_2} dV \leq \int_\Omega |f|^2 e^{-\tilde{\varphi}} dV < \infty,$$

we obtain $f \in L^2_{(p,q+1)}(\Omega, \varphi_2)$. Thus by Remark 2.1, there exists $u \in L^2_{(p,q)}(\Omega, \varphi_1)$ such that $\bar{\partial} u = f$. It follows from Lemma 2.8 that $u \in L^2_{(p,q)}(\Omega, \text{loc})$. □

Lemma 2.22 *Let $f \in \mathcal{D}(\mathbf{R}^N)$. Then*

$$|f(x)| \leq \int_{\mathbf{R}^N} \left| \frac{\partial^N f}{\partial t_1 \cdots \partial t_N}(t) \right| dV(t)$$

for every $x \in \mathbf{R}^N$.

Proof. For $x = (x_1, \cdots, x_N) \in \mathbf{R}^n$, we have

$$|f(x)| = \left| \int_{-\infty}^{x_1} \cdots \int_{-\infty}^{x_N} \frac{\partial^N}{\partial t_1 \cdots \partial t_N} f(t_1, \cdots, t_N) dt_N \cdots dt_1 \right|$$
$$\leq \int_{\mathbf{R}^N} \left| \frac{\partial^N}{\partial t_1 \cdots \partial t_N} f(t) \right| dV(t).$$

□

Definition 2.14 Let f be a locally integrable function in $\mathbf{R}^N$. For a multi-index $\alpha = (\alpha_1, \cdots, \alpha_N)$, where each α_j is a nonnegative integer, define

$$|\alpha| = \alpha_1 + \cdots + \alpha_N$$

and

$$\left(\frac{\partial}{\partial x}\right)^\alpha f = \frac{\partial^{|\alpha|} f}{\partial x_1^{\alpha_1} \cdots \partial x_N^{\alpha_N}}.$$

The $\bar{\partial}$ Problem in Pseudoconvex Domains

Lemma 2.23 *Let f be a locally integrable function in $\mathbf{R}^N$ with compact support. Suppose*

$$\left(\frac{\partial}{\partial x}\right)^\alpha f \in L^2(\mathbf{R}^N)$$

for all multi-indices α with $|\alpha| \leq N+1$. Then f is continuous almost everywhere.

Proof. Choose a function $\Phi \in \mathcal{D}(\mathbf{R}^N)$ such that

$$\int_{\mathbf{R}^N} \Phi \, dV = 1, \quad \operatorname{supp}(\Phi) \subset\subset \{x \in \mathbf{R}^N \mid |x| < 1\}.$$

We define

$$\Phi_\delta(x) = \delta^{-N}\Phi(x/\delta), \quad f_\varepsilon = f * \Phi_\varepsilon.$$

Then we have

$$\left\|\left(\frac{\partial}{\partial x}\right)^\alpha f_\varepsilon\right\|_{L^2} = \left\|\frac{\partial^\alpha f}{\partial x^\alpha} * \Phi_\varepsilon\right\|_{L^2} \leq \left\|\frac{\partial^\alpha f}{\partial x^\alpha}\right\|_{L^2}.$$

Hence there exist positive constants c_1 and c_2 such that

$$\left|\frac{\partial}{\partial x_j} f_\varepsilon(x)\right| \leq \int_{\mathbf{R}^N} \left|\frac{\partial^{N+1} f_\varepsilon(x)}{\partial x_1 \cdots \partial x_j^2 \cdots \partial x_N}\right| dx$$

$$\leq c_1 \left\|\frac{\partial^{N+1} f}{\partial x_1 \cdots \partial x_j^2 \cdots \partial x_N}\right\|_{L^2} \leq c_2.$$

Consequently, there exists a constant $c_3 > 0$ such that

$$|f_\varepsilon(x) - f_\varepsilon(y)| = \left|\sum_{j=1}^N \frac{\partial f_\varepsilon}{\partial x_j}(x + \theta y)(x_j - y_j)\right| \leq c_3 \|x - y\|,$$

which means that $\{f_\varepsilon\}$ is equicontinuous. On the other hand, there exists a constant $c_4 > 0$ such that

$$|f_\varepsilon(x)| \leq \int_{\mathbf{R}^n} \left|\frac{\partial^N}{\partial x_1 \cdots \partial x_N} f_\varepsilon(x)\right| dV(x) \leq c_4 \left\|\frac{\partial^N f}{\partial x_1 \cdots \partial x_N}\right\|_{L^2}.$$

Hence $\{f_\varepsilon\}$ are uniformly bounded. Using the Ascoli-Arzela theorem, one can choose a subsequence $\{f_{\varepsilon_j}\}$ of $\{f_\varepsilon\}$ which converges uniformly to $\tilde{f}$ on every compact subset of Ω. Thus we have $\|f_{\varepsilon_j} - \tilde{f}\|_{L^2} \to 0$. Since

$\|f_{\varepsilon_j} - f\|_{L^2} \to 0$, we have $f = \tilde{f}$ almost everywhere. Since $\tilde{f}$ is continuous, f is continuous almost everywhere. □

Theorem 2.10 *Let f be a locally integrable function in $\mathbf{R}^N$ with compact support. Suppose*

$$\left(\frac{\partial}{\partial x}\right)^\alpha f \in L^2(\mathbf{R}^N)$$

for all multi-indices α with $|\alpha| \leq N + k + 1$. Then there exists a function $h \in C^k(\mathbf{R}^N)$ such that $f = h$ almost everywhere.

Proof. We prove Theorem 2.10 by induction on k. In case $k = 0$, Theorem 2.10 follows from Lemma 2.23. We assume that $k \geq 1$ and Theorem 2.10 has already been proved for $k - 1$. For $1 \leq j \leq N$, we have

$$\frac{\partial^\alpha}{\partial x^\alpha}\left(\frac{\partial f}{\partial x_j}\right) \in L^2(\mathbf{R}^N) \qquad (|\alpha| \leq N + k).$$

Hence by the inductive hypothesis, there exists $\tilde{f}_j \in C^{k-1}(\mathbf{R}^N)$ such that $\frac{\partial f}{\partial x_j} = \tilde{f}_j$ almost everywhere. On the other hand, by Lemma 2.23 there exists a continuous function h such that $f = h$ almost everwhere. Hence in the sense of distributions, we obtain $\tilde{f}_j = \frac{\partial f}{\partial x_j} = \frac{\partial h}{\partial x_j}$. Thus h is partially differentiable and satisfies $\frac{\partial h}{\partial x_j} = \tilde{f}_j$ for $1 \leq j \leq N$, which means that $h \in C^k(\mathbf{R}^N)$. □

Corollary 2.2 *Let $\Omega \subset \mathbf{R}^N$ be an open set. Suppose that f is a locally integrable function in Ω and that*

$$\left(\frac{\partial}{\partial x}\right)^\alpha f \in L^2_{loc}(\Omega) \qquad (|\alpha| = 0, 1, 2, \cdots).$$

Then there exists a function $\tilde{f} \in C^\infty(\Omega)$ such that $f = \tilde{f}$ almost everywhere.

Proof. Fix a function $\eta \in \mathcal{D}(\Omega)$. We set $\eta f = h$. Then we have

$$\left(\frac{\partial}{\partial x}\right)^\alpha h \in L^2(\mathbf{R}^N) \qquad (|\alpha| = 0, 1, 2, \cdots).$$

It follows from Theorem 2.10 that there exists $\tilde{h} \in C(\mathbf{R}^N)$ such that $h = \tilde{h}$ almost everywhere. It follows from Theorem 2.10 that for every k there exists $\varphi \in C^k(\mathbf{R}^N)$ such that $h = \varphi$ almost everywhere. Thus, in the sense of distributions, we have $\frac{\partial \tilde{h}}{\partial x_j} = \frac{\partial \varphi}{\partial x_j}$, which means that $\tilde{h}$ is partially differentiable and satisfies $\frac{\partial \tilde{h}}{\partial x_j} = \frac{\partial \varphi}{\partial x_j}$. Thus we have $\tilde{h} \in C^k(\mathbf{R}^N)$. Since

k is arbitrary, $\tilde{h} \in C^\infty(\mathbf{R}^N)$. If we choose a sequence $\{K_n\}$ of compact subsets of Ω such that $K_n \subset \overset{\circ}{K}_{n+1}$, $\overset{\infty}{\underset{n=1}{\cup}} K_n = \Omega$, then f is of class C^∞ in $K_n - A_n$, where each set A_n is of Lebesgue measure 0. Since $A = \cup_{n=1}^\infty A_n$, f is of class C^∞ almost everywhere. $\square$

Lemma 2.24 *Let φ and ψ be as in Theorem 2.7. We set $\varphi_1 = \varphi - 2\psi$, $\varphi_2 = \varphi - \psi$. If $f \in L^2_{(p,q+1)}(\Omega, \varphi_2)$ satisfies $\bar{\partial} f = 0$, then there exists a unique u which satisfies*

$$Tu = f, \quad u \in (\operatorname{Ker} T)^\perp, \quad u \in L^2_{(p,q)}(\Omega, \varphi_1).$$

Proof. First we show that $\operatorname{Ker} T$ is a closed subset. Let $Tu_\nu = 0$, $u_\nu \to 0$. Then $(u_\nu, Tu_\nu) \to (u, 0)$. By Theorem 2.5, T is a closed operator and satisfies $u \in \mathcal{D}_T$, $(u, 0) = (u, Tu)$. Hence $Tu = 0$, which means that $u \in \operatorname{Ker} T$. Consequently, $\operatorname{Ker} T$ is closed. Since $\operatorname{Ker} S = \mathcal{R}_T$, there exists $\alpha \in L^2_{(p,q)}(\Omega, \varphi_1)$ such that $\bar{\partial} \alpha = f$. Since $\operatorname{Ker} T$ is closed, α can be written

$$\alpha = \alpha_1 + \alpha_2 \quad (\alpha_1 \in \operatorname{Ker} T, \ \alpha_2 \in (\operatorname{Ker} T)^\perp).$$

Define

$$\mathcal{P}\alpha = \alpha_1, \quad \alpha - \mathcal{P}\alpha = u.$$

Then $u \in (\operatorname{Ker} T)^\perp$ and $\bar{\partial} u = \bar{\partial} \alpha = f$, which shows that u is the desired solution. Next we asuume u^* also satisfies the conditions of the lemma. Then $u - u^* \in (\operatorname{Ker} T)^\perp$, and $T(u - u^*) = 0$, which means that $u - u^* \in \operatorname{Ker} T$. Thus $u = u^*$. $\square$

Definition 2.15 Let $\Omega \subset \mathbf{C}^n$ be an open set. For a nonnegative integer s, define the Sobolev space $W^s(\Omega)$ of order s by

$$W^s(\Omega) = \left\{ f \mid \left(\frac{\partial}{\partial z}\right)^\mu \left(\frac{\partial}{\partial \bar{z}}\right)^\eta f \in L^2(\Omega), \ |\mu| + |\eta| \leq s \right\}.$$

Further, we define for $f \in W^s_{(p,q)}(\Omega)$

$$\|f\|^2_{W^s(\Omega)} = \underset{\substack{|\alpha|=p \\ |\beta|=q}}{{\sum}'} \sum_{|\mu|+|\eta|\leq s} \left\| \left(\frac{\partial}{\partial z}\right)^\mu \left(\frac{\partial}{\partial \bar{z}}\right)^\eta f_{\alpha,\beta} \right\|^2_{L^2(\Omega)}.$$

Lemma 2.25 *For $f \in \mathcal{D}(\mathbf{C}^n)$, we have*

$$\left\| \frac{\partial f}{\partial z_j} \right\|_{L^2} = \left\| \frac{\partial f}{\partial \bar{z}_j} \right\|_{L^2} \quad (j = 1, 2, \cdots, n).$$

Proof. Using the integration by parts we have

$$\int_{\mathbf{C}^n} \left|\frac{\partial f}{\partial \bar{z}_j}\right|^2 dV = \int_{\mathbf{C}^n} \frac{\partial f}{\partial \bar{z}_j} \frac{\partial \bar{f}}{\partial z_j} dV = -\int_{\mathbf{C}^n} f \frac{\partial}{\partial \bar{z}_j}\left(\frac{\partial \bar{f}}{\partial z_j}\right) dV$$

$$= -\int_{\mathbf{C}^n} f \frac{\partial}{\partial z_j}\left(\frac{\partial \bar{f}}{\partial \bar{z}_j}\right) dV = \int_{\mathbf{C}^n} \frac{\partial f}{\partial z_j} \frac{\partial \bar{f}}{\partial \bar{z}_j} dV$$

$$= \int_{\mathbf{C}^n} \left|\frac{\partial f}{\partial z_j}\right|^2 dV.$$

□

Lemma 2.26 *Suppose that $f \in L^2(\mathbf{C}^n)$ has a compact support and that*

$$\frac{\partial f}{\partial \bar{z}_j} \in L^2(\mathbf{C}^n) \qquad (j = 1, \cdots, n).$$

Then $f \in W^1(\mathbf{C}^n)$. Moreover, we have

$$\left\|\frac{\partial f}{\partial z_j}\right\|_{L^2} = \left\|\frac{\partial f}{\partial \bar{z}_j}\right\|_{L^2} \qquad (j = 1, 2, \cdots, n).$$

Proof. For $\varepsilon > 0$, we set $f_\varepsilon = f * \Phi_\varepsilon$. Then $f_\varepsilon \in \mathcal{D}(\mathbf{C}^n)$. In $L^2(\mathbf{C}^n)$

$$\frac{\partial f_\varepsilon}{\partial \bar{z}_j} = \frac{\partial f}{\partial \bar{z}_j} * \Phi_\varepsilon \to \frac{\partial f}{\partial \bar{z}_j} \qquad (\varepsilon \to 0).$$

It follows from Lemma 2.25 that

$$\left\|\frac{\partial f_\varepsilon}{\partial z_j} - \frac{\partial f_\delta}{\partial z_j}\right\|_{L^2} = \left\|\frac{\partial f_\varepsilon}{\partial \bar{z}_j} - \frac{\partial f_\delta}{\partial \bar{z}_j}\right\|_{L^2} \to 0.$$

Hence $\{\partial f_\varepsilon/\partial z_j\}$ is a Cauchy sequence. Thus there exists $g \in L^2(\mathbf{C}^n)$ such that

$$\frac{\partial f_\varepsilon}{\partial z_j} \to g$$

as $\varepsilon \to 0$. For $\psi \in \mathcal{D}(\mathbf{C}^n)$

$$\left(\frac{\partial f_\varepsilon}{\partial z_j}, \psi\right) = -\left(f_\varepsilon, \frac{\partial \psi}{\partial z_j}\right) \to -\left(f, \frac{\partial \psi}{\partial z_j}\right) = \left(\frac{\partial f}{\partial z_j}, \psi\right),$$

which means that $\partial f/\partial z_j = g$. Hence we have $\partial f/\partial z_j \in L^2(\mathbf{C}^n)$, and by Lemma 2.25

$$\left\|\frac{\partial f_\varepsilon}{\partial \bar{z}_j}\right\|_{L^2} = \left\|\frac{\partial f_\varepsilon}{\partial z_j}\right\|_{L^2}.$$

Letting $\varepsilon \to 0$, we obtain
$$\left\|\frac{\partial f}{\partial \bar{z}_j}\right\|_{L^2} = \left\|\frac{\partial f}{\partial z_j}\right\|_{L^2}.$$
□

Definition 2.16 For $f \in L^2_{(p,q+1)}(\Omega)$ with $f = \sum'_{\alpha,\beta} f_{\alpha,\beta} dz^\alpha \wedge d\bar{z}^\beta$, define
$$T^* f := (-1)^{p-1} \sum_{\alpha,\gamma}' \sum_{j=1}^n \frac{\partial f_{\alpha,j\gamma}}{\partial z_j} dz^\alpha \wedge d\bar{z}^\gamma.$$

Lemma 2.27 Suppose that f is a differential form in $L^2_{(p,q+1)}(\Omega)$ with compact support and that $\bar{\partial} f \in L^2_{(p,q+2)}(\Omega)$ and $T^* f \in L^2_{(p,q)}(\Omega)$. Then $f \in W^1_{(p,q+1)}(\Omega)$.

Proof. In case $f \in \mathcal{D}_{(p,q+1)}(\Omega)$, we set $\psi = 0$ and $\varphi = 0$ in the proof of Theorem 2.8. Then
$$\int_\Omega \sum_{\alpha,\beta}' \sum_{j=1}^n \left|\frac{\partial f_{\alpha,\beta}}{\partial \bar{z}_j}\right|^2 dV \leq 2\|T^* f\|^2 + \|\bar{\partial} f\|^2. \tag{2.11}$$

In the general case, setting $f * \Phi_\delta = f_\delta$ and applying (2.11) to $f_\delta - f_\varepsilon$, we obtain
$$\|T^* f_\delta - T^* f_\varepsilon\|^2 + \|\bar{\partial} f_\delta - \bar{\partial} f_\varepsilon\|^2 \to 0$$
as $\varepsilon, \delta \to 0$. Consequently,
$$\sum_{\alpha,\beta}' \int_\Omega \sum_{j=1}^n \left|\frac{\partial (f_{\alpha,\beta} * \Phi_\delta)}{\partial \bar{z}_j} - \frac{\partial (f_{\alpha,\beta} * \Phi_\varepsilon)}{\partial \bar{z}_j}\right|^2 dV \to 0$$
as $\varepsilon, \delta \to 0$. Then there exists $\lambda_{\alpha,\beta} \in L^2(\Omega)$ such that in $L^2(\Omega)$,
$$\frac{\partial (f_{\alpha,\beta} * \Phi_\delta)}{\partial \bar{z}_j} \to \lambda_{\alpha,\beta}$$
as $\delta \to 0$. Hence we have $\partial f_{\alpha,\beta}/\partial \bar{z}_j = \lambda_{\alpha,\beta}$ in the sense of distributions, and hence $\partial f_{\alpha,\beta}/\partial \bar{z}_j \in L^2(\Omega)$. By Lemma 2.26, $\partial f_{\alpha,\beta}/\partial z_j \in L^2(\Omega)$, and hence $f \in W^1_{(p,q+1)}(\Omega)$. □

Theorem 2.11 Let $\Omega \subset \mathbf{C}^n$ be a pseudoconvex open set, and let $0 \leq s \leq \infty$. If $f \in W^s_{(p,q+1)}(\Omega, loc)$ and $\bar{\partial} f = 0$, then there exists a solution u of $\bar{\partial} u = f$ such that $u \in W^s_{(p,q)}(\Omega, loc)$.

Proof. In case $q = 0$. By Theorem 2.9, there exists a solution $u = \sum'_\alpha u_\alpha dz^\alpha$ of $\bar{\partial} u = f$ such that $u \in L^2_{(p,q)}(\Omega, \text{loc})$. Hence we have

$$\frac{\partial u_\alpha}{\partial \bar{z}_j} = f_{\alpha,j} \in W^s(\Omega, \text{loc}).$$

Suppose that $u \in W^\theta_{(p,0)}(\Omega, \text{loc})$ for some θ with $0 \leq \theta \leq s$ (if $\theta = 0$, then $u \in W^\theta_{(p,0)}(\Omega, \text{loc})$). For $\eta \in \mathcal{D}(\Omega)$ we have

$$\frac{\partial(\eta u_\alpha)}{\partial \bar{z}_j} = \eta f_{\alpha,j} + \frac{\partial \eta}{\partial \bar{z}_j} u_\alpha \in W^\theta.$$

For $|\mu| + |\nu| \leq \theta$,

$$\frac{\partial}{\partial \bar{z}_j} \left\{ \left(\frac{\partial}{\partial z_j}\right)^\mu \left(\frac{\partial}{\partial \bar{z}_j}\right)^\nu (\eta u_\alpha) \right\} = \left(\frac{\partial}{\partial z_j}\right)^\mu \left(\frac{\partial}{\partial \bar{z}_j}\right)^\nu \left(\frac{\partial}{\partial \bar{z}_j}(\eta u_\alpha)\right) \in L^2.$$

By Lemma 2.26, $\eta u_\alpha \in W^{\theta+1}$. By the inductive hypothesis on θ, we have $\eta u_\alpha \in W^{s+1}$. Thus we have $u \in W^{s+1}(\Omega, \text{loc})$.

In case $q \geq 1$. Since $f \in L^2_{(p,q+1)}(\Omega, \text{loc})$, by Lemma 2.9 there exists $\lambda \in C^\infty(\Omega)$ such that $f \in L^2_{(p,q+1)}(\Omega, \lambda)$. Suppose φ satisfies the condition of Theorem 2.7 for $\rho(z) = \psi + \lambda$. Then for $\varphi_1 = \varphi - 2\psi$ and $\varphi_2 = \varphi - \psi$, $f \in L^2_{(p,q+1)}(\Omega, \varphi_2)$, and hence by Lemma 2.24, there exists $u \in L^2_{(p,q)}(\Omega, \varphi_1)$ such that $\bar{\partial} u = f$, $u \in (\text{Ker } T)^\perp$. Since $\mathcal{R}_T = \text{Ker } S$, $\mathcal{R}_T$ is closed. By Lemma 2.6, $\mathcal{R}_{T^*}$ is closed. By Lemma 2.7,

$$u \in (\text{Ker } T)^\perp = \overline{\mathcal{R}_{T^*}} = \mathcal{R}_{T^*},$$

which means that $u = T^* v$ ($v \in L^2_{(p,q+1)}(\Omega, \varphi_2)$). For $v = \sum'_{\alpha,\beta} v_{\alpha,\beta} dz^\alpha \wedge d\bar{z}^\beta$, we set

$$\delta v = (-1)^{p-1} T^* v = \sum'_{\alpha,\gamma} \sum_{j=1}^n \frac{\partial f_{\alpha,j\gamma}}{\partial z_j} dz^\alpha \wedge d\bar{z}^\gamma.$$

Since $e^{-\varphi_1} u = (-1)^{p-1} \delta(e^{-\varphi_2} v)$, we obtain

$$\delta(e^{-\varphi_1} u) = (-1)^{p-1} \sum'_{\alpha,L} \sum_{k=1}^n \sum_{j=1}^n \frac{\partial(v_{\alpha,jkL} e^{-\varphi_2})}{\partial z_k \partial z_j} dz^\alpha \wedge d\bar{z}^L = 0.$$

Consequently,

$$0 = \sum_{\alpha,\gamma}{}' \sum_{j=1}^n \frac{\partial(e^{-\varphi_1} u_{\alpha,j\gamma})}{\partial z_j} dz^\alpha \wedge d\bar{z}^\gamma$$

$$= \sum_{\alpha,\gamma}{}' \sum_{j=1}^n \frac{\partial u_{\alpha,j\gamma}}{\partial z_j} dz^\alpha \wedge d\bar{z}^\gamma - \sum_{\alpha,\gamma}{}' \sum_{j=1}^n \frac{\partial \varphi_1}{\partial z_j} u_{\alpha,j\gamma} dz^\alpha \wedge d\bar{z}^\gamma.$$

Let $0 \leq \theta \leq s$. Suppose $u \in W_{(p,q)}^\theta(\Omega, \text{loc})$. When $\theta = 0$, it is true. Fix $\eta \in \mathcal{D}(\Omega)$. Then

$$\bar{\partial}(\eta u) = \bar{\partial}\eta \wedge u + \eta f \in W_{(p,q+1)}^\theta(\Omega).$$

On the other hand we have

$$(-1)^{p-1} T^*(\eta u)$$

$$= \sum_{\alpha,\gamma}{}' \sum_{j=1}^n \frac{\partial(\eta u_{\alpha,j\gamma})}{\partial z_j} dz^\alpha \wedge d\bar{z}^\gamma$$

$$= \eta \sum_{\alpha,\gamma}{}' \sum_{j=1}^n \frac{\partial \varphi_1}{\partial z_j} u_{\alpha,j\gamma} dz^\alpha \wedge d\bar{z}^\gamma + \sum_{\alpha,\gamma}{}' \sum_{j=1}^n \frac{\partial \eta}{\partial z_j} u_{\alpha,j\gamma} dz^\alpha \wedge d\bar{z}^\gamma,$$

which implies that

$$T^*(\eta u) \in W_{(p,q-1)}^\theta(\Omega).$$

Suppose $|\mu| + |\nu| \leq \theta$. Then

$$\bar{\partial}\left(\left(\frac{\partial}{\partial z}\right)^\mu \left(\frac{\partial}{\partial \bar{z}}\right)^\nu (\eta u)\right)$$

$$= \sum_{\alpha,\gamma}{}' \left(\frac{\partial}{\partial z}\right)^\mu \left(\frac{\partial}{\partial \bar{z}}\right)^\nu (\bar{\partial}(\eta u))_{\alpha,\gamma} dz^\alpha \wedge d\bar{z}^\gamma \in L_{(p,q+1)}^2(\Omega).$$

Similarly,

$$T^*\left(\left(\frac{\partial}{\partial z}\right)^\mu \left(\frac{\partial}{\partial \bar{z}}\right)^\nu (\eta u)\right) \in L_{(p,q-1)}^2(\Omega).$$

By Lemma 2.27 we have

$$\left(\frac{\partial}{\partial z}\right)^\mu \left(\frac{\partial}{\partial \bar{z}}\right)^\nu (\eta u) \in W_{(p,q)}^1(\Omega),$$

which implies that $\eta u \in W_{(p,q)}^{\theta+1}(\Omega)$. By the inductive hypothesis on θ, $\eta u \in W_{(p,q)}^{s+1}(\Omega)$, and hence $u \in W_{(p,q)}^{s+1}(\Omega, \text{loc})$. □

Corollary 2.3 *Let $\Omega \subset \mathbf{C}^n$ be a pseudoconvex domain. Suppose f is a $C^\infty(p, q+1)$ form in Ω with $\bar\partial f = 0$. Then there exists a $C^\infty(p,q)$ form u in Ω such that $\bar\partial u = f$.*

Proof. Since $C^\infty(\Omega) \subset W_{\text{loc}}^\infty(\Omega)$, we have $f \in W_{(p,q+1)}^\infty(\Omega, \text{loc})$. Then by Theorem 2.11, there exists $u \in W_{(p,q)}^\infty(\Omega, \text{loc})$ such that $\bar\partial u = f$. By Corollary 2.2, there exists a C^∞ (p,q) form $\tilde u$ on Ω such that $u = \tilde u$ almost everywhere. Thus we have $\bar\partial \tilde u = f$. □

Definition 2.17 A metric space is called separable if it has a countable everywhere dense subset.

Definition 2.18 Let H be a Hilbert space. We say that a sequence $\{x_n\}$ in H converges weakly to $x \in H$ if

$$(x_n, y) \to (x, y)$$

for every $y \in H$.

Lemma 2.28 *Every bounded sequence in a separable Hilbert space contains a weakly convergent subsequence.*

Proof. Suppose $\{x_n\}$ is a bounded sequence in a separable Hilbert space H. Then there exists a constant M such that $\|x_n\| \leq M$ for all n. Since H is separable, there is a countable dense subset $\{z_n\}$. We have

$$|(x_n, z_1)| \leq \|x_n\| \|z_1\| \leq M\|z_1\|.$$

Thus $\{(x_n, z_1)\}$ is bounded. Using Bolzano-Weierstrass theorem, there is a subsequence $\{x_n^{(1)}\}$ of $\{x_n\}$ such that $\{(x_n^{(1)}, z_1)\}$ converges. Since

$$|(x_n^{(1)}, z_2)| \leq \|x_n^{(1)}\| \|z_2\| \leq M\|z_2\|,$$

there exists a subsequence $\{x_n^{(2)}\}$ of $\{x_n^{(1)}\}$ such that $\{(x_n^{(2)}, x_2)\}$ converges. Repeating this process, we have sequences $\{x_n^{(k)}\}$ such that

(1) $\{x_n^{(k+1)}\}$ is a subsequence of $\{x_n^{(k)}\}$.
(2) $\{(x_n^{(k)}, z_j)\}$ converge for $j = 1, \cdots, k$.

Thus $\{(x_n^{(n)}, z)\}$ converge for $z = z_1, z_2, \cdots$. We set $y_n = x_n^{(n)}$, and for $\varepsilon > 0$, we set $\delta = \min\{\varepsilon/(3M), \varepsilon/3\}$. If $w_1, w_2 \in H$, $\|w_1 - w_2\| < \delta$, then

$$|(y_n, w_1) - (y_n. w_2)| \leq \|y_n\| \|w_1 - w_2\| < \frac{\varepsilon}{3}.$$

Since for any $z \in H$, there exists z_{n_0} such that $\|z - z_{n_0}\| < \delta$. Hence we have

$$|(y_m, z) - (y_n, z)| \leq |(y_m, z) - (y_m, z_{n_0})| + |(y_m, z_{n_0}) - (y_n, z_{n_0})|$$
$$+ |(y_n, z_{n_0}) - (y_n, z)|$$
$$\leq \frac{\varepsilon}{3} + |(y_m, z_{n_0}) - (y_n, z_{n_0})| + \frac{\varepsilon}{3}.$$

Since $\{(y_n, z_{n_0})\}$ converges, there exists a positive integer N such that if $m, n \geq N$, then

$$|(y_m, z_{n_0}) - (y_n, z_{n_0})| < \frac{\varepsilon}{3}.$$

Hence if $m, n \geq N$, then

$$|(y_m, z) - (y_n, z)| < \varepsilon,$$

which implies that $\{(y_n, z)\}$ converges. Define $\varphi(z) = \lim_{n \to \infty}(z, y_n)$. Then φ is a bounded linear functional on H. It is evident that φ is linear. There exists a positive integer N_1 such that if $n \geq N_1$, then

$$|\varphi(z) - (y_n, z)| < 1 \quad (z \in H),$$

which implies that

$$|\varphi(z)| < 1 + M \quad (z \in H, \|z\| = 1).$$

Hence φ is bounded. Using the Riesz representation theorem, there exists $y \in H$ such that

$$\varphi(z) = (z, y) \quad (z \in H).$$

Then we have

$$(y, z) - (y_n, z) = \overline{\varphi(z) - (z, y_n)} \to 0 \quad (n \to \infty),$$

which implies that $\{y_n\}$ converges weakly to y. □

Theorem 2.12 Let $\Omega \subset \mathbf{C}^n$ be a pseudoconvex open set. Suppose that $\varphi \in C^2(\Omega)$ is a real-valued function and that there exists a continuous positive function c in Ω such that

$$c(z) \sum_{j=1}^{n} |w_j|^2 \leq \sum_{j,k=1}^{n} \frac{\partial^2 \varphi}{\partial z_j \partial \bar{z}_k}(z) w_j \bar{w}_k \quad (z \in \Omega, \ w \in \mathbf{C}^n).$$

If $g \in L^2_{(p,q+1)}(\Omega, \varphi)$ satisfies $\bar{\partial} g = 0$, then there exists $u \in L^2_{(p,q)}(\Omega, \varphi)$ such that $\bar{\partial} u = g$, and

$$\int_\Omega |u|^2 e^{-\varphi} dV \leq 2 \int_\Omega \frac{|g|^2}{c} e^{-\varphi} dV, \qquad (2.12)$$

provided that the right side is finite.

Proof. There exists a C^∞ strictly plurisubharmonic function ρ in Ω such that for any real number a,

$$\Omega_a = \{z \in \Omega \mid \rho(z) < a\} \subset\subset \Omega.$$

Fix a. We choose a sequence $\{K_j\}_{j=0}^\infty$ of compact subsets of Ω with the following properties:

$$\Omega_{a+1} \subset K_0, \quad K_j \subset K_{j+1}^\circ, \quad \bigcup_{j=1}^\infty K_j = \Omega.$$

Let $\eta_j \in C_c^\infty(\Omega)$ be functions such that $\eta_j = 1$ on K_{j-1}, $\mathrm{supp}(\eta_j) \subset K_j$. Define

$$\psi(z) = \begin{cases} \sum_{k=1}^n \left|\frac{\partial \eta_j}{\partial \bar{z}_k}\right|^2 & (z \in K_j - K_{j-1}) \\ 0 & (z \in K_0) \end{cases}.$$

Then $\psi \in C^\infty(\Omega)$. Moreover, we have $\psi(z) = 0$ for $z \in \Omega_{a+1}$, and

$$e^{\psi(z)} \geq \psi(z) = \sum_{k=1}^n \left|\frac{\partial \eta_j}{\partial \bar{z}_k}\right|^2 \quad (j = 1, 2, \cdots).$$

Since ρ is strictly plurisubharmonic in Ω, there exists a positive continuous function m in Ω such that

$$\sum_{j,k=1}^\infty \frac{\partial^2 \rho}{\partial z_j \partial \bar{z}_k} w_j \bar{w}_k \geq m(z) |w|^2.$$

By Lemma 2.21, there exists a convex increasing function χ in $\mathbf{R}$ such that $\chi(t) = 0$ for $-\infty < t < a$, and

$$\chi(\rho(z)) \geq 2\psi(z), \quad \chi'(\rho(z)) \geq \frac{2|\partial \psi|^2}{m(z)}.$$

for all $z \in \Omega$. We set

$$\varphi' = \varphi + \chi \circ \rho, \quad \varphi_j = \varphi' + (j-3)\psi \quad (j=1,2,3).$$

Then
$$\varphi_2 - \varphi \geq \psi \geq 0, \quad 2\varphi_2 - \varphi - \varphi' = \chi \circ \rho - 2\psi \geq 0.$$

Repeating the proof of Theorem 2.7, we obtain

$$\sum_{j,k=1}^n \frac{\partial^2 \varphi'}{\partial z_j \partial \bar{z}_k} w_j \bar{w}_k = \sum_{j,k=1}^\infty \frac{\partial^2 \varphi}{\partial z_j \partial \bar{z}_k} w_j \bar{w}_k + \chi''(\rho(z)) \left| \sum_{j=1}^n \frac{\partial \rho}{\partial z_j}(z) w_j \right|^2$$

$$+ \chi'(\rho(z)) \sum_{j,k=1}^n \frac{\partial^2 \rho}{\partial z_j \partial \bar{z}_k}(z) w_j \bar{w}_k$$

$$\geq (2|\partial \psi|^2 + c)|w|^2.$$

By applying the proof of Theorem 2.8, we have for $f \in \mathcal{D}_{(p,q+1)}(\Omega)$

$$2\|T^* f\|_1^2 + \|Sf\|_3^2$$

$$= \int_\Omega \sum_{\alpha,\gamma}{}' \sum_{j,k=1}^n \left\{ \frac{\partial^2 \varphi'}{\partial z_j \partial \bar{z}_k} \right\} f_{\alpha,j\gamma} \overline{f_{\alpha,k\gamma}} e^{-\varphi'} dV$$

$$- 2 \int_\Omega |\partial \psi|^2 \left(\sum_{\alpha,\gamma}{}' \sum_{j=1}^n |f_{\alpha,j\gamma}|^2 \right) e^{-\varphi'} dV$$

$$\geq \int_\Omega \sum_{\alpha,\gamma}{}' (2|\partial \psi|^2 + c) \sum_{j=1}^n |f_{\alpha,j\gamma}|^2 e^{-\varphi'} dV$$

$$- 2 \int_\Omega |\partial \psi|^2 \left(\sum_{\alpha,\gamma}{}' \sum_{j=1}^n |f_{\alpha,j\gamma}|^2 \right) e^{-\varphi'} dV$$

$$\geq \int_\Omega \sum_{\alpha,\gamma}{}' \sum_{j=1}^n c|f_{\alpha,j\gamma}|^2 e^{-\varphi'} dV \geq \int_\Omega c|f|^2 e^{-\varphi'} dV.$$

Consequently,

$$\int_\Omega c|f|^2 e^{-\varphi'} \leq 2\|T^* f\|_1^2 + \|Sf\|_3^2 \quad (f \in \mathcal{D}_{T^*} \cap \mathcal{D}_S).$$

We set

$$\int_\Omega \frac{|g|^2}{c} e^{-\varphi} dV = A^2.$$

Then using the Schwarz inequality, we have for $f \in \mathcal{D}_{T^*} \cap \mathcal{D}_S$

$$|(g,f)_2|^2 \le \left(\int_\Omega \frac{|g|^2}{c} e^{-\varphi} dV\right)\left(\int_\Omega c|f|^2 e^{\varphi - 2\varphi_2} dV\right)$$

$$\le A^2 \int_\Omega c|f|^2 e^{-\varphi'} dV \le A^2 \left(2\|T^*f\|_1^2 + \|Sf\|_3^2\right).$$

Now we show that

$$|(g,f)_2| \le \sqrt{2}A\|T^*f\|_1 \qquad (f \in \mathcal{D}_{T^*}). \tag{2.13}$$

If $Sf = 0$, then (2.13) is trivial. Suppose $f \in (\operatorname{Ker} S)^\perp$. Since $S \circ T = 0$, we have $\mathcal{R}_T \subset \operatorname{Ker} S$, and hence $f \in (\mathcal{R}_T)^\perp$. Then by Lemma 2.7 we have $T^*f = 0$. Hence we have

$$\int_\Omega |g|^2 e^{-\varphi_2} dV \le \int_\Omega |g|^2 e^{-\varphi} dV < \infty,$$

which implies that $g \in L^2_{(p,q+1)}(\Omega, \varphi_2)$. Since $\bar{\partial} g = 0$, we obtain $g \in \operatorname{Ker} S$, and hence $(g,f)_2 = 0$. Let $f \in \mathcal{D}_{T^*}$. Then f can be uniquely expressed by

$$f = f_1 + f_2 \qquad (f_1 \in \operatorname{Ker} S, \ f_2 \in (\operatorname{Ker} S)^\perp).$$

Then

$$|(g,f)_2| = |(g,f_1)_2| \le \sqrt{2}A\|T^*f_1\| = \sqrt{2}A\|T^*f\|,$$

which implies that (2.13) holds. Now we define a linear functional Φ on $\mathcal{R}_{T^*}$ by $\Phi(T^*f) = (f,g)_2$. If $T^*f = T^*f'$, then (2.13) implies that $f = f'$. Thus Φ is well defined. Applying the Hahn-Banach theorem, Φ can be extended to a bounded linear functional on $L^2_{(p,q)}(\Omega, \varphi_1)$. Using the Riesz representation theorem, there exists $u_a \in L^2_{(p,q)}(\Omega, \varphi_1)$ such that

$$(f,g)_2 = \Phi(T^*f) = (T^*f, u_a)_1, \quad \|\Phi\| = \|u_a\|_1.$$

It follows from (2.13) that

$$\|\Phi\| = \sup_{f \in \mathcal{D}_{T^*}} \frac{|(f,g)_2|}{\|T^*f\|} \le \sqrt{2}A,$$

which implies that

$$\int_\Omega |u_a|^2 e^{-\varphi_1} dV \le 2A^2.$$

Further we have

$$|(T^*f, u_a)_1| \leq \|g\|_2 \|f\|_2,$$

which implies that by Lemma 2.5 $u_a \in \mathcal{D}_{(T^*)^*} = \mathcal{D}_T$. Hence we have

$$(f,g)_2 = (T^*f, u_a) = (f, Tu_a) \qquad (f \in \mathcal{D}_{T^*}).$$

Thus we obtain $Tu_a = g$. Let $\{u_{a_j}\}$ be a subsequence of $\{u_a\}$ such that $a_j \to \infty$. By Lemma 2.28, we can choose a subsequence of $\{u_{a_j}\}$ which converges weakly in $L^2_{(p,q)}(\Omega, \varphi_1)$. Hence we may assume that $\{u_{a_j}\}$ converges weakly. For any real number α, $\{u_{a_j}\}$ converges weakly in $H_\alpha = L^2_{(p,q)}(\Omega_\alpha, \varphi)$. We set $\lim_{j \to \infty} u_{a_j} = u$. If $a_j > \alpha$, then $\varphi = \varphi_1$ on Ω_α. Hence we have

$$\int_{\Omega_\alpha} |u_{a_j}|^2 e^{-\varphi} dV \leq 2A^2.$$

Using the equality

$$(u_{a_j}, u)_{H_\alpha} = (u_{a_j} - u, u)_{H_\alpha} + (u, u)_{H_\alpha},$$

we obtain

$$\|u\|^2_{H_\alpha} \leq \sqrt{2}A \|u\|_{H_\alpha} + |(u_{a_j} - u, u)_{H_\alpha}|,$$

which implies that $\|u\|_{H_\alpha} \leq \sqrt{2}A$. Namely, for any α,

$$\int_{\Omega_\alpha} |u|^2 e^{-\varphi} dV \leq 2A^2.$$

For $f \in \mathcal{D}_{T^*}$, we have

$$(g,f)_2 = (Tu_{a_j}, f)_2 = (u_{a_j}, T^*f) \to (u, T^*f) = (Tu, f),$$

which implies that $Tu = g$. □

Now we are going to prove L^2 estimates for the $\bar{\partial}$ problem in pseudoconvex domains obtained by Hörmander [HR2].

Theorem 2.13 (L^2 **estimates**) *Let $\Omega \subset \mathbf{C}^n$ be a pseudoconvex open set and let φ be a plurisubharmonic function in Ω. If $g \in L^2_{(p,q+1)}(\Omega, \varphi)$ satisfies $\bar{\partial} g = 0$, then there exists a solution $u \in L^2_{(p,q)}(\Omega, loc)$ of the equation $\bar{\partial} u = g$ such that*

$$\int_\Omega |u|^2 e^{-\varphi} (1 + |z|^2)^{-2} dV \leq \int_\Omega |g|^2 e^{-\varphi} dV. \qquad (2.14)$$

Proof. First we assume that $\varphi \in C^2(\Omega)$. We set $\varphi' = \varphi + 2\log(1+|z|^2)$. Then

$$\sum_{j,k=1}^n \frac{\partial^2 \varphi'}{\partial z_j \partial \bar{z}_k} w_j \bar{w}_k \geq 2 \sum_{j,k=1}^n \frac{\partial^2}{\partial z_j \partial \bar{z}_k} \{\log(1+|z|^2)\} w_j \bar{w}_k$$

$$= \frac{2}{(1+|z|^2)^2} \left(|w|^2 (1+|z|^2) - |\sum_{j=1}^n \bar{z}_j w_j|^2 \right)$$

$$\geq 2(1+|z|^2)^{-2} |w|^2.$$

Then (2.14) follows from Theorem 2.12. In the general case, there exists a C^∞ strictly plurisubharmonic function ρ in Ω such that for any real number a

$$\Omega_a = \{z \in \Omega \mid \rho(z) < a\} \subset\subset \Omega.$$

Suppose that $\Phi \in \mathcal{D}(\mathbf{C}^n)$ is a function depending only on $|z_1|, \cdots, |z_n|$ and that $\Phi = 0$ on $|z| \geq 1$, $0 \leq \Phi$ and $\int \Phi dV = 1$. Define

$$\varphi_\varepsilon(z) = \int_\Omega \varphi(z - \varepsilon \zeta) \Phi(\zeta) dV(\zeta).$$

Then φ_ε is a C^∞ plurisubharmonic function in $\{z \in \mathbf{C}^n \mid d(z, \Omega^c) > \varepsilon\}$, and $\varphi_\varepsilon \downarrow \varphi$ if $\varepsilon \downarrow 0$. We choose $a(\varepsilon)$ with the properties that $a(\varepsilon) \to \infty$ if $\varepsilon \to 0$, and that φ_ε is a C^∞ plurisubharmonic function in $\Omega_{a(\varepsilon)}$. By using the result of the C^2 case, there exists $u_\varepsilon \in L^2_{(p,q)}(\Omega_{a(\varepsilon)}, \varphi_\varepsilon)$ such that $\bar{\partial} u_\varepsilon = g$ in $\Omega_{a(\varepsilon)}$, and

$$\int_{\Omega_{a(\varepsilon)}} |u_\varepsilon|^2 e^{-\varphi_\varepsilon} (1+|z|^2)^{-2} dV \leq \int_{\Omega_{a(\varepsilon)}} |g|^2 e^{-\varphi_\varepsilon} dV \leq \int_\Omega |g|^2 e^{-\varphi} dV.$$

Fix a real number α. We choose $\delta > 0$ such that $a(\delta) > \alpha$. Then there exists a constant $c_1 > 0$ such that

$$c_1 \leq e^{-\varphi_\delta}(1+|z|^2)^{-2} \qquad (z \in \Omega_\alpha),$$

which implies that for $\varepsilon > \delta$,

$$\int_{\Omega_\alpha} |u_\varepsilon|^2 dV \leq \frac{1}{c_1} \int_\Omega |g|^2 e^{-\varphi} dV.$$

Hence we can choose a subsequence $\{u_{\varepsilon_j}\}$ of $\{u_\varepsilon\}$ which converges weakly in $L^2_{(p,q)}(\Omega_\alpha)$. Let $\{\alpha_k\}$ be a sequence such that $\alpha_k \to \infty$. Since we can choose a subsequence $\{u_{j,k}\}$ of $\{u_{\varepsilon_j}\}$ which converges weakly in $\{\Omega_{\alpha_k}\}$, $\{u_{k,k}\}$ converges weakly to u in $L^2_{(p,q)}(\Omega, \text{loc})$. Hence we have for $\varepsilon > \delta$,

$$\int_{\Omega_\alpha} |u|^2 e^{-\varphi_\varepsilon}(1+|z|^2)^{-2} dV \leq \int_\Omega |g|^2 e^{-\varphi} dV.$$

Letting $\varepsilon \to 0$ we have

$$\int_{\Omega_\alpha} |u|^2 e^{-\varphi}(1+|z|^2)^{-2} dV \leq \int_\Omega |g|^2 e^{-\varphi} dV.$$

Since $\bar{\partial} u = g$, and α is arbitrary, we have (2.14). □

Theorem 2.14 *Let $\Omega \subset \mathbf{C}^n$ be a pseudoconvex open set. Define*

$$\omega = \Omega \cap \{z_n = 0\}.$$

Suppose $f : \omega \to \mathbf{C}$ is holomorphic in $\tilde{\omega}$, where

$$\tilde{\omega} = \{(z_1, \cdots, z_{n-1}) \in \mathbf{C}^{n-1} \mid (z_1, \cdots, z_{n-1}, 0) \in \omega\}.$$

Then there exists a holomorphic function $F : \Omega \to \mathbf{C}$ such that $F|_\omega = f$.

Proof. Let $\pi : \mathbf{C}^n \to \mathbf{C}^{n-1}$ be the projection such that

$$\pi(z_1 \cdots, z_n) = (z_1 \cdots, z_{n-1}).$$

If $B = \{z \in \Omega \mid \pi(z) \notin \tilde{\omega}\}$, then, ω and B are closed subset of Ω, and $\omega \cap B = \phi$. Then there exists a function $\psi \in C^\infty(\Omega)$ such that $\psi = 1$ on an open subset of Ω which contains ω, and $\psi = 0$ on an open subset of Ω which contains B. Define

$$F(z) = \psi(z) f(\pi(z), 0) + z_n v(z),$$

where v is a C^∞ function in Ω. We will determine v later to satisfy $\bar{\partial} F = 0$. Then F is a C^∞ function on Ω and

$$\bar{\partial} F(z) = \bar{\partial} \psi(z) f(\pi(z), 0) + z_n \bar{\partial} v(z).$$

If v satisfies

$$\bar{\partial} v(z) = \frac{(-\bar{\partial}\psi(z)) f(\pi(z), 0)}{z_n}, \tag{2.15}$$

then $0 = \bar{\partial} F$. Since $\bar{\partial}\psi(z) = 0$ in a neighborhood of ω, the right side of (2.15) is of class C^∞ in Ω. Further, the right side of (2.15) is $\bar{\partial}$ closed. Therefore, by Corollary 2.3 there exists $v \in C^\infty(\Omega)$ which satisfies (2.15). Thus F is holomorphic in Ω. Since $F|_\omega = f$, F is the desired function. □

Theorem 2.15 *Let $\Omega \subset \mathbf{C}^n$ be an open set. Suppose that for every $f \in C^\infty_{(0,q+1)}$ ($0 \leq q \leq n-2$) with $\bar\partial f = 0$, there exists $u \in C^\infty_{(0,q)}(\Omega)$ such that $\bar\partial u = f$. Then Ω is a domain of holomorphy.*

Proof. We prove Theorem 2.15 by induction on n. If $n = 1$, then the theorem is true since every open set is a domain of holomorphy. Assume that the theorem has already been proved for $n-1$ dimensions. In order to prove that the domain $\Omega \subset \mathbf{C}^n$ is a domain of holomorphy, it is sufficient to show that for every open convex set $D \subset \Omega$ such that some boundary point z^0 of D is on $\partial\Omega$, there is a holomorphic function in Ω which cannot be continued holomorphically to a neighboorhood of z^0. (For, if a holomorphic function in Ω can be extended holomorphically to a neighborhood G of $w \subset \partial\Omega$, then there exists $r > 0$ such that $B(w, 2r) \subset G$. For $z_1 \in B(w,r) \cap \Omega$, there exists $\delta (0 < \delta \leq r)$ such that $B(z_1, \delta) \subset \Omega$, $\partial B(z_1, \delta) \cap \partial\Omega \neq \phi$. Then every holomorphic function on Ω can be extended to a neighborhood of $\partial B(z_1, \delta) \cap \partial\Omega$, which is a contradiction.) We choose a coordinate system such that $z_0 = 0$ and $D_0 = \{z_n = 0\} \cap D$ is not empty. Since D is convex, 0 is a boundary point of D_0, and hence a boundary point of $\omega = \{z \mid z \in \Omega, z_n = 0\}$. Suppose $f \in C^\infty_{(0,q+1)}(\omega)$ and $\bar\partial f = 0$. For ω, let ψ is the function in the proof of Theorem 2.14. Let $i : \omega \to \Omega$ be the inclusion mapping. Let $\pi(z_1, \cdots, z_n) = (z_1, \cdots, z_{n-1})$. For $z \in \Omega$ and $v \in C^\infty_{(0,q+1)}(\Omega)$, we set

$$F(z) = \psi(z)\pi^* f(z) - z_n v(z).$$

Then $F \in C^\infty_{(0,q+1)}(\Omega)$. We have

$$\bar\partial F = \bar\partial\psi \wedge \pi^* f - z_n \bar\partial v.$$

Hence, if we choose v such that

$$\bar\partial v = \frac{\bar\partial\psi \wedge \pi^* f}{z_n}, \qquad (2.16)$$

then $\bar\partial F = 0$. Since the right side of (2.16) belongs to $C^\infty_{(0,q+1)}(\Omega)$, and $\bar\partial$ closed, by the assumption, there exists $v \in C^\infty_{(0,q)}(\Omega)$ satisfying (2.16). Thus there exists $F \in C^\infty_{(0,q+1)}(\Omega)$ such that $\bar\partial F = 0$ and $i^* F = f$. By the assumption, there exists $U \in C^\infty_{(0,q)}(\Omega)$ such that $\bar\partial U = F$. If we set $u = i^* U$, then $u \in C_{(0,q)}(\omega)$, and

$$\bar\partial u = \bar\partial i^* U = i^* \bar\partial U = i^* u = f.$$

By the inductive hypothesis, ω is a domain of holomorphy. Hence there exists a holomorphic function h in ω which cannot be extended holomorphically to a neighborhood of $\pi(z_0)$. On the other hand, by Theorem 2.14 there exists a holomorphic function H on Ω such that $H|_\omega = h$. Since H cannot be extended holomorphically in a neighborhood of z_0, Ω is a domain of holomorphy. Hence Theorem 2.15 holds for n. □

Corollary 2.4 *(Levi's problem) Let $\Omega \subset \mathbf{C}^n$ be an open set. Then Ω is pseudoconvex if and only if Ω is a domain of holomorphy.*

Proof. Corollary 2.4 follows from Corollary 1.6, Corollary 2.3 and Theorem 2.15. □

Lemma 2.29 *Suppose that p is a C^∞ strictly plurisubharmonic function in Ω and that for every $c \in \mathbf{R}$*

$$K_c = \{z \in \Omega \mid p(z) \le c\} \subset\subset \Omega.$$

Then every holomorphic function in a neighborhood of K_0 can be approximated in L^2 norm on K_0 by functions in $\mathcal{O}(\Omega)$.

Proof. Let u be a holomorphic function in a neighborhood of K_0. By the Hahn-Banach theorem, it is sufficient to show that if φ is a bounded linear functional on $L^2(K_0)$ which vanishes on $\mathcal{O}(\Omega)$, then $\varphi(u) = 0$. By the Riesz representation theorem, there exists $v \in L^2(K_0)$ such that

$$\varphi(x) = (x, v) = \int_{K_0} x \bar{v} dV.$$

Hence it is sufficient to show that

$$\int_{K_0} f \bar{v} dV = 0 \quad (f \in \mathcal{O}(\Omega)) \implies \int_{K_0} u \bar{v} dV = 0.$$

By setting $v = 0$ outside of K_0, we extend v to the function on Ω. If f satisfies the equation $\bar{\partial} f = 0$ on Ω, then f is holomorphic in Ω. Hence, by the assumption,

$$(f, v e^{\varphi_1})_1 = \int_{K_0} f \bar{v} e^{\varphi_1} e^{-\varphi_1} dV = \int_{K_0} f \bar{v} dV = 0,$$

which means that $v e^{\varphi_1} \in (\operatorname{Ker} T)^\perp$. Suppose $\mathcal{R}_T = \operatorname{Ker} S$. Then $\operatorname{Ker} S$ is closed, and hence $\mathcal{R}_T$ is closed. By Lemma 2.6, $\mathcal{R}_{T^*}$ is closed. By Lemma

2.7, we have $(\operatorname{Ker} T)^\perp = \mathcal{R}_{T^*}$, and hence $ve^{\varphi_1} \in \mathcal{R}_{T^*} = T^*(\mathcal{D}_{T^*})$. By Theorem 2.4, we have

$$\|f\|_2 \le c\|T^*f\|_1 \qquad (f \in \mathcal{R}_T \cap \mathcal{D}_{T^*}).$$

Next we show that $T^*(\mathcal{R}_T \cap \mathcal{D}_{T^*}) = T^*(\mathcal{D}_{T^*})$. Let $u \in \mathcal{D}_{T^*}$. Then u can be uniquely expressed by

$$u = u_1 + u_2 \qquad (u_1 \in \mathcal{R}_T \cap \mathcal{D}_{T^*},\ u_2 \in (\mathcal{R}_T)^\perp \cap \mathcal{D}_{T^*}).$$

Since $(\mathcal{R}_T)^\perp = \operatorname{Ker} T^*$, we obtain

$$T^*(u) = T^*(u_1 + u_2) = T^* u_1.$$

Hence we have $T^*(\mathcal{R}_T \cap \mathcal{D}_{T^*}) = T^*(\mathcal{D}_{T^*})$, which means that there exists $f \in \mathcal{R}_T \cap \mathcal{D}_{T^*}$ such that

$$ve^{\varphi_1} = T^*f, \quad \|f\|_2 \le c\|T^*f\|_1,$$

that is, if $f = \sum_{j=1}^n f_j d\bar{z}_j$, then we have

$$ve^{\varphi_1} = -e^{\varphi_1} \sum_{j=1}^n \frac{\partial(e^{-\varphi_2} f_j)}{\partial z_j}.$$

We set $g = fe^{-\varphi_2}$. Then we have

(1) $v = -\sum_{j=1}^n \dfrac{\partial g_j}{\partial z_j}$

(2) $\displaystyle\int_\Omega \sum_{j=1}^n |g_j|^2 e^{\varphi_2} dV \le c^2 \int_\Omega |v|^2 e^{\varphi_1} dV.$

Define

$$\rho_\nu(t) = \begin{cases} \nu e^{-1/t^2} & (t > 0) \\ 0 & (t \le 0) \end{cases}, \qquad \lambda_\nu(t) = \int_{-\infty}^t \rho_\nu(t) dt.$$

Let χ be the function in the proof of Theorem 2.7. Define

$$\chi_\nu = \chi + \lambda_\nu,$$

and

$$\varphi_1 = \chi_\nu \circ p - 2\psi, \quad \varphi_2 = \chi_\nu \circ p - \psi, \quad \varphi_3 = \chi_\nu \circ p,$$

where ψ is the function defined in Lemma 2.13. By Corollary 2.1, $\operatorname{Ker} S = \mathcal{R}_T$. Hence for φ_i ($i = 1, 2, 3$) we can construct g_ν satisfying (1), (2). χ_ν

satisfies that $\chi_\nu(t) = \chi(t)$ for $t \leq 0$, and $\chi_\nu(t) \uparrow \infty$ $(\nu \to \infty)$ for $t > 0$. Using (2), we have

$$\int_\Omega \sum_{j=1}^n |g_j^\nu|^2 \exp(\chi_\nu \circ p - \psi) dV \leq c^2 \int_\Omega |v|^2 \exp(\chi_\nu \circ p - 2\psi) dV$$

$$= c^2 \int_{K_0} |v|^2 \exp(\chi \circ p - 2\psi) dV$$

$$\leq C_1.$$

Thus we obtain

$$\int_\Omega \sum_{j=1}^n |g_j^\nu|^2 \exp(\chi_1 \circ p - \psi) dV \leq C_1,$$

which means that $\{g^\nu\}$ is a bounded sequence in $L_{(0,1)}(\Omega, \psi - \chi_1 \circ p)$. Hence we can choose a subsequence $\{g^{\nu_k}\}$ of $\{g^\nu\}$ which converges weakly. Let $g^{\nu_k} \to g$. For $s_2 > s_1 > 0$, we set

$$M = \max_{x \in K_{s_2}} \psi(x).$$

Since $\exp(\chi_\nu \circ p - \psi) \geq \exp(\chi_\nu(s_1) - M)$ on $K_{s_2} - K_{s_1}$, we have

$$\int_{K_{s_2} - K_{s_1}} \sum_{j=1}^n |g_j^\nu|^2 dV \leq C_1 \exp(M - \chi_\nu(s_1)) \to 0 \quad (\nu \to \infty).$$

Thus $\{g_j^\nu\}$ converges 0 in $L^2(K_{s_2} - K_{s_1})$, which means that $\{g_j^\nu\}$ converges to 0 in $\Omega - K_0$ in the sense of distributions. Hence, $g = 0$ on $\Omega - K_0$. v is written in the following form

$$v = -\sum_{j=1}^n \frac{\partial g_j^\nu}{\partial z_j}.$$

Letting $\nu \to \infty$ we have in the sense of distributions

$$v = -\sum_{j=1}^n \frac{\partial g_j}{\partial z_j}.$$

Hence we have for $u \in \mathcal{D}(\Omega)$

$$\int_\Omega u \bar{v} dV = -\int_\Omega \sum_{j=1}^n u \frac{\partial \bar{g}_j}{\partial \bar{z}_j} dV = \int_\Omega \sum_{j=1}^n \frac{\partial u}{\partial \bar{z}_j} \bar{g}_j dV.$$

Since $g = 0$ outside of K_0, we have for every holomorphic function in a neighborhood of K_0

$$\int_\Omega u\bar{v} dV = \int_{K_0} \sum_{j=1}^n \frac{\partial u}{\partial \bar{z}_j} \bar{g}_j dV = 0,$$

which completes the proof of Lemma 2.29. □

Theorem 2.16 *Let $\Omega \subset \mathbf{C}^n$ be a pseudoconvex open set and let K be a compact subset of Ω, ω a neighborhood of $\hat{K}_\Omega^P$. Then there exists a function $u \in C^\infty(\Omega)$ such that*

(a) u is strictly plurisubharmonic in Ω.
(b) $u < 0$ in K, $u > 0$ in $\Omega \cap \omega^c$.
(c) $\{z \in \Omega \mid u(z) < c\} \subset\subset \Omega$ for every $c \in \mathbf{R}$.

Proof. By Theorem 1.15, there exists a C^∞ strictly plurisubharmonic function u_0 on Ω such that for every real number c,

$$\{z \in \Omega \mid u_0(z) < c\} \subset\subset \Omega.$$

Without loss of generality, we may assume that $u_0 < 0$ in K. Define

$$K' = \{z \in \Omega \mid u_0(z) \leq 2\}, \quad L = \{z \mid z \in \Omega \cap \omega^c, u_0(z) \leq 0\}.$$

Then $L \subset K'$, and K' and L are compact. Since $z \notin \hat{K}_\Omega^P$ for each $z \in L$, there exists $g \in P(\Omega)$ such that

$$|g(z)| > \sup_K \|g\|.$$

Let d be such that $|g(z)| > d > \sup_K \|g\|$. We set $\tilde{g} = g - d$. Then $\tilde{g}(z) > 0$, $\tilde{g} < 0$ in K. There exists a C^∞ plurisubharmonic function $g_\varepsilon(z)$ defined in an open set W with $K' \subset W \subset \Omega$ such that $g_\varepsilon \downarrow \tilde{g}$ ($\varepsilon \to 0$). For any sufficiently small $\varepsilon > 0$, $g_\varepsilon > 0$ in some neighborhood of z, $g_\varepsilon < 0$ in K. Since L is compact, by the Heine-Borel theorem, there exist finitely many C^∞ strictly plurisubharmonic functions $\varphi_1, \cdots, \varphi_k$ in W such that if we define $\varphi = \max(\varphi_1, \cdots, \varphi_k)$, then φ is a continuous plurisubharmonic function in W, $\varphi > 0$ in some neighborhood of L and $\varphi < 0$ in K. We set

$$c = \sup_{K'} \varphi > 0.$$

Define

$$v(z) = \begin{cases} \sup(\varphi(z), cu_0(z)) & (u_0(z) < 2) \\ cu_0(z) & (u_0(z) > 1) \end{cases}.$$

If $1 < u_0(z) < 2$, then $z \in K'$, and hence $\varphi(z) \leq c < cu_0(z)$. Then
$$\sup(\varphi(z), cu_0(z)) = cu_0(z),$$
which implies that v is a continuous plurisubharmonic function in Ω. Further, $v < 0$ in K. If $u_0(z) \leq 0$ for $z \in \Omega \cap \omega^c$, then $z \in L$, and hence $v(z) = \varphi(z) > 0$. If $u_0(z) > 0$, then $v(z) \geq cu_0(z) > 0$, which implies that $v(z) > 0$ for $z \in \Omega \cap \omega^c$. Thus v is a continuous plurisubharmonic function in Ω satisfying (b) and (c). We set
$$\Omega_c = \{z \in \Omega \mid v(z) < c\},$$
and
$$v_j(z) = \int_{\Omega_{j+1}} \frac{v(\zeta)\lambda((z-\zeta)/\varepsilon)}{\varepsilon^{2n}} dV(\zeta) + \varepsilon|z|^2,$$
where λ is the function defined in Theorem 1.14. If we choose $\varepsilon > 0$ sufficiently small, then $v_j \in C^\infty(\mathbf{C}^n)$. Further, $v_j > v$, and v_j is strictly plurisubharmonic in a neighborhood of $\bar{\Omega}_j$. Choosing ε sufficiently small, we may assume that in K, $v_0 < 0$ and $v_1 < 0$. Further, $v_j < v + 1$ in Ω_j ($j = 1, 2, \cdots$). We choose a convex function $\chi \in C^\infty(\mathbf{R})$ such that $\chi(t) = 0$ for $t < 0$, $\chi'(t) > 0$ for $t > 0$. Then $\chi(v_j + 1 - j)$ is strictly plurisubharmonic in a neighborhood of $\bar{\Omega}_j \setminus \Omega_{j-1}$. We choose a_j ($j = 1, 2, \cdots$) such that if we set
$$u_m = v_0 + \sum_{j=1}^m a_j \chi(v_j + 1 - j),$$
then u_m is strictly plurisubharmonic in a neighborhood of $\bar{\Omega}$, and $u_m > v$. If $i > j$, $k > j$, then $u_k = u_i$ in Ω_j, which implies that $u = \lim_{i \to \infty} u_i$ exists. Since u is a C^∞ plurisubharmonic function in Ω, $u = v_0 < 0$ in K and $u > v$ in Ω, u satisfies (a), (b) and (c). $\square$

Theorem 2.17 *Let $\Omega \subset \mathbf{C}^n$ be a pseudoconvex open set and K a compact subset of Ω with $K = \hat{K}_\Omega^P$. Then every holomorphic function in a neighborhood of K can be approximated uniformly on K by functions in $\mathcal{O}(\Omega)$. (Since $\hat{K}_\Omega^P \subset \hat{K}_\Omega^\mathcal{O}$, Theorem 2.17 holds for every compact set K with $K = \hat{K}_\Omega^\mathcal{O}$.)*

Proof. Let u be holomorphic in a neighborhood ω of K. By Theorem 2.16, there exists a C^∞ strictly plurisubharmonic function p in Ω such that p satisfies the assumption in Lemma 2.29, and if we set $K_c = \{z \in \Omega \mid p(z) < c\}$, p satisfies $K \subset K_0^\circ \subset K_0 \subset \omega$. By Lemma 2.29, there exist $u_j \in \mathcal{O}(\Omega)$

such that $u_j \to u$ in $L^2(K_0)$. Using Corollary 1.3, $u_j - u$ converges to 0 uniformly on K. □

2.3 The Ohsawa-Takegoshi Extension Theorem

Let $\Omega \subset\subset \mathbf{C}^n$ be a pseudoconvex domain and let $H = \{z \in \mathbf{C}^n \mid z_n = 0\}$. Then Ohsawa and Takegoshi [OHT] proved that every L^2 holomorphic function in $H \cap \Omega$ can be extended to an L^2 holomorphic function in Ω. The proof given here is based on the proof of Jarnicki-Pflug [JP].

Let H^j, $j = 0, 1, 2$, be Hilbert spaces. Let $\mathcal{D}_j$ be dense subsets of H^j, $j = 0, 1$, respectively. Let

$$T : \mathcal{D}_0 \to H^1, \quad S : \mathcal{D}_1 \to H^2$$

be closed linear operators such that $ST = 0$. Let $L : H^1 \to H^1$ be a linear bijection satisfying

$$(Lx, x)_1 \geq 0 \quad (x \in H^1). \tag{2.17}$$

Then we have the following lemma.

Lemma 2.30 *Suppose*

$$|(Lv, v)_1| \leq \|T^*v\|_0^2 + \|Sv\|_2^2,$$

for every $v \in \mathcal{D}_{T^*} \cap \mathcal{D}_S$. *Then for* $g \in \operatorname{Ker} S$, *there exists* $u \in \mathcal{D}_T$ *with the following properties:*

$$Tu = g, \quad \|u\|_0^2 \leq |(L^{-1}g, g)_1|.$$

Proof. It follows from (2.17) that

$$(L(x+y), x+y)_1 = (x+y, L(x+y))_1,$$

$$(L(x+iy), x+iy)_1 = (x+iy, L(x+iy))_1,$$

which implies that

$$(Lx, y)_1 + (Ly, x)_1 = (x, Ly)_1 + (y, Lx)_1,$$

$$-(Lx, y)_1 + (Ly, x)_1 = -(x, Ly)_1 + (y, Lx)_1.$$

Thus we obtain
$$(Lx, y)_1 = (x, Ly)_1 \quad (x, y \in H^1).$$

It follows from (2.17) that for $t \in \mathbf{C}$ we obtain
$$(L(x + ty)_1, x + ty)_1 \geq 0.$$

Hence for every real number t,
$$(L(x + (Lx, y)_1 ty)_1, x + (Lx, y)_1 ty)_1 \geq 0,$$

which implies that for every real number t,
$$(Lx, x)_1 + 2|(Lx, y)_1|^2 t + |(Lx, y)_1|^2 (Ly, y)_1 t^2 \geq 0.$$

Hence we have
$$|(Lx, y)_1|^2 \leq (Lx, x)_1 (Ly, y)_1 \quad (x, y \in H^1).$$

Since L is bijective, there exists $\tilde{g} \in H^1$ such that $L\tilde{g} = g$. Thus for $v \in \mathcal{D}_{T^*} \cap \operatorname{Ker} S$, we have
$$|(v, g)_1|^2 = |(v, L\tilde{g})_1|^2 \leq (Lv, v)_1 (L\tilde{g}, \tilde{g})_1$$
$$\leq (L\tilde{g}, \tilde{g})_1 (\|T^*v\|_0^2 + \|Sv\|_2^2) = (L\tilde{g}, \tilde{g})_1 \|T^*v\|^2.$$

Since $(v, g)_1 = 0$ for $v \in \mathcal{D}_{T^*} \cap (\operatorname{Ker} S)^\perp$, we have for $v \in \mathcal{D}_{T^*}$,
$$|(v, g)_1|^2 \leq (L\tilde{g}, \tilde{g})_1 \|T^*v\|_0^2. \tag{2.18}$$

Hence if we define a bounded linear functional $\varphi : \mathcal{R}_{T^*} \to \mathbf{C}$ by $\varphi(T^*v) = (v, g)_1$, then by the Hahn-Banach theorem, φ is extended to a bounded linear functional on H^0. By the Riesz representaion theorem, there exists $u_0 \in H^0$ such that
$$\varphi(w) = (w, u_0)_0, \quad \|\varphi\| = \|u_0\|_0 \quad (w \in H^0).$$

It follows from (2.18) that
$$|\varphi(T^*v)| = |(g, v)_1| \leq \sqrt{(L\tilde{g}, \tilde{g})_1} \|T^*v\|_0,$$

which implies that $\|\varphi\|^2 \leq (L\tilde{g}, \tilde{g})_1$. Consequently,
$$\|u_0\|_0^2 \leq (L\tilde{g}, \tilde{g})_1.$$

On the other hand we have

$$\varphi(T^*v) = (T^*v, u_0)_0 = (v, g)_1 \quad (v \in \mathcal{D}_{T^*}). \tag{2.19}$$

Hence by (2.19) we have $|(T^*v, u_0)_0| \leq \|v\|_1 \|g\|_1$ for $v \in \mathcal{D}_{T^*}$, which implies that $u_0 \in \mathcal{D}_{T^{**}} = \mathcal{D}_T$. By (2.19), $(v, g)_1 = (v, Tu_0)$ for $v \in \mathcal{D}_{T^*}$, which implies that $Tu_0 = g$. $\square$

Let $\Omega \subset \mathbf{C}^n$ be a bounded pseudoconvex domain with C^2 boundary. Then there exist a neighborhood U of $\partial \Omega$ and a C^2 plurisubharmonic function ρ in U such that

$$U \cap \Omega = \{z \in U \mid \rho(z) < 0\}.$$

We assume that $|d\rho(z)| = 1$ for $z \in \partial\Omega$. Further, we assume that φ is a C^2 plurisubharmonic function in a neighborhood $\widetilde{\Omega}$ of $\overline{\Omega}$. For $l \in (0,1)$, define $\widetilde{\chi} \in C^\infty(\mathbf{R})$ such that (see Exercise 2.4)

$$\widetilde{\chi}(t) = \begin{cases} 1 & (t \leq l) \\ 0 & (t \geq 1) \end{cases}, \quad |\widetilde{\chi}'| \leq \frac{2}{1-l}.$$

For $0 < \varepsilon < \frac{1}{2}$, define

$$\chi_\varepsilon(z) = \widetilde{\chi}\left(\frac{|z_n|^2}{\varepsilon^2}\right).$$

Further, for $f \in \mathcal{O}(\widetilde{\Omega})$, define

$$g_\varepsilon(z) = \bar{\partial}\left(\frac{\chi_\varepsilon(z)f(z)}{z_n}\right).$$

Then g_ε is a $\bar{\partial}$ closed C^∞ $(0,1)$ form on $\widetilde{\Omega}$. We have

$$\int_\Omega |g_\varepsilon(z)|^2 e^{-\varphi(z)} dV(z) = \frac{1}{\varepsilon^4} \int_{\Omega_\varepsilon} |f(z)|^2 \left|\widetilde{\chi}'\left(\frac{|z_n|^2}{\varepsilon^2}\right)\right|^2 e^{-\varphi(z)} dV(z),$$

where

$$\Omega_\varepsilon = \{z \in \Omega \mid l\varepsilon^2 \leq |z_n|^2 \leq \varepsilon^2\},$$

and dV is the Lebesgue measure in $\mathbf{C}^n$. We choose $A > 1$ such that $\Omega \subset \mathbf{C}^{n-1} \times \{z_n \mid |z_n| < A/2\}$. Define

$$\gamma_\varepsilon(z) = \frac{1}{\varepsilon^2 + |z_n|^2}, \quad \eta_\varepsilon(z) = \log(A^2 \gamma_\varepsilon(z)).$$

Then $z \in \Omega$, and for $\varepsilon \in (0, 1/2)$, $\eta_\varepsilon(z) \geq \log 2$. Define

$$\sigma(z) = \frac{|z|^2}{\log 2}, \quad \psi = \varphi + \sigma.$$

Then we have

$$\eta_\varepsilon(z) \sum_{j,k=1}^n \frac{\partial^2 \sigma}{\partial z_j \partial \bar{z}_k}(z) w_j \bar{w}_k = \eta_\varepsilon(z) \frac{|w|^2}{\log 2} \geq |w|^2$$

for $z \in \Omega$, $w \in \mathbf{C}^n$, $\varepsilon \in (0, 1/2)$. Consequently,

$$\eta_\varepsilon(z) \sum_{j,k=1}^n \frac{\partial^2 \psi}{\partial z_j \partial \bar{z}_k}(z) w_j \bar{w}_k \geq |w|^2 \quad (z \in \Omega, w \in \mathbf{C}^n). \tag{2.20}$$

For $0 \leq \varepsilon < 1/2$, define

$$\alpha_\varepsilon = \begin{cases} 1 & (\varepsilon = 0) \\ \eta_\varepsilon + \gamma_\varepsilon & (\varepsilon > 0) \end{cases}.$$

We set

$$H^0 = L^2_{(0,0)}(\Omega, \psi), \quad H^1 = L^2_{(0,1)}(\Omega, \psi), \quad H^2 = L^2_{(0,2)}(\Omega, \psi),$$

and

$$T_\varepsilon(u) = \bar{\partial}(\sqrt{\alpha_\varepsilon} u), \quad S_\varepsilon = \sqrt{\alpha_\varepsilon} \bar{\partial}, \quad T = T_0, \quad S = S_0.$$

Then we have

$$\mathcal{D}_{T_\varepsilon} = \mathcal{D}_T, \quad \mathcal{D}_{S_\varepsilon} = \mathcal{D}_S, \quad \mathcal{D}_{T^*_\varepsilon} = \mathcal{D}_{T^*}.$$

Now we define a linear operator $L_\varepsilon : H^1 \to H^1$ by

$$L_\varepsilon \left(\sum_{j=1}^{n-1} v_j d\bar{z}_j + v_n d\bar{z}_n \right) = \sum_{j=1}^{n-1} v_j d\bar{z}_j + \frac{\varepsilon^2}{(\varepsilon^2 + |z_n|^2)^2} v_n d\bar{z}_n.$$

Then $L_\varepsilon : H^1 \to H^1$ is bijective and satisfies

$$(L_\varepsilon(x), x)_1 \geq 0,$$

for every $x \in H^1$.

Lemma 2.31 Let $v = \sum_{j=1}^{n} v_j d\bar{z}_j \in C^2_{(0,1)}(\widetilde{\Omega})$. Then $v \in \mathcal{D}_{T^*_\varepsilon}$ if and only if

$$\sum_{j=1}^{n} v_j(z) \frac{\partial \rho}{\partial z_j}(z) = 0 \qquad (z \in \partial\Omega).$$

Proof. Suppose $v = \sum_{j=1}^{n} v_j d\bar{z}_j \in C^2_{(0,1)}(\widetilde{\Omega}) \cap \mathcal{D}_{T^*_\varepsilon}$. Then

$$(u, T^*v)_0 = (Tu, v)_1 \qquad (u \in \mathcal{D}_T),$$

which means that

$$T^*v = -\sum_{j=1}^{n} e^\psi \frac{\partial}{\partial z_j}(v_j e^{-\psi}).$$

We set

$$\tilde{v}(z) = \sum_{j=1}^{n} v_j(z) \frac{\partial \rho}{\partial z_j}(z).$$

Suppose there exists $z^0 \in \partial\Omega$ such that $\tilde{v}(z^0) \neq 0$. We may assume that $\operatorname{Re} \tilde{v} > 0$ in some neighborhood W of z^0. We choose a function $\tilde{u} \in C^\infty_c(\mathbf{C}^n)$ with the properties that $\tilde{u} \geq 0$, $\tilde{u}(z^0) > 0$, $\operatorname{supp}(\tilde{u}) \subset W$. Since $\tilde{u} \in \mathcal{D}_T$, it follows from Green's theorem (Theorem 2.1) that

$$(\tilde{u}, T^*v)_1 = (T\tilde{u}, v)_2 = \int_\Omega \sum_{j=1}^{n} \frac{\partial \tilde{u}}{\partial \bar{z}_j} v_j e^{-\psi} dV$$

$$= -\int_\Omega \tilde{u} \sum_{j=1}^{n} e^\psi \frac{\partial(v_j e^{-\psi})}{\partial \bar{z}_j} e^{-\psi} dV + \int_{\partial\Omega} \sum_{j=1}^{n} \frac{\partial \rho}{\partial \bar{z}_j} \tilde{u} v_j e^{-\psi} dS$$

$$= (\tilde{u}, T^*v)_1 + \int_{\partial\Omega} \tilde{u}\tilde{v} e^{-\psi} dS,$$

which implies that

$$\int_{\partial\Omega} \tilde{u}\tilde{v} e^{-\psi} dS = 0.$$

This contradicts the choice of $\tilde{v}$ and $\tilde{u}$. Thus we have $\tilde{v}|_{\partial\Omega} = 0$. Similarly we can prove the sufficiency. □

For $u \in \mathcal{D}_{T^*}$ and $v \in \mathcal{D}_T$, we have

$$(v, T_\varepsilon^* u)_0 = (T_\varepsilon v, u)_1 = (\bar\partial(\sqrt{\alpha_\varepsilon} v), u)_1 = (v, \sqrt{\alpha_\varepsilon} T^* u)_0,$$

which implies that $T_\varepsilon^* u = \sqrt{\alpha_\varepsilon} T^* u$. Hence, for $u = \sum_{k=1}^{n} u_k d\bar z_k \in C^2_{(0,1)}(\widetilde\Omega) \cap \mathcal{D}_{T_\varepsilon^*}$,

$$T_\varepsilon^* u = -\sqrt{\alpha_\varepsilon} e^\psi \sum_{j=1}^{n} \frac{\partial}{\partial z_j}(u_j e^{-\psi}).$$

Theorem 2.18 *For $0 < \varepsilon < 1/2$ and $u \in C^2_{(0,1)}(\widetilde\Omega) \cap \mathcal{D}_{T_\varepsilon^*}$, we have*

$$(L_\varepsilon u, u) \leq \|T_\varepsilon^* u\|_0^2 + \|S_\varepsilon u\|_2^2.$$

Proof. Using Green's theorem (Theorem 2.1), we have

$$\|T_\varepsilon^* u\|_0^2 + \|S_\varepsilon u\|_2^2$$
$$= (\alpha_\varepsilon T^* u, T^* u)_0 + (\alpha_\varepsilon Su, Su)_2$$
$$= (\gamma_\varepsilon T^* u, T^* u)_0 + (\gamma_\varepsilon Su, Su)_2 + (\bar\partial(\eta_\varepsilon T^* u), u)_1$$
$$+ \int_\Omega \eta_\varepsilon \sum_{j<k} \left(\frac{\partial u_k}{\partial \bar z_j} - \frac{\partial u_j}{\partial \bar z_k}\right) \overline{\left(\frac{\partial u_k}{\partial \bar z_j} - \frac{\partial u_j}{\partial \bar z_k}\right)} e^{-\psi} dV$$

$$= (\gamma_\varepsilon T^* u, T^* u)_0 + (\gamma_\varepsilon Su, Su)_2 + (\bar\partial(\eta_\varepsilon T^* u), u)_1$$
$$+ \int_\Omega \eta_\varepsilon \sum_{j,k=1}^{n} \left(\frac{\partial u_k}{\partial \bar z_j} - \frac{\partial u_j}{\partial \bar z_k}\right) \overline{\frac{\partial u_k}{\partial \bar z_j}} e^{-\psi} dV$$
$$= (\gamma_\varepsilon T^* u, T^* u)_0 + (\gamma_\varepsilon Su, Su)_2 + (\bar\partial(\eta_\varepsilon T^* u), u)_1$$
$$- \int_\Omega \sum_{j,k=1}^{n} \frac{\partial}{\partial z_j}\left\{\eta_\varepsilon \left(\frac{\partial u_k}{\partial \bar z_j} - \frac{\partial u_j}{\partial \bar z_k}\right) e^{-\psi}\right\} \bar u_k dV$$
$$+ \int_{\partial\Omega} \eta_\varepsilon \sum_{j,k=1}^{n} \left(\frac{\partial u_k}{\partial \bar z_j} - \frac{\partial u_j}{\partial \bar z_k}\right) \frac{\partial \rho}{\partial z_j} \bar u_k e^{-\psi} dS.$$

Since $\sum_{j=1}^{n} u_j \frac{\partial \rho}{\partial z_j} = 0$ on $\partial\Omega$, there exists a C^1 function Θ in a neighborhood of $\partial\Omega$ such that

$$\sum_{j=1}^{n} u_j \frac{\partial \rho}{\partial z_j} = \Theta \rho. \qquad (2.21)$$

Differentiating (2.21) with respect to $\bar{z}_k$, we have on $\partial\Omega$

$$\sum_{j=1}^{n}\left(\frac{\partial u_j}{\partial \bar{z}_k}\frac{\partial \rho}{\partial z_j}+u_j\frac{\partial^2 \rho}{\partial \bar{z}_k \partial z_j}\right)=\rho\frac{\partial \Theta}{\partial \bar{z}_k}+\Theta\frac{\partial \rho}{\partial \bar{z}_k}=\Theta\frac{\partial \rho}{\partial \bar{z}_k}. \quad (2.22)$$

If we multiply by $\bar{u}_k$ and add, we obtain on $\partial\Omega$

$$\sum_{j,k=1}^{n}\bar{u}_k\left(\frac{\partial u_j}{\partial \bar{z}_k}\frac{\partial \rho}{\partial z_j}+u_j\frac{\partial^2 \rho}{\partial \bar{z}_k \partial z_j}\right)=\Theta\overline{\sum_{k=1}^{n}\frac{\partial \rho}{\partial z_k}u_k}=0.$$

Consequently,

$$\int_{\partial\Omega}\eta_\varepsilon\sum_{j,k=1}^{n}\left(\frac{\partial u_k}{\partial \bar{z}_j}-\frac{\partial u_j}{\partial \bar{z}_k}\right)\frac{\partial \rho}{\partial z_j}u_k e^{-\psi}dS$$

$$=\int_{\partial\Omega}\eta_\varepsilon\sum_{j,k=1}^{n}\frac{\partial u_k}{\partial \bar{z}_j}\frac{\partial \rho}{\partial z_j}\bar{u}_k e^{-\psi}dS+\int_{\partial\Omega}\sum_{j,k=1}^{n}\eta_\varepsilon\bar{u}_k u_j\frac{\partial^2 \rho}{\partial z_j\partial \bar{z}_k}e^{-\psi}dS$$

$$\geq\int_{\partial\Omega}\eta_\varepsilon\sum_{j,k=1}^{n}\frac{\partial u_k}{\partial \bar{z}_j}\frac{\partial \rho}{\partial z_j}\bar{u}_k e^{-\psi}dS$$

$$=\int_{\Omega}\sum_{j,k=1}^{n}\eta_\varepsilon\frac{\partial u_k}{\partial \bar{z}_j}\overline{\frac{\partial u_k}{\partial \bar{z}_j}}e^{-\psi}dV+\int_{\Omega}\sum_{j,k=1}^{n}\bar{u}_k\frac{\partial}{\partial z_j}\left(\eta_\varepsilon\frac{\partial u_k}{\partial \bar{z}_j}e^{-\psi}\right)dV$$

$$\geq\int_{\Omega}\sum_{j,k=1}^{n}\bar{u}_k\frac{\partial}{\partial z_j}\left(\eta_\varepsilon\frac{\partial u_k}{\partial \bar{z}_j}e^{-\psi}\right)dV.$$

Thus if we use a representation

$$\|T_\varepsilon^* u\|_0^2+\|S_\varepsilon u\|_2^2=(\gamma_\varepsilon T^* u, T^* u)_0+(\gamma_\varepsilon S u, S u)_2+(*),$$

then

$$(*)\geq(\eta_\varepsilon TT^* u, u)_1+\int_{\Omega}\sum_{j=1}^{n}\frac{\partial \eta_\varepsilon}{\partial \bar{z}_j}T^*(u)\bar{u}_j e^{-\psi}dV$$

$$-\int_{\Omega}\sum_{j,k=1}^{n}\frac{\partial \eta_\varepsilon}{\partial z_j}\left(\frac{\partial u_k}{\partial \bar{z}_j}-\frac{\partial u_j}{\partial \bar{z}_k}\right)e^{-\psi}\bar{u}_k dV$$

$$-\int_{\Omega}\sum_{j,k=1}^{n}\eta_\varepsilon\frac{\partial}{\partial z_j}\left\{\left(\frac{\partial u_k}{\partial \bar{z}_j}-\frac{\partial u_j}{\partial \bar{z}_k}\right)e^{-\psi}\right\}\bar{u}_k dV$$

$$+\int_{\Omega}\sum_{j,k=1}^{n}\bar{u}_k\frac{\partial}{\partial z_j}\left(\eta_\varepsilon\frac{\partial u_k}{\partial \bar{z}_j}e^{-\psi}\right)dV.$$

Taking into account that

$$(\eta_\varepsilon TT^*u, u)_1 = \int_\Omega \sum_{k=1}^n \eta_\varepsilon \frac{\partial}{\partial \bar{z}_k}(T^*u)\bar{u}_k e^{-\psi}dV$$

$$= \int_\Omega \eta_\varepsilon \sum_{j,k=1}^n \left(\frac{\partial^2 \psi}{\partial z_j \partial \bar{z}_k}u_j - \frac{\partial^2 u_j}{\partial z_j \partial \bar{z}_k} + \frac{\partial \psi}{\partial z_j}\frac{\partial u_j}{\partial \bar{z}_k}\right)\bar{u}_k e^{-\psi}dV,$$

we obtain

$$(*) \geq \int_\Omega \sum_{j=1}^n \frac{\partial \eta_\varepsilon}{\partial \bar{z}_j} T^*(u)\bar{u}_j e^{-\psi}dV - \int_\Omega \sum_{j,k=1}^n \frac{\partial \eta_\varepsilon}{\partial z_j}\left(\frac{\partial u_k}{\partial \bar{z}_j} - \frac{\partial u_j}{\partial \bar{z}_k}\right)e^{-\psi}\bar{u}_k dV$$

$$+ \int_\Omega \sum_{j,k=1}^n \left(\eta_\varepsilon u_j \bar{u}_k \frac{\partial^2 \psi}{\partial z_j \partial \bar{z}_k} + \frac{\partial \eta_\varepsilon}{\partial z_j}\frac{\partial u_k}{\partial \bar{z}_j}\bar{u}_k\right)e^{-\psi}dV$$

$$= \int_\Omega \sum_{j=1}^n \frac{\partial \eta_\varepsilon}{\partial \bar{z}_j} T^*(u)\bar{u}_j e^{-\psi}dV$$

$$+ \int_\Omega \sum_{j,k=1}^n \left(\eta_\varepsilon u_j \frac{\partial^2 \psi}{\partial z_j \partial \bar{z}_k} + \frac{\partial \eta_\varepsilon}{\partial z_j}\frac{\partial u_j}{\partial \bar{z}_k}\right)\bar{u}_k e^{-\psi}dV.$$

Since $u \in \mathcal{D}_{T^*}$, we have

$$\int_\Omega \sum_{j,k=1}^n \frac{\partial \eta_\varepsilon}{\partial z_j}\frac{\partial u_j}{\partial \bar{z}_k}\bar{u}_k e^{-\psi}dV$$

$$= -\int_\Omega \sum_{j,k=1}^n \left(\frac{\partial^2 \eta_\varepsilon}{\partial z_j \partial \bar{z}_k}u_j \bar{u}_k e^{-\psi} + \frac{\partial \eta_\varepsilon}{\partial z_j}u_j \frac{\partial}{\partial \bar{z}_k}(\bar{u}_k e^{-\psi})\right)dV$$

$$+ \int_{\partial\Omega} \sum_{j,k=1}^n \frac{\partial \eta_\varepsilon}{\partial z_j}\frac{\partial \rho}{\partial \bar{z}_k}u_j \bar{u}_k e^{-\psi}dS$$

$$= -\int_\Omega \sum_{j,k=1}^n \left(\frac{\partial^2 \eta_\varepsilon}{\partial z_j \partial \bar{z}_k}u_j \bar{u}_k e^{-\psi} + \frac{\partial \eta_\varepsilon}{\partial z_j}u_j \frac{\partial}{\partial \bar{z}_k}(\bar{u}_k e^{-\psi})\right)dV.$$

Consequently,

$$(*) \geq \int_\Omega \sum_{j=1}^n \frac{\partial \eta_\varepsilon}{\partial \bar{z}_j} T^*(u) \bar{u}_j e^{-\psi} dV + \int_\Omega \sum_{j,k=1}^n \eta_\varepsilon \frac{\partial^2 \psi}{\partial z_j \partial \bar{z}_k} u_j \bar{u}_k e^{-\psi} dV$$
$$- \int_\Omega \sum_{j,k=1}^n \left(\frac{\partial^2 \eta_\varepsilon}{\partial z_j \partial \bar{z}_k} u_j \bar{u}_k e^{-\psi} + \frac{\partial \eta_\varepsilon}{\partial z_j} u_j \frac{\partial}{\partial \bar{z}_k} (\bar{u}_k e^{-\psi}) \right) dV$$
$$= \int_\Omega \sum_{j=1}^n \frac{\partial \eta_\varepsilon}{\partial \bar{z}_j} T^*(u) \bar{u}_j e^{-\psi} dV + \int_\Omega \sum_{j,k=1}^n \eta_\varepsilon \frac{\partial^2 \psi}{\partial z_j \partial \bar{z}_k} u_j \bar{u}_k e^{-\psi} dV$$
$$- \int_\Omega \sum_{j,k=1}^n \frac{\partial^2 \eta_\varepsilon}{\partial z_j \partial \bar{z}_k} u_j \bar{u}_k e^{-\psi} dV + \int_\Omega \sum_{j=1}^n \frac{\partial \eta_\varepsilon}{\partial z_j} \overline{T^*(u)} u_j e^{-\psi} dV$$
$$= \int_\Omega \sum_{j,k=1}^n \eta_\varepsilon \frac{\partial^2 \psi}{\partial z_j \partial \bar{z}_k} u_j \bar{u}_k e^{-\psi} dV - \int_\Omega \sum_{j,k=1}^n \frac{\partial^2 \eta_\varepsilon}{\partial z_j \partial \bar{z}_k} u_j \bar{u}_k e^{-\psi} dV$$
$$+ 2\operatorname{Re} \int_\Omega \sum_{j=1}^n \frac{\partial \eta_\varepsilon}{\partial z_j} \overline{T^*(u)} u_j e^{-\psi} dV.$$

Using the inequality

$$\left| \sum_{j=1}^n \frac{\partial \eta_\varepsilon}{\partial z_j} u_j \overline{T^*(u)} \right| = \left| -\frac{\bar{z}_n}{\varepsilon^2 + |z_n|^2} u_n \overline{T^*(u)} \right| \leq \frac{|z_n|^2 |u_n|^2 + |T^*(u)|^2}{2(\varepsilon^2 + |z_n|^2)},$$

we obtain

$$(*) \geq \int_\Omega \sum_{j,k=1}^n \eta_\varepsilon \frac{\partial^2 \psi}{\partial z_j \partial \bar{z}_k} u_j \bar{u}_k e^{-\psi} dV - \int_\Omega \sum_{j,k=1}^n \frac{\partial^2 \eta_\varepsilon}{\partial z_j \partial \bar{z}_k} u_j \bar{u}_k e^{-\psi} dV$$
$$- \int_\Omega \frac{|z_n|^2 |u_n|^2}{\varepsilon^2 + |z_n|^2} e^{-\psi} dV - \int_\Omega \frac{|T^*(u)|^2}{\varepsilon^2 + |z_n|^2} e^{-\psi} dV.$$

Consequently,

$$\|T_\varepsilon^* u\|_0^2 + \|S_\varepsilon u\|_2^2$$
$$\geq \int_\Omega \gamma_\varepsilon (|T^* u|^2 + |S_\varepsilon u|^2) e^{-\psi} dV$$
$$+ \int_\Omega \sum_{j,k=1}^n \eta_\varepsilon \frac{\partial^2 \psi}{\partial z_j \partial \bar{z}_k} u_j \bar{u}_k e^{-\psi} dV$$
$$- \int_\Omega \sum_{j,k=1}^n \frac{\partial^2 \eta_\varepsilon}{\partial z_j \partial \bar{z}_k} u_j \bar{u}_k e^{-\psi} dV$$
$$- \int_\Omega \frac{|z_n|^2 |u_n|^2}{\varepsilon^2 + |z_n|^2} e^{-\psi} dV - \int_\Omega \frac{|T^*(u)|^2}{\varepsilon^2 + |z_n|^2} e^{-\psi} dV.$$

It follows from (2.20) that

$$\|T_\varepsilon^* u\|_0^2 + \|S_\varepsilon u\|_2^2$$
$$\geq \int_\Omega \sum_{j,k=1}^n \eta_\varepsilon \frac{\partial^2 \psi}{\partial z_j \partial \bar{z}_k} u_j \bar{u}_k e^{-\psi} dV$$
$$- \int_\Omega \sum_{j,k=1}^n \frac{\partial^2 \eta_\varepsilon}{\partial z_j \partial \bar{z}_k} u_j \bar{u}_k e^{-\psi} dV$$
$$- \int_\Omega \frac{|z_n|^2 |u_n|^2}{\varepsilon^2 + |z_n|^2} e^{-\psi} dV$$
$$= \int_\Omega \sum_{j,k=1}^n \eta_\varepsilon \frac{\partial^2 \psi}{\partial z_j \partial \bar{z}_k} u_j \bar{u}_k e^{-\psi} dV$$
$$+ \int_\Omega \gamma_\varepsilon |u_n|^2 (\varepsilon^2 \gamma_\varepsilon - |z_n|^2) e^{-\psi} dV$$
$$\geq \int_\Omega \left(\sum_{j=1}^{n-1} u_j^2 + \frac{\varepsilon^2 |u_n|^2}{(\varepsilon^2 + |z_n|^2)^2} \right) e^{-\psi} dV$$
$$= (L_\varepsilon u, u)_1,$$

which completes the proof of Theorem 2.18. □

The following theorem was proved by Hörmander [HR1]. We omit the proof.

Theorem 2.19 *For $f = \sum_{j=1}^n f_j d\bar{z}_j \in \mathcal{D}_{T^*} \cap \mathcal{D}_S$, there exists a sequence $\{f_\nu\}$ with the following properties:*

(a) $f_\nu \in L^2_{(0,1)}(\Omega, \psi)$.
(b) If $f_\nu = \sum_{\nu=1}^n f_{\nu,j} d\bar{z}_j$, then $f_{\nu,j} \in C^2(\overline{\Omega})$.
(c) $\sum_{j=1}^n f_{\nu,j} \frac{\partial \rho}{\partial z_j}|_{\partial\Omega} = 0$, that is, $f_\nu \in \mathcal{D}_{T^*}$.
(d) $\|f - f_\nu\|_1 + \|Sf_\nu - Sf\|_2 + \|T^* f_\nu - T^* f\|_0 \to 0 \quad (\nu \to \infty)$.

Corollary 2.5 *For $g_\varepsilon = \bar{\partial}(\chi_\varepsilon f / z_n)$, there exists $u_\varepsilon \in H^1$ such that $T_\varepsilon u_\varepsilon = g_\varepsilon$, and*

$$\int_\Omega |u_\varepsilon|^2 e^{-\psi} dV \le \frac{4}{(1-l)^2 \varepsilon^6} \int_{\Omega_\varepsilon} (\varepsilon^2 + |z_n|^2)^2 |f|^2 e^{-\psi} dV.$$

Proof. Using Theorem 2.18 and Theorem 2.19, for $0 < \varepsilon < 1/2$ and $u \in \mathcal{D}_{S_\varepsilon} \cap \mathcal{D}_{T^*_\varepsilon}$, we have

$$(L_\varepsilon u, u)_1 \le \|T^*_\varepsilon u\|_0^2 + \|S_\varepsilon u\|_2^2.$$

By Lemma 2.30, there exists $u_\varepsilon \in \mathcal{D}_T$ such that

$$T_\varepsilon u_\varepsilon = g_\varepsilon, \quad \|u_\varepsilon\|_0 \le |(L_\varepsilon^{-1} g_\varepsilon, g_\varepsilon)_1|.$$

On the other hand we have

$$L_\varepsilon^{-1} g_\varepsilon = \frac{(\varepsilon^2 + |z_n|^2)^2}{\varepsilon^2} \frac{\partial \chi_\varepsilon}{\partial \bar{z}_n} \frac{f}{z_n} d\bar{z}_n,$$

which implies that

$$|(L_\varepsilon^{-1} g_\varepsilon, g_\varepsilon)_1| \le \int_\Omega \frac{(\varepsilon^2 + |z_n|^2)^2}{\varepsilon^2} \left|\widetilde{\chi}'\left(\frac{|z_n|^2}{\varepsilon^2}\right)\right|^2 \left(\frac{|z_n|}{\varepsilon^2}\right)^2 \left|\frac{f}{z_n}\right|^2 e^{-\psi} dV$$

$$\le \frac{4}{(1-l)^2} \int_{\Omega_\varepsilon} \frac{(\varepsilon^2 + |z_n|^2)^2}{\varepsilon^6} |f|^2 e^{-\psi} dV.$$

□

We set

$$F_\varepsilon = \chi_\varepsilon f - \sqrt{\alpha_\varepsilon} z_n u_\varepsilon.$$

Since $\bar{\partial} F_\varepsilon = 0$, F_ε is holomorphic in Ω. Moreover we have $F_\varepsilon|_{H \cap \Omega} = f$. We set $\hat{\Omega}_\varepsilon = \{z \in \Omega \mid |z_n| \le \varepsilon\}$. Then it follows from Minkowski's inequality that

$$\|F_\varepsilon\|_0 := \left(\int_\Omega |F_\varepsilon|^2 e^{-\psi} dV\right)^{1/2}$$
$$\leq \left(\int_{\hat{\Omega}_\varepsilon} |\chi_\varepsilon|^2 |f|^2 e^{-\psi} dV\right)^{\frac{1}{2}} + \left(\int_\Omega |z_n|^2 |\alpha_\varepsilon| |u_\varepsilon|^2 e^{-\psi} dV\right)^{\frac{1}{2}}$$
$$\leq \left(\int_{\hat{\Omega}_\varepsilon} |\chi_\varepsilon|^2 |f|^2 e^{-\psi} dV\right)^{\frac{1}{2}} + \sup_{z\in\Omega} |z_n|\sqrt{|\alpha_\varepsilon|} \left(\int_\Omega |u_\varepsilon|^2 e^{-\psi} dV\right)^{\frac{1}{2}}.$$

Since there exists a constant $B > 0$, such that

$$|z_n|\sqrt{|\alpha_\varepsilon|} \leq \sqrt{|z_n|^2 \log\left(\frac{A^2}{\varepsilon^2 + |z_n|^2}\right) + 1} \leq B.$$

It follows from Corollary 2.5 that

$$\left(\int_\Omega |F_\varepsilon|^2 e^{-\psi} dV\right)^{1/2} \leq \left(\int_{\hat{\Omega}_\varepsilon} |\chi_\varepsilon|^2 |f|^2 e^{-\psi} dV\right)^{\frac{1}{2}} \quad (2.23)$$
$$+ \frac{2B}{(1-l)\varepsilon^3}\left(\int_{\Omega_\varepsilon} (\varepsilon^2 + |z_n|^2)^2 |f|^2 e^{-\psi} dV\right)^{\frac{1}{2}}.$$

The first term in the right side of (2.23) converges to 0 as $\varepsilon \to 0$. In order to investigate the second term in the right side of (2.23), we need the following lemma.

Lemma 2.32 *For $\varphi \in C^\infty(\overline{\Omega})$, we have*

$$\lim_{\varepsilon\to 0+} \int_{\Omega_\varepsilon} \frac{\varphi(z)}{(|z_n|^2+\varepsilon)^2} dV(z) = (1-l)\pi \int_{\{z_n=0\}\cap\Omega} \varphi(z) dV_{n-1}(z),$$

where dV and dV_{n-1} are the Lebesgue measures in $\mathbf{C}^n$ and $\mathbf{C}^{n-1}$, respectively.

Proof. Let $0 < \varepsilon \leq 1/2$. If we choose ε sufficiently small, then there exist a constant $\alpha > 0$ and compact sets $E^{(\varepsilon)}$, $F^{(\varepsilon)} \subset \mathbf{C}^{n-1}$ with the following properties:

$$E^{(\varepsilon)} \times \{\sqrt{l\varepsilon} \leq |z_n| \leq \varepsilon\} \subset \Omega_\varepsilon \subset F^{(\varepsilon)} \times \{\sqrt{l\varepsilon} \leq |z_n| \leq \varepsilon\} \quad (2.24)$$

and

$$\mu(F^{(\varepsilon)} - E^{(\varepsilon)}) \leq \alpha\varepsilon, \quad (2.25)$$

where μ is the Lebesgue measure in $\mathbf{C}^{n-1}$. We set $z' = (z_1, \cdots, z_{n-1})$, $z = (z', z_n)$. We define τ by $\tau(z) = \varphi(z) - \varphi(z', 0)$. Then there exists a constant $C > 0$ such that $|\tau(z)| \leq C|z_n|$. On the other hand we have

$$\int_{\sqrt{l\varepsilon} \leq |z_n| \leq \varepsilon} \frac{dx_n dy_n}{(|z_n|^2 + \varepsilon)^2} = 2\pi \int_{\sqrt{l\varepsilon}}^{\varepsilon} \frac{rdr}{(r^2 + \varepsilon)^2}$$
$$= \frac{(1-l)\pi}{(l\varepsilon + 1)(\varepsilon + 1)} \to (1-l)\pi,$$

as $\varepsilon \to 0$. Hence we obtain

$$\lim_{\varepsilon \to 0+} \int_{\Omega_\varepsilon} \frac{\varphi(z)}{(|z_n|^2 + \varepsilon)^2} dV(z) = \lim_{\varepsilon \to 0+} \int_{\Omega_\varepsilon} \frac{\varphi(z', 0)}{(|z_n|^2 + \varepsilon)^2} dV(z)$$
$$= \lim_{\varepsilon \to 0+} (1-l)\pi \int_{E(\varepsilon)} \varphi(z', 0) dV_{n-1}(z')$$
$$= (1-l)\pi \int_{\{z_n=0\} \cap \Omega} \varphi(z', 0) dV_{n-1}(z'),$$

which completes the proof of Lemma 2.32. □

Since $\varepsilon^2 \geq (\varepsilon^2 + |z_n|^2)/2$ and $\varepsilon \geq (\varepsilon + |z_n|^2)/2$ in D_ε, it follows from Lemma 2.32 that

$$\frac{1}{\varepsilon^6} \int_{\Omega_\varepsilon} (\varepsilon^2 + |z_n|^2)^2 |f|^2 e^{-\psi} dV$$
$$\leq 16 \int_{\Omega_\varepsilon} \frac{|f|^2 e^{-\psi}}{(\varepsilon + |z_n|^2)^2} dV$$
$$\to 16(1-l)\pi \int_{H \cap \Omega} |f(z', 0)|^2 e^{-\psi(z', 0)} dV_{n-1}$$
$$\leq 16(1-l)\pi \sup_{z \in H \cap \Omega} e^{-\sigma(z)} \int_{H \cap \Omega} |f(z', 0)|^2 e^{-\psi(z', 0)} dV_{n-1}.$$

Consequently,

$$\limsup_{\varepsilon \to 0} \int_\Omega |F_\varepsilon|^2 e^{-\psi} dV \leq C \int_{H \cap \Omega} |f(z', 0)|^2 e^{-\psi(z', 0)} dV_{n-1}, \qquad (2.26)$$

where $C = (64B^2\pi)/(1-l) \sup_{z \in H \cap \Omega} e^{-\sigma(z)}$.

The following lemma is well known. So we omit the proof (See Excercise 2.3).

Lemma 2.33 *(Montel's theorem)* Let $\{u_k\}$ be a sequence of holomorphic functions in Ω which are uniformly bounded on every compact subset of Ω. Then there exists a subsequence $\{u_{k_j}\}$ of $\{u_k\}$ which converges uniformly on every compact subset of Ω.

Lemma 2.34 Let Ω be a bounded pseudoconvex domain in $\mathbf{C}^n$ with C^2 boundary whose defining function ρ satisfies $|d\rho| = 1$ on $\partial\Omega$. Then there exists a constant $C > 0$ such that for every holomorphic function f in $H \cap \Omega$, there exists a holomorphic function F in Ω which satisfies $F|_{H \cap \Omega} = f$ and

$$\int_\Omega |F|^2 e^{-\psi} dV \le C \int_{H \cap \Omega} |f(z',0)|^2 e^{-\psi(z',0)} dV_{n-1}.$$

Proof. Lemma 2.34 follows from Lemma 2.33 and (2.26). □

In order to prove the Ohsawa-Takegoshi extension theorem we need the following lemma.

Lemma 2.35 Let $\Omega \subset\subset \mathbf{C}^n$ be a strictly pseudoconvex domain with C^3 boundary. Then there exist a neighborhood U of $\partial\Omega$ and a C^2 strictly plurisubharmonic function $\tilde{\rho}$ in U such that

$$U \cap \Omega = \{z \in U \mid \tilde{\rho}(z) < 0\}, \quad |d\tilde{\rho}(z)| = 1 \ (z \in \partial\Omega).$$

Proof. By the definition of the strictly pseudoconvex domain, there exist a neighborhood V of $\partial\Omega$ and a strictly plurisubharmonic function ρ in V such that

$$V \cap \Omega = \{z \in V \mid \rho(z) < 0\}, \quad d\rho(z) \ne 0 \ (z \in \partial\Omega).$$

We may assume that $d\rho(z) \ne 0$ in V. If we set $\rho_1(z) = \rho(z)/|d\rho(z)|$, then for $z \in \partial\Omega$, $w \in \mathbf{C}^n - \{0\}$ with $\sum_{j=1}^n \frac{\partial \rho_1}{\partial z_j}(z)w_j = 0$, we have

$$\sum_{j,k=1}^n \frac{\partial^2 \rho_1}{\partial z_j \partial \bar{z}_k}(z) w_j \bar{w}_k > 0.$$

For $A > 0$, we set

$$\tilde{\rho}(z) = \rho_1(z) e^{A\rho_1(z)},$$

where we will determine A later. Then we have $|d\tilde{\rho}| = 1$ on $\partial\Omega$. Let $P \in \partial\Omega$. Then we obtain

$$\sum_{j,k=1}^n \frac{\partial^2 \tilde{\rho}}{\partial z_j \partial \bar{z}_k}(P) w_j \bar{w}_k = \frac{\partial^2 \rho_1}{\partial z_j \partial \bar{z}_k}(P) w_j \bar{w}_k + \left| \sum_{j=1}^n \frac{\partial \rho_1}{\partial z_j}(P) w_j \right|^2 (A + A^2).$$

Define
$$X = \{w \mid |w| = 1, \sum_{j,k=1}^{n} \frac{\partial^2 \rho_1}{\partial z_j \partial \bar{z}_k}(P) w_j \bar{w}_k \leq 0\}.$$

Then X is compact, and
$$X \subset \{w \mid |w| = 1, \sum_{j=1}^{n} \frac{\partial \rho_1}{\partial z_j}(P) w_j \neq 0\}.$$

Hence $|\sum_{j=1}^{n} \frac{\partial \rho_1}{\partial z_j}(P) w_j|$ has the minimum $m > 0$ in X. We set
$$A = \frac{-\min_{w \in X} \sum_{j,k=1}^{n} \frac{\partial^2 \rho_1}{\partial z_j \partial \bar{z}_k}(P) w_j \bar{w}_k}{m^2} + 1.$$

Then for $w \in X$,
$$\sum_{j,k=1}^{n} \frac{\partial^2 \tilde{\rho}}{\partial z_j \partial \bar{z}_k}(P) w_j \bar{w}_k \geq m^2 > 0.$$

In case $|w| = 1$ and $w \notin X$, we have
$$\frac{\partial^2 \rho_1}{\partial z_j \partial \bar{z}_k}(P) w_j \bar{w}_k > 0.$$

Hence for $|w| = 1$, we obtain
$$\sum_{j,k=1}^{n} \frac{\partial^2 \tilde{\rho}}{\partial z_j \partial \bar{z}_k}(P) w_j \bar{w}_k > 0. \tag{2.27}$$

For each $P \in \partial\Omega$, there exists $A = A(P) > 0$ and a neighborhood $W(P)$ of P such that (2.27) holds for $z \in W(P)$. Thus there exist a constant A and a neighborhood U ($U \subset V$) of $\partial\Omega$ such that for $z \in U$ and $|w| = 1$,
$$\sum_{j,k=1}^{n} \frac{\partial^2 \tilde{\rho}}{\partial z_j \partial \bar{z}_k}(z) w_j \bar{w}_k > 0,$$

which implies that $\tilde{\rho}$ is strictly plurisubharmonic in U. □

Now we are going to prove the Ohsawa-Takegoshi extension theorem.

Theorem 2.20 *(Ohsawa-Takegoshi extension theorem [OHT]) Let $\Omega \subset \mathbf{C}^n$ be a bounded pseudoconvex domain and let $H = \{z \in \mathbf{C}^n \mid z_n = 0\}$. Suppose φ is plurisubharmonic in Ω. Then there exists a constant*

$C > 0$ such that for every holomorphic function f in $H \cap \Omega$, there exists a holomorphic function F in Ω which satisfies $F|_{H \cap \Omega} = f$ and

$$\int_\Omega |F|^2 e^{-\varphi} dV \leq C \int_{H \cap \Omega} |f|^2 e^{-\varphi} dV_{n-1}.$$

Proof. We choose an increasing sequence $\{\Omega_j\}$ of strictly pseudoconvex domains in $\mathbf{C}^n$ with C^∞ boundary such that $\overline{\Omega}_j$ are compact subsets of Ω and $\cup_{j=1}^\infty \Omega_j = \Omega$. By Lemma 2.35, we can choose the defining functions ρ_j for Ω_j with the properties that $|d\rho_j| = 1$ on $\partial \Omega_j$ for $j = 1, 2, \cdots$. Let $\{\varphi_j\}$ be a sequence of C^∞ plurisubharmonic functions on $\overline{\Omega}_j$ with $\varphi_j \downarrow \varphi$ (Such a sequence $\{\varphi_j\}$ exists by Theorem 1.15 and Theorem 2.16). By Theorem 2.14, we may assume that f is holomorphic in Ω. Suppose

$$\int_{H \cap \Omega} |f|^2 e^{-\varphi} dV_{n-1} = M < \infty.$$

It follows from Lemma 2.34 that there exist holomorphic functions F_j in Ω_j such that $F_j|_{H \cap \Omega_j} = f$ and

$$\int_{\Omega_j} |F_j|^2 e^{-\varphi_j} dV \leq C \int_{H \cap \Omega_j} |f(z', 0)|^2 e^{-\varphi_j(z', 0)} dV(z') \leq CM.$$

Let $K \subset \Omega$ be a compact set. Then there exists a positive integer N such that $K \subset \Omega_j$ for $j \geq N$. If we set $L_N = \min_{\overline{\Omega}_N} e^{-\varphi_N}$, then it follows from Corollary 1.3 that

$$CM \geq \int_{\Omega_j} |F_j|^2 e^{-\varphi_j} dV \geq L_N \int_{\Omega_N} |F_j|^2 dV \geq L_N \widetilde{C} \sup_K |F_j|$$

for $j \geq N$. Hence $\{F_j\}$ is uniformly bounded on every compact subset of Ω, and hence by the Montel theorem (Lemma 2.33), we can choose a subsequence $\{F_{k_j}\}$ of $\{F_j\}$ which converges uniformly on every compact subset of Ω. We set $\lim_{j \to \infty} F_{k_j} = F$. Then F is holomorphic in Ω and $F|_{H \cap \Omega} = f$. Moreover we have

$$\int_K |F|^2 e^{-\varphi} dV = \lim_{j \to \infty} \int_K |F_{k_j}|^2 e^{-\varphi_{k_j}} dV$$
$$\leq \lim_{j \to \infty} \int_{\Omega_{k_j}} |F_{k_j}|^2 e^{-\varphi_{k_j}} dV \leq CM.$$

$\square$

Berndtsson [BR2] improved Ohsawa-Takegoshi extension theorem as follows. We omit the proof.

Theorem 2.21 Let Ω be a bounded pseudoconvex domain in $\mathbf{C}^n$ and let φ be plurisubharmonic in Ω. Let $M = \{z \in \Omega \mid h(z) = 0\}$ be a hypersurface defined by a holomorphic function bounded by 1 in Ω. Then, for any holomorphic function, f, on M there is a holomorphic function F in Ω such that $F = f$ on M and

$$\int_\Omega |F|^2 e^{-\varphi} dV \leq 4\pi \int_M |f|^2 \frac{e^{-\varphi}}{|\partial h|^2} dV_M,$$

where dV_M is the surface measure on M.

Remark 2.2 Siu [SI2] proved that the constant C in Theorem 2.20 can be chosen to be $\frac{64}{9}\pi \left(1 + \frac{1}{4e}\right)^{1/2}$, provided $\Omega \subset \{z \in \mathbf{C}^n \mid |z_n| \leq 1\}$.

Exercises

2.1 Prove Theorem 2.2 for any (p, q).

2.2 Show that if a family $\mathcal{F} = \{f_\lambda \mid \lambda \in \Lambda\}$ of holomorphic functions in a domain $\Omega \subset \mathbf{C}^n$ is uniformly bounded in Ω, then it is equicontinuous on every compact subset of Ω.

2.3 Prove Lemma 2.33.

2.4 Let $a > 0$ and $c > 1$. Construct a function f which satisfies the following conditions:

(a) $f \in C^\infty(\mathbf{R})$, $0 \leq f(x) \leq 1$ $(x \in \mathbf{R})$.
(b) $f(x) = 1$ $(x \leq 0)$, $f(x) = 0$ $(x \geq a)$.
(c) $|f'(x)| \leq \frac{c}{a}$.

Chapter 3

Integral Formulas for Strictly Pseudoconvex Domains

In this chapter we study integral formulas for differential forms on bounded domains in $\mathbf{C}^n$ with smooth boundary. Using integral formulas, we prove Hölder estimates for the $\bar{\partial}$ problem in strictly pseudoconvex domains with smooth boundary. Moreover, by following Henkin-Leiterer [HER] and Henkin [HEN3] we prove bounded and continuous extensions of holomorphic functions from submanifolds in general position of strictly pseudoconvex domains in $\mathbf{C}^n$ with smooth boundary. We also study H^p and C^k extensions of holomorphic functions from submanifolds of strictly pseudoconvex domains in $\mathbf{C}^n$ with smooth boundary. Next we prove Fefferman's mapping theorem [FEF] concerning biholomorphic mappings between strictly pseudoconvex domains with smooth boundary. The proof of Fefferman's mapping theorem given here is based on integral formulas for strictly pseudoconvex domains obtained by Henkin-Leiterer [HER] and the method developed by Range [RAN2].

3.1 The Homotopy Formula

Let $\Omega \subset \mathbf{R}^n$ be an open set and let f be a differential form with degree s on Ω. Then f is expressed by

$$f(x) = \sum_{1 \leq i_1 < \cdots < i_s \leq n} f_{i_1 \cdots i_s}(x) dx_{i_1} \wedge \cdots \wedge dx_{i_s} \qquad (x \in \Omega).$$

Definition 3.1 Define

$$|f(x)| = \left\{ \sum_{1 \leq i_1 < \cdots < i_s \leq n} |f_{i_1 \cdots i_s}(x)|^2 \right\}^{\frac{1}{2}} \qquad (x \in \Omega).$$

Let $z_1, \cdots, z_n$ be the coordinate system in $\mathbf{C}^n$. Then a (p,q) form f in Ω is expressed by

$$f = {\sum}'_{\substack{|I|=p \\ |J|=q}} f_{I,J} dz^I \wedge d\bar{z}^J,$$

where $f_{I,J}$ are functions on Ω, and $I = (i_1, \cdots, i_p)$, $J = (j_1, \cdots, j_q)$ are multi-indices with $1 \leq i_\nu, j_\mu \leq n$, and that

$$dz^I = dz_{i_1} \wedge \cdots \wedge dz_{i_p}, \quad d\bar{z}^J = d\bar{z}_{j_1} \wedge \cdots \wedge d\bar{z}_{j_q}.$$

Further, $\sum'$ means that the summation is performed only over strictly increasing multi-indices such that $i_1 < \cdots < i_p$, $j_1 < \cdots < j_q$. For a continuous function f in Ω, we define

$$\partial f = \sum_{j=1}^n \frac{\partial f}{\partial z_j} dz_j, \quad \bar{\partial} f = \sum_{j=1}^n \frac{\partial f}{\partial \bar{z}_j} d\bar{z}_j,$$

$$df = \partial f + \bar{\partial} f,$$

where the differentiation of functions is in the sense of distributions. For a differential form $f = \sum'_{I,J} f_{I,J} dz^I \wedge d\bar{z}^J$, define

$$\partial f = {\sum}'_{I,J} \partial f_{I,J} \wedge dz^I \wedge d\bar{z}^J,$$

$$\bar{\partial} f = {\sum}'_{I,J} \bar{\partial} f_{I,J} \wedge dz^I \wedge d\bar{z}^J,$$

$$df = \partial f + \bar{\partial} f.$$

If f is a (p,q) form, then ∂f is a $(p+1, q)$ form and $\bar{\partial} f$ is a $(p, q+1)$ form.

We can prove easily the following lemma. We omit the poof.

Lemma 3.1 $\partial^2 f = \bar{\partial}^2 f = 0$.

Definition 3.2 Let $D \subset \mathbf{C}^n$ and $G \subset \mathbf{C}^m$ be open sets and let $h = (h_1, \cdots, h_m) : D \to \mathbf{C}^m$ be a holomorphic mapping, and $h(D) \subset G$. Then for a differential form $f = {\sum}'_{I,J} f_{I,J} dz^I \wedge d\bar{z}^J$ on G, the pullback $h^* f$ of f with respect to h is defined by

$$h^* f = {\sum}'_{I,J} (f_{I,J} \circ h) dh^I \wedge d\bar{h}^J,$$

where for $I = (i_1, \cdots, i_p)$ and $J = (j_1, \cdots, j_q)$, we define

$$dh^I = dh_{i_1} \wedge \cdots \wedge dh_{i_p}, \quad d\bar{h}^J = d\bar{h}_{j_1} \wedge \cdots \wedge d\bar{h}_{j_q}.$$

Since $dh_j = \partial h_j$ and $d\bar{h}_j = \bar{\partial}\bar{h}_j$, h^*f is a (p,q) form in D if f is (p,q) form in G. Further, we have

$$\partial h^* f = h^* \partial f, \quad \bar{\partial} h^* f = h^* \bar{\partial} f.$$

Definition 3.3 Let X be a real C^1 manifold of dimension n and let $u : X \to \mathbf{C}^n$ and $v : X \to \mathbf{C}^n$ be C^1 mappings. Define

$$\omega(u) = du_1 \wedge \cdots \wedge du_n$$

$$\omega'(v) = \sum_{j=1}^{n} (-1)^{j+1} v_j dv_1 \wedge \cdots \widehat{_j} \cdots \wedge dv_n,$$

where $u = (u_1, \cdots, u_n)$, $v = (v_1, \cdots, v_n)$, and $\widehat{_j}$ means that dv_j is omitted. Further, we define

$$<v, u> := \sum_{j=1}^{n} v_j u_j.$$

Lemma 3.2 Let X be a C^∞ real manifold and let $u : X \to \mathbf{C}^n$ and $v : X \to \mathbf{C}^n$ be C^1 mappings. Then we have

$$d\left(\frac{\omega'(v) \wedge \omega(u)}{<u,v>^n}\right) = 0,$$

provided $<u(x), v(x)> \neq 0$ for $x \in X$.

Proof. Since $d\omega(u) = 0$, we have

$$d\left(\frac{\omega'(v) \wedge \omega(u)}{<u,v>^n}\right) = \frac{d\omega'(v) \wedge \omega(u)}{<v,u>^n} - \frac{d(<v,u>^n) \wedge \omega'(v) \wedge \omega(u)}{<v,u>^{2n}}.$$

Moreover, we have $d\omega'(v) = n\omega(v)$, and

$$d<v,u>^n \wedge \omega'(v) \wedge \omega(u)$$
$$= n<v,u>^{n-1} \sum_{j=1}^{n}(v_j du_j + u_j dv_j) \wedge \omega'(v) \wedge \omega(u)$$
$$= n<v,u>^n \omega(v) \wedge \omega(u).$$

$\square$

Lemma 3.3 Let $\zeta_j = x_j + ix_{j+n}$ for $j = 1, \cdots, n$ and let $z \in \mathbf{C}^n$. Then
$$d_\zeta(\omega'(\bar{\zeta} - \bar{z}) \wedge \omega(\zeta)) = n(2i)^n dx_1 \wedge \cdots \wedge dx_{2n}.$$

Proof. Since
$$d\bar{\zeta}_j \wedge d\zeta_j = (dx_j - idx_{j+n}) \wedge (dx_j + idx_{j+n}) = 2i dx_j \wedge dx_{j+n},$$
we obtain
$$d_\zeta \left(\sum_{j=1}^n (-1)^{j+1} (\bar{\zeta}_j - \bar{z}_j) \bigwedge_{k \neq j} d\bar{\zeta}_k \bigwedge_{s=1}^n d\zeta_s \right) = n \bigwedge_{k=1}^n d\bar{\zeta}_k \bigwedge_{s=1}^n d\zeta_s$$
$$= n(2i)^n \bigwedge_{j=1}^n dx_j.$$
□

Definition 3.4 Let $\mathcal{A}$ be an algebra whose elements are functions or a differential forms. Let $A = (a_{ij})_{i,j=1}^n$ be an $n \times n$ matrix with entries $a_{ij} \in \mathcal{A}$. Define
$$\det A = \sum_\sigma \operatorname{sgn}(\sigma) a_{\sigma(1),1} \cdots a_{\sigma(n),n},$$
where $\sum_\sigma$ means that the summation is performed over all permutations σ of $\{1, \cdots, n\}$.

Definition 3.5 Define
$$Z = \{a \in \mathcal{A} \mid a \wedge b = b \wedge a \text{ for all } b \in \mathcal{A}\}.$$

Then Z is the subalgebra of $\mathcal{A}$ which consists of functions and differential forms of even degree.

Lemma 3.4 Suppose $z_i \in Z$ for $i = 1, 2, \cdots, n$. Then we have
$$\begin{vmatrix} a_{11} & \cdots & z_1 \wedge b_k & \cdots & z_1 \wedge b_s & \cdots & a_{1n} \\ a_{21} & \cdots & z_2 \wedge b_k & \cdots & z_2 \wedge b_s & \cdots & a_{2n} \\ \vdots & & \vdots & & \vdots & & \vdots \\ a_{n1} & \cdots & z_n \wedge b_k & \cdots & z_n \wedge b_s & \cdots & a_{nn} \end{vmatrix} = 0, \qquad (3.1)$$

where $b_k \in \mathcal{A}$ is contained in the k-th column, and b_s is contained in the s-th column.

Proof. We denote by $\det A$ the determinant in (3.1). For a permutation $\tau = (ks)$, we obtain

$$\begin{aligned}\det A &= \sum_\sigma \operatorname{sgn}(\sigma) a_{\sigma(1)1} \cdots z_{\sigma(k)} b_k \cdots z_{\sigma(s)} b_s \cdots a_{\sigma(n)n} \\ &= \sum_\sigma \operatorname{sgn}(\sigma) a_{\sigma(1)1} \cdots z_{\sigma(s)} b_k \cdots z_{\sigma(k)} b_s \cdots a_{\sigma(n)n} \\ &= \operatorname{sgn}(\tau) \sum_\sigma \operatorname{sgn}(\sigma\tau) a_{\sigma\tau(1)1} \cdots z_{\sigma\tau(k)} b_k \cdots z_{\sigma\tau(s)} b_s \cdots a_{\sigma\tau(n)n} \\ &= -\sum_\sigma \operatorname{sgn}(\sigma) a_{\sigma(1)1} \cdots z_{\sigma(k)} b_k \cdots z_{\sigma(s)} b_s \cdots a_{\sigma(n)n} = -\det A,\end{aligned}$$

which means that $\det A = 0$. □

Lemma 3.5 *For a C^1 function ψ on X,*

$$\omega'(\psi v) = \psi^n \omega'(v).$$

Proof. $\omega'(v)$ can be written

$$\omega'(v) = \frac{1}{(n-1)!} \begin{vmatrix} v_1 & dv_1 & \cdots & dv_1 \\ \vdots & \vdots & & \vdots \\ v_n & dv_n & \cdots & dv_n \end{vmatrix}.$$

Since $v_1, \cdots, v_n \in Z$, it follows from Lemma 3.4 that

$$\begin{aligned}\omega'(\psi v) &= \det(\psi v, d(\psi v), \cdots, d(\psi v)) \\ &= \det(\psi v, \psi dv + v d\psi, \cdots, \psi dv + v d\psi) \\ &= \det(\psi v, \psi dv, \cdots, \psi dv) \\ &= \psi^n \det(v, dv, \cdots, dv) = \psi^n \omega'(v).\end{aligned}$$

□

Let $\Omega \subset \mathbf{C}^n$ be a bounded domain and let f be a bounded $(0,1)$ form in Ω. For $z \in \Omega$, define

$$(B_\Omega f)(z) := \frac{(n-1)!}{(2\pi i)^n} \int_{\zeta \in \Omega} f(\zeta) \wedge \frac{\omega'_\zeta(\bar\zeta - \bar z) \wedge \omega(\zeta)}{|\zeta - z|^{2n}}, \tag{3.2}$$

where

$$\omega(\zeta) := d\zeta_1 \wedge \cdots \wedge d\zeta_n, \quad \omega'_\zeta(\bar\zeta - \bar z) := \sum_{j=1}^n (-1)^{j+1} (\bar\zeta_j - \bar z_j) \bigwedge_{k \neq j} d\bar\zeta_k.$$

Let $\Omega \subset \mathbf{C}^n$ be a bounded domain with C^1 boundary, and let f be a bounded function on $\partial\Omega$. For $z \in \Omega$, define

$$(B_{\partial\Omega} f)(z) := \frac{(n-1)!}{(2\pi i)^n} \int_{\zeta \in \partial\Omega} f(\zeta) \frac{\omega'_\zeta(\bar{\zeta} - \bar{z}) \wedge \omega(\zeta)}{|\zeta - z|^{2n}}. \tag{3.3}$$

For $(z, \zeta) \in \mathbf{C}^n \times \mathbf{C}^n$, define

$$\frac{\omega_{z,\zeta}(\bar{\zeta} - \bar{z}) \wedge \omega(\zeta)}{|\zeta - z|^{2n}} := \sum_{j=1}^n (-1)^{j+1} \frac{\bar{\zeta}_j - \bar{z}_j}{|\zeta - z|^{2n}} \bigwedge_{k \neq j} (d\bar{\zeta}_k - d\bar{z}_k) \wedge \omega(\zeta).$$

Let f be a bounded differential form in Ω. Define

$$(B_\Omega f)(z) := \frac{(n-1)!}{(2\pi i)^n} \int_{\zeta \in \Omega} f(\zeta) \wedge \frac{\omega'_{z,\zeta}(\bar{\zeta} - \bar{z}) \wedge \omega(\zeta)}{|\zeta - z|^{2n}}. \tag{3.4}$$

If f is a $(0,1)$ form, then (3.4) coincides with (3.2). Further, if f is a function, then $f(\zeta) \omega'_{z,\zeta}(\bar{\zeta} - \bar{z}) \wedge \omega(\zeta)$ is at most of degree $2n - 1$ with respect to ζ, and hence $B_\Omega f = 0$.

Let $\Omega \subset \mathbf{C}^n$ be a bounded domain with C^1 boundary, and let f be a bounded differential form on $\partial\Omega$. Define

$$(B_{\partial\Omega} f)(z) := \frac{(n-1)!}{(2\pi i)^n} \int_{\zeta \in \partial\Omega} f(\zeta) \frac{\omega'_{z,\zeta}(\bar{\zeta} - \bar{z}) \wedge \omega(\zeta)}{|\zeta - z|^{2n}}. \tag{3.5}$$

If f is a function, then (3.5) coincides with (3.3).

Definition 3.6 Let $\Omega \subset \mathbf{R}^n$ be an open set and let $\alpha = (\alpha_1, \cdots, \alpha_n)$ be a multi-index, where each α_j is a nonnegative integer. Define

$$|\alpha| = \alpha_1 + \cdots + \alpha_n,$$

$$\partial^\alpha = \frac{\partial^{|\alpha|}}{\partial x_1^{\alpha_1} \cdots \partial x_n^{\alpha_n}}.$$

For $f \in C^k(\Omega)$, define the C^k-norm $|f|_{k,\Omega}$ of f in Ω by

$$|f|_{k,\Omega} = \sum_{|\alpha| \leq k} \sup_{x \in \Omega} |\partial^\alpha f(x)|.$$

$|f|_{0,\Omega}$ is denoted by $|f|_\Omega$. Then $\{f \in C^k(\Omega) \mid |f|_{k,\Omega} < \infty\}$ is a Banach space with respect to this norm. For $f \in C(\Omega)$ and $0 < \alpha < 1$, define the

α-Lipschitz norm (or the α-Hölder norm) $|f|_{\alpha,\Omega}$ by

$$|f|_{\alpha,\Omega} = |f|_{0,\Omega} + \sup_{z,\zeta \in \Omega, z \neq \zeta} \frac{|f(z) - f(\zeta)|}{|\zeta - z|^\alpha}.$$

We set

$$\Lambda_\alpha(\Omega) = \{f \in C(\Omega) \mid |f|_{\alpha,\Omega} < \infty\}.$$

We call $\Lambda_\alpha(\Omega)$ the Lipschitz space (or the Hölder space) of order α. A function $f \in \Lambda_\alpha(\Omega)$ is bounded and uniformly continuous in Ω.

Lemma 3.6 *Let $\Omega \subset \mathbf{R}^n$ be an open set. Then*

(a) For $0 < \alpha < 1$, $\Lambda_\alpha(\Omega)$ is a Banach space.
(b) If $f \in \Lambda_\alpha(\Omega)$ then f is continuous on $\overline{\Omega}$.

Proof. (a) Choose a sequence $\{f_n\}$ such that $f_n \in \Lambda_\alpha(\Omega)$, $|f_n - f_m|_{\alpha,\Omega} \to 0$. Since

$$\sup_{x \in \Omega} |f_n(x) - f_m(x)| \to 0 \qquad (n, m \to \infty),$$

there exists $f \in C(\Omega)$ such that $\lim_{n \to \infty} f_n(x) = f(x)$. On the other hand, for $\varepsilon > 0$, there exists a positive integer N such that if $n, m \geq N$, then

$$|f_n(x) - f_m(x)| < \frac{\varepsilon}{2}$$

for all $x \in \Omega$, and

$$\frac{|f_n(x) - f_m(x) - (f_n(y) - f_m(y))|}{|x - y|^\alpha} < \frac{\varepsilon}{2}$$

for $x \neq y$. Letting $m \to \infty$, we have

$$|f_n - f|_{\alpha,\Omega} \leq \varepsilon.$$

Consequently,

$$|f|_{\alpha,\Omega} \leq \varepsilon + |f_n|_{\alpha,\Omega} < \infty.$$

Thus $f \in \Lambda_\alpha(\Omega)$ and $f_n \to f$. Hence $\Lambda_\alpha(\Omega)$ is a Banach space.

(b) Suppose $a \in \partial\Omega$ and $z_n \in \Omega$ such that $z_n \to a$. Then $\{f(z_n)\}$ converges. Let $\lim_{n \to \infty} f(z_n) := f(a)$. Then for any sequence $\{w_n\}$ with $w_n \in \Omega$, $w_n \to a$, we have $f(w_n) \to f(a)$. Next suppose $z_n \in \overline{\Omega}$ and $z_n \to a$,

$a \in \partial\Omega$. Then there exists $\{w_n\} \subset \Omega$ such that $f(z_n) - f(w_n) \to 0$ and $w_n \to a$. Then

$$|f(z_n) - f(a)| \le |f(z_n) - f(w_n)| + |f(w_n) - f(a)| \to 0,$$

which implies that f is continuous on $\overline{\Omega}$. □

Lemma 3.7 *Let $\Omega \subset \mathbf{C}^n$ be a bounded domain. Then For a bounded differential form f in Ω, $B_\Omega f \in \Lambda_\alpha(\Omega)$ for every $0 < \alpha < 1$. In particular, $B_\Omega f$ is continuous on $\overline{\Omega}$. Moreover, there exists a constant $C_\alpha > 0$ such that for every bounded function f in Ω, $\|B_\Omega f\|_{\alpha,\Omega} \le C_\alpha \|f\|_{0,\Omega}$.*

Proof. By the definition of $B_\Omega f$, there exists a constant $C_1 > 0$ such that for $z, \xi \in \Omega$

$$\|B_\Omega f(z) - B_\Omega f(\xi)\| \le C_1 |f|_{0,\Omega} \sum_{j=1}^{n} \int_{\zeta \in \Omega} \left| \frac{\bar{\zeta}_j - \bar{z}_j}{|\zeta - z|^{2n}} - \frac{\bar{\zeta}_j - \bar{\xi}_j}{|\zeta - \xi|^{2n}} \right| dV(\zeta).$$

Thus it is sufficient to show that there exists a constant $C_2 > 0$ and $R > 0$ such that for $t, s \in \mathbf{R}^n$ with $|t|, |s| \le R$,

$$\int_{x \in \mathbf{R}^n, |x| < R} \left| \frac{x_1 - t_1}{|x - t|^n} - \frac{x_1 - s_1}{|x - s|^n} \right| dV(x) \le C_2 |t - s| |\log |t - s||.$$

Now we have

$$\int_{|x-t| \le |t-s|/2} \left| \frac{x_1 - t_1}{|x - t|^n} - \frac{x_1 - s_1}{|x - s|^n} \right| dV(x)$$

$$\le \int_{|x-t| \le |t-s|/2} |x - t|^{1-n} dV_n(x)$$

$$+ \int_{|x-t| \le |t-s|/2} \left| \frac{x_1 - s_1}{|x - s|^n} \right| dV(x).$$

If $|x - t| \le |t - s|/2$, then

$$|x_1 - s_1| \le |x - t| + |t - s| \le \frac{3}{2}|t - s|, \quad |x - s| \ge \frac{1}{2}|t - s|.$$

Hence there exists a constant $C_3 > 0$ such that

$$\int_{|x-t| \le |t-s|/2} \left| \frac{x_1 - s_1}{|x - s|^n} \right| dV(x) \le C_3 |t - s|^{1-n} \int_{|x-t| \le |t-s|/2} dV(x).$$

Using the polar coordinate system, there exists a constant $C_4 > 0$ such that

$$\int_{|x-t| \le |t-s|/2} \left| \frac{x_1 - t_1}{|x - t|^n} - \frac{x_1 - s_1}{|x - s|^n} \right| dV_n(x) \le C_4 |t - s|.$$

Similarly, there exists a constant $C_5 > 0$ such that

$$\int_{|x-s|\leq |t-s|/2} \left| \frac{x_1 - t_1}{|x-t|^n} - \frac{x_1 - s_1}{|x-s|^n} \right| dV_n(x) \leq C_5 |t-s|.$$

On the other hand, we have

$$\left| \frac{x_1 - t_1}{|x-t|^n} - \frac{x_1 - s_1}{|x-s|^n} \right|$$

$$= \frac{(x_1 - t_1)|x-s|^n - (x_1 - s_1)|x-t|^n}{|x-t|^n |x-s|^n}$$

$$= \frac{(x_1 - t_1)(|x-s|^n - |x-t|^n) - (t_1 - s_1)|x-t|^n}{|x-t|^n |x-s|^n}$$

$$\leq ||x-s| - |x-t|| \sum_{\nu=0}^{n-1} \frac{|x-s|^\nu |x-t|^{n-\nu}}{|x-s|^n |x-t|^n} + \frac{|t-s|}{|x-s|^n}$$

$$\leq n|t-s| \left(\max \left\{ \frac{1}{|x-s|^n}, \frac{1}{|x-t|^n} \right\} + \frac{1}{|x-s|^n} \right)$$

$$\leq 2n|t-s| \left(\frac{1}{|x-s|^n} + \frac{1}{|x-t|^n} \right).$$

We set

$$A = \{ x \in \mathbf{R}^n \mid |x-t| \geq \frac{|t-s|}{2}, \ |x-s| \geq \frac{|t-s|}{2}, \ |x| < R \}.$$

Then there exist constants C_6, C_7 and C_8 such that

$$\int_A \left| \frac{x_1 - t_1}{|x-t|^n} - \frac{x_1 - s_1}{|x-s|^n} \right| dV(x) \leq C_6 \int_{|t-s|/2 \leq |y| \leq 2R} \frac{dV(y)}{|y|^n}$$

$$\leq C_7 |t-s| \int_{|t-s|/2}^{2R} \frac{dr}{r}$$

$$\leq C_8 |t-s| |\log |t-s||,$$

which completes the proof of Lemma 3.7. □

Theorem 3.1 *(Bochner-Martinelli formula) Let $\Omega \subset \mathbf{C}^n$ be a bounded domain with C^1 boundary and let f be a continuous function on $\overline{\Omega}$ such that $\bar{\partial} f$ is continuous on $\overline{\Omega}$, where the differentiation means in the sense of distributions. Then*

$$f(z) = (B_{\partial \Omega} f)(z) - (B_\Omega \bar{\partial} f)(z) \qquad (z \in \Omega).$$

Proof. For $z \in \Omega$, we set

$$\varphi(\zeta) = \frac{(n-1)!}{(2\pi i)^n} \frac{\omega'_\zeta(\bar\zeta - \bar z) \wedge \omega(\zeta)}{|\zeta - z|^{2n}}.$$

By Lemma 3.2, we have $d\varphi(\zeta) = 0$ for $\zeta \in \Omega - \{z\}$. Then we have

$$d(f(\zeta)\varphi(\zeta)) = \bar\partial f(\zeta) \wedge \varphi(\zeta)$$

for $\zeta \in \Omega - \{z\}$. For any sufficiently small $\varepsilon > 0$, we set $\Omega_\varepsilon = \{\zeta \in \Omega \mid |\zeta - z| > \varepsilon\}$. It follows from Stokes' theorem that

$$\int_{|\zeta-z|=\varepsilon} f(\zeta)\varphi(\zeta) = \int_{\partial\Omega} f(\zeta)\varphi(\zeta) - \int_{\Omega_\varepsilon} \bar\partial f(\zeta) \wedge \varphi(\zeta). \qquad (3.6)$$

The left side in (3.6) can be rewritten as

$$\int_{|\zeta-z|=\varepsilon} f(\zeta)\varphi(\zeta) = f(z)\int_{|\zeta-z|=\varepsilon} \varphi(\zeta) + \int_{|\zeta-z|=\varepsilon} (f(\zeta) - f(z))\varphi(\zeta).$$

It follows from Stokes' theorem that

$$\int_{|\zeta-z|=\varepsilon} \varphi(\zeta) = \frac{(n-1)!}{(2\pi i)^n} \int_{|\zeta-z|=\varepsilon} \frac{\omega'_\zeta(\bar\zeta - \bar z) \wedge \omega(\zeta)}{\varepsilon^{2n}}$$
$$= \frac{(n-1)!}{(2\pi i)^n} \frac{1}{\varepsilon^{2n}} \int_{|\zeta-z|<\varepsilon} d(\omega'_\zeta(\bar\zeta - \bar z) \wedge \omega(\zeta))$$
$$= \frac{n!}{\varepsilon^{2n} \pi^n} \int_{|\zeta-z|<\varepsilon} dx_1 \wedge \cdots \wedge dx_{2n} = 1.$$

On the other hand we have

$$\int_{|\zeta-z|=\varepsilon} (f(\zeta) - f(z))\chi(\zeta)$$
$$= \frac{(n-1)!}{(2\pi i)^n \varepsilon^{2n-1}} \int_{|\zeta-z|=\varepsilon} (f(\zeta) - f(z))\frac{\omega'_\zeta(\bar\zeta - \bar z) \wedge \omega(\zeta)}{|\zeta - z|}.$$

Since $(\omega'_\zeta(\bar\zeta - \bar z) \wedge \omega(\zeta))/|\zeta - z|$ is bounded on Ω, there exists a constant $C > 0$ such that

$$\left| \int_{|\zeta-z|=\varepsilon} (f(\zeta) - f(z))\varphi(\zeta) \right| \le C \max_{|\zeta-z|=\varepsilon} |f(\zeta) - f(z)| \to 0 \qquad (\varepsilon \to 0).$$

Letting $\varepsilon \to 0$ in (3.6), we obtain the desired formula. $\qquad\square$

Corollary 3.1 *Let $\Omega \subset \mathbf{C}^n$ be a bounded domain with C^1 boundary and let f be continuous on $\overline{\Omega}$ and holomorphic in Ω. Then*

$$f(z) = (B_{\partial\Omega}f)(z) \qquad (z \in \Omega).$$

Proof. Corollary 3.1 follows easily from Theorem 3.1. □

Next we prove the Koppelman formula which is a generalization of the Bochner-Martinelli formula to differential forms.

Theorem 3.2 **(Koppelman formula)** *Let $\Omega \subset \mathbf{C}^n$ be a bounded domain with C^2 boundary and let f be a continuous $(0,q)$ form on $\overline{\Omega}$, $0 \le q \le n$, such that $\bar{\partial}f$ is continuous on $\overline{\Omega}$. Then*

$$(-1)^q f(z) = (B_{\partial\Omega}f)(z) - (B_\Omega \bar{\partial}f)(z) + (\bar{\partial}B_\Omega f)(z) \qquad (z \in \Omega). \qquad (3.7)$$

Proof. If $q = 0$, then $B_\Omega f = 0$. Hence (3.7) is the Bochner-Martinelli formula. Let $1 \le q \le n$. By Lemma 3.7, $B_\Omega f$, $B_\Omega \bar{\partial}f$ and $B_{\partial\Omega}f$ are all continuous in Ω. Hence if we can prove the equation

$$\bar{\partial}B_\Omega f = (-1)^q f - B_{\partial\Omega}f + B_\Omega \bar{\partial}f$$

in the sense of distributions, then $\bar{\partial}B_\Omega f$ is continuous in Ω, which means (3.7). Since $(B_\Omega f)(z)$ is a $(0, q-1)$ form, we have

$$\bar{\partial}(B_\Omega f \wedge v) = \bar{\partial}B_\Omega f \wedge v + (-1)^{q-1} B_\Omega f \wedge \bar{\partial}v$$

for every $(n, n-q)$ form v in Ω. Hence it is sufficient to show that

$$(-1)^q \int_\Omega B_\Omega f \wedge \bar{\partial}v = (-1)^q \int_\Omega f \wedge v - \int_\Omega B_{\partial\Omega}f \wedge v + \int_\Omega B_\Omega \bar{\partial}f \wedge v \qquad (3.8)$$

for $v \in \mathcal{D}_{(n,n-q)}(\Omega)$. We set

$$\varphi(z,\zeta) = \frac{(n-1)!}{(2\pi i)^n} \frac{\omega'_{z,\zeta}(\bar{\zeta} - \bar{z}) \wedge \omega(\zeta)}{|\zeta - z|^{2n}}.$$

Then it follows from (3.8) that

$$(-1)^q \int_{(\zeta,z) \in \Omega \times \Omega} f(\zeta) \wedge \varphi(z,\zeta) \wedge \bar{\partial}v(z) = (-1)^q \int_\Omega f(z) \wedge v(z)$$
$$- \int_{(\zeta,z) \in \partial\Omega \times \Omega} f(\zeta) \wedge \varphi(z,\zeta) \wedge v(z) + \int_{\Omega \times \Omega} \bar{\partial}f(\zeta) \wedge \varphi(z,\zeta) \wedge v(z).$$

We set
$$\Phi(z,\zeta) = \frac{(n-1)!}{(2\pi i)^n} \frac{\omega'_{z,\zeta}(\bar\zeta - \bar z) \wedge \omega_{z,\zeta}(\zeta - z)}{|\zeta - z|^{2n}},$$
where
$$\omega_{z,\zeta}(\zeta - z) = \bigwedge_{j=1}^{n} (d\zeta_j - dz_j).$$

By Lemma 3.2, we have $d_{z,\zeta}\Phi(z,\zeta) = 0$ ($z \neq \zeta$). Since $\Phi(z,\zeta) - \varphi(z,\zeta)$ contains one of $dz_1, \cdots, dz_n$, we have
$$(\Phi(z,\zeta) - \varphi(z,\zeta)) \wedge v(z) = 0.$$

Hence we obtain
$$d_{z,\zeta}(\varphi(z,\zeta) \wedge v(z)) = d_{z,\zeta}(\Phi(z,\zeta) \wedge v(z)) = (-1)^{2n-1}\Phi(z,\zeta) \wedge dv(z)$$
$$= -\varphi(z,\zeta) \wedge \bar\partial v(z).$$

Consequently,
$$d_{z,\zeta}(f(\zeta) \wedge \varphi(z,\zeta) \wedge v(z)) = \bar\partial f(\zeta) \wedge \varphi(z,\zeta) \wedge v(z) - (-1)^q f(\zeta) \wedge \varphi(z,\zeta) \wedge \bar\partial v(z).$$

For $\varepsilon > 0$, we set
$$U_\varepsilon = \{(\zeta, z) \in \mathbf{C}^n \times \mathbf{C}^n \mid |\zeta - z| < \varepsilon\}.$$

Since $\mathrm{supp}(v) \subset\subset \Omega$, we have for any sufficiently small $\varepsilon > 0$,
$$\partial(\Omega \times \Omega \setminus U_\varepsilon) \cap (\mathbf{C}^n \times \mathrm{supp}(v)) = (\partial\Omega \times \Omega \cup \partial U_\varepsilon) \cap (\mathbf{C}^n \times \mathrm{supp}(v)).$$

It follows from Stokes' theorem that
$$\int_{\partial\Omega \times \Omega} f(\zeta) \wedge \varphi(z,\zeta) \wedge v(z) - \int_{\partial U_\varepsilon} f(\zeta) \wedge \varphi(z,\zeta) \wedge v(z)$$
$$= \int_{\Omega \times \Omega \setminus U_\varepsilon} \bar\partial f(\zeta) \wedge \varphi(z,\zeta) \wedge v(z)$$
$$- (-1)^q \int_{\Omega \times \Omega \setminus U_\varepsilon} f(\zeta) \wedge \varphi(z,\zeta) \wedge \bar\partial v(z).$$

Hence, if we prove the equality
$$\lim_{\varepsilon \to 0} \int_{\partial U_\varepsilon} f(\zeta) \wedge \varphi(z,\zeta) \wedge v(z) = (-1)^q \int_\Omega f(z) \wedge v(z),$$

then we have (3.8). We define a holomorphic mapping $T : \mathbf{C}^n \times \mathbf{C}^n \to \mathbf{C}^n \times \mathbf{C}^n$ by $T(\xi, z) = (z + \xi, z)$. If we set $S_\varepsilon = \{\xi \in \mathbf{C}^n \mid |\xi| = \varepsilon\}$, then $T(S_\varepsilon \times \mathbf{C}^n) = \partial U_\varepsilon$. Since $\omega_{z,\xi}(z+\xi) \wedge v(z) = \omega(\xi) \wedge v(z)$, we obtain

$$T^*(f(\zeta) \wedge \varphi(z,\zeta) \wedge v(z)) = \sum_I f_I(z+\xi) d(\bar{z}+\bar{\xi})^I \frac{(n-1)!}{(2\pi i)^n} \frac{\omega'(\bar{\xi}) \wedge \omega(\xi)}{|\xi|^{2n}} \wedge v(z).$$

Since S_ε is real $2n - 1$ dimensional, we have on S_ε

$$d(\bar{z} + \bar{\xi})^I \wedge \omega'(\bar{\xi}) \wedge \omega(\xi) = d\bar{z}^I \wedge \omega'(\bar{\xi}) \wedge \omega(\xi).$$

Further, using the equation $d\bar{z}^I \wedge \omega'(\bar{\xi}) \wedge \omega(\xi) = (-1)^q \omega'(\bar{\xi}) \wedge \omega(\xi) \wedge d\bar{z}^I$, we have

$$\int_{\partial U_\varepsilon} f(\zeta) \wedge \varphi(z,\zeta) \wedge v(z)$$
$$= (-1)^q \int_{z \in \mathbf{C}^n} \sum_I \left[\frac{(n-1)!}{(2\pi i)^n} \int_{\xi \in S_\varepsilon} f_I(z+\xi) \frac{\omega'(\bar{\xi}) \wedge \omega(\xi)}{|\xi|^{2n}} \right] d\bar{z}^I \wedge v(z).$$

By the Bochner-Martinelli formula, the term in brackets is equal to

$$f_I(z) + \frac{(n-1)!}{(2\pi i)^n} \int_{\xi \in S_\varepsilon} (f_I(z+\xi) - f_I(z)) \frac{\omega'(\bar{\xi}) \wedge (\xi)}{|\xi|^{2n}}. \quad (3.9)$$

Using the same method as the proof of Theorem 3.1, (3.9) converges to f_I, which completes the proof of Theorem 3.2. $\square$

Definition 3.7 (**Leray map**) Let $\Omega \subset \mathbf{C}^n$ be a bounded open set. A C^1 mapping $w : \Omega \times \partial\Omega \to \mathbf{C}^n$ is called a Leray map for Ω if

$$< w(z,\zeta), \zeta - z > \neq 0$$

for all $(z, \zeta) \in \Omega \times \partial\Omega$.

Let $w(z, \zeta)$ be a Leray map for Ω and let Ω have a C^1 boundary. Define

$$\eta(z, \zeta, \lambda) = (1 - \lambda) \frac{w(z, \zeta)}{< w(z,\zeta), \zeta - z >} + \lambda \frac{\bar{\zeta} - \bar{z}}{|\zeta - z|^2}, \quad (3.10)$$

for $z \in \Omega$, $1 \leq \lambda \leq 1$ and $\zeta \in \partial\Omega$. Further, we define

$$\omega'_\zeta(w(z,\zeta)) := \sum_{j=1}^n (-1)^{j+1} w_j(z,\zeta) \bigwedge_{k \neq j} \bar{\partial}_\zeta w_k(z,\zeta)$$

and

$$\omega'_{\zeta,\lambda}(\eta(z,\zeta,\lambda)) := \sum_{j=1}^{n}(-1)^{j+1}\eta_j(z,\zeta,\lambda) \bigwedge_{k\neq j} d_{\zeta,\lambda}\eta_k(z,\zeta,\lambda),$$

where

$$\eta(z,\zeta,\lambda) = (\eta_1(z,\zeta,\lambda),\cdots,\eta_n(z,\zeta,\lambda)), \quad d_{\zeta,\lambda} = d_\zeta + d_\lambda.$$

For a bounded function f on $\partial\Omega$, define

$$(L_{\partial\Omega}f)(z) := \frac{(n-1)!}{(2\pi i)^n}\int_{\zeta\in\partial\Omega} f(\zeta)\frac{\omega'_\zeta(w(z,\zeta))\wedge\omega(\zeta)}{<w(z,\zeta),\zeta-z>^n} \quad (z\in\Omega). \quad (3.11)$$

For a bounded $(0,1)$ form f on $\partial\Omega$ and $z\in\Omega$, define

$$(R_{\partial\Omega}f)(z) := \frac{(n-1)!}{(2\pi i)^n}\int_{\substack{\zeta\in\partial\Omega\\ 0\leq\lambda\leq 1}} f(\zeta)\wedge\omega'_{\zeta,\lambda}(\eta(z,\zeta,\lambda))\wedge\omega(\zeta). \quad (3.12)$$

Remark 3.1 If $w(z,\zeta) = \bar\zeta - \bar z$, then $L_{\partial\Omega} = B_{\partial\Omega}$ and $\eta(z,\zeta,\lambda) = w(z,\zeta)$, which implies that $d_\lambda\eta_k(z,\zeta,\lambda) = 0$, and hence $R_{\partial\Omega} = 0$.

Definition 3.8 Let $\Omega\subset\mathbf{C}^n$ be a bounded domain with C^1 boundary and let $w(z,\zeta)$ be a Leray map for Ω. For $(z,\zeta,\lambda)\in\Omega\times\partial\Omega\times[0,1]$, define

$$\omega'_{z,\zeta}(w(z,\zeta)) := \sum_{j=1}^{n}(-1)^{j+1}w_j(z,\zeta)\bigwedge_{k\neq j}\bar\partial_{z,\zeta}w_k(z,\zeta)$$

and

$$\omega'_{z,\zeta,\lambda}(\eta(z,\zeta,\lambda)) = \sum_{j=1}^{n}(-1)^{j+1}\eta_j(z,\zeta,\lambda)\bigwedge_{k\neq j}(\bar\partial_{z,\zeta}+d_\lambda)\eta_k(z,\zeta,\lambda),$$

where $\bar\partial_{z,\zeta} = \bar\partial_z + \bar\partial_\zeta$.

For a bounded differential form f on $\partial\Omega$ and $z\in\Omega$, define

$$(L_{\partial\Omega}f)(z) := \frac{(n-1)!}{(2\pi i)^n}\int_{\zeta\in\partial\Omega}f(\zeta)\wedge\frac{\omega'_{z,\zeta}(w(z,\zeta))\wedge\omega(\zeta)}{<w(z,\zeta),\zeta-z>^n} \quad (3.13)$$

and

$$(R_{\partial\Omega}f)(z) = \frac{(n-1)!}{(2\pi i)^n}\int_{\substack{\zeta\in\partial\Omega\\ 0\leq\lambda\leq 1}}f(\zeta)\wedge\omega'_{z,\zeta,\lambda}(\eta(z,\zeta,\lambda))\wedge\omega(\zeta). \quad (3.14)$$

If f is a function, then (3.13) coincides with (3.11). Further, if f is a $(0,1)$ form, (3.14) coincides with (3.12).

Remark 3.2 If $w(z,\zeta) = \bar{\zeta} - \bar{z}$, then $L_{\partial\Omega} = B_{\partial\Omega}$, $R_{\partial\Omega} = 0$.

Theorem 3.3 *(Leray formula)* Let $\Omega \subset \mathbf{C}^n$ be a bounded domain with C^1 boundary. If f is a continuous function on $\overline{\Omega}$ such that $\bar{\partial}f$ is continuous on $\overline{\Omega}$, then

$$f(z) = (L_{\partial\Omega}f)(z) - (R_{\partial\Omega}\bar{\partial}f)(z) - (B_\Omega\bar{\partial}f)(z) \qquad (z \in \Omega).$$

Proof. Fix $z \in \Omega$. Since

$$<\eta(z,\zeta,\lambda), \zeta - z> = 1 \qquad (\zeta \in \partial\Omega,\ \lambda \in [0,1]),$$

it follows from Lemma 3.2 that $d_{\zeta,\lambda}[\omega'_{\zeta,\lambda}(\eta(z,\zeta,\lambda)) \wedge \omega(\zeta)] = 0$. Hence we have

$$d_{\zeta,\lambda}[f(\zeta)\omega'_{\zeta,\lambda}(\eta(z,\zeta,\lambda)) \wedge \omega(\zeta)] = \bar{\partial}f(\zeta) \wedge \omega'_{\zeta,\lambda}(\eta(z,\zeta,\lambda)) \wedge \omega(\zeta).$$

By applying Stokes' theorem to the equation $f(\zeta)\omega'_{\zeta,\lambda}(\eta(z,\zeta,\lambda)) \wedge \omega(\zeta)$ on $\partial\Omega \times [0,1]$, we have

$$\int_{\partial(\partial\Omega \times [0,1])} f(\zeta)\omega'_{\zeta,\lambda}(\eta(z,\zeta,\lambda)) \wedge \omega(\zeta)$$
$$= \int_{\partial\Omega \times [0,1]} \bar{\partial}f(\zeta)\omega'_{\zeta,\lambda}(\eta(z,\zeta,\lambda)) \wedge \omega(\zeta).$$

On the other hand we have

$$\partial(\partial\Omega \times [0,1]) = \partial\Omega \times \{1\} - \partial\Omega \times \{0\}.$$

Since we have equalities

$$\omega'_{\zeta,\lambda}(\eta(z,\zeta,\lambda)) \wedge \omega(\zeta)|_{(\zeta,\lambda) \in \partial\Omega \times \{0\}} = \frac{\omega'_\zeta(w(z,\zeta)) \wedge \omega(\zeta)}{<w(z,\zeta), \zeta - z>^n}$$

and

$$\omega'_{\zeta,\lambda}(\eta(z,\zeta,\lambda)) \wedge \omega(\zeta)|_{(\zeta,\lambda) \in \partial\Omega \times \{1\}} = \frac{\omega'_\zeta(\bar{\zeta} - \bar{z}) \wedge \omega(\zeta)}{|\zeta - z|^{2n}},$$

we obtain $R_{\partial\Omega}\bar{\partial}f = L_{\partial\Omega}f - B_{\partial\Omega}f$. Together with the Bochner-Martinelli formula, we have the desired formula. $\square$

Corollary 3.2 Let $\Omega \subset \mathbf{C}^n$ be a bounded domain with C^1 boundary and let $w(z,\zeta)$ be a Leray map for Ω. If f is continuous on $\overline{\Omega}$ and holomorphic in Ω, then

$$f(z) = (L_{\partial\Omega}f)(z) \qquad (z \in \Omega). \tag{3.15}$$

Proof. Corollary 3.2 follows easily from Theorem 3.3. □

Definition 3.9 (3.15) is called the Cauchy-Fantappiè formula.

Now we are going to prove the Koppelman-Leray formula. The Koppelman-Leray formula is a generalization of the Leray formula to differential forms. The Koppelman formula and Koppelman-Leray formula are called the homotopy formula (see Lieb-Michel [LIM] and Range [RAN2]).

Theorem 3.4 *(Koppelman-Leray formula)* Let $\Omega \subset \mathbf{C}^n$ be a bounded domain with C^1 boundary and let $w(z,\zeta)$ be a C^2 Leray map for Ω. Suppose f is a continuous $(0,q)$ $(0 \leq q \leq n)$ form on $\overline{\Omega}$ such that $\bar{\partial}f$ is also continuous on $\overline{\Omega}$. Then

$$(-1)^q f = L_{\partial\Omega} f - (R_{\partial\Omega} + B_\Omega)\bar{\partial}f + \bar{\partial}(R_{\partial\Omega} + B_\Omega)f. \qquad (3.16)$$

In particular, if $q = 0$, then by degree reasons $R_{\partial\Omega}f = B_\Omega f = 0$, and hence $f = L_{\partial\Omega} f - (R_{\partial\Omega} + B_\Omega)\bar{\partial}f$.

Proof. By definition, $L_{\partial\Omega}f$ and $R_{\partial\Omega}f$ are continuous in Ω. Further, by Lemma 3.7 and Theorem 3.3, $B_\Omega f$, $B_\Omega \bar{\partial}f$ and $\bar{\partial}B_\Omega f$ are all continuous in Ω. $\bar{\partial}R_{\partial\Omega}f$ is also continuous in Ω since we can perform the differentiation under the integral sign. By the Koppelman formula, it is sufficient to show that

$$\bar{\partial}R_{\partial\Omega}f = B_{\partial\Omega}f - L_{\partial\Omega}f + R_{\partial\Omega}\bar{\partial}f.$$

If $q = 0$, then (3.16) is the Leray formula. Let $1 \leq q \leq n$. It is sufficient to show that

$$\int_\Omega (\bar{\partial}R_{\partial\Omega}f)(z) \wedge v(z) = \int_\Omega (B_{\partial\Omega}f)(z) \wedge v(z) - \int_\Omega (L_{\partial\Omega}f)(z) \wedge v(z)$$
$$+ \int_\Omega (R_{\partial\Omega}\bar{\partial}f)(z) \wedge v(z)$$

for $v \in \mathcal{D}_{(n,n-q)}(\Omega)$. For simplicity, we set $\omega = \omega(\zeta)$, $\tilde{\omega} = \omega_{z,\zeta}(\zeta - z)$. Define

$$\theta = \frac{(n-1)!}{(2\pi i)^n} \sum_{j=1}^n (-1)^{j+1} \eta_j(z,\zeta,\lambda) \underset{k \neq j}{\wedge} d_{z,\zeta,\lambda} \eta_k(z,\zeta,\lambda)$$

and

$$\tilde{\theta} = \frac{(n-1)!}{(2\pi i)^n} \sum_{j=1}^n (-1)^{j+1} \eta_j(z,\zeta,\lambda) \underset{k \neq j}{\wedge} (\bar{\partial}_{z,\zeta} + d_\lambda)\eta_k(z,\zeta,\lambda).$$

Since $<\eta(z,\zeta,\lambda), \zeta-z> = 1$, it follows from Lemma 3.2 that $d_{z,\zeta,\lambda}(\theta \wedge \widetilde{\omega}) = 0$. Consequently,

$$d_{z,\zeta,\lambda}(\theta \wedge \omega) \wedge v(z) = d_{z,\zeta,\lambda}(\theta \wedge \widetilde{\omega}) \wedge v(z) = 0.$$

Since $\partial_\zeta(\theta \wedge \omega) = 0$, we have $(\bar\partial_{z,\zeta} + d_\lambda + \partial_z)(\theta \wedge \omega) \wedge v(z) = 0$. Hence we obtain

$$[(\bar\partial_{z,\zeta} + d_\lambda)(\tilde\theta \wedge \omega) + \partial_z(\tilde\theta \wedge \omega) + (\bar\partial_{z,\zeta} + d_\lambda + \partial_z)((\theta - \tilde\theta) \wedge \omega)] \wedge v(z) = 0.$$

Since the second and the third terms in the bracket contain one of $dz_1, \cdots, dz_n$, we obtain

$$(\bar\partial_{z,\zeta} + d_\lambda)(\tilde\theta \wedge \omega) \wedge v(z) = 0.$$

Hence we have

$$(\bar\partial_\zeta + d_\lambda)(\tilde\theta \wedge \omega) \wedge v(z) = -\bar\partial_z(\tilde\theta \wedge \omega) \wedge v(z).$$

Consequently,

$$\begin{aligned}d_{\zeta,\lambda}(f \wedge \tilde\theta \wedge \omega) \wedge v(z) &= (\bar\partial_\zeta + d_\lambda)(f \wedge \tilde\theta \wedge \omega) \wedge v(z) \\ &= \{\bar\partial f \wedge \tilde\theta \wedge \omega - \bar\partial_z(f \wedge \tilde\theta \wedge \omega)\} \wedge v(z).\end{aligned}$$

It follows from Stokes' theorem that

$$\int_{z \in \Omega} \left\{ \int_{(\zeta,\lambda) \in \partial(\partial\Omega \times [0,1])} f \wedge \tilde\theta \wedge \omega \right\} \wedge v(z)$$
$$= \int_{z \in \Omega} \left\{ \int_{(\zeta,\lambda) \in \partial\Omega \times [0,1]} \bar\partial f \wedge \tilde\theta \wedge \omega - \bar\partial_z \int_{(\zeta,\lambda) \in \partial\Omega \times [0,1]} f \wedge \tilde\theta \wedge \omega \right\} \wedge v$$
$$= \int_\Omega (R_{\partial\Omega}\bar\partial f)(z) \wedge v(z) - \int_\Omega (\bar\partial_z R_{\partial\Omega} f)(z) \wedge v(z).$$

On the other hand we have

$$\partial(\partial\Omega \times [0,1]) = (-1)^{2n-1}\partial\Omega \times \partial[0,1] = -\partial\Omega \times \{1\} + \partial\Omega \times \{0\},$$

$$\tilde\theta \wedge \omega|_{\lambda=0} = \frac{(n-1)!}{(2\pi i)^n} \frac{\omega'(w(z,\zeta)) \wedge \omega}{<w(z,\zeta), \zeta-z>^n}$$

and

$$\tilde\theta \wedge \omega|_{\lambda=1} = \frac{(n-1)!}{(2\pi i)^n} \frac{\omega'(\bar\zeta - \bar z) \wedge \omega}{|\zeta - z|^{2n}}.$$

Consequently,

$$\int_{z\in\Omega}\left\{\int_{(\zeta,\lambda)\in\partial(\partial\Omega\times[0,1])} f\wedge\tilde{\theta}\wedge\omega\right\}\wedge v(z)$$
$$= \int_\Omega (L_{\partial\Omega}f)(z)\wedge v(z) - \int_\Omega (B_{\partial\Omega}f)(z)\wedge v(z),$$

which completes the proof of Theorem 3.4. □

Corollary 3.3 *Let $\Omega\subset\mathbf{C}^n$ be a bounded domain with C^1 boundary and let $w(z,\zeta)$ be a Leray map for Ω such that $w(\cdot,\zeta)$ is holomorphic in Ω for fixed ζ. Define for $1\leq q\leq n$*

$$T_q = (-1)^q(R_{\partial\Omega} + B_\Omega).$$

Let f be a continuous $(0,q)$ form on $\overline{\Omega}$ such that $\bar\partial f$ is also continuous on $\overline{\Omega}$. Then

$$f = \bar\partial T_q f + T_{q+1}\bar\partial f.$$

Moreover, if $\bar\partial f = 0$, then $u = T_q f$ is a continuous solution of the equation $\bar\partial u = f$ in Ω.

Proof. By definition, we have

$$(L_{\partial\Omega}f)(z)$$
$$= \frac{(n-1)!}{(2\pi i)^n}\int_{\partial\Omega} f(\zeta)\sum_{j=1}^n (-1)^{j+1}\frac{w_j(z,\zeta)}{<w(z,\zeta),\zeta-z>^n}$$
$$\times \bigwedge_{k\neq j}\bar\partial_{z,\zeta}w_k(z,\zeta)\wedge\omega(\zeta).$$

If $q\geq 1$, then one of $\bar\partial_z w_k(z,\zeta)$, $k=1,\cdots,n$, is contained in each term in the right side of the above equality. Hence $L_{\partial\Omega}f(z) = 0$. By Theorem 3.4, we have $f = \bar\partial T_q f + T_{q+1}\bar\partial f$. It is trivial that $R_{\partial\Omega}f$ is C^∞ in Ω. By Lemma 3.7, $B_\Omega f$ is C^α in Ω. □

Next we will construct a Leray map for a strictly convex domain with C^2 boundary.

Definition 3.10 *Let $\Omega\subset\subset\mathbf{R}^n$ be a domain such that $\Omega = \{x\mid \rho(x) < 0\}$, where ρ is a C^2 defining function for Ω defined in a neighborhood of $\overline{\Omega}$. We say that Ω is strictly convex if*

$$\sum_{j,k=1}^n \frac{\partial^2\rho}{\partial x_j\partial x_k}(p)w_j w_k > 0$$

for every p and $0 \neq w = (w_1, \cdots, w_n) \in \mathbf{R}^n$ satisfying

$$p \in \partial\Omega, \quad \sum_{j=1}^n \frac{\partial \rho}{\partial x_j}(p) w_j = 0.$$

We can prove the following lemma by using the same method as the proof of Corollary 1.4. We omit the proof.

Lemma 3.8 *Let $\Omega \subset\subset \mathbf{R}^n$ be a strictly convex domain. Then there exists a defining function ρ for Ω such that*

$$\sum_{j,k=1}^n \frac{\partial^2 \rho}{\partial x_j \partial x_k}(p) w_j w_k \geq C|w|^2 \quad (p \in \partial\Omega, w \in \mathbf{R}^n). \tag{3.17}$$

Lemma 3.9 *Every strictly convex domain is a strictly pseudoconvex domain.*

Proof. Let Ω be a strictly convex domain. Then $\Omega = \{z \in \mathbf{C}^n \mid \rho(z) < 0\}$, where ρ satisfies (3.17). Let $p \in \partial\Omega$. Using Taylor's formula, we have for $t = (t_1, \cdots, t_n) \in \mathbf{C}^n$

$$\rho(p+t) = \rho(p) + 2\mathrm{Re}\left(\sum_{j=1}^n \frac{\partial \rho}{\partial \zeta_j}(p) t_j + \frac{1}{2}\sum_{j,k=1}^n \frac{\partial^2 \rho}{\partial \zeta_j \partial \zeta_k}(p) t_j t_k\right)$$
$$+ \sum_{j,k=1}^n \frac{\partial^2 \rho}{\partial \zeta_j \partial \bar\zeta_k}(p) t_j \bar t_k + o(|t|^2).$$

We set

$$Q_p(t) = \frac{1}{2}\sum_{j,k=1}^n \frac{\partial^2 \rho}{\partial \zeta_j \partial \zeta_k}(p) t_j t_k$$

$$L_p(t) = \sum_{j,k=1}^n \frac{\partial^2 \rho}{\partial \zeta_j \partial \bar\zeta_k}(p) t_j \bar t_k.$$

Then for $t_j = x_j + i x_{n+j}$,

$$\frac{1}{2}\sum_{j,k=1}^n \frac{\partial^2 \rho}{\partial x_j \partial x_k}(p) x_j x_k = 2\mathrm{Re} Q_p(t) + L_p(t).$$

Since Ω is strictly convex, we obtain

$$2\mathrm{Re} Q_p(t) + L_p(t) > 0 \quad (0 \neq t \in \mathbf{C}^n).$$

Since
$$Q_p(it) = -Q_p(t), \quad L_p(it) = L_p(t),$$
we obtain
$$-2\mathrm{Re}Q_p(t) + L_p(t) > 0 \qquad (0 \neq t \in \mathbf{C}^n).$$
Hence $L_p(t) > 0$ for $t \neq 0$, which means that Ω is strictly pseudoconvex. $\square$

Definition 3.11 Let $\Omega_1, \Omega_2 \subset \mathbf{C}^n$ be open sets. A holomorphic mapping $f : \Omega_1 \to \Omega_2$ is called biholomorphic if $f : \Omega_1 \to \Omega_2$ is bijective and $f^{-1} : \Omega_2 \to \Omega_1$ is a holomorphic mapping. (It follows from Corollary 5.4 that if a holomorphic mapping $f : \Omega_1 \to \Omega_2$ is bijective, then $f : \Omega_1 \to \Omega_2$ is biholomorphic.)

Lemma 3.10 *(Narashimhan's lemma) Suppose $\Omega \subset\subset \mathbf{C}^n$ is a strictly pseudoconvex domain with C^2 boundary and $p \in \partial\Omega$. Then there exist a neighborhood U of p and a biholomorphic mapping φ in U such that $\varphi(U\cap\Omega)$ is a strictly convex domain.*

Proof. There are a neighborhood W of $\partial\Omega$ and a strictly plurisubharmonic function ρ in W such that $\Omega\cap W = \{z \in W \mid \rho(z) < 0\}$. By Corollary 1.4, there exists a constant $C > 0$ such that

$$\sum_{j,k=1}^n \frac{\partial^2 \rho}{\partial z_j \partial \bar{z}_k}(p) w_j \bar{w}_k \geq C|w|^2 \qquad (w \in \mathbf{C}^n).$$

We choose a coordinate system such that

$$p = 0, \quad \left(\frac{\partial \rho}{\partial z_1}(p), \cdots, \frac{\partial \rho}{\partial z_n}(p)\right) = (1, 0, \cdots, 0).$$

Using Taylor's formula, we have

$$\rho(w) = 2\mathrm{Re}\left(\sum_{j=1}^n \frac{\partial \rho}{\partial z_j}(p) w_j + \frac{1}{2}\sum_{j,k=1}^n \frac{\partial^2 \rho}{\partial z_j \partial z_k}(p) w_j w_k\right)$$
$$+ \sum_{j,k=1}^n \frac{\partial^2 \rho}{\partial z_j \partial \bar{z}_k}(p) w_j \bar{w}_k + o(|w|^2).$$

We set

$$w'_1 = w_1 + \frac{1}{2}\sum_{j,k=1}^n \frac{\partial^2 \rho}{\partial z_j \partial z_k}(p) w_j w_k$$
$$w'_j = w_j \quad (j=2,\cdots,n).$$

Then $w' = \varphi(w)$ is a biholomorphic mapping in a neighborhood of 0 (see Corollary 5.3). Further, we have

$$\rho \circ \varphi^{-1}(w') = 2\mathrm{Re}\,w'_1 + \frac{1}{2}\sum_{j,k=1}^n \frac{\partial^2 \rho \circ \varphi^{-1}}{\partial z_j \partial \bar{z}_k}(0) w'_j \bar{w}'_k + o(|w'|^2).$$

Thus there exists a neighborhood U of p such that $\rho \circ \varphi^{-1}$ is strictly convex in U. Hence $\varphi(U \cap \Omega)$ is strictly convex. $\square$

Lemma 3.11 *Let $\Omega \subset\subset \mathbf{R}^n$ be a strictly convex domain with a C^2 defining function ρ. Then Ω has a C^2 boundary, that is, ρ satisfies $d\rho(x) \neq 0$ for all $x \in \partial\Omega$.*

Proof. Suppose there exists a point $p = (a_1,\cdots,a_n) \in \partial\Omega$ such that $d\rho(p) = \rho(p) = 0$. Using Taylor's formula we obtain

$$\rho(x) = \sum_{j,k=1}^n \frac{\partial^2 \rho(p)}{\partial x_j \partial x_k}(x_j - a_j)(x_k - a_k) + o(|x-p|^2)$$
$$\geq \alpha |x-p|^2 - o(|x-p|^2).$$

Hence there exists $r > 0$ such that $\rho(x) > 0$ for $0 < |x - p| < r$, which contradicts that p is a boundary point of Ω. $\square$

Lemma 3.12 *Suppose $\Omega \subset\subset \mathbf{R}^n$ is a strictly convex domain with C^2 boundary. Then Ω is geometrically convex, that is, if $P_1, P_2 \in \Omega$ and $0 \leq \lambda \leq 1$, then $\lambda P_1 + (1-\lambda)P_2 \in \Omega$. Moreover, if $P_1, P_2 \in \overline{\Omega}$, $P_1 \neq P_2$ and $0 < \lambda < 1$, then $\lambda P_1 + (1-\lambda)P_2 \in \Omega$.*

Proof. Suppose Lemma 3.12 does not hold. Then there exist λ_0 with $0 < \lambda_0 < 1$ and $P_1, P_2 \in \Omega$ such that $\lambda P_1 + (1-\lambda)P_2 \in \overline{\Omega}$ for $0 \leq \lambda \leq 1$, $P = \lambda_0 P_1 + (1-\lambda_0)P_2 \in \partial\Omega$ and $P_1 - P_2$ is contained in the tangent space to $\partial\Omega$ at P. We set $\varphi(\lambda) = \rho(\lambda P_1 + (1-\lambda)P_2)$. Then we have $\varphi(\lambda_0) = \rho(P) = 0$, $\varphi'(\lambda_0) = 0$, We set $P_1 = (a_1,\cdots,a_n)$ and $P_2 = (b_1,\cdots,b_n)$. By Taylor's formula,

$$\varphi(\lambda) = \sum_{j,k=1}^n \frac{\partial^2 \rho}{\partial x_j \partial x_k}(P)(a_j - b_j)(a_k - b_k)(\lambda - \lambda_0)^2 + o(|\lambda - \lambda_0|^2)$$

$$\geq \alpha|P_1 - P_2|^2(\lambda - \lambda_0)^2 + o(|\lambda - \lambda_0|^2).$$

Thus there exists $r > 0$ such that if $0 < |\lambda - \lambda_0| \leq r$, then $\varphi(\lambda) > 0$, which is a contradiction. One can prove the latter half similarly. $\square$

Lemma 3.13 *Let $\Omega \subset\subset \mathbf{R}^n$ be a domain with C^2 boundary and let ρ be a defining function for Ω. Then Ω is geometrically convex if and only if*

$$\sum_{j,k=1}^{n} \frac{\partial^2 \rho}{\partial x_j \partial x_k}(P) w_j w_k \geq 0 \tag{3.18}$$

for all $P \in \partial\Omega$ and $w = (w_1, \cdots, w_n) \in \mathbf{R}^n$ with $\sum_{j=1}^{n} \frac{\partial \rho}{\partial x_j}(P) w_j = 0$.

Proof. Let Ω be geometrically convex. Suppose (3.18) does not hold. Then there exist points $P \in \partial\Omega$ and $w \in \mathbf{R}^n$ such that

$$\sum_{j=1}^{n} \frac{\partial \rho}{\partial x_j}(P) w_j = 0,$$

and

$$\sum_{j,k=1}^{n} \frac{\partial^2 \rho}{\partial x_j \partial x_k}(P) w_j w_k = -2k < 0.$$

We may assume that $P = 0$, $\operatorname{grad} \rho(P) = (0, \cdots, 0.1)$. We set $Q = tw + \varepsilon(0, \cdots, 0, 1)$, where $\varepsilon > 0$ and $t \in \mathbf{R}$ will be determined later. Using Taylor's formula we have

$$\rho(Q) = \rho(0) + \sum_{j=1}^{n} \frac{\partial \rho}{\partial x_j}(0) Q_j + \frac{1}{2} \sum_{j,k=1}^{n} \frac{\partial^2 \rho}{\partial x_j \partial x_k}(0) Q_j Q_k + o(|Q|^2)$$

$$= \varepsilon + \frac{t^2}{2} \sum_{j,k=1}^{n} \sum_{j,k=1}^{n} \frac{\partial^2 \rho}{\partial x_j \partial x_k}(0) w_j w_k + O(\varepsilon^2) + O(\varepsilon t) + o(t^2; \varepsilon^2)$$

$$= \varepsilon - kt^2 + R(\varepsilon, t),$$

where $R(\varepsilon, t)$ satisfies $|R(\varepsilon, t)| \leq ct^2 + C\varepsilon^2$, and we can make c sufficiently small if C is sufficiently large. We choose ε so small that $\varepsilon \gg \varepsilon^2$. If $t = 0$, then $\rho(Q) > 0$. On the other hand, if $|t| > \sqrt{2\varepsilon/k}$, then $\rho(Q) < 0$. For t with $|t| > \sqrt{2\varepsilon/k}$, we set $Q_1 = tw + \varepsilon(0, \cdots, 0, 1)$ and $Q_2 = -tw + \varepsilon(0, \cdots, 0, 1)$. Then $Q_1, Q_2 \in \Omega$ and $(Q_1 + Q_2)/2 = \varepsilon(0, \cdots, 0, 1) \notin \overline{\Omega}$. This contradicts that Ω is geometrically convex.

Conversely, assume that (3.18) holds. Suppose $0 \in \Omega$. For $\varepsilon > 0$ and a positive integer M, we set $\rho_\varepsilon(x) = \rho(x) + \varepsilon |x|^{2M}/M$, $\Omega_\varepsilon = \{x \mid \rho_\varepsilon(x) < 0\}$.

Then $\Omega = \cup_{\varepsilon>0}\Omega_\varepsilon$. If $\varepsilon > 0$ is sufficiently small, then Ω_ε is strictly convex, and hence geometrically convex. Hence Ω is geometrically convex. □

Lemma 3.14 *Let $\Omega \subset\subset \mathbf{C}^n$ be a strictly convex domain. Then there exist a neighborhood U of $\partial\Omega$ and constants $\varepsilon > 0$ and $\beta > 0$ such that for $\zeta \in U$ and $z \in \mathbf{C}^n$ with $|\zeta - z| \leq \varepsilon$, we have*

$$2\mathrm{Re} <\partial\rho(\zeta), \zeta - z> \geq \rho(\zeta) - \rho(z) + \beta|\zeta - z|^2,$$

where

$$\partial\rho(\zeta) = \left(\frac{\partial\rho(\zeta)}{\partial\zeta_1}, \cdots, \frac{\partial\rho(\zeta)}{\partial\zeta_n}\right).$$

Proof. For $\zeta_j = \xi_j + i\xi_{j+n}$ and $z_j = x_j + ix_{j+n}$,

$$2\mathrm{Re} <\partial\rho(\zeta), \zeta - z> = \sum_{j=1}^{2n} \frac{\partial\rho}{\partial x_j}(\zeta)(\xi_j - x_j).$$

Hence by Taylor's formula, we have

$$\rho(z) = \rho(\zeta) - 2\mathrm{Re} <\partial\rho(\zeta), \zeta - z>$$
$$+ \frac{1}{2}\sum_{j,k=1}^{2n} \frac{\partial^2\rho}{\partial x_j \partial x_k}(\zeta)(\xi_j - x_j)(\xi_k - x_k) + o(|\zeta - z|^2).$$

Consequently, if we choose $\varepsilon > 0$ sufficiently small, then we have for $\zeta \in \partial\Omega$ with $|z - \zeta| \leq \varepsilon$

$$2\mathrm{Re} <\partial\rho(\zeta), \zeta - z> \geq \rho(\zeta) - \rho(z) + \frac{\alpha}{4}|\zeta - z|^2.$$

□

Theorem 3.5 *Let $\Omega \subset\subset \mathbf{C}^n$ be a strictly convex domain with C^2 boundary. Then $2\partial\rho(\zeta)$ is a Leray map for Ω.*

Proof. Let $z \in \Omega$ and $\zeta \in \partial\Omega$. By Lemma 3.14, there exists $\varepsilon > 0$ such that

$$\mathrm{Re} <2\partial\rho(\zeta), \zeta - z> \geq -\rho(z) > 0,$$

provided $|\zeta - z| \leq \varepsilon$. Let $|\zeta - z| > \varepsilon$. If we set

$$z_\varepsilon = \left(1 - \frac{\varepsilon}{|\zeta - z|}\right)\zeta + \frac{\varepsilon}{|\zeta - z|}z.$$

Then by Lemma 3.12, $z_\varepsilon \in \Omega$. Since $|\zeta - z_\varepsilon| = \varepsilon$, we have

$$\text{Re} < 2\rho(\zeta), \zeta - z > = 2\text{Re}\frac{|\zeta - z|}{\varepsilon} < \partial\rho(\zeta), \zeta - z_\varepsilon >> 0,$$

which implies that

$$< 2\rho(\zeta), \zeta - z > \neq 0 \qquad ((z, \zeta) \in \Omega \times \partial\Omega).$$

Hence $2\partial\rho(\zeta)$ is a Leray map for Ω. □

Corollary 3.4 *Let $\Omega \subset\subset \mathbf{C}^n$ be a strictly convex domain with C^2 boundary and let f be a continuous $(0, q)$ form, $1 \leq q \leq n$, on $\overline{\Omega}$ such that $\bar{\partial}f = 0$ in Ω. Let $w(z, \zeta) = 2\partial\rho(\zeta)$ be a Leray map, where ρ is a defining function for Ω. Then*

$$u = (-1)^q (R_{\partial\Omega} f + B_\Omega f)$$

is a continuous solution of the equation $\bar{\partial}u = f$.

Proof. Corollary 3.4 follows from Theorem 3.5 and Corollary 3.3. □

Let $\Omega \subset\subset \Omega$ be a strictly pseudoconvex domain with C^∞ boundary. In 3.2, we will construct a Leray map $w(z, \zeta)$ for Ω which is of class C^∞ in a neighborhood of $\overline{\Omega} \times \partial\Omega$ depending holomorphically on z for ζ fixed. In order to show that $w(z, \zeta)$ is of class C^∞ on $\overline{\Omega} \times \partial\Omega$, we need the following Theorem 3.6 and Theorem 3.7. We begin with Lemma 3.15.

Lemma 3.15 *Let $\Omega \subset \mathbf{C}^n$ be a pseudoconvex open set. Define*

$$M_k = \{z \in \mathbf{C}^n \mid z_1 = \cdots = z_k = 0\}$$

for $1 \leq k \leq n$. If f is holomorphic in Ω, and $f(z) = 0$ for $z \in M_k \cap \Omega$, then there exist holomorphic functions $f_1, \cdots f_k$ in Ω such that

$$f(z) = \sum_{j=1}^{k} z_j f_j(z) \qquad (z \in \Omega).$$

Proof. We prove the lemma by induction on k. When $k = 1$, we set $f_1(z) = f(z)/z_1$. Assume that Lemma 3.15 has already been proved for $k - 1$. Suppose f is holomorphic in Ω and $f(z) = 0$ for $z \in M_k \cap \Omega$. Since $\Omega \cap M_1$ is a pseudoconvex open set in M_1, by the inductive hypothesis there exist holomorphic functions $\tilde{f}_2, \cdots, \tilde{f}_k$ in $\Omega \cap M_1$ such that

$$f(z) = \sum_{j=2}^{k} z_j \tilde{f}_j(z_2, \cdots, z_n) \qquad (z \in \Omega \cap M_1).$$

By Theorem 2.14, there exist holomorphic functions f_j ($j = 2, \cdots, k$) in Ω such that $f_j(z) = \tilde{f}_j(z_2, \cdots, z_n)$ in $\Omega \cap M_1$. If we set

$$f_1(z) := \frac{1}{z_1}(f(z) - \sum_{j=2}^{k} z_j f_j(z)) \qquad (z \in \Omega),$$

then f_1 is holomorphic in Ω, which completes the proof of Lemma 3.15. $\square$

Lemma 3.16 *Let $\Omega \subset \mathbf{C}^n$ be a pseudoconvex open set and let f be a holomorphic function in Ω. Then there exist holomorphic functions $f_1, \cdots, f_n$ in $\Omega \times \Omega$ such that*

$$f(w) - f(z) = \sum_{j=1}^{n} (w_j - z_j) f_j(w, z)$$

for $(w, z) \in \Omega \times \Omega$.

Proof. We set $g(w, z) = f(w) - f(z)$. Then g is holomorphic in $\Omega \times \Omega$. By a change of variables $z_i^* = w_i - z_i$, $z_{n+i}^* = z_i$ for $i = 1, \cdots, n$, we have $g(w, z) = 0$ in $M_n = \{(w, z) \in \Omega \times \Omega \mid w = z\}$. We set $z^* = (z_1^*, \cdots, z_{2n}^*)$ and $h(z^*) = g(w, z)$. By Lemma 3.15, there exist holomorphic functions $h_1, \cdots, h_n$ in $\Omega \times \Omega$ such that

$$h(z^*) = \sum_{j=1}^{n} z_j^* h_j(z^*).$$

We set $f_j(w, z) = h_j(w_1 - z_1, \cdots, w_n - z_n, z_1, \cdots, z_n)$. Then f_j, $j = 1, \cdots, n$, are holomorphic in $\Omega \times \Omega$ and satisfy the equality

$$f(w) - f(z) = \sum_{j=1}^{n} (w_j - z_j) f_j(w, z).$$

$\square$

The following two theorems were proved by Range [RAN2] and used to show the smoothness of the Leray map for strictly pseudoconvex domains.

Theorem 3.6 *Let $\Omega \subset\subset \mathbf{C}^n$ be a pseudoconvex domain and let $K \subset\subset \Omega$ be compact. Then there exist neighborhoods V_0 and V of K, and a C^∞ function $\Phi(z, \zeta)$ in $V_0 \times \partial V$ with the following properties:*

(a) $V_0 \subset\subset V \subset\subset \Omega$.
(b) V has C^∞ boundary.
(c) $\Phi(z, \zeta)$ is holomorphic with respect to z for fixed ζ.
(d) $\Phi(z, \zeta) \neq 0$ for every $(z, \zeta) \in V_0 \times \partial V$.

(e) There exist C^∞ functions $w_j(z,\zeta)$, $j=1,\cdots,n$, in $V_0 \times \partial V$ such that $w_j(z,\zeta)$ are holomorphic with respect to z and satisfy

$$\Phi(z,\zeta) = \sum_{j=1}^n w_j(z,\zeta)(\zeta_j - z_j).$$

Proof. We may assume that $K = \widehat{K}_\Omega^{\mathcal{O}}$. Let $\omega \subset\subset \Omega$ be a neighborhood of K. By Theorem 1.18, there exist $h_k \in \mathcal{O}(\Omega)$, $1 \le k \le N$, such that if

$$A = \{z \in \omega \mid |h_k(z)| < 1, k = 1, \cdots, N\},$$

then $K \subset A \subset\subset \omega$. Let Δ^N be the unit polydisc in $\mathbf{C}^N$ and $H = (h_1,\cdots,h_N) : \Omega \to \Delta^N$. Since $H(K)$ is a compact subset of Δ^N, there exists a convex neighborhood U of $H(K)$ such that ∂U is C^∞ boundary of U and $U \subset\subset \Delta^N$. Let $U = \{t \in \Delta^N \mid \rho(t) < 0\}$. Then we have

$$< \partial\rho(\eta), \eta - t > \ne 0 \qquad ((t,\eta) \in U \times \partial U).$$

If we define $\Phi : A \times A \to \mathbf{C}$ by

$$\Phi(z,\zeta) = \sum_{k=1}^N \frac{\partial \rho}{\partial \eta_k}(H(\zeta))(h_k(\zeta) - h_k(z)),$$

then $\zeta \in (H|_A)^{-1}(\partial U)$ and $\Phi(z,\zeta) \ne 0$ for $z \in K$. By the continuity, there exists a neighborhood V of K such that $V_0 \subset\subset V \subset\subset A$, V has a C^∞ boundary and

$$\Phi(z,\zeta) \ne 0 \qquad ((z,\zeta) \in V_0 \times \partial V).$$

By Lemma 3.16, there exist $Q_{j,k} \in \mathcal{O}(G \times G)$ such that

$$h_k(\zeta) - h_k(z) = \sum_{j=1}^n Q_{jk}(z,\zeta)(\zeta_j - z_j).$$

We set

$$w_j(z,\zeta) = \sum_{k=1}^N \frac{\partial \rho}{\partial \eta_k}(H(\zeta))Q_{jk}(z,\zeta).$$

for $j = 1, \cdots, n$. Then we obtain

$$\Phi(z,\zeta) = \sum_{j=1}^n w_j(z,\zeta)(\zeta_j - z_j),$$

which completes the proof of Theorem 3.6. □

Theorem 3.7 *Let $\Omega \subset\subset \mathbf{C}^n$ be a pseudoconvex domain and let K be a compact subset of Ω. Then there exist neighborhoods V_0 and V of K with $V_0 \subset\subset V \subset\subset \Omega$, and a continuous linear operator $T_q : C_{(0,q)}(\overline{V}) \to C_{(0,q-1)}(V_0)$ ($1 \le q \le n$) with the following properties:*

(a) for $k = 0, 1, 2, \cdots$, $T_q f \in C^k_{(0,q-1)}(V_0)$ if $f \in C^k_{(0,q)}(\overline{V})$.
(b) If $\bar{\partial} f = 0$ on V, then $\bar{\partial} T_q f = f$ on V_0.

Proof. For $f \in C^k_{(0,q)}(\overline{V})$ and $z \in V_0$, we set

$$(R_{\partial V} f)(z) = \frac{(n-1)!}{(2\pi i)^n} \int_{\substack{\zeta \in \partial V \\ 0 \le \lambda \le 1}} f(\zeta) \wedge \omega'_{z,\zeta,\lambda}(\eta(z,\zeta,\lambda)) \wedge \omega(\zeta),$$

where $\eta(z,\zeta,\lambda)$ is defined by using $w(z,\zeta)$ and $\Phi(z,\zeta)$ in Theorem 3.6. Define

$$T_q = (-1)^q (R_{\partial V} + B_V).$$

Then it follows from theorem 3.4 that

$$f(z) = (\bar{\partial} T_q f)(z) + (T_{q+1} \bar{\partial} f)(z) \qquad (z \in V_0),$$

which completes the proof of Theorem 3.7. □

3.2 Hölder Estimates for the $\bar{\partial}$ Problem

Let $\Omega \subset\subset \mathbf{C}^n$ be a strictly pseudoconvex domain with C^∞ boundary. Then there exist a neighborhood U of $\partial \Omega$ and a strictly plurisubharmonic C^∞ function ρ in U such that

$$U \cap \Omega = \{z \in U \mid \rho(z) < 0\}, \quad d\rho(z) \ne 0 \ (z \in \partial \Omega).$$

Results in 3.2, 3.3 and 3.4 are still valid under the assumption that the boundary $\partial \Omega$ is of class C^2. However, because of Fefferman's mapping theorem, we assume that the boundary $\partial \Omega$ is of class C^∞.

Definition 3.12 For $\zeta \in U$ and $z \in \mathbf{C}^n$, we define the Levi polynomial $F(z,\zeta)$ by

$$F(z,\zeta) = 2 \sum_{j=1}^n \frac{\partial \rho}{\partial \zeta_j}(\zeta)(\zeta_j - z_j) - \sum_{j,k=1}^n \frac{\partial^2 \rho}{\partial \zeta_j \partial \zeta_k}(\zeta)(\zeta_j - z_j)(\zeta_k - z_k).$$

$F(z,\zeta)$ is a C^∞ function in $\mathbf{C}^n \times U$ and holomorphic with respect to z. By Taylor's formula, there exist a constant $\beta > 0$ and $\varepsilon > 0$ such that for $\zeta \in U$ with $|z - \zeta| \leq 2\varepsilon$,

$$\operatorname{Re} F(z,\zeta) \geq \rho(\zeta) - \rho(z) + \beta|\zeta - z|^2. \tag{3.19}$$

If we choose $\varepsilon > 0$ sufficiently small, then for $\zeta \in \partial\Omega$, we have

$$\{z \in \mathbf{C}^n \mid |\zeta - z| \leq 3\varepsilon\} \subset U.$$

If $\varepsilon \leq |\zeta - z| \leq 2\varepsilon$, then by (3.19) we have

$$\operatorname{Re} F(z,\zeta) \geq \rho(\zeta) - \rho(z) + \beta\varepsilon^2 \qquad (\zeta, z \in U).$$

We choose a neighborhood $U_1 \subset U$ of $\partial\Omega$ such that $|\rho(\zeta)| \leq \beta\varepsilon^2/3$ for $\zeta \in U_1$, and $\{z \mid |z - \zeta| \leq 2\varepsilon\} \subset U$ for $\zeta \in U_1$. We set $V_{\overline{\Omega}} = \Omega \cup U_1$. Then for $(z,\zeta) \in V_{\overline{\Omega}} \times U_1$ with $|z - \zeta| \leq 2\varepsilon$, we have $z, \zeta \in U$, and

$$\operatorname{Re} F(z,\zeta) \geq \frac{\beta\varepsilon^2}{3}.$$

Hence we can define $\log F(z,\zeta)$ for $\varepsilon \leq |\zeta - z| \leq 2\varepsilon$, $(z,\zeta) \in V_{\overline{\Omega}} \times U_1$. Choose a function $\chi \in C^\infty(\mathbf{C}^n \times \mathbf{C}^n)$ with the properties that $0 \leq \chi \leq 1$, and

$$\chi(z,\zeta) = \begin{cases} 1 & (|\zeta - z| \leq 5\varepsilon/4) \\ 0 & (|\zeta - z| \geq 7\varepsilon/4) \end{cases}.$$

For $(z,\zeta) \in V_{\overline{\Omega}} \times U_1$, define

$$f(z,\zeta) = \begin{cases} \bar{\partial}_z[\chi(\zeta - z)\log F(z,\zeta)] & (\varepsilon \leq |\zeta - z| \leq 2\varepsilon) \\ 0 & (\text{otherwise}) \end{cases}.$$

Then we have $f \in C^\infty_{(0,1)}(V_{\overline{\Omega}} \times U_1)$ and $\bar{\partial}_z f = 0$. By Theorem 3.7, there exists a neighborhood U_2 of $\partial\Omega$ with $U_2 \subset\subset U_1$ such that if we set $U_{\overline{\Omega}} = \Omega \cup U_2$, then $U_{\overline{\Omega}} \subset\subset V_{\overline{\Omega}}$. It follows from Theorem 3.7 that there exists a continuous linear operator $T_1 : C^\infty_{(0,1)}(V_{\overline{\Omega}}) \to C^\infty(U_{\overline{\Omega}})$ such that $\bar{\partial}_z T_1(f(\cdot,\zeta))(z) = f(z,\zeta)$ for $z \in U_{\overline{\Omega}}$. Define $u(z,\zeta) = T_1(f(\cdot,\zeta))(z)$. Then $u \in C^\infty(U_{\overline{\Omega}} \times U_2)$ and $\bar{\partial}_z u = f$. For $(z,\zeta) \in U_{\overline{\Omega}} \times U_2$, we define

$$M(z,\zeta) := e^{-u(z,\zeta)},$$

$$\Phi(z,\zeta) := \begin{cases} F(z,\zeta)M(z,\zeta) & (|\zeta - z| \leq \varepsilon) \\ \exp[\chi(\zeta - z)\log F(z,\zeta) - u(z,\zeta)] & (|\zeta - z| \geq \varepsilon) \end{cases}.$$

Then we have the following theorem.

Theorem 3.8 $\Phi(z,\zeta)$ satisfies the following:

(a) $\Phi(z,\zeta)$ is a C^∞ function in $U_{\overline{\Omega}} \times U_2$.
(b) $\Phi(z,\zeta)$ is holomorphic with respect to $z \in U_{\overline{\Omega}}$.
(c) $\Phi(z,\zeta) \neq 0$ for $(z,\zeta) \in U_{\overline{\Omega}} \times U_2$ with $|\zeta - z| \geq \varepsilon$.
(d) There exists a C^∞ function $M(z,\zeta) \neq 0$ in $U_{\overline{\Omega}} \times U_2$ such that

$$\Phi(z,\zeta) = F(z,\zeta)M(z,\zeta) \qquad ((z,\zeta) \in U_{\overline{\Omega}} \times U_2, \ |\zeta - z| \leq \varepsilon).$$

Proof. (a) holds since $u(z,\zeta)$ is C^∞ in $U_{\overline{\Omega}} \times U_2$. If $|z - \zeta| \leq \varepsilon$, then

$$\bar{\partial}_z \Phi(z,\zeta) = F(z,\zeta) e^{-u} \bar{\partial}_z(-u) = -F(z,\zeta) e^{-u} f = 0.$$

If $\varepsilon \leq |z - \zeta| \leq 2\varepsilon$, then

$$\bar{\partial}_z \Phi = \exp[\chi \log F(z,\zeta) - u(z,\zeta)] \bar{\partial}_z \{\chi(\zeta - z) \log F(z,\zeta) - u(z,\zeta)\} = 0.$$

If $2\varepsilon \leq |z - \zeta|$, then

$$\bar{\partial}_z \Phi(z,\zeta) = e^{-u} \bar{\partial}_z(-u(z,\zeta)) = -e^{-u} f = 0,$$

which implies that $\Phi(z,\zeta)$ is holomorphic with respect to z. This proves (b). (c) and (d) follow from the definition of $\Phi(z,\zeta)$. □

It follows from (3.19) that for $(z,\zeta) \in U_{\overline{\Omega}} \times U_1$ with $\varepsilon \leq |\zeta - z| \leq 2\varepsilon$, we have

$$\operatorname{Re} F(z,\zeta) - 2\rho(\zeta) \geq -\rho(\zeta) - \rho(z) + \beta |\zeta - z|^2 \geq \frac{\beta \varepsilon^2}{3}.$$

Hence we can define $\log(F(z,\zeta) - 2\rho(\zeta))$ for $(z,\zeta) \in U_{\overline{\Omega}} \times U_2$ with $\varepsilon \leq |\zeta - z| \leq 2\varepsilon$, . For $(z,\zeta) \in U_{\overline{\Omega}} \times U_2$, define

$$\tilde{f}(z,\zeta) = \begin{cases} \bar{\partial}_z[\chi(\zeta - z) \log(F(z,\zeta) - 2\rho(\zeta))] & (\varepsilon \leq |\zeta - z| \leq 2\varepsilon) \\ 0 & (\text{otherwise}) \end{cases}.$$

Then $\bar{\partial}_z \tilde{f} = 0$. It follows from Theorem 3.7 that there exists a C^∞ function $\tilde{u}(z,\zeta)$ in $U_{\overline{\Omega}} \times U_2$ such that $\bar{\partial}_z \tilde{u} = \tilde{f}$. In particular, if $\zeta \in \partial\Omega$, then $\tilde{f}(z,\zeta) = f(z,\zeta)$. Hence we obtain that $\tilde{u}(z,\zeta) = u(z,\zeta)$ for $\zeta \in \partial\Omega$. Define

$$\widetilde{M}(z,\zeta) = e^{-\tilde{u}(z,\zeta)},$$

$$\widetilde{\Phi}(z,\zeta) = \begin{cases} (F(z,\zeta) - 2\rho(\zeta)) \widetilde{M}(z,\zeta) & (|\zeta - z| \leq \varepsilon) \\ \exp(\chi \log(F(z,\zeta) - 2\rho(\zeta)) - \tilde{u}(z,\zeta)) & (|\zeta - z| \geq \varepsilon) \end{cases}.$$

Then we have the following theorem which is proved in the same way as the proof of Theorem 3.8. So we omit the proof.

Theorem 3.9 $\widetilde{\Phi}(z,\zeta)$ satisfies the following:

(a) $\widetilde{\Phi}(z,\zeta)$ is a C^∞ function in $U_{\overline{\Omega}} \times U_2$.
(b) $\widetilde{\Phi}(z,\zeta)$ is holomorphic with respect to $z \in U_{\overline{\Omega}}$.
(c) $\widetilde{\Phi}(z,\zeta) \neq 0$ for $(z,\zeta) \in U_{\overline{\Omega}} \times U_2$ with $|\zeta - z| \geq \varepsilon$.
(d) There exists a C^∞ function $\widetilde{M}(z,\zeta) \neq 0$ in $U_{\overline{\Omega}} \times U_2$ such that

$$\widetilde{\Phi}(z,\zeta) = (F(z,\zeta) - 2\rho(\zeta))\widetilde{M}(z,\zeta) \qquad ((z,\zeta) \in U_{\overline{\Omega}} \times U_2, \ |\zeta - z| \leq \varepsilon).$$

(e) $\widetilde{\Phi}(z,\zeta) = \Phi(z,\zeta)$ for $\zeta \in \partial\Omega$.

In particular, it follows from (c) and (d) that $\widetilde{\Phi}(z,\zeta) \neq 0$ for $(z,\zeta) \in \overline{\Omega} \times \overline{\Omega} \setminus (\partial\Omega \times \partial\Omega)$.

Lemma 3.17 Let $\Omega \subset\subset \mathbf{C}^n$ be a strictly pseudoconvex domain with C^∞ boundary and let $W = \{z_n = 0\} \cap U_{\overline{\Omega}}$. Let a function $f \in C^\infty(W \times U_2)$ satisfy $f(\cdot,\zeta) \in \mathcal{O}(W)$. Then there exist an open set V_0 with $\overline{\Omega} \subset V_0 \subset U_{\overline{\Omega}}$ and a function $F \in C^\infty(V_0 \times U_2)$ such that $F(\cdot,\zeta) \in \mathcal{O}(V_0)$ $(\zeta \in U_2)$, and

$$F((z',0),\zeta) = f((z',0),\zeta) \qquad ((z',0) \in \{z_n = 0\} \cap V_0, \zeta \in U_2).$$

Proof. We define a mapping $\pi : \mathbf{C}^n \to \mathbf{C}^n$ by $\pi(z',z_n) = (z',0)$. Since W and $U_{\overline{\Omega}} - \pi^{-1}(W)$ are closed disjoint subsets of $U_{\overline{\Omega}}$, there exists a function $\chi \in C^\infty(U_{\overline{\Omega}})$ with the properties that $\chi = 1$ in an open subset of $U_{\overline{\Omega}}$ containing W, and $\chi = 0$ in an open subset of $U_{\overline{\Omega}}$ containing $U_{\overline{\Omega}} - \pi^{-1}(W)$. We set

$$\alpha(z,\zeta) = \frac{\bar{\partial}_z \{\chi(z) f(\pi(z),\zeta)\}}{z_n}.$$

It follows from the definition of χ that $\alpha \in C^\infty_{(0,1)}(U_{\overline{\Omega}} \times U_2)$. By Theorem 3.7, there exist an open set V, V_0 ($\overline{\Omega} \subset V_0 \subset\subset V \subset U_{\overline{\Omega}}$), and a continuous linear operator $T_1 : C^\infty_{(0,1)}(\overline{V}) \to C^\infty(V_0)$ such that $\bar{\partial} T_1(\alpha(\cdot,\zeta))(z) = \alpha(z,\zeta)$ for $z \in V_0$. Define $g(z,\zeta) = T_1(\alpha(\cdot,\zeta))(z)$. Then $g \in C^\infty(V_0 \times U_2)$. If we set $F(z,\zeta) = \chi(z) f(\pi(z),\zeta) - z_n g(z,\zeta)$, then $F(\cdot,\zeta)$ is holomorphic in V_0. $F \in C^\infty(V_0 \times U_2)$ and $F(\pi(z),\zeta) = f(\pi(z),\zeta)$, which completes the proof of Lemma 3.17. □

The following lemma is the parametrized version of Lemma 3.15.

Lemma 3.18 Let $\Omega \subset\subset \mathbf{C}^n$ be a strictly pseudoconvex domain with C^∞ boundary. Define

$$M_k = \{z \in \mathbf{C}^n \mid z_1 = \cdots = z_k = 0\}$$

for $1 \leq k \leq n$. If $f(z,\zeta)$ is of class C^∞ in $U_{\overline{\Omega}} \times U_2$ and holomorphic with respect to $z \in U_{\overline{\Omega}}$ for $\zeta \in U_2$ fixed, and $f(z,\zeta) = 0$ for $z \in M_k \cap U_{\overline{\Omega}}$, then there exist an open set V_0 ($\overline{\Omega} \subset V_0 \subset U_{\overline{\Omega}}$) and C^∞ functions $f_1(z,\zeta), \cdots f_k(z,\zeta)$ in $V_0 \times U_2$ which are holomorphic with respect to $z \in \Omega$ for ζ fixed such that

$$f(z,\zeta) = \sum_{j=1}^{k} z_j f_j(z,\zeta) \qquad ((z,\zeta) \in V_0 \times U_2).$$

Proof. Lemma 3.18 follows from Lemma 3.17 and the proof of Lemma 3.15. □

Lemma 3.19 Let $\Omega \subset \mathbf{C}^n$ be a strictly pseudoconvex domain with C^∞ boundary and let $f(z,\zeta)$ be of class C^∞ in $U_{\overline{\Omega}} \times U_2$ and holomorphic with respect to $z \in U_{\overline{\Omega}}$ for $\zeta \in U_2$ fixed. Then there exist an open set V_0 ($\overline{\Omega} \subset V_0 \subset U_{\overline{\Omega}}$) and C^∞ functions $f_1, \cdots, f_n$ in $V_0 \times V_0 \times U_2$ which are holomorphic with respect to $(z,w) \in V_0 \times V_0$ for $\zeta \in U_2$ fixed such that

$$f(w,\zeta) - f(z,\zeta) = \sum_{j=1}^{n}(w_j - z_j)f_j(w,z,\zeta)$$

for $(w,z,\zeta) \in V_0 \times V_0 \times U_2$.

Proof. Lemma 3.19 follows from Lemma 3.18 and the proof of Lemma 3.16. □

Theorem 3.10 Let $\Omega \subset\subset \mathbf{C}^n$ be a strictly pseudoconvex domain with C^∞ boundary. Then there exist a neighborhood V_0 of $\overline{\Omega}$ and C^∞ functions $w_j(z,\zeta)$ for $j = 1, \cdots, n$ in $V_0 \times U_2$ such that

$$\Phi(z,\zeta) = \sum_{j=1}^{n}(z_j - \zeta_j)w_j(z,\zeta) \qquad ((z,\zeta) \in V_0 \times U_2).$$

Moreover, $w_j(z,\zeta)$, $j = 1, \cdots, n$, are holomorphic with respect to $z \in V_0$ for ζ fixed.

Proof. By Lemma 3.19 there exist a neighborhood V_0 and functions $f_j \in C^\infty(V_0 \times V_0 \times U_2)$ such that $f_j(z,w,\zeta)$ are holomorphic with respect to $(z,w) \in V_0 \times V_0$ for ζ fixed and

$$\Phi(z,\zeta) - \Phi(w,\zeta) = \sum_{j=1}^{n}(z_j - w_j)f_j(z,w,\zeta).$$

We set $w = \zeta$. Because of $\Phi(\zeta, \zeta) = 0$, we have

$$\Phi(z, \zeta) = \sum_{j=1}^{n}(z_j - \zeta_j) f_j(z, \zeta, \zeta).$$

If we set $w_j(z, \zeta) = f_j(z, \zeta, \zeta)$, then each $w_j(z, \zeta)$ satisfies the desired conditions. □

Let $\Omega \subset\subset \mathbf{C}^n$ be a strictly pseudoconvex domain with C^∞ boundary and $w(z, \zeta)$ be the Leray map for Ω. For any $f \in \mathcal{O}(\Omega)$ which is continuous on $\overline{\Omega}$, the Cauchy-Fantappié formula (3.15) is expressed by

$$f(z) = (L_{\partial\Omega} f)(z) = \frac{(n-1)!}{(2\pi i)^n} \int_{\zeta \in \partial\Omega} f(\zeta) \frac{\omega'_\zeta(w(z,\zeta)) \wedge \omega(\zeta)}{<w(z,\zeta), \zeta - z>^n}. \quad (3.20)$$

Definition 3.13 Let Ω be a strictly pseudoconvex domain in $\mathbf{C}^n$ with C^∞ boundary. The kernel of the Cauchy-Fantappié formula (3.20) is called the Henkin-Ramirez kernel.

We need the following lemma in order to prove $\frac{1}{2}$-Hölder estimate for the $\bar{\partial}$ problem in Ω. We omit the proof (see Exercise 3.1).

Lemma 3.20 Let Ω be a bounded domain in $\mathbf{R}^n$ with C^1 boundary. Suppose $f \in C^1(\Omega)$ and that for some $0 < \alpha < 1$ there exists a constant $C > 0$ such that $\|df(x)\| \leq C[\mathrm{dist}(x, \partial\Omega)]^{\alpha-1}]$ for all $x \in \Omega$. Then $f \in \Lambda_\alpha(\Omega)$.

Now we are going to prove the $\frac{1}{2}$-Hölder estimate for the $\bar{\partial}$ problem in strictly pseudoconvex domains with smooth boundary.

Theorem 3.11 Let Ω be a strictly pseudoconvex domain in $\mathbf{C}^n$ with C^∞ boundary. Let w be a Leray map defined in Theorem 3.10. Then for any bounded differential form f on $\partial\Omega$

$$\|R_{\partial\Omega} f\|_{1/2,\Omega} \leq C \|f\|_{0,\Omega}.$$

Proof. By definition $R_{\partial\Omega}$ is expressed by

$$(R_{\partial\Omega})(z) = \int_{\partial\Omega \times [0,1]} f \wedge \sum_{s=0}^{n-2} p_s \det\nolimits_{1,1,n-s-2,s} \left(\frac{w}{\Phi}, \frac{\bar\zeta - \bar z}{|\zeta - z|^2}, \frac{\bar\partial_\zeta w}{\Phi}, \frac{d\bar\zeta - d\bar z}{|\zeta - z|^2} \right) \wedge d\lambda \wedge \omega(\zeta).$$

Hence the coefficients of the form $R_{\partial\Omega}f$ are linear combinations of the integrals of the following type

$$E(z) = \int_{\partial\Omega} \frac{\psi}{\Phi^{n-s-1}|\zeta - z|^{2s+2}} \bigwedge_{j \neq m} d\bar{\zeta}_j \wedge \omega(\zeta),$$

where ψ satisfies

$$|\psi| \leq C|\zeta - z|.$$

By Lemma 3.20, it is sufficient to prove that, for $j = 1, \cdots, n$,

$$\left|\frac{\partial E(z)}{\partial z_j}\right|, \left|\frac{\partial E(z)}{\partial \bar{z}_j}\right| \leq C\|f\|_{0,\Omega}|\rho(z)|^{-1/2}.$$

Therefore it is sufficient to show that for every $\xi \in \partial\Omega$, there are a neighborhood U and a constant $C > 0$ such that

$$\int_{\partial\Omega \cap U} \frac{d\sigma_{2n-1}}{|\Phi(z,\zeta)|^{n-s-1}|\zeta - z|^{2s+2}} \leq C|\rho(z)|^{-1/2}$$

and

$$\int_{\partial\Omega \cap U} \frac{d\sigma_{2n-1}}{|\Phi(z,\zeta)|^{n-s}|\zeta - z|^{2s+1}} \leq C|\rho(z)|^{-1/2},$$

where $d\sigma_{2n-1}$ is the surface measure on $\partial\Omega$. We can choose a local coordinate system $t = (t_1, \cdots, t_{2n-1})$ in $U \cap \partial\Omega$ such that $t_1 = \operatorname{Im}\Phi(z,\zeta)$ and $|t| \approx |z - \zeta|$. It follows from (3.19) that

$$\int_{\partial\Omega \cap U} \frac{d\sigma_{2n-1}}{|\Phi(z,\zeta)|^{n-s}|\zeta - z|^{2s+1}} \leq C \int_{|t|<R} \frac{dt_1 \cdots dt_{2n-1}}{(|t_1| + |t|^2 + |\rho(z)|)^{n-s}|t|^{2s+1}},$$

where R is some positive constant. We set $t' = (t_2, \cdots, t_{2n-1})$. Then

$$\int_{\partial\Omega \cap U} \frac{d\sigma_{2n-1}}{|\Phi(z,\zeta)|^{n-s}|\zeta - z|^{2s+1}} \leq \int_{|t'|<R} \frac{dt_2 \cdots dt_{2n-1}}{(|t'|^2 + |\rho(z)|)^{n-s-1}|t'|^{2s+1}}$$

$$\leq C \int_0^R \frac{r^{2n-3}dr}{(r^2 + |\rho(z)|)^{n-s-1}r^{2s+1}}$$

$$\leq C|\rho(z)|^{-1/2}.$$

In the same way we obtain

$$\int_{\partial\Omega \cap U} \frac{d\sigma_{2n-1}}{|\Phi(z,\zeta)|^{n-s-1}|\zeta - z|^{2s+2}} \leq C|\rho(z)|^{-1/2}. \qquad \square$$

Corollary 3.5 *Let Ω be a strictly pseudoconvex domain in $\mathbf{C}^n$ with smooth boundary. For every continuous $(0,q)$-form f on $\overline{D}$ such that $\bar{\partial} f = 0$ in Ω, $1 \leq q \leq n$,*

$$u = (-1)^q(R_{\partial\Omega} f + B_\Omega f)$$

is a continuous solution of $\bar{\partial} u = f$ such that $\|u\|_{1/2,\Omega} \leq C\|f\|_{0,\Omega}$.

Proof. Corollary 3.5 follows from Lemma 3.7, Corollary 3.3 and Theorem 3.11. □

Example 3.1 (E. M. Stein) For $\alpha > \frac{1}{2}$, there exist a strictly pseudoconvex domain Ω and a continuous function f on $\overline{\Omega}$ such that the equation $\bar{\partial} u = f$ does not have any solution satisfying $u \in \Lambda_\alpha(\Omega)$.

Proof. Let $\Omega = \{(z_1, z_2) \in \mathbf{C}^n \mid |z_1|^2 + |z_2|^2 < 1\}$. We set

$$f(z_1, z_2) = \begin{cases} \frac{d\bar{z}_2}{\log(z_1-1)} & ((z_1, z_2) \in \overline{\Omega}\setminus\{(1,0)\}) \\ 0 & ((z_1, z_2) = (1,0)) \end{cases}.$$

Then f is a C^∞ $(0,1)$ form in $\overline{\Omega}\setminus\{(1,0)\}$. Since $\log(z_1-1) \to \infty$ as $z_1 \to \infty$, f is continuous on $\overline{\Omega}$. Further, $\bar{\partial} f = 0$ in Ω. Suppose there exists $u \in \Lambda_\alpha(\Omega)$ such that $\bar{\partial} u = f$ for $\alpha > 1/2$. Then $(u - \bar{z}_2)/\log(z_1 - 1)$ is holomorphic in Ω. Let ε be such that $0 < 2\varepsilon < 1$. We set

$$C_1 = \{(z_1, z_2) \in \mathbf{C}^2 \mid z_1 = 1 - \varepsilon, |z_2| = \sqrt{\varepsilon}\},$$

$$C_2 = \{(z_1, z_2) \in \mathbf{C}^2 \mid z_1 = 1 - 2\varepsilon, |z_2| = \sqrt{\varepsilon}\}.$$

Then $C_1, C_2 \subset \Omega$. By the Cauchy integral formula we have

$$\int_{|z_2|=\sqrt{\varepsilon}} u(1-\varepsilon, z_2) dz_2 = \int_{|z_2|=\sqrt{\varepsilon}} \frac{\bar{z}_2 dz_2}{\log(-\varepsilon)} = \frac{2\pi i}{\log(-\varepsilon)}.$$

$$\int_{|z_2|=\sqrt{\varepsilon}} u(1-2\varepsilon, z_2) dz_2 = \frac{2\pi i}{\log(-2\varepsilon)}.$$

Since $u \in \Lambda_\alpha(\Omega)$, there exists a constant $C > 0$ such that for every $0 < 2\varepsilon < 1$, we have

$$\left| \frac{1}{\log(-\varepsilon)} - \frac{1}{\log(-2\varepsilon)} \right| \leq C\varepsilon^{\alpha-1/2}.$$

Consequently, for any $0 < 2\varepsilon < 1$ we have

$$\log 2 = |\log(-2\varepsilon) - \log(-\varepsilon)| \leq C\varepsilon^{\alpha-1/2}|\log(-\varepsilon)\log(-2\varepsilon)|,$$

which is a contradiction. □

As an application of Theorem 3.11, we have the following lemma.

Lemma 3.21 *Let $\Omega \subset\subset \mathbf{C}^n$ be a strictly pseudoconvex domain with C^∞ boundary and let f be a holomorphic function in Ω that is continuous on $\overline{\Omega}$. Then f can be approximated uniformly on $\overline{\Omega}$ by functions holomorphic in a neighborhood of $\overline{\Omega}$.*

Proof. Let $\mathcal{U} = \{U_i \mid i = 1, \cdots, N\}$ be a finite open cover of $\overline{\Omega}$. Choose $\chi_j \in C_c^\infty(U_j)$ such that $\sum_{j=1}^N \chi_j = 1$ on $\overline{\Omega}$. Define

$$f_j = L_{\partial\Omega}(\chi_j f),$$

where w is the Leray map defined in Theorem 3.10. By Corollary 3.2 we have

$$f = \sum_{j=1}^N f_j$$

and each f_j is holomorphic in some neighborhood of $\overline{\Omega} \setminus (\partial\Omega \cap U_j)$. By Theorem 3.3 we have

$$f_j = \chi_j f + R_{\partial\Omega}(f\bar{\partial}\chi_j) + B_\Omega(f\bar{\partial}\chi_j).$$

It follows from Lemma 3.7 that $B_\Omega(f\bar{\partial}\chi_j)$ is continuous on $\overline{\Omega}$. By Theorem 3.11 we have

$$\|R_{\partial\Omega} f\|_{1/2,\Omega} \leq C\|f\|_{0,\Omega}.$$

Hence $R_{\partial\Omega}^w f$ is continuous on $\overline{\Omega}$. Hence each f_j is continuous on $\overline{\Omega}$. It is sufficient to show that each f_j can be approximated uniformly on $\overline{\Omega}$ by functions holomorphic in a neighborhood of $\overline{\Omega}$. The required approximation can be obtained by a shift in the direction of the normal vector of $\partial\Omega$ at some point in $\partial\Omega \cap U_j$. □

Remark 3.3 *Lemma 3.21 was first proved by Lieb [LI1]. The above proof is due to Henkin-Leiterer [HER].*

Suppose f is a continuous $(0, q)$ form on $\overline{\Omega}$ such that $\bar{\partial} f = 0$. We set $T_q = (-1)^q (R_{\partial\Omega} + B_\Omega)$. By (3.16) or by Corollary 3.3 we have $f = \bar{\partial} T_q f$, which means that $T_q f$ is a solution of the $\bar{\partial}$ problem. Using $T_q f$, Henkin [HEN2], Grauert-Lieb [GRL], Lieb [LI2; LI3], Kerzman [KER], Ovrelid [OV], Henkin-Romanov [HEV] and Krantz [KR1] obtained L^p and Hölder estimates for the $\bar{\partial}$ problem in strictly pseudoconvex domains with smooth boundary. We proved $\frac{1}{2}$-Hölder estimate for the $\bar{\partial}$ problem in strictly pseudoconvex domains with smooth boundary in Corollary 3.5. In 4.2 we will prove L^p estimates for the $\bar{\partial}$ problem in strictly pseudoconvex domains with smooth boundary by applying the Berndtsson-Andersson formula. Bruna and Burgués [BRG] obtained $\frac{1}{2}$-Hölder and L^p estimates for the $\bar{\partial}$ problem in strictly pseudoconvex domains with nonsmooth boundary using the Berndtsson-Andersson formula. Siu [SI1] and Lieb-Range [LIR] studied the differentiability for solutions of the $\bar{\partial}$ problem in strictly pseudoconvex domains with smooth boundary. Moreover, in the finite intersection of strictly pseudoconvex domains with smooth boundary, Michel [MIC] and Michel-Perotti [MIP] obtained C^k estimates, and Range-Siu [RAS] and Menini [MEN] obtained L^p and Hölder estimates for the $\bar{\partial}$ problem. Menini used the Bendtsson-Andersson formula. Range [RAN1], Diederich-Fornaess-Wiegerinck [DIK] and Chen-Krantz-Ma [CHK] obtained Hölder and L^p estimates for the $\bar{\partial}$ problem in real or complex ellipsoids. Bruna-Castillo [BRJ], Polking [POL] and Range [RAN4] obtained Hölder and L^p estimates for the $\bar{\partial}$ problem in some convex domains. S.C. Chen [CH] and Z. Chen [CHE] investigated the real analyticity for solutions of the $\bar{\partial}$ problem in certain convex domains. Fischer-Lieb [FIL], Ho [Ho1], Schmalz [SCH] and Ma [MA] investigated the $\bar{\partial}$ problem in q-convex domains. Fleron [FLE], Ho [Ho2] and Verdera [VER] obtained Hölder and uniform estimates for the $\bar{\partial}$ problem in some domains. On the other hand, Kohn [KON] proved the global regurality for solutions of the $\bar{\partial}$ problem in pseudoconvex domains in $\mathbf{C}^n$ with smooth boundary, that is, if Ω is a pseudoconvex domain in $\mathbf{C}^n$ with C^∞ boundary and f is a C^∞ $(0,1)$ form on $\overline{\Omega}$ with $\bar{\partial} f = 0$, then there exists a C^∞ function u on $\overline{\Omega}$ such that $\bar{\partial} u = f$ (see D'Angelo [DA]). Fefferman-Kohn [FEK] studied Hölder estimates for the $\bar{\partial}$ problem in pseudoconvex domains of finite type in $\mathbf{C}^2$ with smooth boundary. Range [RAN3] also investigated the $\bar{\partial}$ problem in pseudoconvex domains of finite type in $\mathbf{C}^2$ with smooth boundary using the homotopy formula.

3.3 Bounded and Continuous Extensions

Let $\Omega \subset\subset \mathbf{C}^n$ be a strictly pseudoconvex domain with smooth boundary and let X be a submanifold in a neighborhood of $\overline{\Omega}$ which intersects $\partial\Omega$ transversally. In 1972, Henkin [HEN3] proved that every bounded holomorphic function in $V = X \cap \Omega$ can be extended to a bounded holomorphic function in Ω. Moreover, Henkin [HEN3] proved that if f is holomorphic in V that is continuous on $\overline{V}$, then f can be extended to a holomorphic function in Ω that is continuous on $\overline{\Omega}$. In 1984, Henkin and Leiterer [HER] extended Henkin's results to strictly pseudoconvex domains with non-smooth boundary in a Stein manifold without assuming the transversality. In 3.3 and 3.4, we only treat the smooth domain and assume the transversality. We prove first the bounded extension from complex hypersurfaces by following the method of Henkin-Leiterer [HER], and then the continuous and bounded extensions from submanifolds by following the method of Henkin [HEN3].

Let Ω be a strictly pseudoconvex domain in $\mathbf{C}^n$ with C^∞ boundary. Then there exists a neighborhood U of $\partial\Omega$ and a strictly plurisubharmonic C^∞ function ρ in U such that

$$U \cap \Omega = \{z \in U \mid \rho(z) < 0\}, \quad d\rho(z) \neq 0 \ (z \in \partial\Omega).$$

Let U_2 be the open set in Theorem 3.8. We choose $\varepsilon_0 > 0$ such that $\{\zeta \in U \mid |\rho(\zeta)| < 2\varepsilon_0\} \subset\subset U_2$. Let $\chi \in \mathcal{D}(\mathbf{C}^n)$ be a function with the following properties:

(a) $0 \leq \chi \leq 1$.
(b) $\chi(\zeta) = 1$ for $\zeta \in U$ with $\rho(\zeta) \geq -\varepsilon_0$.
(c) $\chi(\zeta) = 0$ for $\zeta \in (\Omega - U) \cup \{\zeta \in U \mid \rho(\zeta) \leq -2\varepsilon_0\}$.

Define

$$\omega_\zeta\left(\frac{\chi(\zeta)w(z,\zeta)}{\widetilde{\Phi}(z,\zeta)}\right) = \bigwedge_{j=1}^n d_\zeta\left(\frac{\chi(\zeta)w_j(z,\zeta)}{\widetilde{\Phi}(z,\zeta)}\right). \tag{3.21}$$

By Theorem 3.9 (d), the differential form in (3.21) is continuous with respect to $(z,\zeta) \in \Omega \times \overline{\Omega}$.

Definition 3.14 For an L^1 function f in Ω, define

$$L_\Omega f(z) := \frac{n!}{(2\pi i)^n} \int_\Omega f(\zeta)\omega_\zeta\left(\frac{\chi(\zeta)w(z,\zeta)}{\widetilde{\Phi}(z,\zeta)}\right) \wedge \omega(\zeta) \quad (z \in \Omega).$$

Since $w(z,\zeta)$ and $\widetilde{\Phi}(z,\zeta)$ are holomorphic with respect to z, $L_\Omega f$ is holomorphic in Ω.

Definition 3.15 For $0 \leq \lambda \leq 1$, define

$$\tilde{\eta}_j(z,\zeta,\lambda) := (1-\lambda)\frac{\chi(\zeta)w_j(z,\zeta)}{\widetilde{\Phi}(z,\zeta)} + \lambda\frac{\bar{\zeta}_j - \bar{z}_j}{|\zeta - z|^2}, \quad \tilde{\eta} = (\tilde{\eta}_1, \cdots, \tilde{\eta}_n),$$

and
$$\omega(\tilde{\eta}(z,\zeta,\lambda)) := \bigwedge_{j=1}^{n}(\bar{\partial}_{z,\zeta} + d_\lambda)\tilde{\eta}_j(z,\zeta,\lambda).$$

Definition 3.16 For an L^1 function f in Ω, define
$$R_\Omega f(z) := \frac{n!}{(2\pi i)^n}\int_{(\zeta,\lambda)\in\Omega\times[0,1]} f(\zeta)\wedge\omega(\tilde{\eta}(z,\zeta,\lambda))\wedge\omega(\zeta) \qquad (z\in\Omega).$$

Then $R_\Omega f(z)$ is continuous in Ω. If f is a $(0,q)$ form, then $R_\Omega f$ is a $(0,q-1)$ form. In particular, if f is a function, then $R_\Omega f = 0$.

Theorem 3.12 Let $\Omega \subset\subset \mathbf{C}^n$ be a strictly pseudoconvex domain with C^∞ boundary.

(a) Suppose f is an L^1 function in Ω such that $\bar{\partial}f$ is an L^1 form in Ω. Then
$$f(z) = L_\Omega f(z) + R_\Omega \bar{\partial}f(z) \qquad (z\in\Omega).$$

(b) Suppose f is an $L^1(0,q)$ $(1\leq q\leq n)$ form in Ω such that $\bar{\partial}f$ is an L^1 form in Ω. Then
$$f(z) = \bar{\partial}R_\Omega f(z) + R_\Omega\bar{\partial}f(z) \qquad (z\in\Omega).$$

Proof. Let $0 \leq q \leq n$. Suppose f is a continuous $(0,q)$ form on $\overline{\Omega}$ such that $\bar{\partial}f$ is continuous on $\overline{\Omega}$. For $z\in\Omega$, $\zeta\in\overline{\Omega}\setminus\{z\}$ and $0 \leq \lambda \leq 1$, define
$$\omega'_{z,\zeta,\lambda}(\tilde{\eta}(z,\zeta,\lambda)) := \sum_{j=1}^{n}(-1)^{j+1}\tilde{\eta}_j(z,\zeta,\lambda)\bigwedge_{k\neq j}(\bar{\partial}_{z,\zeta} + d_\lambda)\tilde{\eta}_k(z,\zeta,\lambda).$$

Then $\omega_{z,\zeta,\lambda}(\tilde{\eta}(z,\zeta,\lambda))$ is continuous for (z,ζ,λ) with $z\in\Omega$, $\zeta\in\overline{\Omega}\setminus\{z\}$, $0 \leq \lambda \leq 1$. Since $\tilde{\Phi}(z,\zeta) \neq 0$ for $(z,\zeta)\in\Omega\times\overline{\Omega}$, each term of $\omega_{z,\zeta,\lambda}(\tilde{\eta}(z,\zeta,\lambda))$ involving $d\lambda$ is equal to $O(|\zeta-z|^{-(2n-2)})$. Hence
$$\int_{(\zeta,\lambda)\in\Omega\times[0,1]} f(\zeta)\wedge\omega'_{z,\zeta,\lambda}(\tilde{\eta}(z,\zeta,\lambda))\wedge\omega(\zeta), \qquad (z\in\Omega)$$
is differentiable with respect to $z\in\Omega$. Differentiating under the integral sign and taking into account that $\dim_{\mathbf{R}}(\Omega\times[0,1])$ is odd, we have
$$\bar{\partial}_z\int_{\Omega\times[0,1]} f(\zeta)\wedge\omega'_{z,\zeta,\lambda}(\tilde{\eta}(z,\zeta,\lambda))\wedge\omega(\zeta)$$
$$= -\int_{\Omega\times[0,1]} \bar{\partial}_z[f(\zeta)\wedge\omega'_{z,\zeta,\lambda}(\tilde{\eta}(z,\zeta,\lambda))\wedge\omega(\zeta)],$$

and
$$(\partial_{z,\zeta} + d_\lambda)\omega'_{z,\zeta,\lambda}(\tilde{\eta}(z,\zeta,\lambda)) = n\omega(\tilde{\eta}(z,\zeta,\lambda)).$$

Consequently,
$$\begin{aligned}
&d_{\zeta,\lambda}[f(\zeta) \wedge \omega'_{z,\zeta,\lambda}(\tilde{\eta}(z,\zeta,\lambda)) \wedge \omega(\zeta)] \\
&= \bar{\partial}_\zeta f(\zeta) \wedge \omega'_{z,\zeta,\lambda}(\tilde{\eta}(z,\zeta,\lambda)) \wedge \omega(\zeta) \\
&\quad + (-1)^q n f(\zeta) \wedge \omega(\tilde{\eta}(z,\zeta,\lambda)) \wedge \omega(\zeta) \\
&\quad - \bar{\partial}_z[f(\zeta) \wedge \omega'_{z,\zeta,\lambda}(\tilde{\eta}(z,\zeta,\lambda)) \wedge \omega(\zeta)].
\end{aligned}$$

Since
$$\begin{aligned}
\partial(\Omega \times [0,1]) &= \partial\Omega \times [0,1] + (-1)^{\dim_{\mathbf{R}}\Omega} \times \partial([0,1]) \\
&= \partial\Omega \times [0,1] - \Omega \times \{0\} + \Omega \times \{1\},
\end{aligned}$$

it follows from Stokes' theorem that
$$\int_{\Omega \times [0,1]} \bar{\partial}_\zeta f(\zeta) \wedge \omega'_{z,\zeta,\lambda}(\tilde{\eta}(z,\zeta,\lambda)) \wedge \omega(\zeta) + (-1)^q \frac{(2\pi i)^n}{(n-1)!} R_\Omega f(z)$$
$$- \int_{\Omega \times [0,1]} \bar{\partial}_z[f(\zeta) \wedge \omega'_{z,\zeta,\lambda}(\tilde{\eta}(z,\zeta,\lambda)) \wedge \omega(\zeta)]$$

$$= \int_{\partial\Omega \times [0,1]} f(\zeta) \wedge \omega'_{z,\zeta,\lambda}(\tilde{\eta}(z,\zeta,\lambda)) \wedge \omega(\zeta)$$
$$- \int_{\Omega \times \{0\}} f(\zeta) \wedge \omega'_{z,\zeta,\lambda}(\tilde{\eta}(z,\zeta,\lambda)) \wedge \omega(\zeta)$$
$$+ \int_{\Omega \times \{1\}} f(\zeta) \wedge \omega'_{z,\zeta,\lambda}(\tilde{\eta}(z,\zeta,\lambda)) \wedge \omega(\zeta).$$

If $\zeta \in \partial\Omega$, then $\chi(\zeta) = 1$, and hence $\tilde{\eta}(z,\zeta,\lambda) = \eta(z,\zeta,\lambda)$. Thus we obtain
$$\int_{\partial\Omega \times [0,1]} f(\zeta) \wedge \omega'_{z,\zeta,\lambda}(\tilde{\eta}(z,\zeta,\lambda)) \wedge \omega(\zeta) = \frac{(2\pi i)^n}{(n-1)!} R_{\partial\Omega} f(z).$$

On the other hand we have
$$\int_{\Omega \times \{1\}} f(\zeta) \wedge \omega'_{z,\zeta,\lambda}(\tilde{\eta}(z,\zeta,\lambda)) \wedge \omega(\zeta) = \frac{(2\pi i)^n}{(n-1)!} B_\Omega f(z).$$

Since $\eta(z,\zeta,0) = \chi(\zeta)w(z,\zeta)/\widetilde{\Phi}(z,\zeta)$ is holomorphic with respect to z, we have

$$\int_{\Omega\times\{0\}} f(\zeta) \wedge \omega'_{z,\zeta,\lambda}(\tilde{\eta}(z,\zeta,\lambda)) \wedge \omega(\zeta) = \int_{\Omega} f(\zeta) \wedge \omega'_\zeta\left(\frac{\chi(\zeta)w(z,\zeta)}{\widetilde{\Phi}(z,\zeta)}\right) \wedge \omega(\zeta).$$

We set $T_q = (-1)^q(R_{\partial\Omega} + B_\Omega)$. For $z \in \Omega$, we have

$$R_\Omega f(z) = T_q f(z)$$

$$+ (-1)^{q+1}\frac{(n-1)!}{(2\pi i)^n}\left[\bar{\partial}_z \int_{\Omega\times[0,1]} f(\zeta) \wedge \omega'_{z,\zeta,\lambda}(\tilde{\eta}(z,\zeta,\lambda)) \wedge \omega(\zeta)\right.$$

$$+ \int_\Omega f(\zeta) \wedge \omega'_\zeta\left(\frac{\chi(\zeta)w(z,\zeta)}{\widetilde{\Phi}(z,\zeta)}\right) \wedge \omega(\zeta)$$

$$\left.+ \int_{\Omega\times[0,1]} \bar{\partial}_\zeta f(\zeta) \wedge \omega'_{z,\zeta,\lambda}(\tilde{\eta}(z,\zeta,\lambda)) \wedge \omega(\zeta)\right].$$

In the above equality, if we replace f by $\bar{\partial}f$, then we have

$$R_\Omega \bar{\partial} f(z) = T_{q+1}\bar{\partial} f(z) + (-1)^q \frac{(n-1)!}{(2\pi i)^n} \times$$

$$\left[\bar{\partial}_z \int_{\Omega\times[0,1]} \bar{\partial}_\zeta f(\zeta) \wedge \omega'_{z,\zeta,\lambda}(\tilde{\eta}(z,\zeta,\lambda)) \wedge \omega(\zeta)\right.$$

$$\left.+ \int_\Omega \bar{\partial}_\zeta f(\zeta) \wedge \omega'_\zeta\left(\frac{\chi(\zeta)w(z,\zeta)}{\widetilde{\Phi}(z,\zeta)}\right) \wedge \omega(\zeta)\right] \quad (3.22)$$

for $z \in \Omega$. By degree reasons we have for $q \geq 1$

$$\int_\Omega f(\zeta) \wedge \omega'_\zeta\left(\frac{\chi(\zeta)w(z,\zeta)}{\widetilde{\Phi}(z,\zeta)}\right) \wedge \omega(\zeta) = 0. \quad (3.23)$$

Since $w(z,\zeta)$ and $\widetilde{\Phi}(z,\zeta)$ are holomorphic with respect to z, we have

$$\bar{\partial}_z \int_\Omega f(\zeta) \wedge \omega'_\zeta\left(\frac{\chi(\zeta)w(z,\zeta)}{\widetilde{\Phi}(z,\zeta)}\right) \wedge \omega(\zeta) = 0. \quad (3.24)$$

It follows from (3.22), (3.23) and (3.24) that

$$\bar{\partial} R_\Omega f(z) = \bar{\partial} T_q f(z)$$

$$+(-1)^{q+1}\frac{(n-1)!}{(2\pi i)^n}\bar{\partial}_z\int_{\Omega\times[0,1]}\bar{\partial}_\zeta f(\zeta)\wedge\omega'_{z,\zeta,\lambda}(\tilde\eta(z,\zeta,\lambda))\wedge\omega(\zeta). \qquad (3.25)$$

First we prove (b). Suppose $q \geq 1$. Since the last integral in the right side of (3.22) equals 0 by degree reasons, it follows from (3.22) and (3.25) that

$$R_\Omega\bar{\partial}f + \bar{\partial}R_\Omega f = \bar{\partial}T_q f + T_{q+1}\bar{\partial}f.$$

By Corollary 3.3, we have $f = \bar{\partial}T_q f + T_{q+1}\bar{\partial}f$, which proves (b).

Next we prove (a). By Theorem 3.4, we have

$$f = L_{\partial\Omega}f + T_1\bar{\partial}f.$$

Since $\tilde\Phi(z,\zeta) = \Phi(z,\zeta)$ for $\zeta \in \partial\Omega$, we have

$$L_{\partial\Omega}f(z) = \frac{(n-1)!}{(2\pi i)^n}\int_{\partial\Omega}f(\zeta)\omega'_\zeta\left(\frac{\chi(\zeta)w(z,\zeta)}{\tilde\Phi(z,\zeta)}\right)\wedge\omega(\zeta).$$

Since

$$d_\zeta\omega'_\zeta\left(\frac{\chi(\zeta)w(z,\zeta)}{\tilde\Phi(z,\zeta)}\right) = n\omega_\zeta\left(\frac{\chi(\zeta)w(z,\zeta)}{\tilde\Phi(z,\zeta)}\right),$$

it follows from Stokes' theorem that

$$L_{\partial\Omega}f(z) = \frac{(n-1)!}{(2\pi i)^n}\left[\int_\Omega \bar{\partial}_\zeta f(\zeta)\omega'_\zeta\left(\frac{\chi(\zeta)w(z,\zeta)}{\tilde\Phi(z,\zeta)}\right)\wedge\omega(\zeta)\right] + L_\Omega f(z). \qquad (3.26)$$

Since $q = 0$, we obtain by degree reasons

$$\int_{\Omega\times[0,1]}\bar{\partial}_\zeta f(\zeta)\wedge\omega'_{z,\zeta,\lambda}(\tilde\eta(z,\zeta,\lambda))\wedge\omega(\zeta) = 0.$$

Hence we have together with (3.22) and (3.26)

$$R_\Omega\bar{\partial}f(z) = T_1\bar{\partial}f(z) + L_{\partial\Omega}f(z) - L_\Omega f(z).$$

This proves (a). In the general case, Theorem 3.12 is proved using the fact that f and $\bar{\partial}f$ can be approximated in L^1 norm by continuous functions with compact support. □

Definition 3.17 Let $\Omega \subset\subset \mathbf{C}^n$ be a strictly pseudoconvex domain with C^∞ boundary. Define

$$X := \{z \in \mathbf{C}^n \mid z_n = 0\}.$$

For $\zeta = (\zeta_1, \cdots, \zeta_n) \in \mathbf{C}^n$, define $\zeta' = (\zeta_1, \cdots, \zeta_{n-1})$. Further, we define

$$\partial_{\zeta'} := \sum_{j=1}^{n-1} \frac{\partial}{\partial \zeta_j} d\zeta_j, \quad \bar{\partial}_{\zeta'} := \sum_{j=1}^{n-1} \frac{\partial}{\partial \bar{\zeta}_j} d\bar{\zeta}_j,$$

$$d_{\zeta'} := \bar{\partial}_{\zeta'} + \partial_{\zeta'}, \quad \omega_{\zeta'}(\zeta) = d\zeta_1 \wedge \cdots \wedge d\zeta_{n-1},$$

and

$$(w'(z,\zeta)) := (w_1(z,\zeta), \cdots, w_{n-1}(z,\zeta)),$$

where $w(z,\zeta) = (w_1(z,\zeta), \cdots, w_n(z,\zeta))$ is the Leray map defined in Theorem 3.10. Define

$$\omega_{\zeta'}\left(\frac{\chi(\zeta)(w(z,\zeta))'}{\widetilde{\Phi}(z,\zeta)}\right) := \bigwedge_{j=1}^{n-1} \bar{\partial}_{\zeta'}\left(\frac{\chi(\zeta) w_j(z,\zeta)}{\widetilde{\Phi}(z,\zeta)}\right).$$

By Theorem 3.9, there exists a neighborhood $U_{\partial\Omega \setminus X}$ of $\partial\Omega \setminus X$ such that

$$\widetilde{\Phi}(z,\zeta) \neq 0 \quad (\zeta \in X \cap \overline{\Omega}, \ z \in \Omega \cup U_{\partial\Omega \setminus X}).$$

We set $V = X \cap \Omega$. For $f \in \mathcal{O}(V) \cap L^1(V)$ and $z \in \Omega \cup U_{\partial\Omega \setminus \overline{V}}$, define

$$Ef(z) := \frac{(n-1)!}{(2\pi i)^{n-1}} \int_V f(\zeta) \omega_{\zeta'}\left(\frac{\chi(\zeta)(w(z,\zeta))'}{\widetilde{\Phi}(z,\zeta)}\right) \wedge \omega_{\zeta'}(\zeta). \quad (3.27)$$

The following theorem follows from Theorem 3.12.

Theorem 3.13 *Let $f \in \mathcal{O}(V) \cap L^1(V)$. Then Ef is holomorphic in $\Omega \cup U_{\partial\Omega \setminus \overline{V}}$ and satisfies*

$$Ef(z) = f(z) \quad (z \in V).$$

Definition 3.18 Let $\Omega \subset\subset \mathbf{C}^n$ be a domain with C^∞ boundary. Suppose there exist a neighborhood $\widetilde{\Omega}$ of $\overline{\Omega}$ and a C^∞ function ρ in $\widetilde{\Omega}$ such that $\Omega = \{z \in \widetilde{\Omega} \mid \rho(z) < 0\}$. Let X be a k dimensional complex submanifold in a neighborhood of $\overline{\Omega}$. We set $V = X \cap \Omega$. Let $P \in \partial V$. Then there exist a neighborhood $U^{(P)}$ of P and a holomorphic coordinate system $f_1^{(P)}, \cdots, f_n^{(P)}$ in $U^{(P)}$ such that

$$U^{(P)} \cap X = \{z \in U^{(P)} \mid f_1^{(P)}(z) = \cdots = f_{n-k}^{(P)}(z) = 0\}.$$

We say that X intersects $\partial\Omega$ transversally if

$$df_1^{(P)}(P) \wedge \cdots \wedge df_{n-k}^{(P)}(P) \wedge d\rho(P) \neq 0$$

for every point $P \in \partial\Omega \cap X$. Moreover, in this case we say that the submanifold $X \cap \Omega$ of Ω is a submanifold in general position of Ω.

In what follows we assume that X intersects $\partial\Omega$ transversally.

Definition 3.19 Let U_2 be the neighborhood of $\partial\Omega$ in Theorem 3.8. For $(z,\zeta) \in \Omega \times U_2$, define

$$\Phi^*(z,\zeta) = \Phi(\zeta,z), \quad {}^*w(z,\zeta) = -w(\zeta,z),$$

$$^*w'(z,\zeta) = ({}^*w_1(z,\zeta), \cdots, {}^*w_{n-1}(z,\zeta)).$$

The following lemma was proved by Henkin-Leiterer [HER]. In their proof the transversality is not assumed. For simplicity, we assume that X intersects $\partial\Omega$ transversally in the following lemma.

Lemma 3.22 *If f is a bounded holomorphic function in $V = X \cap \Omega$, then*

$$Ef(z)$$
$$= z_n \frac{(-1)^{n-1}}{(2\pi i)^{n-1}} \int_{X \cap \Omega} f(\zeta) det_{1,n-1}\left(\frac{{}^*w(z,\zeta)}{\Phi^*(z,\zeta)}, \bar{\partial}_{\zeta'} \frac{\chi(\zeta)w(z,\zeta)}{\tilde{\Phi}(z,\zeta)}\right) \wedge \omega_{\zeta'}(\zeta)$$

for $z \in \partial\Omega \backslash \overline{V}$.

Proof. By applying the expansion formula of the determinant to the n-th column, we have

$$(-1)^n det_{1,n-1}\left(\frac{{}^*w}{\Phi^*}, \bar{\partial}_{\zeta'} \frac{\chi w}{\tilde{\Phi}}\right)$$

$$= (-1)^n \begin{vmatrix} \frac{{}^*w_1}{\Phi^*} & \frac{{}^*w_2}{\Phi^*} & \cdots & \frac{{}^*w_n}{\Phi^*} \\ \bar{\partial}_{\zeta'} \frac{\chi w_1}{\tilde{\Phi}} & \bar{\partial}_{\zeta'} \frac{\chi w_2}{\tilde{\Phi}} & \cdots & \bar{\partial}_{\zeta'} \frac{\chi w_n}{\tilde{\Phi}} \\ \cdots & \cdots & \cdots & \cdots \\ \bar{\partial}_{\zeta'} \frac{\chi w_1}{\tilde{\Phi}} & \bar{\partial}_{\zeta'} \frac{\chi w_2}{\tilde{\Phi}} & \cdots & \bar{\partial}_{\zeta'} \frac{\chi w_n}{\tilde{\Phi}} \end{vmatrix}$$

$$= -\frac{{}^*w_n}{\Phi^*} det_{n-1}\left(\bar{\partial}_{\zeta'} \frac{\chi w'}{\tilde{\Phi}}\right)$$

$$+ (n-1)\bar{\partial}_{\zeta'} \frac{\chi w_n}{\tilde{\Phi}} \wedge det_{1,n-2}\left(\frac{{}^*w'}{\Phi^*}, \bar{\partial}_{\zeta'} \frac{\chi w'}{\tilde{\Phi}}\right).$$

We have by the definition of the determinant

$$det_{n-1}\left(\bar{\partial}_{\zeta'} \frac{\chi w'}{\tilde{\Phi}}\right) = (n-1)! \bigwedge_{j=1}^{n-1} \bar{\partial}_{\zeta'} \frac{\chi w'_j}{\tilde{\Phi}} = (n-1)! \omega_{\zeta'}\left(\frac{\chi w'}{\tilde{\Phi}}\right). \quad (3.28)$$

For $\zeta \in V$ we have
$$\sum_{j=1}^{n} \frac{{}^*w_j(z,\zeta)(\zeta_j - z_j)}{\Phi^*(z,\zeta)} = -1.$$

Consequently,
$$-z_n \frac{{}^*w_n}{\Phi^*} = -1 - \sum_{j=1}^{n-1}(\zeta_j - z_j)\frac{{}^*w_j}{\Phi^*}. \qquad (3.29)$$

Since
$$\sum_{j=1}^{n} \frac{\chi w_j(\zeta_j - z_j)}{\widetilde{\Phi}} + \frac{\chi \Phi}{\widetilde{\Phi}} = 0,$$

it follows that for $\zeta \in V$
$$z_n \bar{\partial}_{\zeta'} \frac{\chi w_n}{\widetilde{\Phi}} = \sum_{j=1}^{n-1}(\zeta_j - z_j)\bar{\partial}_{\zeta'}\frac{\chi w_j}{\widetilde{\Phi}} + \bar{\partial}_{\zeta'}\frac{\chi \Phi}{\widetilde{\Phi}}.$$

Therefore, together with (3.28) and (3.29), we obtain for $\zeta \in V$,
$$z_n(-1)^n \det{}_{1,n-1}\left(\frac{{}^*w}{\Phi^*}, \bar{\partial}_{\zeta'}\frac{\chi w}{\widetilde{\Phi}}\right)$$
$$= \left(-1 - \sum_{j=1}^{n-1}(\zeta_j - z_j)\frac{{}^*w_j}{\Phi^*}\right)(n-1)!\omega_{\zeta'}\left(\frac{\chi w'}{\widetilde{\Phi}}\right)$$
$$+ (n-1)\sum_{j=1}^{n-1}(\zeta_j - z_j)\bar{\partial}_{\zeta'}\frac{\chi w_j}{\widetilde{\Phi}} \wedge \det{}_{1,n-2}\left(\frac{{}^*w'}{\Phi^*}, \bar{\partial}_{\zeta'}\frac{\chi w'}{\widetilde{\Phi}}\right)$$
$$+ (n-1)\bar{\partial}_{\zeta'}\frac{\chi \Phi}{\widetilde{\Phi}} \wedge \det{}_{1,n-2}\left(\frac{{}^*w'}{\Phi^*}, \bar{\partial}_{\zeta'}\frac{\chi w'}{\widetilde{\Phi}}\right).$$

On the other hand we have
$$\bar{\partial}_{\zeta'}\frac{\chi w_j}{\widetilde{\Phi}} \wedge \det{}_{1,n-2}\left(\frac{{}^*w'}{\Phi^*}, \bar{\partial}_{\zeta'}\frac{\chi w'}{\widetilde{\Phi}}\right)$$
$$= \bar{\partial}_{\zeta'}\frac{\chi w_j}{\widetilde{\Phi}} \wedge \sum_{\sigma(1)=j}\operatorname{sgn}(\sigma)\frac{{}^*w_{\sigma(1)}}{\Phi^*}\bar{\partial}_{\zeta'}\frac{\chi w_{\sigma(2)}}{\widetilde{\Phi}} \wedge \cdots \wedge \bar{\partial}_{\zeta'}\frac{\chi w_{\sigma(n-1)}}{\widetilde{\Phi}}$$

$$= (n-2)! \frac{^*w_j}{\Phi^*} \bar{\partial}_{\zeta'} \frac{\chi w_1}{\widetilde{\Phi}} \wedge \cdots \wedge \bar{\partial}_{\zeta'} \frac{\chi w_{n-1}}{\widetilde{\Phi}}$$

$$= (n-2)! \frac{^*w_j}{\Phi^*} \omega_{\zeta'} \left(\frac{\chi w'}{\widetilde{\Phi}} \right).$$

Hence we have for $\zeta \in V$

$$z_n (-1)^n \det{}_{1,n-1} \left(\frac{^*w(z,\zeta)}{\Phi^*(z,\zeta)}, \bar{\partial}_{\zeta'} \frac{\chi(\zeta) w(z,\zeta)}{\widetilde{\Phi}(z,\zeta)} \right)$$

$$= -(n-1)! \omega_{\zeta'} \left(\frac{\chi w'}{\widetilde{\Phi}} \right)$$

$$+ (n-1) \bar{\partial}_{\zeta'} \frac{\chi \Phi}{\widetilde{\Phi}} \wedge \det{}_{1,n-2} \left(\frac{^*w'}{\Phi^*}, \bar{\partial}_{\zeta'} \frac{\chi w'}{\widetilde{\Phi}} \right).$$

Now we set

$$I(z) = \int_V f(\zeta) \bar{\partial}_{\zeta'} \frac{\chi \Phi}{\widetilde{\Phi}} \wedge \det{}_{1,n-2} \left(\frac{^*w'}{\Phi^*}, \bar{\partial}_{\zeta'} \frac{\chi w'}{\widetilde{\Phi}} \right) \wedge \omega_{\zeta'}(\zeta).$$

Since $^*w(z,\zeta)$, $\Phi^*(z,\zeta)$ and $f(\zeta)$ are holomorphic with respect to ζ, it follows from Stokes' theorem that

$$I(z) = \int_V \bar{\partial}_{\zeta'} \left\{ f(\zeta) \frac{\chi \Phi}{\widetilde{\Phi}} \wedge \det{}_{1,n-2} \left(\frac{^*w'}{\Phi^*}, \bar{\partial}_{\zeta'} \frac{\chi w'}{\widetilde{\Phi}} \right) \right\} \wedge \omega_{\zeta'}(\zeta)$$

$$= \int_{\partial V} f(\zeta) \frac{\chi \Phi}{\widetilde{\Phi}} \det{}_{1,n-2} \left(\frac{^*w'}{\Phi^*}, \bar{\partial}_{\zeta'} \frac{\chi w'}{\widetilde{\Phi}} \right) \wedge \omega_{\zeta'}(\zeta).$$

Since $\Phi(z,\zeta) = \widetilde{\Phi}(z,\zeta)$ and $\chi(\zeta) = 1$ for $\zeta \in \partial \Omega$, it follows from Stokes' theorem that

$$I(z) = \int_V \bar{\partial}_{\zeta'} \left\{ f(\zeta) \det{}_{1,n-2} \left(\frac{^*w'}{\Phi^*}, \bar{\partial}_{\zeta'} \frac{\chi w'}{\widetilde{\Phi}} \right) \right\} \wedge \omega_{\zeta'}(\zeta) = 0,$$

which completes the proof of Lemma 3.22. □

Lemma 3.23 *There exists a constant $C > 0$ such that for all $z \in \partial \Omega \setminus \overline{V}$ the following estimates hold:*

(a)

$$\int_{\zeta \in V} \frac{dV_{n-1}(\zeta)}{|\zeta - z|^{2n-1}} \leq \frac{C}{|z_n|}.$$

(b)
$$\int_{\zeta \in V \cap U_2} \frac{dV_{n-1}(\zeta)}{|\widetilde{\Phi}(z,\zeta)||\Phi^*(z,\zeta)||\zeta - z|^{2n-4}} \leq \frac{C}{|z_n|}.$$

(c)
$$\int_{\zeta \in V \cap U_2} \frac{dV_{n-1}(\zeta)}{|\widetilde{\Phi}(z,\zeta)|^2|\Phi^*(z,\zeta)||\zeta - z|^{2n-5}} \leq \frac{C}{|z_n|},$$

where dV_{n-1} denotes the Lebesgue measure on $\mathbb{C}^{n-1}$.

Proof. In what follows we denote by C any constant which depends only on Ω and V.

(a) we have
$$\int_{\zeta \in V} \frac{dV_{n-1}(\zeta)}{|\zeta - z|^{2n-1}} \leq \int_V \frac{dV_{n-1}(\zeta)}{(|z_n|^2 + |\zeta' - z'|^2)^2 |\zeta' - z'|^{2n-5}}.$$

We set $\zeta_j - z_j = t_j + it_{j+n-1}$ for $j = 1, \cdots, n-1$. Then
$$\int_{\zeta \in V} \frac{dV_{n-1}(\zeta)}{|\zeta - z|^{2n-1}} \leq \int_{|t| \leq C} \frac{dt_1 \cdots dt_{2n-2}}{(|z_n|^2 + |t|^2)^2 |t|^{2n-5}}$$
$$\leq \int_{r \leq C} \frac{r^{2n-3} dr}{(|z_n|^2 + r^2)^2 r^{2n-5}} \leq \frac{C}{|z_n|}.$$

This proves (a).

(b) We set $\zeta' = (\zeta_1, \cdots, \zeta_{n-1})$ and $z' = (z_1, \cdots, z_{n-1})$. Let $z^0 \in \partial V$. We may assume that $(\partial \rho / \partial z_1)(z^0) \neq 0$. Let U be a neighborhood of z^0 such that $(\partial \rho / \partial z_1)(z) \neq 0$ for $z \in \overline{U}$. For $z, \zeta \in U$, define
$$t_{2j-1}(\zeta) = \operatorname{Re}(\zeta_j - z_j), \quad t_{2j}(\zeta) = \operatorname{Im}(\zeta_j - z_j) \quad j = 2, \cdots, n-1,$$
$$t_1(\zeta) = \rho(\zeta) - \rho(z), \quad t_2(\zeta) = \operatorname{Im} \Phi(z, \zeta).$$

Then
$$\frac{\partial t_2}{\partial x_{2j}}(z) = -\frac{1}{2} \frac{\partial \rho}{\partial x_{2j-1}}(z), \quad \frac{\partial t_2}{\partial x_{2j-1}}(z) = \frac{1}{2} \frac{\partial \rho}{\partial x_{2j}}(z).$$

Consequently,
$$\frac{\partial(t_1, \cdots, t_{2n-2})}{\partial(x_1, \cdots, x_{2n-2})} = 2 \left| \frac{\partial \rho}{\partial \zeta_1} \right|^2 \neq 0$$

Hence $t_1, \cdots, t_{2n-2}$ form a local coordinate system in U.

Since $\rho(z) = 0$, it follows from (3.19) that

$$|\Phi^*(z,\zeta)| \geq |\Phi(\zeta,z)| \geq C|F(\zeta,z)| \geq C(|t_1| + |\zeta - z|^2),$$

$$|\widetilde{\Phi}(z,\zeta)| \geq C(|t_1| + |\zeta - z|^2).$$

Hence we have

$$\int_{\zeta \in X \cap \Omega \cap U_2} \frac{dV_{n-1}(\zeta)}{|\widetilde{\Phi}(z,\zeta)||\Phi^*(z,\zeta)||\zeta - z|^{2n-4}}$$
$$\leq \int_{|t| \leq C} \frac{dt_1 \cdots dt_{2n-2}}{(|z_n|^2 + |t_1| + |t|^2)^2 |t|^{2n-4}}.$$

We set $t' = (t_2, \cdots, t_{2n-2})$. Then

$$\int_{|t| \leq C} \frac{dt_1 \cdots dt_{2n-2}}{(|z_n|^2 + |t_1| + |t|^2)^2 |t|^{2n-4}} \leq \int_{|t'| \leq C} \frac{dt_2 \cdots dt_{2n-2}}{(|z_n|^2 + |t'|^2)|t'|^{2n-4}}$$
$$\leq \frac{C}{|z_n|} \int_0^\infty \frac{dy}{1+y^2} \leq \frac{C}{|z_n|}.$$

This proves (b).

(c) We have

$$|\Phi^*(z,\zeta)| = |\Phi(\zeta,z)| \geq C(|t_1| + |t_2| + |\zeta - z|^2),$$

$$|\widetilde{\Phi}(z,\zeta)| \geq C(|t_1| + |t_2| + |\zeta - z|^2).$$

We set $t'' = (t_3, \cdots, t_{2n-2})$. Then we have

$$\int_{\zeta \in V \cap U_2} \frac{dV_{n-1}(\zeta)}{|\widetilde{\Phi}(z,\zeta)|^2 |\Phi^*(z,\zeta)||\zeta - z|^{2n-5}}$$
$$\leq \int_{|t| \leq C} \frac{dt_1 \cdots dt_{2n-2}}{(|z_n|^2 + |t_1| + |t_2| + |t|^2)^3 |t|^{2n-5}}$$
$$\leq \int_{|t''| \leq C} \frac{dt_3 \cdots dt_{2n-2}}{(|z_n|^2 + |t''|^2)^3 |t''|^{2n-5}}$$
$$\leq \frac{C}{|z_n|}.$$

This proves (c). □

Theorem 3.14 *Let $\Omega \subset\subset \mathbf{C}^n$ be a strictly pseudoconvex domain with C^∞ boundary and let $X = \{z \in \mathbf{C}^n \mid z_n = 0\}$, $V = \Omega \cap X$. Assume that X*

intersects $\partial\Omega$ transversally. If f is a bounded holomorphic function in V, then there exists a constant $C > 0$ such that

$$|Ef(z)| \leq C \sup_{\zeta \in V} |f(\zeta)|$$

for $z \in \partial\Omega \backslash \overline{V}$.

Proof. We set $U_3 = \{z \in U \mid |\rho(z)| < \varepsilon_0\}$. Since $\chi = 1$ in U_3, we have

$$\det{}_{1,n-1}\left(\frac{{}^*w}{\Phi^*}, \bar{\partial}_{\zeta'}\frac{\chi w}{\widetilde{\Phi}}\right) = \det{}_{1,n-1}\left(\frac{{}^*w}{\Phi^*}, \frac{\bar{\partial}_{\zeta'}w}{\widetilde{\Phi}} - w\frac{\bar{\partial}_{\zeta'}\widetilde{\Phi}}{\widetilde{\Phi}^2}\right).$$

It follows from Lemma 3.4 that any determinant which contains $w\frac{\bar{\partial}_{\zeta'}\widetilde{\Phi}}{\widetilde{\Phi}^2}$ in two columns equals 0. Hence we have

$$\det{}_{1,n-1}\left(\frac{{}^*w}{\Phi^*}, \bar{\partial}_{\zeta'}\frac{\chi w}{\widetilde{\Phi}}\right)$$
$$= c_1 \det{}_{1,n-1}\left(\frac{{}^*w}{\Phi^*}, \frac{\bar{\partial}_{\zeta'}w}{\widetilde{\Phi}}\right) + c_2 \det{}_{1,1,n-2}\left(\frac{{}^*w}{\Phi^*}, w\frac{\bar{\partial}_{\zeta'}\widetilde{\Phi}}{\widetilde{\Phi}^2}, \frac{\bar{\partial}_{\zeta'}w}{\widetilde{\Phi}}\right).$$

Since

$${}^*w(z,\zeta) + w(z,\zeta) = -w(\zeta,z) + w(z,\zeta) = O(|\zeta - z|),$$

we have

$$\det{}_{1,1,n-2}\left(\frac{{}^*w}{\Phi^*}, w\frac{\bar{\partial}_{\zeta'}\widetilde{\Phi}}{\widetilde{\Phi}^2}, \frac{\bar{\partial}_{\zeta'}w}{\widetilde{\Phi}}\right)$$
$$= \det{}_{1,1,n-2}\left(\frac{O(|\zeta - z|)}{\Phi^*}, w\frac{\bar{\partial}_{\zeta'}\widetilde{\Phi}}{\widetilde{\Phi}^2}, \frac{\bar{\partial}_{\zeta'}w}{\widetilde{\Phi}}\right).$$

It follows from (3.19) that

$$|\widetilde{\Phi}(z,\zeta)| \geq \alpha|\zeta - z|^2, \quad |\Phi^*(z,\zeta)| \geq \alpha|\zeta - z|^2.$$

Hence we obtain

$$\left|\det{}_{1,n-1}\left(\frac{{}^*w}{\Phi^*}, \frac{\bar{\partial}_{\zeta'}w}{\widetilde{\Phi}}\right)\right| \leq \frac{C}{|\Phi^*||\widetilde{\Phi}|^{n-1}} \leq \frac{C}{|\Phi^*||\widetilde{\Phi}||\zeta - z|^{2n-4}}$$

and

$$\left|\det{}_{1,1,n-2}\left(\frac{O(|\zeta-z|)}{\Phi^*}, w\frac{\bar{\partial}_{\zeta'}\widetilde{\Phi}}{\widetilde{\Phi}^2}, \frac{\bar{\partial}_{\zeta'}w}{\widetilde{\Phi}}\right)\right|$$
$$\leq \frac{C|\zeta-z|}{|\Phi^*||\widetilde{\Phi}|^n} \leq \frac{C}{|\Phi^*||\widetilde{\Phi}|^2|\zeta-z|^{2n-5}}.$$

Using Lemma 3.23, we have the desired inequality. □

Theorem 3.15 *Let $\Omega \subset\subset \mathbf{C}^n$ be a strictly pseudoconvex domain with C^∞ boundary and let $X = \{z \in \mathbf{C}^n \mid z_n = 0\}$, $V = \Omega \cap X$. Assume that X intersects $\partial\Omega$ transversally. If f is a bounded holomorphic function in V, then Ef is a bounded holomorphic function in Ω satisfying $Ef = f$ on V. Moreover, there exists a constant $C > 0$ such that*

$$\sup_{z\in\Omega}|F(z)| \leq C\sup_{z\in V}|f(z)|.$$

Proof. By Theorem 3.13, Ef is holomorphic in $\Omega \cup U_{\partial\Omega\setminus X}$. Let $X_a = \{z \in \mathbf{C}^n \mid z_n = a\}$. If $a \neq 0$, then Ef is holomorphic in the closure of $\Omega \cap X_a$. Hence by the maximum principle, $|Ef|$ has the maximum in $\partial(\Omega \cap X_a)$. It follows from Theorem 3.14 that $|Ef(z)| \leq C\|f\|_V$ for $z \in \partial(\Omega \cap X_a)$, which means that $|Ef(z)| \leq C\|f\|_V$ for $z \in \overline{\Omega}\setminus X$. Since Ef is holomorphic in Ω, $|Ef(z)| \leq C\|f\|_V$ for $z \in \overline{\Omega}\setminus(\partial\Omega \cap X)$. Hence Ef is bounded in Ω. □

Next we prove bounded and continuous extensions of holomorphic functions from submanifolds in general position of strictly pseudoconvex domains in $\mathbf{C}^n$ with smooth boundary by the method of Henkin [HEN3]. For simplicity, we assume that the codimension of submanifolds is one. The general case can be proved in the same way.

Definition 3.20 Let D be an open set in a complex manifold. We denote by $H^\infty(D)$ the Banach space of all bounded holomorphic functions in D. We also denote by $A(D)$ the Banach space of all continuous functions on $\overline{D}$ that are holomorphic in D.

Let $\Omega \subset\subset \mathbf{C}^n$ be a strictly pseudoconvex domain with C^∞ boundary and let H be a holomorphic function in a neighborhood $\widetilde{\Omega}$ of $\overline{\Omega}$. We set $X = \{z \in \widetilde{\Omega} \mid H(z) = 0\}$ and $V = X \cap \Omega$. Assume that X intersects $\partial\Omega$ transversally. By Lemma 3.16, there are holomorphic functions $h_1, \cdots, h_n$ in a neighborhood of $\overline{\Omega} \times \overline{\Omega}$ such that for $z, \zeta \in \overline{\Omega}$

$$H(z) - H(\zeta) = \sum_{i=1}^n (z_i - \zeta_i) h_i(z, \zeta).$$

By Theorem 3.10 there exist C^∞ functions $w_1(z,\zeta), \cdots, w_n(z,\zeta)$ in a neighborhood of $\overline{\Omega} \times \partial\Omega$ holomorphic with respect to z such that

$$\Phi(z,\zeta) = \sum_{j=1}^n w_j(z,\zeta)(z_j - \zeta_j),$$

where $\Phi(z,\zeta)$ is the function defined in Theorem 3.8. Define

$$\alpha(z,\zeta) = -(-1)^{n(n+1)/2} \det(w_j, h_j, \overbrace{\bar{\partial}_\zeta w_j, \cdots, \bar{\partial}_\zeta w_j}^{n-2}),$$

$$\beta(\zeta) = \left(\sum_{j=1}^n \left|\frac{\partial H}{\partial \zeta_j}(\zeta)\right|^2\right)^{-2} \det\left(\frac{\partial H}{\partial \zeta_j}(\zeta), \overbrace{d\zeta_j, \cdots, d\zeta_j}^{n-1}\right).$$

Define

$$K(z,\zeta) = \alpha(z,\zeta) \wedge \beta(\zeta).$$

Then Stout [STO] and Hatziafratis [HAT1] proved the following theorem. We omit the proof.

Theorem 3.16 *Let $f \in A(V)$. Then*

$$f(z) = \int_{\partial V} f(\zeta) \frac{K(z,\zeta)}{\Phi(z,\zeta)^{n-1}}$$

for all $z \in V$.

Remark 3.4 *Stout [STO] obtained the integral formula on submanifolds of one codimension, and then Hatziafratis [HAT1] extended the integral formula obtained by Stout to the formula on submanifolds of any codimension.*

In what follows, we prove bounded and continuous extensions by following Henkin [HEN3]. Let $\Omega = \{z \mid \rho(z) < 0\}$ be a strictly convex domain in $\mathbf{C}^n$ with C^∞ boundary and let $X = \{z_n = 0\}$, $V = X \cap \Omega$. Assume that X intersects $\partial\Omega$ transversally. We may assume that $0 \in \Omega$. Let f be a bounded holomorphic function in V. It follows from Fatou's theorem (see Stein [STE]) that there is a bounded measurable function f^* on ∂V such that

$$f^*(\zeta) = \lim_{\theta \uparrow 1} f(\theta\zeta)$$

for almost all $\zeta \in \partial V$. Then we have the following lemma.

Lemma 3.24 *For $z \in V$, we have*
$$f(z) = \int_{\zeta \in \partial V} f^*(\zeta) \frac{K(z,\zeta)}{\Phi(z,\zeta)^{n-1}}.$$

Proof. Let $0 < \theta < 1$. We set $F(z) = f(\theta z)$ for $z \in \overline{V}$. Then F is holomorphic in a neighborhood of $\overline{V}$. We fix $z_0 \in V$. It follows from Theorem 3.16 that
$$F(z_0) = \int_{\zeta \in \partial V} F(\zeta) \frac{K(z_0,\zeta)}{\Phi(z_0,\zeta)^{n-1}}.$$

Consequently,
$$f(\frac{z_0}{\theta}) = \int_{\zeta \in \partial V} f(\theta\zeta) \frac{K(z_0,\zeta)}{\Phi(z_0,\zeta)^{n-1}}.$$

By Lebesgue's dominated convergence theorem, we may pass the limit under the integral sign as $\theta \to 1$ in the above equality. $\square$

Definition 3.21 Let f be a bounded holomorphic function in V and let f^* be the boundary value of f on ∂V. For $z \in \Omega$, define
$$E_1 f(z) = \int_{\partial V} f^*(\zeta) \frac{K(z,\zeta)}{\Phi(z,\zeta)^{n-1}}. \tag{3.30}$$

Then $E_1 f$ is holomorphic in a neighborhood of $\overline{\Omega} \setminus \partial V$ and $E_1 f|_V = f$. If $f \in A(V)$, then $f|_{\partial V} = f^*$.

Let $\Omega = \{z \mid \rho(z) < 0\}$ be a strictly convex domain in $\mathbb{C}^n$ with C^∞ boundary and let $X = \{z_n = 0\}$, $V = X \cap \Omega$. Assume that X intersects $\partial \Omega$ transversally. We fix a point $z^* \in \partial V$. Suppose
$$\frac{\partial \rho}{\partial z_1}(z^*) \neq 0.$$

Then there exists a constant $\sigma_1 > 0$ such that $\frac{\partial \rho}{\partial z_1}(z) \neq 0$ for all $z \in \overline{B}(z^*, \sigma_1)$.

In this setting, we prove Lemma 3.25, Lemma 3.26, Lemma 3.27, Theorem 3.17 and Theorem 3.18.

Lemma 3.25 *For $z \in B(z^*, \sigma_1)$, we consider a system of equations for $\zeta^* = (\zeta_1^*, \cdots, \zeta_n^*)$ of the following form:*
$$\sum_{i=1}^n \frac{\partial \rho}{\partial \zeta_i}(\zeta^*)(\zeta_i^* - z_i) = 0, \tag{3.31}$$

$$\zeta_i^* = z_i \quad (i = 2, \cdots, n-1), \quad \zeta_n^* = 0. \tag{3.32}$$

Then there exist positive constants σ_2, γ_1 and γ_2 depending only on Ω and V, such that for any $z \in B(z^*, \sigma_2)$ there exists a unique solution $\zeta^* = \zeta^*(z)$ of the system of equations (3.31) and (3.32) which belongs to the set $B(z^*, \sigma_2) \cap X$. Moreover, the point ζ^* has the following properties:

$$|z - \zeta^*|^2 \leq \frac{1}{\gamma_1} \{\rho(z) - \rho(\zeta^*)\},$$

$$|z - \zeta^*|^2 \geq |z_n|^2 \geq \gamma_2 \{\rho(z) - \rho(\zeta^*)\},$$

$$\zeta^* = z \quad \text{for} \quad z \in B(z^*, \sigma_2) \cap X.$$

Proof. (3.31) can be written

$$\zeta_1^* = z_1 + \frac{\partial \rho}{\partial z_n}(\zeta^*) \left(\frac{\partial \rho}{\partial z_1}(\zeta^*)\right)^{-1} z_n.$$

Define

$$g(\zeta) = \frac{\partial \rho}{\partial z_n}(\zeta) \left(\frac{\partial}{\partial z_1}(\zeta)\right)^{-1}.$$

We choose $\sigma_2 > 0$ so small that $|dg(\zeta)||z_n| \leq 1/2$ for $z, \zeta \in B(z^*, \sigma_2)$. Define $\{\zeta^{(j)}\}$ by recurrence such that

$$\zeta_1^{(1)} = z_1,$$
$$\zeta^{(j)} = (\zeta_1^{(j)}, z_2, \cdots, z_{n-1}, 0),$$
$$\zeta_1^{(j+1)} = z_1 + g(\zeta^{(j)}) z_n.$$

Then $|\zeta_1^{(j)} - \zeta_1^{(j-1)}| \leq \frac{1}{2}|\zeta_1^{(j-1)} - \zeta_1^{(j-2)}|$, and hence $\{\zeta^{(j)}\}$ converges. Let $\lim_{j \to \infty} \zeta^{(j)} = \zeta^*$. Then ζ^* satisfies (3.31) and (3.32). The strict convexity of ρ yields for some positive constants γ_1 and C_1

$$\rho(\zeta^*) - \rho(z) + \gamma_1 |\zeta^* - z|^2 \leq 2\mathrm{Re} \sum_{j=1}^n \frac{\partial \rho}{\partial \zeta_j}(\zeta^*)(\zeta_j^* - z_j) = 0$$

$$\rho(\zeta^*) - \rho(z) + C_1 |\zeta^* - z|^2 \geq 2\mathrm{Re} \sum_{j=1}^n \frac{\partial \rho}{\partial \zeta_j}(\zeta^*)(\zeta_j^* - z_j) = 0.$$

Since $\zeta_1^* = z_1 + g(\zeta^*)z_n$, there exists a constant C_2 such that $|\zeta_1^* - z_1| \leq C_2|z_n|$. Hence ζ^* satisfies the desired inequalities. If there are two solutions ζ^* and $\tilde\zeta^*$. Then we have

$$|\zeta_1^* - \tilde\zeta_1^*| \leq \|dg\| |\zeta_1^* - \tilde\zeta_1^*| |z_n| \leq 1/2 |\zeta_1^* - \tilde\zeta_1^*|,$$

which implies that $\zeta^* = \tilde\zeta^*$. □

Lemma 3.26 *For any $z^* \in \partial V$ and any $z \in (\partial\Omega \backslash \partial V) \cap B(z^*, \sigma_2)$ we have*

$$\left| \frac{d(E_1 f)(\zeta^* + \lambda(z - \zeta^*))}{d\lambda} \right|_{\lambda=1} \leq C \sup_{\zeta \in V} |f(\zeta)|$$

where $\zeta^* = \zeta^*(z)$ and σ_2 are from Lemma 3.25 and the constant C depends only on Ω and V.

Proof. Since

$$\sum_{j=1}^n \frac{\partial\rho}{\partial\zeta_i}(\zeta^*)(\zeta_i^* - z_i) = 0$$

and

$$\frac{\partial \Phi}{\partial z_i}(z, \zeta^*) = -2 \frac{\partial\rho}{\partial\zeta_i}(\zeta^*) + O(|\zeta^* - z|),$$

we have

$$\left| \sum_{i=1}^n \frac{\partial \Phi}{\partial z_i}(z, \zeta)(\zeta_i^* - z_i) \right| = \left| \sum_{i=1}^n \left(\frac{\partial \Phi}{\partial z_i}(z, \zeta) + 2\frac{\partial\rho}{\partial\zeta_i}(\zeta^*) \right)(\zeta_i^* - z_i) \right|$$

$$\leq \left| \sum_{i=1}^n \left(\frac{\partial \Phi}{\partial z_i}(z, \zeta) - \frac{\partial \Phi}{\partial z_i}(z, \zeta^*) + O(|\zeta^* - z|) \right)(\zeta_i^* - z_i) \right|$$

$$\leq C\varepsilon(|\zeta - z| + \varepsilon),$$

where $\varepsilon = |z_n|$. Then we have

$$\left| \frac{d(E_1 f)(\zeta^* + \lambda(z - \zeta^*))}{d\lambda} \right|_{\lambda=1} = \left| \frac{d(E_1 f)(z + \lambda(z - \zeta^*))}{d\lambda} \right|_{\lambda=0}$$

$$\leq C \int_{\partial V \cap B(z^*, \sigma_2)} \frac{|f^*(\zeta)||z - \zeta^*|}{|\Phi(z, \zeta)|^n} d\sigma_{2n-3}(\zeta)$$

$$+ C \int_{\partial V \cap B(z^*, \sigma_2)} \frac{|f^*(\zeta)|\varepsilon(|\zeta - z| + \varepsilon)}{|\Phi(z, \zeta)|^n} d\sigma_{2n-3}(\zeta).$$

We choose a local coordinate system $(t_1, t_2, \cdots, t_{2n})$ in $B(z^*, \sigma_2)$ such that $t_1 + it_2 = \rho(\zeta) - \rho(z) + i\mathrm{Im}\,\Phi(z, \zeta)$. Then we obtain

$$\left|\frac{d(E_1 f)(\zeta^* + \lambda(z - \zeta^*))}{d\lambda}\right|_{\lambda=1} \leq C \sup_{\zeta \in V} |f(\zeta)|.$$

$\square$

Theorem 3.17 *If $f \in H^\infty(V)$, then $E_1 f \in H^\infty(\Omega)$. Moreover, $E_1 : H^\infty(V) \to H^\infty(\Omega)$ defines a bounded linear operator.*

Proof. Let $\sigma > 0$. Let f be a bounded holomorphic function in V. Since $E_1 f$ is holomorphic in $\overline{\Omega} \setminus \partial V$, it is sufficient to prove that

$$\sup_{z \in \partial\Omega \setminus \partial V} |E_1 f(z)| \leq C \sup_{\zeta \in V} |f(\zeta)|.$$

It is easily proved that $\sup_{z \in \partial\Omega \setminus (\partial V)_\sigma} |E_1 f(z)| \leq C \sup_{\zeta \in V} |f(\zeta)|$, where $(\partial V)_\sigma$ is the σ-neighborhood of ∂V. Therefore, it is sufficient to show that

$$\sup_{z \in \{(\partial V)_\sigma \setminus \partial V\} \cap \partial\Omega} |E_1 f(z)| \leq C \sup_{\zeta \in V} |f(\zeta)|.$$

Let $z \in \{(\partial V)_\sigma \setminus \partial V\} \cap \partial\Omega$. We set

$$\Delta = \{\lambda \in \mathbb{C} \mid z(\lambda) = \zeta^* + \lambda(z - \zeta^*) \in \Omega\}.$$

Then Δ is a convex domain containing $\lambda = 0$ since Ω is convex. Since $\rho(z) = 0$, we have by Lemma 3.25

$$\frac{\varepsilon^2}{\gamma_1} \leq -\rho(\zeta^*) \leq \frac{\varepsilon^2}{\gamma_2}.$$

If $\lambda \in \partial\Delta$, then $z(\lambda) \in \partial\Omega$, and hence $\rho(z(\lambda)) = 0$. Then

$$|z(\lambda) - \zeta^*| \leq \frac{1}{\sqrt{\gamma_1}}\sqrt{\rho(z) - \rho(\zeta^*)} \leq \frac{\varepsilon}{\sqrt{\gamma_1 \gamma_2}}$$

for $\lambda \in \partial\Delta$. Consequently,

$$|z(\lambda) - z^*| \leq |z(\lambda) - \zeta^*| + |\zeta^* - z^*|$$

$$\leq \frac{\sigma}{\sqrt{\gamma_1 \gamma_2}} + \frac{\sigma_2}{4}.$$

We impose the further assumption that the constant $\sigma < \sigma_2 \sqrt{\gamma_1 \gamma_2}/4$. Then $|z(\lambda) - z^*| < \sigma_2/2$. Then $\zeta^* = \zeta^*(z)$ satisfies (3.31) and (3.32) for $z(\lambda)$ with

$\lambda \in \partial\Delta$, and hence $\zeta(z(\lambda)) = \zeta^*(z)$ for any $\lambda \in \partial\Delta$. Moreover, it follows from Lemma 3.25 that

$$\frac{\varepsilon}{\sqrt{\gamma_1\gamma_2}}|\lambda|\varepsilon \geq \frac{|\lambda|}{\sqrt{\gamma_1}}\sqrt{\rho(z)-\rho(\zeta^*)} \geq |\lambda||z-\zeta^*| = |z(\lambda)-\zeta^*| \geq \varepsilon$$

for $\lambda \in \partial\Delta$. Consequently,

$$|\lambda| \geq \sqrt{\gamma_1\gamma_2} \quad \text{for any} \quad \lambda \in \partial\Delta.$$

Since

$$\left.\frac{d(E_1f)(\zeta^*+t(z(\lambda)-\zeta^*))}{dt}\right|_{t=1} = \frac{d(E_1f)(\zeta^*+\lambda(z-\zeta^*))}{d\lambda}\lambda,$$

we obtain for some constant $C_1 > 0$ and $C_2 > 0$,

$$\left|\frac{d(E_1f)(\zeta^*+\lambda(z-\zeta^*))}{d\lambda}\right| \leq \frac{C_1}{|\lambda|}\sup_{\zeta \in V}|f(\zeta)| \leq C_2 \sup_{\zeta \in V}|f(\zeta)|$$

for every $\lambda \in \partial\Delta$. Since the function $d(E_1f)(\zeta^*+\lambda(z-\zeta^*))/d\lambda$ is holomorphic for all $\lambda \in \overline{\Delta}$, we have for some constant $C_3 > 0$

$$\sup_{\lambda \in \Delta}\left|\frac{d(E_1f)(\zeta^*+\lambda(z-\zeta^*))}{d\lambda}\right| \leq C_3 \sup_{\zeta \in V}|f(\zeta)|.$$

Consequently, there exists a constant $C_4 > 0$ such that

$$|E_1f(z) - E_1f(\zeta^*)| = \left|\int_0^1 \frac{d}{d\lambda}E_1f(\zeta^*+\lambda(z-\zeta^*))d\lambda\right| \leq C_4 \sup_{\zeta \in V}|f(\zeta)|.$$

Since $\zeta^* \in V$, we have $E_1f(\zeta^*) = f(\zeta^*)$. Hence there exists a constant $C_5 > 0$ such that

$$\sup_{z \in \{(\partial V)_\sigma \setminus \partial V\} \cap \partial\Omega} |E_1f(z)| \leq C_5 \sup_{\zeta \in V}|f(\zeta)|,$$

which completes the proof of Theorem 3.17. $\square$

Lemma 3.27 *Let V' be a domain with smooth boundary in X such that $\overline{V} \subset V'$. We denote by C any positive constant which depends only on Ω, V and V'. Then for $z \in \Omega$ and $\varepsilon = |z_n|$,*

(a)

$$\int_{V' \setminus V} \frac{1}{|\Phi(z,\zeta)|^n} dV_{n-1}(\zeta) \leq C|\log\varepsilon|.$$

(b)
$$\int_{V'\setminus V} \frac{|z-\zeta|}{|\Phi(z,\zeta)|^{n+1}} dV_{n-1}(\zeta) \leq C\frac{1}{\varepsilon}.$$

(c)
$$\int_{V'\setminus V} \frac{|z-\zeta|^2}{|\Phi(z,\zeta)|^{n+1}} dV_{n-1}(\zeta) \leq C|\log \varepsilon|.$$

Proof. We may assume that $|z - \zeta| < \varepsilon$, where ε is the constant in Theorem 3.8. By contracting V' if necessary, it follows from (3.19) that

$$2\mathrm{Re}\Phi(z,\zeta) \geq \rho(\zeta) - \rho(z) + \beta|\zeta - z|^2$$

for $(z,\zeta) \in \{\overline{\Omega} \times (V'\setminus V)\} \cap \{(z,\zeta) \mid |\zeta - z| < \varepsilon\}$. We can choose a local coordinate system $t = (t_1, \cdots, t_{2n-2})$ such that $\rho(\zeta) = t_1$, $\mathrm{Im}\,\Phi(z,\zeta) = t_2$. We set $t' = (t_3, \cdots, t_{2n-2})$. Then

(a)
$$\int_{V'\setminus V} \frac{1}{|\Phi(z,\zeta)|^n} dV_{n-1}(\zeta) \leq C \int_{|t|\leq R} \frac{dt_1 \cdots dt_{2n-2}}{(|t'|^2 + \varepsilon^2 + |t_1| + |t_2|)^n}$$
$$\leq C \int_{|t'|\leq R} \frac{dt_3 \cdots dt_{2n-2}}{(|t'|^2 + \varepsilon^2)^{n-2}}$$
$$\leq C \int_0^R \frac{r^{2n-5}}{(r^2 + \varepsilon^2)^{n-2}} dr$$
$$\leq C|\log \varepsilon|.$$

(b)
$$\int_{V'\setminus V} \frac{|z-\zeta|}{|\Phi(z,\zeta)|^{n+1}} dV_{n-1}(\zeta)$$
$$\leq C \int_{V'\setminus V} \frac{1}{(|z-\zeta|^2 + |\mathrm{Im}\,\Phi(z,\zeta)| + |\rho(\zeta)|)^{n+(1/2)}} dV_{n-1}(\zeta)$$
$$\leq C \int_{|t'|\leq R} \frac{dt_3 \cdots dt_{2n-2}}{(|t'|^2 + \varepsilon^2)^{n-(3/2)}} \leq C\frac{1}{\varepsilon}.$$

(c)
$$\int_{V'\setminus V} \frac{|z-\zeta|^2}{|\Phi(z,\zeta)|^{n+1}} dV_{n-1}(\zeta) \leq C \int_{|t'|\leq R} \frac{dt_3 \cdots dt_{2n-2}}{(|t'|^2 + \varepsilon^2)^{n-2}} \leq C|\log \varepsilon|.$$

$\square$

Theorem 3.18 *If $f \in A(V)$, then $E_1 f \in A(\Omega)$. Moreover, $E_1 f : A(V) \to A(\Omega)$ defines a bounded linear operator.*

Proof. Let $z^* \in \partial V$ and let $z \in (\overline{\Omega} \setminus \overline{V}) \cap B(z^*, \sigma_2)$. Let ζ^* be the solution for the system of equations (3.31) and (3.32). We set $z(\theta) = \zeta^* + \theta(z - \zeta^*)$ for $0 \leq \theta \leq 1$. Then $z(\theta)_n = \theta z_n$. Further, $\zeta^* = \zeta^*(z)$ also satisfies the system of equations (3.31) and (3.32) for $z(\theta)$ instead of z. By the uniqueness of the solution, we have $\zeta^*(z) = \zeta^*(z(\theta))$. Let V' be a domain with smooth boundary in X such that $\overline{V} \subset V'$. Then

$$E_1 f(z) = \int_{\partial V} \frac{f(\zeta) K(z,\zeta)}{\Phi(z,\zeta)^{n-1}}$$
$$= \int_{\partial V'} \frac{f(\zeta) K(z,\zeta)}{\Phi(z,\zeta)^{n-1}}$$
$$- \int_{V' \setminus V} f(\zeta) \bar{\partial}_\zeta \left(\frac{K(z,\zeta)}{\Phi(z,\zeta)^{n-1}} \right).$$

Define

$$F_1(z) = \int_{(V' \setminus V) \cap B(z^*, \sigma_2)} f(\zeta) \bar{\partial}_\zeta \left(\frac{K(z,\zeta)}{\Phi(z,\zeta)^{n-1}} \right).$$

By Theorem 3.8, we may assume that $\Phi(z,\zeta) = F(z,\zeta) M(z,\zeta)$ on $B(z^*, \sigma_2) \times B(z^*, \sigma_2)$, where $F(z,\zeta)$ is the Levi polynomial. Then F_1 is expressed by

$$F_1(z) = \int_{(V' \setminus V) \cap B(z^*, \sigma)} f(\zeta) \frac{A(z,\zeta)}{\Phi(z,\zeta)^{n-1}}$$
$$+ \int_{(V' \setminus V) \cap B(z^*, \sigma)} f(\zeta) \frac{\sum_{j=1}^n (\zeta_j - z_j) B_j(z,\zeta)}{\Phi(z,\zeta)^n},$$

where $A(z,\zeta)$ and $B_j(z,\zeta)$, $1 \leq j \leq n$, are C^∞ $(2n-2)$ forms on $\overline{\Omega} \times \overline{V}$. Then it follows from Lemma 3.27 that there exist constants C_1, C_2 and C_3 such that

$$\left| \frac{dF_1(\zeta^* + \lambda(z - \zeta^*))}{d\lambda} \bigg|_{\lambda=1} \right| \leq C_1 |f|_{\overline{V}} \int_{(V' \setminus V) \cap B(z^*, \sigma)} \frac{\varepsilon}{|\Phi(z,\zeta)|^n} dV_{n-1}(\zeta)$$

$$+ C_2 |f|_{\overline{V}} \int_{(V' \setminus V) \cap B(z^*, \sigma)} \frac{\varepsilon |\zeta - z|(|\zeta - z| + \varepsilon)}{|\Phi(z,\zeta)|^{n+1}} dV_{n-1}(\zeta) \leq C_3 \varepsilon |\log \varepsilon| |f|_{\overline{V}},$$

for any point $z^* \in \partial V, z \in (\overline{\Omega} \setminus \partial V) \cap B(z^*, \sigma_2)$, where $\varepsilon = |z_n|$.

Then
$$|F_1(z) - F_1(\zeta^*)| = \left|\int_0^1 \frac{d}{d\theta} F_1(\zeta^* + \theta(z-\zeta^*))d\theta\right|$$
$$= \left|\int_0^1 \frac{dF_1(\zeta^* + \lambda\theta(z-\zeta^*))}{d\lambda}\bigg|_{\lambda=1} d\theta\right|$$
$$= \left|\int_0^1 \frac{1}{\theta}\frac{dF_1(\zeta^* + \lambda(z(\theta)-\zeta^*))}{d\lambda}\bigg|_{\lambda=1} d\theta\right|$$
$$\le C_4 \int_0^1 \varepsilon|\log\varepsilon\theta| d\theta \sup_{\zeta\in\overline{V}} |f(\zeta)|.$$

Hence we obtain
$$|E_1 f(z) - E_1 f(\zeta^*)| \le C_5 \sigma_2 |\log\sigma_2| \sup_{\zeta\in\overline{V}} |f(\zeta)|$$

We may assume that $f \in C^1(\overline{V})$. Since $\zeta^* \in \overline{V}$, we obtain
$$|E_1 f(z) - E_1 f(z^*)| = |E_1 f(z) - E_1 f(\zeta^*)| + |f(\zeta^*) - f(z^*)|$$
$$\le |E_1 f(z) - E_1 f(\zeta^*)| + C_6\{|\zeta^* - z| + |z - z^*|\}$$
$$\le C_7 \sigma_2 |\log\sigma_2| \sup_{\zeta\in\overline{V}} |f(\zeta)|.$$

Consequently, $\lim_{z\to z^*} E_1 f(z) = f(z^*)$. □

Lemma 3.28 *Let Ω be a strictly pseudoconvex domain with C^∞ boundary in $\mathbf{C}^n$ and let X be a closed submanifold of codimension one in a neighborhood $\widetilde{\Omega}$ of $\overline{\Omega}$. Let Ω' be a pseudoconvex domaim such that $\overline{\Omega} \subset \Omega' \subset \overline{\Omega}' \subset \widetilde{\Omega}$. We set $V = \Omega \cap X$ and $V' = \Omega' \cap X$.*

Assume that X intersects $\partial\Omega$ transversally. Let $\zeta \in \partial V$. It follows from Theorem 5.20 (a) (Cartan theorem A) that there exist $\sigma > 0$ and holomorphic functions $F_1, \cdots, F_q \in \Gamma(\Omega', \mathcal{F}_{V'})$ such that $\mathcal{F}_{V'}$ is generated by $F_1, \cdots, F_q$ in $B(\zeta, \sigma)$. Then there exist constants $\sigma > \sigma_1 > \delta > 0$ with the following properties:

(a) *For some integer q_1 with $1 \le q_1 \le q$ and some integers $m_1, \cdots, m_{n-2}$ from the set $\{1, \cdots, n\}$ the mapping*
$$\varphi(z) = (z_{m_1} - \zeta_{m_1}, \cdots, z_{m_{n-2}} - \zeta_{m_{n-2}}, F(z,\zeta), F_{q_1}(z))$$
is a biholomorphic mapping of the ball $B(\zeta, \sigma_1) \subset \Omega'$ onto a neighborhood W_ζ of 0, where $F(z,\zeta)$ is the Levi polynomial defined in Definition 3.12.

(b) There exists a strictly convex domain $U_\zeta \subset\subset W_\zeta$ such that

$$\Omega \cap B(\zeta, \delta) \subset \varphi^{-1}(U_\zeta) \subset \Omega,$$

where $U_\zeta = \{w \in W_\zeta \mid \rho_\zeta(w) < 0\}$, and ρ_ζ is a real-valued C^2 function in the domain W_ζ that is strictly convex in a neighborhood of $\overline{U}_\zeta$.

Proof. There exists q_1 $(1 \leq q_1 \leq q)$ such that the equation

$$\sum_{j=1}^{m} \frac{\partial F_{q_1}}{\partial z_j}(\zeta)(z_j - \zeta_j) = 0$$

defines a $(n-1)$ dimensional analytic plane tangent to V' at the point $z = \zeta$. Since X and $\partial\Omega$ intersect transversally, the equations

$$\sum_{j=1}^{m} \frac{\partial F_{q_1}}{\partial z_j}(\zeta)(z_j - \zeta_j) = 0,$$

$$\sum_{j=1}^{n} \frac{\partial F}{\partial z_j}(\zeta, \zeta)(z_j - \zeta_j) = 2\sum_{j=1}^{n} \frac{\partial \rho}{\partial z_j}(\zeta)(z_j - \zeta_j) = 0$$

define a $(n-2)$ dimensional analytic plane if $\zeta \in \partial V$. Therefore we can choose numbers $m_1, \cdots, m_{n-2}$ so that

$$\varphi(z) = \{z_{m_1} - \zeta_{m_1}, \cdots, z_{m_{n-2}} - \zeta_{m_{n-2}}, F(z,\zeta), F_{q_1}(z))$$

has a non-zero Jacobian at the point $z = \zeta$. By the implicit function theorem there exists $\sigma > 0$ such that the mapping φ of the ball $B(\zeta, \sigma)$ onto some domain W_ζ containing 0 has the inverse mapping φ^{-1} (see Corollary 5.3). Define $\tilde{\rho}_\zeta(w) = \rho(\varphi^{-1}(w))$. By Taylor's formula we have

$$\rho(z) = \text{Re}\left(2\sum_{j=1}^{n} \frac{\partial \rho}{\partial z_j}(\zeta)(z_j - \zeta_j) + \sum_{j,k=1}^{n} \frac{\partial^2 \rho}{\partial z_j \partial z_k}(\zeta)(z_j - \zeta_j)(z_k - \zeta_k)\right)$$

$$+ \sum_{j,k=1}^{n} \frac{\partial^2 \rho}{\partial z_j \partial \bar{z}_k}(\zeta)(z_j - \zeta_j)(\bar{z}_k - \bar{\zeta}_k) + o(|z - \zeta|^2).$$

Since $F(z,\zeta) = w_{n-1}$ and

$$z_i - \zeta_i = z_i(w) - z_i(0) = \sum_{\nu=1}^{n} \frac{\partial z_i}{\partial w_\nu}(0) w_\nu + o(|w|),$$

we obtain

$$\tilde{\rho}_\zeta(w) = \mathrm{Re} w_{n-1} + \sum_{i,j=1}^n \frac{\partial^2 \rho}{\partial z_i \partial \bar{z}_j}(\zeta) \left(\sum_{\nu=1}^n \frac{\partial z_i}{\partial w_\nu}(0) w_\nu\right)$$

$$\times \left(\sum_{\nu=1}^n \frac{\partial \bar{z}_j}{\partial \bar{w}_\nu}(0) \bar{w}_\nu\right) + o(|w|^2).$$

$\tilde{\rho}_\zeta$ is strictly convex in a neighborhood $U_1 (\subset U_0)$ of $w = 0$. Define

$$t_\zeta(w) = \mathrm{Re} \sum_{i=1}^n \frac{\partial \tilde{\rho}_\zeta}{\partial w_i}(0) w_i.$$

The equation $t_\zeta(w) = 0$ defines the real tangent plane to the boundary of the convex domain $U_2 := \{w \in U_1 \mid \tilde{\rho}_\zeta(w) < 0\}$ at the point $w = 0$. Since U_2 is strictly convex near 0, there exists $\varepsilon > 0$ such that if we define $U_3 = \{w \in U_1 \mid \tilde{\rho}_\zeta(w) < 0, t_\zeta(w) > -\varepsilon\}$, then $U_3 \subset\subset U_1$. Define

$$\chi(t) = \begin{cases} 0 & (t \geq -\frac{\varepsilon}{2}) \\ (t + \frac{\varepsilon}{2})^4 & (t \leq -\frac{\varepsilon}{2}) \end{cases}.$$

Then χ is of class C^2 in $\mathbf{R}$. We choose a constant $A > 0$ in such a way that

$$\sup_{\zeta \partial V} \sup_{w \in U_1} |\tilde{\rho}_\zeta(w)| < A\chi(-\varepsilon).$$

Define

$$\rho_\zeta(w) = \tilde{\rho}_\zeta(w) + A\chi(t_\zeta(w)).$$

Since

$$\sum_{j,k=1}^{2n} \frac{\partial^2}{\partial u_j \partial u_k}[\chi(t_\zeta(w))] u_j u_k = \frac{1}{4}\chi''(t_\zeta(w)) \left(\sum_{j=1}^{2n} \frac{\partial \tilde{\rho}_\zeta}{\partial u_j}(0) u_j\right)^2 \geq 0,$$

$\rho_\zeta(w)$ is strictly convex in U_1. Then $U_\zeta = \{w \in U_1 \mid \rho_\zeta(w) < 0\}$ is a strictly convex domain in U_1 with $U_\zeta \subset\subset U_1$. Define $G_\zeta = \{z \in B(\zeta, \sigma) \mid \varphi(z) \in U_\zeta\}$. If we choose $\delta > 0$ sufficiently small, then we obtain $\Omega \cap B(\zeta, \delta) \subset G_\zeta \subset \Omega$. □

Lemma 3.29 *Let L_Ω and R_Ω be the integral operators defined in Definition 3.14 and Definition 3.16, respectively.*

(a) If f is a bounded function in Ω, then $R_\Omega f$ is continuous on $\overline{\Omega}$.

(b) If f is a bounded holomorphic function in Ω and if φ is a C^1 function in $\mathbf{C}^n$, then $L_\Omega(f\varphi)$ is bounded in Ω.

Proof. (a) We write $R_\Omega f$ in the following form

$$R_\Omega f(z) = \int_\Omega f(\zeta) H(z,\zeta) dV(\zeta),$$

where $dV(\zeta)$ is the Lebesgue measure on Ω. Then

$$|H(z,\zeta)| \leq C \left| \int_0^1 \left| \omega\left(1 - \lambda \frac{\chi(\zeta) w(z,\zeta)}{\widetilde{\Phi}(z,\zeta)} + \lambda \frac{\bar{\zeta} - \bar{z}}{|z-\zeta|^2}\right) \right| d\lambda \right|$$

$$\leq C \left(\frac{1}{|\widetilde{\Phi}|^2 |\zeta - z|^{2n-3}} + \frac{1}{|\widetilde{\Phi}||\zeta - z|^{2n-2}} + \frac{1}{|\zeta - z|^{2n-1}} \right).$$

For $\varepsilon > 0$, there exists $\delta > 0$ such that for any $w \in \overline{\Omega}$

$$\int_{B(w,3\delta) \cap \Omega} |H(w,\zeta)| dV(\zeta) < \varepsilon.$$

Let

$$K_\delta = \{(w,\zeta) \in \overline{\Omega} \times \partial\Omega \mid |w - \zeta| \geq \delta\}.$$

Since $H(w,\zeta)$ is continuous on the compact set K_δ, we can choose $\delta > \delta_1 > 0$ such that for any point w with $|z - w| < \delta_1$,

$$|H(z,\zeta) - H(w,\zeta)| < \varepsilon$$

for all ζ satisfying $|\zeta - z| \geq 2\delta$. For $|z - w| < \delta_1$, we have

$$|R_\Omega f(z) - R_\Omega f(w)| = \left| \int_\Omega f(\zeta)(H(z,\zeta) - H(w,\zeta)) dV(\zeta) \right|$$

$$\leq \left| \int_{\Omega \setminus B(z,2\delta)} (H(z,\zeta) - H(w,\zeta)) dV(\zeta) \right|$$

$$+ \int_{\Omega \cap B(z,2\delta)} |f(\zeta)| |H(z,\zeta)| dV(\zeta)$$

$$+ \int_{\Omega \cap B(w,3\delta)} |f(\zeta)| |H(w,\zeta)| dV(\zeta)$$

$$\leq \varepsilon \int_{\Omega \setminus B(z,2\delta)} |f(\zeta)| dV(\zeta) + \varepsilon \|f\| + \varepsilon \|f\| < C\varepsilon.$$

Hence $R_\Omega f$ is continuous on $\overline{\Omega}$.

(b) Since $\bar{\partial}(f\varphi) = f\bar{\partial}\varphi$, $\bar{\partial}(f\varphi)$ is bounded in Ω. It follows from Theorem 3.12 that

$$\varphi f = L_\Omega(\varphi f) + R_\Omega(\bar{\partial}(\varphi f)),$$

which means that $L_\Omega(\varphi f)$ is bounded in Ω. □

Lemma 3.30 *Let $\Omega \subset\subset \mathbf{C}^n$ be a strictly pseudoconvex domain with C^∞ boundary and let $U_j \subset \mathbf{C}^n$ ($j = 1, \cdots, N$) be open sets such that $\overline{\Omega} \subset \cup_{j=1}^N U_j$. Then any $f \in H^\infty(\Omega)$ admits a decomposition $f = \sum_{j=1}^N f_j$, where every f_j is bounded and holomorphic in some neighborhood of $\overline{\Omega} \setminus (\partial\Omega \cap U_j)$. In addition, if f is continuous on $\overline{\Omega}$, then every f_j is continuous on $\overline{\Omega}$.*

Proof. Choose C^∞ functions χ_j in $\mathbf{C}^n$ such that $\sum_{j=1}^N \chi_j = 1$ on $\overline{\Omega}$ and $\mathrm{supp}(\chi_j) \subset U_j$. Define $f_j = L_\Omega(\chi_j f)$. Since $\chi_j = 0$ on $\mathbf{C}^n \setminus U_j$, f_j is bounded and holomorphic in some neighborhood of $\overline{\Omega} \setminus (\partial\Omega \cap U_j)$. By Theorem 3.12

$$f = L_\Omega f = \sum_{j=1}^N L_\Omega(\chi_j f) = \sum_{j=1}^N f_j.$$

Suppose f is continuous on $\overline{\Omega}$. By Theorem 3.12 we have

$$f_j = \chi_j f - R_\Omega(f\bar{\partial}\chi_j).$$

By Lemma 3.29 $R_\Omega(f\bar{\partial}\chi_j)$ is continuous on $\overline{\Omega}$. Hence f_j is continuous on $\overline{\Omega}$. □

The following theorem was proved by Henkin [HEN3] which is a generalization of Lemma 3.30 to submanifolds of strictly pseudoconvex domains in $\mathbf{C}^n$ with smooth boundary.

Theorem 3.19 *Let $\Omega \subset\subset \mathbf{C}^n$ be a strictly pseudoconvex domain with C^∞ boundary and let X be a closed submanifold in a neighborhood $\tilde{\Omega}$ of $\overline{\Omega}$, $V = \Omega \cap X$. Assume that X intersects $\partial\Omega$ transversally. Let $U_j \subset X$ ($j = 1, \cdots, N$) be open sets in X such that $\overline{V} \subset \cup_{j=1}^N U_j$. Then any $f \in H^\infty(V)$ admits a decomposition $f = \sum_{j=1}^N f_j$, where every f_j is bounded and holomorphic in some neighborhood of $\overline{V} \setminus (\partial V \cap U_j)$. In addition, if f is continuous on $\overline{V}$, then every f_j is also continuous on $\overline{V}$.*

Proof. Let Ω' be a strictly psuedoconvex domain such that $\Omega \subset\subset \Omega' \subset\subset \widetilde{\Omega}$. We set $V' = \Omega' \cap X$. Let $\varepsilon > 0$ be given. Let χ_i, $i = 1, \cdots, N$, be real-valued, nonnegative C^∞ functions such that $\sum_{i=1}^N \chi_i = 1$ in a neighborhood of $\overline{V}'$ and the diameter of each set $\operatorname{supp}(\chi_i)$ is less that $\varepsilon/3$. Define

$$\chi_\nu^1 = \sum_{\{i \mid \operatorname{supp}(\chi_i) \cap \operatorname{supp}(\chi_\nu) \neq \phi\}} \chi_i,$$

$$\widetilde{\chi}_\nu = \sum_{\{i \mid \operatorname{supp}(\chi_i) \cap \operatorname{supp}(\chi_\nu^1) = \phi\}} \chi_i.$$

We consider domains for $\nu = 1, \cdots, N$,

$$\Omega_\nu = \{z \in \Omega' \mid \rho(z) - \sum_{i=1}^\nu \lambda_i \chi_i(z) < 0\},$$

$$\widetilde{\Omega}_\nu = \{z \in \Omega' \mid \rho(z) - \sum_{i=1}^{\nu-1} \lambda_i \chi_i(z) - \lambda \widetilde{\chi}_\nu(z) < 0\}.$$

We set $\Omega_0 = \Omega$, $V_\nu = V' \cap \Omega_\nu$ and $\widetilde{V}_\nu = V' \cap \widetilde{\Omega}_\nu$ for $\nu = 0, 1, \cdots, N$. In order to prove Theorem 3.19 we need the following lemma.

Lemma 3.31 *For sufficiently small $\lambda_1, \cdots, \lambda_N > 0$ and for any $\nu = 1, \cdots, N$, there exist bounded operators $L_\nu^0 : H^\infty(V_{\nu-1}) \to H^\infty(V_\nu)$ and $L_\nu^1 : H^\infty(V_{\nu-1}) \to H^\infty(\widetilde{V}_\nu)$ with the following properties:*

(a) $f(z) = (L_\nu^0 f)(z) + (L_\nu^1 f)(z)$ for any $f \in H^\infty(V_{\nu-1})$ and any $z \in V_{\nu-1}$.
(b) $L_\nu^0 f \in A(V_\nu)$ and $L_\nu^1 f \in A(\widetilde{V}_\nu)$ if $f \in A(V_{\nu-1})$.

Proof of Lemma 3.31. Suppose that constants $\lambda_1, \cdots, \lambda_{\nu-1}$ satisfying the conditions of the lemma have already been chosen. We set $U_\nu = \operatorname{supp}(\chi_\nu^1)$. We may assume that $U_\nu \cap \partial V_{\nu-1} \neq \phi$. We fix a point $\zeta^* \in U_\nu \cap \partial V_{\nu-1}$. By Lemma 3.28, there exists a biholomorphic mapping φ of the ball $B(\zeta^*, (3/4)\sigma)$ onto a neighborhood W_{ζ^*} of 0 such that

$$\Omega_{\nu-1} \cap B(\zeta^*, (3/4)\delta) \subset G_{\zeta^*} = \{z \in B(\zeta^*, (3/4)\delta) \mid \rho_{\zeta^*}(\varphi(z)) < 0\},$$

where $\rho_{\zeta^*}(w)$ is strictly convex in a neighborhood of the set $\overline{E}_{\zeta^*} = \{w \in W_{\zeta^*} \mid \rho_{\zeta^*}(w) \leq 0\}$. We set $I_{\zeta^*} = \overline{E}_{\zeta^*} \cap \varphi(V_{\nu-1} \cap G_{\zeta^*})$. For any function $f \in H^\infty(V_{\nu-1})$ and any $z \in G_{\zeta^*} \cap V_{\nu-1}$, it follows from (3.30) that

$$f(z) = f(\varphi^{-1}(w)) = \int_{\zeta \in \partial I_{\zeta^*}} f(\varphi^{-1}(\zeta)) \frac{K(w, \zeta)}{\Phi(w, \zeta)^{n-1}}.$$

We set
$$\chi_\nu^0 = 1 - \chi_\nu^1$$
and
$$R_\nu^\alpha f(z) = \int_{\zeta \in \partial I_{\zeta^*}} f(\varphi^{-1}(\zeta))\chi_\nu^\alpha(\varphi^{-1}(\zeta)) \frac{K(\varphi(z),\zeta)}{\Phi(\varphi(z),\zeta)^{n-1}}$$

for $\alpha = 1, 2$ and $f \in H^\infty(V_{\nu-1})$. Then we have
$$f(z) = (R_\nu^0 f)(z) + (R_\nu^1 f)(z) \qquad (z \in G_{\zeta^*} \cap V_{\nu-1}).$$

We choose $\lambda_\nu < \lambda_0$ sufficiently small. We set
$$V'' = \{z \in V' \mid \rho(z) - \sum_{i=1}^{\nu-1} \lambda_i \chi_i(z) < \lambda_0\} = V_0'' \cup V_1'',$$

where $V_0'' = V'' \cap B(\zeta^*, (3/4)\delta)$ and $V_1'' = V'' \backslash B(\zeta^*, (1/2)\delta)$. Then we have a representation

$$R_\nu^\alpha f(z) = \int_{\zeta \in \partial I_{\zeta^*}} f(\varphi^{-1}(\zeta))\chi_\nu^\alpha(\varphi^{-1}(\zeta)) \frac{K(\varphi(z),\zeta)}{\Phi(\varphi(z),\zeta)^{n-1}}$$

$$= \chi_\nu^\alpha(z)f(z) + \int_{\zeta \in \partial I_{\zeta^*}} f(\varphi^{-1}(\zeta))\{\chi_\nu^\alpha(\varphi^{-1}(\zeta)) - \chi_\nu^\alpha(z)\} \frac{K(\varphi(z),\zeta)}{\Phi(\varphi(z),\zeta)^{n-1}}.$$

We set
$$A_\nu^\alpha(z) = \int_{\zeta \in \partial I_{\zeta^*}} f(\varphi^{-1}(\zeta))\{\chi_\nu^\alpha(\varphi^{-1}(\zeta)) - \chi_\nu^\alpha(z)\} \frac{K(\varphi(z),\zeta)}{\Phi(\varphi(z),\zeta)^{n-1}}.$$

Then we can prove that A_ν^0 is a bounded operator from $H^\infty(V_{\nu-1})$ to $A(V_\nu \cap B(\zeta^*, (3/4)\delta))$, and A_ν^1 is a bounded operator from $H^\infty(V_{\nu-1})$ to $A(\widetilde{V}_\nu \cap B(\zeta^*, (3/4)\delta))$ using the same method as the proof of Lemma 3.29 (a). Therefore, we can prove that R_ν^0 is a bounded operator from $H^\infty(V_{\nu-1})$ to $H^\infty(V_\nu \cap B(\zeta^*, (3/4)\delta))$. On the other hand R_ν^1 is a bounded operator from $H^\infty(V_{\nu-1})$ to $H^\infty(\widetilde{V}_\nu \cap B(\zeta^*, (3/4)\delta))$. If we choose $0 < \chi_\nu < \chi_0$ sufficiently small, then $\chi_\nu^1 = 0$ in a neighborhood of $\overline{V_0'' \cap V_1''}$. Hence R_ν^1 is a bounded operator from $H^\infty(V_{\nu-1})$ to $A(V_0'' \cap V_1'')$. If $f \in A(V_{\nu-1})$, then

$$R_\nu^0 f \in A(V_\nu \cap B(\zeta^*, (3/4)\delta)) \quad \text{and} \quad R_\nu^1 f \in A(\widetilde{V}_\nu \cap B(\zeta^*, (3/4)\delta)).$$

It follows from Theorem 5.26 that for $f \in \mathcal{O}(V_0'' \cap V_1'')$ there exist mappings $T_\nu^\alpha : \mathcal{O}(V_0'' \cap V_1'') \to \mathcal{O}(V_\alpha'')$ such that

$$f = T_\nu^0 f + T_\nu^1 f.$$

For $z \in V_{\nu-1} \cap V_0'' \cap V_1''$, we have

$$f(z) = R_\nu^0 f(z) + R_\nu^1 f(z) = R_\nu^0 f(z) + T_\nu^0(R_\nu^1 f)(z) + T_\nu^1(R_\nu^1 f)(z).$$

We set

$$(L_\nu^0 f)(z) = \begin{cases} (R_\nu^0 f)(z) + (T_\nu^0 \circ R_\nu^1 f)(z) & (z \in V_\nu \cap B(\zeta^*, (3/4)\delta)) \\ f(z) - (T_\nu^1 \circ R_\nu^1 f)(z) & (z \in V_\nu \backslash B(\zeta^*, (1/2)\delta)) \end{cases},$$

$$(L_\nu^1 f)(z) = \begin{cases} (R_\nu^1 f)(z) + (T_\nu^0 \circ R_\nu^1 f)(z) & (z \in \widetilde{V}_\nu \cap B(\zeta^*, (3/4)\delta)) \\ (T_\nu^1 \circ R_\nu^1 f)(z) & (z \in \widetilde{V}_\nu \backslash B(\zeta^*, (1/2)\delta)) \end{cases},$$

Then L_ν^0 and L_ν^1 satisfy conditions (a) and (b), which completes the proof of Lemma 3.31.

End of the proof of Theorem 3.19. We set $L_i = L_i^1 \circ L_{i-1}^0 \circ \cdots \circ L_1^0$ for $i = 1, \cdots, N-1$, and $L_N = L_{N-1}^0 \circ L_{N-2}^0 \circ \cdots L_1^0$. If $f \in H^\infty(V)$, then $L_i f \in H^\infty(\widetilde{V}_i)$ for $i = 1, 2, \cdots, N-1$, and $L_N f \in H^\infty(V_{N-1})$. If $f \in A(V)$, then $L_i f \in A(\widetilde{V}_i)$ for $i = 1, 2, \cdots, N-1$, and $L_N f \in A(V_{N-1})$. Moreover, if $f \in H^\infty(V)$ and $z \in V$, then

$$f(z) = \sum_{i=1}^N L_i f(z).$$

The diameter of the set $\overline{V} \backslash \widetilde{V}_{N-1}$ is less than ε, and the diameter of the set $\overline{V} \backslash V_{N-1}$ is less than $\varepsilon/3$. Theorem 3.19 is proved. □

Now we are going to prove bounded and continuous extensions of holomorphic functions from submanifolds in general position of a strictly pseudoconvex domain in $\mathbf{C}^n$ with C^∞ boundary.

Corollary 3.6 *Let Ω be a strictly pseudoconvex domain in $\mathbf{C}^n$ with C^∞ boundary and let X be a closed submanifold in a neighborhood of $\overline{\Omega}$. Let $V = \Omega \cap X$. Assume that X intersects $\partial \Omega$ transversally. Then for any $f \in H^\infty(V)$ there exists $g \in H^\infty(\Omega)$ such that $g = f$ on V. Moreover, if $f \in A(V)$, then there exists $g \in A(\Omega)$ such that $g = f$ on V.*

Proof. Let f be a bounded holomorphic function in V. Since $\overline{V}$ is compact, there is a biholomorphic mapping $h_\xi : B(\xi, \delta) \to \mathbf{C}^n$ such that $h_\xi(X \cap B(\xi, \delta))$ is the intersection of $h_\xi(B(\xi, \delta))$ with a complex hyperplane in $\mathbf{C}^n$. By Theorem 3.19, it is sufficient to prove the theorem for the case when f has the following property:

There is a point $\xi \in \partial V$ and a strictly pseudoconvex open set $\Omega_0 \subset \mathbf{C}^n$ such that

$$\overline{\Omega} \setminus (\partial \Omega \cap B(\xi, \delta/3)) \subset \Omega_0$$

and f is bounded holomorphic in $X \cap \Omega_0$.

We can choose a strictly pseudoconvex open set $\Omega_\xi \subset \mathbf{C}^n$ such that

$$B(\xi, \delta/3) \cap \Omega_\xi \subset B(\xi, \delta/2) \cap \Omega_0.$$

Then we have

$$\overline{\Omega} \subset \Omega_0 \cup B(\xi, \delta/3).$$

Therefore, we can choose a strictly pseudoconvex open set Ω_1 such that

$$\Omega \subset\subset \Omega_1 \subset\subset \Omega_0 \cup B(\xi, \delta/3).$$

We set

$$U_\xi := B(\xi, \delta/3) \cap \Omega_1, \quad U_0 := \Omega_0 \cap \Omega_1.$$

Then $\{U_0, U_\xi\}$ forms an open covering of Ω_1. By choosing δ sufficiently small, we may assume that $X \cap B(\xi, \delta)$ is a complex hypersurface. It follows from Theorem 3.15 (or Theorem 3.17) that there exists a bounded holomorphic function f_ξ on Ω_ξ such that $f_\xi = f$ on $X \cap \Omega_\xi$. Since Ω_0 is a pseudoconvex domain, there exists a holomorphic function f_0 in Ω_0 such that $f_0 = f$ on $X \cap \Omega_0$. Then $f_0 - f_\xi$ is holomorphic in $\Omega_\xi \cap \Omega_0$ and $f_0 - f_\xi = 0$ on $X \cap \Omega_\xi \cap \Omega_0$. Since $U_\xi \cap U_0 \subset \Omega_\xi \cap \Omega_1 \cap \Omega_0$, it follows from Theorem 5.22 that there exist $\tilde{f}_\xi \in \Gamma(U_\xi, \mathcal{F}_V)$ and $\tilde{f}_0 \in \Gamma(U_0, \mathcal{F}_V)$ such that $f_0 - f_\xi = \tilde{f}_0 - \tilde{f}_\xi$ on $U_\xi \cap U_0$. We set $g := f_0 - \tilde{f}_0$ in U_0 and $g := f_\xi - \tilde{f}_\xi$ in $U_\xi \cap \Omega_\xi$. Then g is holomorphic in $U_0 \cup (U_\xi \cap \Omega_\xi)$ and equals $f_\xi - \tilde{f}_\xi$ in $U_\xi \cap \Omega_\xi$ Therefore, g is bounded and holomorphic in Ω and satisfies $g|_V = f$. If $f \in A(V)$, then we can prove similarly that there exists $g \in A(\Omega)$ such that $g|_V = f$. □

More generally, Henkin-Leiterer [HER] proved bounded and continuous extensions in the case when Ω is a strictly pseudoconvex open set (with not necessarily smooth boundary) in a Stein manifold without assuming that X intersects $\partial\Omega$ transversally. Amar [AMA2] also obtained bounded extensions of holomorphic functions from submanifolds of strictly pseudoconvex domains without assuming the transversality. Using the integral formula obtained by Hatziafratis, Hatziafratis [HAT2] proved the bounded extension of holomorphic functions from submanifolds in general position of strictly convex domains. Fornaess [FOR] investigated the integral formula by embedding strictly pseudoconvex domains into strictly convex domains. Adachi [ADA2; ADA3] proved bounded and continuous extensions from a submanifold V in general position of a weakly pseudoconvex domain Ω under the assumption that ∂V consists of strictly pseudoconvex boundary points of Ω. Using the method of Kerzman-Stein [KES] and the integral formula obtained by Hatziafratis [HAT1], Adachi-Kajimoto [ADK] obtained the holomorphic extension of Lipschitz functions from the boundary. Further, Jakóbczak [JK1; JK2] studied extensions of holomorphic functions in various function spaces.

3.4 H^p and C^k Extensions

In this section we study H^p ($1 \leq p < \infty$) and C^k ($k = 1, 2, \cdots, \infty$) extensions of holomorphic functions from submanifolds in general position of a strictly pseudoconvex domain Ω in $\mathbf{C}^n$ with C^∞ boundary by following the methods of Beatrous [BEA] and Ahern-Schneider [AHS2], respectively.

Let X be a closed submanifold in a neighborhood of $\overline{\Omega}$ and let $V = \Omega \cap X$. Assume that X intersects $\partial\Omega$ transversally. We may assume that $X = \{z_n = 0\}$. For $f \in \mathcal{O}(V) \cap L^p(V)$ ($1 \leq p < \infty$) and $z \in \Omega$, we define

$$Ef(z) = \frac{(n-1)!}{(2\pi i)^{n-1}} \int_V f(\zeta) \omega_{\zeta'} \left(\frac{\chi(\zeta)(w(z,\zeta))'}{\widetilde{\Phi}(z,\zeta)} \right) \wedge \omega_{\zeta'}(\zeta).$$

By Theorem 3.13, Ef is holomorphic in Ω and $Ef(z) = f(z)$ for $z \in V$. There exists a C^∞ $(n-1, n-1)$ form $\eta_0(z, \zeta)$ on $\overline{\Omega} \times \overline{\Omega}$ with respect to ζ such that

$$Ef(z) = \int_V \frac{f(\zeta)\eta_0(z,\zeta)}{\widetilde{\Phi}(z,\zeta)^n}.$$

Then we obtain the following lemma.

Lemma 3.32 *There exists a C^∞ $(n-1, n-1)$ form $\eta(z, \zeta)$ on $\overline{\Omega} \times \overline{\Omega}$ with respect to ζ with the following properties:*

(a) $\eta(\cdot,\zeta)$ is holomorphic in Ω for each $\zeta \in \overline{V}$ fixed.
(b) For $f \in \mathcal{O}(V) \cap L^1(V)$
$$Ef(z) = \int_V \frac{f(\zeta)\rho(\zeta)\eta(z,\zeta)}{\widetilde{\Phi}(z,\zeta)^{n+1}}.$$

Proof. We choose a function $\varphi \in C^\infty(\mathbf{R})$ with the following properties:
$$0 \leq \varphi(t) \leq 1 \ (t \in \mathbf{R}), \quad \varphi(t) = \begin{cases} 1 \ (|t| \leq 1) \\ 0 \ (|t| \geq 2) \end{cases}.$$

For $\varepsilon > 0$, we set
$$\varphi_\varepsilon(z) = \varphi\left(\frac{\rho(z)}{\varepsilon}\right).$$

Then $\varphi_\varepsilon(z) = 1$ if $|\rho(z)| \leq \varepsilon$ and $\varphi_\varepsilon(z) = 0$ if $|\rho(z)| \geq 2\varepsilon$. We choose C^∞ functions ψ_j, $j = 1, \cdots, N$, in U_2 with the properties that $1 = \sum_{j=1}^N \psi_j(z)$ for $z \in U_2$ and there exists a constant $c > 0$ such that if $z \in \mathrm{supp}(\psi_j)$, then there exists a positive integer $k = k(j)$ with $\left|\frac{\partial \rho}{\partial \zeta_k}(z)\right| > c$. Then we have
$$Ef(z) = \sum_{j=1}^N \int_V f(\zeta) \frac{\eta_0(z,\zeta)\psi_j(\zeta)}{\widetilde{\Phi}(z,\zeta)^n}.$$

Since
$$d\bar\zeta_1 \wedge \cdots \wedge d\bar\zeta_{k-1} \wedge \bar\partial\rho \wedge d\bar\zeta_{k+1} \wedge \cdots \wedge d\bar\zeta_{n-1} = \frac{\partial\rho}{\partial\bar\zeta_k} d\bar\zeta_1 \wedge \cdots \wedge d\bar\zeta_{n-1}$$

on $\mathrm{supp}(\psi_i)$, we obtain
$$Ef(z) = \int_V f(\zeta) \frac{\bar\partial\rho(\zeta) \wedge \omega(z,\zeta)}{\widetilde{\Phi}(z,\zeta)^n},$$

where $\omega(z,\zeta)$ is a C^∞ $(n-1, n-2)$ form on $\overline\Omega \times \overline V$, and holomorphic with respect to $z \in \Omega$ for each fixed $\zeta \in \overline V$. Now we have
$$\int_V f(\zeta) \frac{\bar\partial\rho(\zeta) \wedge \omega(z,\zeta)}{\widetilde{\Phi}(z,\zeta)^n} = \int_V f(\zeta) \frac{\bar\partial\rho(\zeta)\varphi_\varepsilon(\zeta) \wedge \omega(z,\zeta)}{\widetilde{\Phi}(z,\zeta)^n}$$
$$+ \int_V f(\zeta) \frac{\bar\partial\rho(\zeta)(1-\varphi_\varepsilon(\zeta)) \wedge \omega(z,\zeta)}{\widetilde{\Phi}(z,\zeta)^n}$$
$$:= I_1^\varepsilon + I_2^\varepsilon.$$

Then $\lim_{\varepsilon \to 0} I_1^\varepsilon = 0$. On the other hand, $1 - \varphi_\varepsilon = 0$ on ∂V, which implies that

$$I_2^\varepsilon = \int_V \bar\partial \left\{ \frac{f(\zeta)\rho(\zeta)(1 - \varphi_\varepsilon) \wedge \omega(z,\zeta)}{\widetilde\Phi(z,\zeta)^n} \right\}$$
$$- \int_V f(\zeta)\rho(\zeta) \bar\partial \left\{ \frac{(1 - \varphi_\varepsilon) \wedge \omega(z,\zeta)}{\widetilde\Phi(z,\zeta)^n} \right\}$$
$$= - \int_V f(\zeta)\rho(\zeta) \frac{[-\widetilde\Phi\bar\partial\varphi_\varepsilon \wedge \omega + \widetilde\Phi(1-\varphi_\varepsilon)\bar\partial\omega - n(1-\varphi_\varepsilon)\bar\partial_\zeta\widetilde\Phi \wedge \omega]}{\widetilde\Phi^{n+1}}.$$

We obtain

$$\left| \int_V f(\zeta)\rho(\zeta)\widetilde\Phi\bar\partial\varphi_\varepsilon \wedge \omega \right|$$
$$= \left| \int_V f(\zeta)\rho(\zeta)\varphi'\left(\frac{\rho(\zeta)}{\varepsilon}\right) \frac{1}{\varepsilon}\widetilde\Phi\bar\partial\rho \wedge \omega \right|$$
$$\leq C \int_{X \cap \{-2\varepsilon \leq \rho \leq -\varepsilon\}} \frac{|f(\zeta)||\rho(\zeta)|}{\varepsilon} dV_{n-1}(\zeta)$$
$$\leq C \int_{X \cap \{-2\varepsilon \leq \rho \leq -\varepsilon\}} |f(\zeta)| dV_{n-1}(\zeta).$$

Consequently,

$$\lim_{\varepsilon \to 0} \int_V f(\zeta)\rho(\zeta)\widetilde\Phi\bar\partial\varphi_\varepsilon \wedge \omega = 0.$$

Thus we have

$$Ef(z) = \lim_{\varepsilon \to 0} I_2^\varepsilon = \int_V \frac{f(\zeta)\rho(\zeta)(\widetilde\Phi\bar\partial\omega - n\omega \wedge \bar\partial_\zeta\widetilde\Phi)}{\widetilde\Phi^{n+1}}.$$

We set

$$\eta(z,\zeta) = \widetilde\Phi(z,\zeta)\bar\partial_\zeta\omega(z,\zeta) - n\omega(z,\zeta) \wedge \bar\partial_\zeta\widetilde\Phi(z,\zeta).$$

Then we have

$$Ef(z) = \int_V \frac{f(\zeta)\rho(\zeta)\eta(z,\zeta)}{\widetilde\Phi(z,\zeta)^{n+1}} \quad (z \in \Omega).$$

$\square$

Definition 3.22 For $z \in \Omega$, we denote by $\delta_X(z)$ the distance from z to X.

Lemma 3.33 For $0 < \varepsilon < 1$ we have

(a) $\displaystyle\int_{\partial\Omega} \frac{\delta_X(z)^{-2\varepsilon}}{|\widetilde{\Phi}(z,\zeta)|^n} d\sigma_{2n-1}(z) \leq C_\varepsilon |\rho(\zeta)|^{-\varepsilon}.$

(b) $\displaystyle\int_{\partial\Omega} \frac{d\sigma_{2n-1}(z)}{|\widetilde{\Phi}(z,\zeta)|^{n+\varepsilon}} \leq C_\varepsilon |\rho(\zeta)|^{-\varepsilon}.$

Proof. We set $\tau(z,\zeta) = \mathrm{Im} F(z,\zeta)$, where $F(z,\zeta)$ is the Levi polynomial. We prove Lemma 3.33 in case $n \geq 3$. We have

$$\int_{\partial\Omega} \frac{\delta_X(z)^{-2\varepsilon}}{\widetilde{\Phi}(z,\zeta)^n} d\sigma_{2n-1}(z)$$
$$\leq C \int_{\mathbf{C}} \int_{\mathbf{C}^{n-2}} \int_0^\infty (|\rho(\zeta)| + \tau(z,\zeta) + |w'|^2 + |w''|^2)^{-n} |w''|^{-2\varepsilon}$$
$$\times d\tau dw' dw''.$$

By the change of variables

$$\tau(z,\zeta) = |\rho(\zeta)| x_1, \quad w' = \sqrt{|\rho(\zeta)|} x', \quad w'' = \sqrt{|\rho(\zeta)|} x'',$$

we obtain

$$\int_{\partial\Omega} \frac{\delta_X(z)^{-2\varepsilon}}{\widetilde{\Phi}(z,\zeta)^n} d\sigma_{2n-1}(z)$$
$$\leq C \int_{\mathbf{C}} \int_{\mathbf{C}^{n-2}} \int_0^\infty (1 + x_1 + |x'|^2 + |x''|^2)^{-n} |(\sqrt{|\rho(\zeta)|})^{-2\varepsilon} |x''|^{-2\varepsilon}$$
$$\times dx_1 dx' dx''$$
$$\leq C|\rho(\zeta)|^{-\varepsilon} \int_{\mathbf{C}} \int_{\mathbf{C}^{n-2}} (1 + |x'|^2 + |x''|^2)^{-n+1} |x''|^{-2\varepsilon} dx' dx''$$
$$\leq C|\rho(\zeta)|^{-\varepsilon} \int_0^\infty \int_0^\infty (1 + r_1^2 + r_2^2)^{-n+1} r_1^{2n-5} r_2^{-2\varepsilon} r_2 dr_1 dr_2.$$

Now we set $r = \lambda\cos\theta$, $r_2 = \lambda\sin\theta$. Then

$$\leq C|\rho(\zeta)|^{-\varepsilon} \int_0^\infty \int_0^{\frac{\pi}{2}} (1 + \lambda^2)^{-n+1} \lambda^{2n-3-2\varepsilon} (\sin\theta)^{1-2\varepsilon} d\lambda d\theta$$
$$\leq C_\varepsilon |\rho(\zeta)|^{-\varepsilon} \int_0^\infty (1 + \lambda^2)^{-n+1} \lambda^{2n-3-2\varepsilon} d\lambda$$
$$\leq C_\varepsilon |\rho(\zeta)|^{-\varepsilon} \int_1^\infty \frac{d\lambda}{\lambda^{1+\varepsilon}} \leq C_\varepsilon |\rho(\zeta)|^{-\varepsilon}.$$

This proves (a). By the same method as the proof of (a), we have

$$\int_{\partial\Omega} \frac{d\sigma_{2n-1}(z)}{|\widetilde{\Phi}(z,\zeta)|^{n+\varepsilon}} \leq C\delta(\zeta)^{-\varepsilon} \int_{\mathbf{C}} \int_{\mathbf{C}^{n-2}} \frac{dx'dx''}{(1+|x'|^2+|x''|^2)^{n-1+\varepsilon}}$$

$$\leq C_\varepsilon \delta(\zeta)^{-\varepsilon} \int_1^\infty \frac{d\lambda}{\lambda^{1+\varepsilon}}$$

$$\leq C_\varepsilon \delta(\zeta)^{-\varepsilon}.$$

This proves (b). □

Lemma 3.34 *For $0 < \varepsilon < 1$, we have*

$$\int_V \frac{|\rho(\zeta)|^{-\varepsilon}}{|\widetilde{\Phi}(z,\zeta)|^n} dV_{n-1}(\zeta) \leq C_\varepsilon (|\rho(z)| + \delta_X(z)^2)^{-\varepsilon}.$$

Proof. We set $\delta(\zeta) = |\rho(\zeta)|$ and $\zeta' = (\zeta_1, \cdots, \zeta_{n-2})$. We may assume that $(\rho(\zeta), \tau(z,\zeta), \zeta')$ forms a real coordinate system in a neighborhood of $\partial\Omega$. Hence we have

$$\int_V \frac{\delta(\zeta)^{-\varepsilon}}{|\widetilde{\Phi}(z,\zeta)|^n} dV_{n-1}(\zeta)$$
$$\leq C \int_{\mathbf{C}^{n-2}} \int_0^\infty \int_0^\infty (\delta(z) + \delta_X(z)^2 + \delta(\zeta) + \tau(z,\zeta) + |\zeta'|^2)^{-n} \delta(\zeta)^{-\varepsilon}$$
$$\times d\delta d\tau d\zeta'.$$

By a change of variables $\delta(\zeta) = (\delta(z) + \delta_X(z)^2)x_1$, $\tau(z,\zeta) = (\delta(z) + \delta_X(z)^2)x_2$ and $\zeta' = \sqrt{(\delta(z) + \delta_X(z)^2}x'$, we obtain

$$\int_V \frac{\delta(\zeta)^{-\varepsilon}}{|\widetilde{\Phi}(z,\zeta)|^n} dV_{n-1}(\zeta)$$
$$\leq C(\delta(z) + \delta_X(z)^2)^{-\varepsilon} \int_{\mathbf{C}^{n-2}} \int_0^\infty \int_0^\infty (1 + x_1 + x_2 + |x'|^2)^{-n} x_1^{-\varepsilon}$$
$$\times dx_1 dx_2 dx'$$
$$\leq C(\delta(z) + \delta_X(z)^2)^{-\varepsilon} \int_{\mathbf{C}^{n-2}} \int_0^\infty (1 + x_1 + |x'|^2)^{-n+1} x_1^{-\varepsilon}$$
$$\times dx_1 dx'$$
$$\leq C(\delta(z) + \delta_X(z)^2)^{-\varepsilon} \int_0^\infty \int_0^\infty (1 + x_1 + r^2)^{-n+1} x_1^{-\varepsilon} r^{2n-5} dx_1 dr.$$

We set $x_1 = y_1^2$. Then

$$\int_V \frac{\delta(\zeta)^{-\varepsilon}}{|\widetilde{\Phi}(z,\zeta)|^n} dV_{n-1}(\zeta)$$
$$\leq C(\delta(z) + \delta_X(z)^2)^{-\varepsilon} \int_0^\infty \int_0^\infty (1 + y_1^2 + r^2)^{-n+1} y_1^{1-2\varepsilon} r^{2n-5} dy_1 dr.$$

We set $y_1 = \lambda \cos\theta$, $r = \lambda \sin\theta$. Then we obtain

$$\int_V \frac{\delta(\zeta)^{-\varepsilon}}{|\widetilde{\Phi}(z,\zeta)|^n} dV_{n-1}(\zeta)$$
$$\leq C(\delta(z) + \delta_X(z)^2)^{-\varepsilon} \int_0^{\frac{\pi}{2}} \int_0^\infty (1 + \lambda^2)^{-n+1} \lambda^{2n-3-2\varepsilon} (\cos\theta)^{1-2\varepsilon} d\lambda d\theta$$
$$\leq C_\varepsilon (\delta(z) + \delta_X(z)^2)^{-\varepsilon} \int_0^\infty (1 + \lambda^2)^{-n+1} \lambda^{2n-3-2\varepsilon} d\lambda$$
$$\leq C_\varepsilon (\delta(z) + \delta_X(z)^2)^{-\varepsilon} \int_1^\infty \frac{d\lambda}{\lambda^{1+2\varepsilon}}$$
$$\leq C_\varepsilon (\delta(z) + \delta_X(z)^2)^{-\varepsilon}.$$

□

Now we are going to prove H^p extensions of L^p holomorphic functions in V.

Theorem 3.20 *Let Ω be a strictly pseudoconvex domain in $\mathbf{C}^n$ with smooth boundary and let X be a submanifold in a neighborhood of $\overline{\Omega}$ which intersects $\partial\Omega$ transversally. Let $V = X \cap \Omega$ and $1 \leq p < \infty$. If $f \in L^p(V)$, then $Ef \in H^p(\Omega)$. Moreover, $E: L^p(V) \to H^p(\Omega)$ is a continuous linear operator.*

Proof. We may assume that $V = \Omega \cap X$, where $X = \{z_n = 0\}$. First we assume that $1 < p < \infty$. Let $q > 1$ satisfy $\frac{1}{p} + \frac{1}{q} = 1$. We set $\delta(z) = |\rho(z)|$. For any sufficiently small $\varepsilon > 0$, by applying the Hölder inequality and Lemma 3.34, we have

$$|Ef(z)| \leq C \int_V \frac{|f(\zeta)|\delta(\zeta)^\varepsilon \delta(\zeta)^{-\varepsilon}}{|\widetilde{\Phi}(z,\zeta)|^n} dV_{n-1}(\zeta)$$
$$\leq \left(\int_V \frac{|f(\zeta)|^p \delta(\zeta)^{\varepsilon p}}{|\widetilde{\Phi}(z,\zeta)|^n} dV_{n-1}(\zeta)\right)^{\frac{1}{p}} \left(\int_V \frac{\delta(\zeta)^{-\varepsilon q}}{|\widetilde{\Phi}(z,\zeta)|^n} dV_{n-1}(\zeta)\right)^{\frac{1}{q}}$$
$$\leq C_\varepsilon \delta_X(z)^{-2\varepsilon} \left(\int_V \frac{|f(\zeta)|^p \delta(\zeta)^{\varepsilon p}}{|\widetilde{\Phi}(z,\zeta)|^n} dV_{n-1}(\zeta)\right)^{\frac{1}{p}}.$$

Hence we obtain

$$|Ef(z)|^p \leq C_\varepsilon \delta_X(z)^{-2\varepsilon p} \left(\int_V \frac{|f(\zeta)|^p \delta(\zeta)^{\varepsilon p}}{|\widetilde{\Phi}(z,\zeta)|^n} dV_{n-1}(\zeta) \right).$$

Using Fubini's theorem and Lemma 3.33, we have

$$\int_{\partial\Omega} |Ef(z)|^p d\sigma_{2n-1}(z)$$

$$\leq C_\varepsilon \int_V |f(\zeta)|^p \delta(\zeta)^{\varepsilon p} \left(\int_{\partial\Omega} \frac{\delta_X(z)^{-2\varepsilon p}}{|\widetilde{\Phi}(z,\zeta)|^n} d\sigma_{2n-1}(z) \right) dV_{n-1}(\zeta)$$

$$\leq C_\varepsilon \int_V |f(\zeta)|^p \delta(\zeta)^{\varepsilon p} \delta(\zeta)^{-\varepsilon p} dV_{n-1}(\zeta)$$

$$= C_\varepsilon \int_V |f(\zeta)|^p dV_{n-1}(\zeta).$$

Thus Theorem 3.20 is proved in case $1 < p < \infty$. Next we prove Theorem 3.20 for $p = 1$. By Lemma 3.32 we have

$$Ef(z) = \int_V \frac{f(\zeta)\rho(\zeta)\eta(z,\zeta)}{\widetilde{\Phi}(z,\zeta)^{n+1}} dV_{n-1}(\zeta),$$

where $\eta(z,\zeta)$ is a C^∞ function on $\overline{\Omega} \times \overline{\Omega}$. Let $0 < \varepsilon < 1$. Then we have

$$|Ef(z)| \leq C \int_V |f(\zeta)| \frac{\delta(\zeta)}{|\widetilde{\Phi}(z,\zeta)|^{n+1}} dV_{n-1} X(\zeta) \leq C_\varepsilon \int_V \frac{|f(\zeta)|\delta(\zeta)^\varepsilon}{|\widetilde{\Phi}(z,\zeta)|^{n+\varepsilon}} dV_{n-1}(\zeta).$$

By Lemma 3.33, we obtain

$$\int_{\partial\Omega} |Ef(z)| d\sigma_{2n-1}(z) \leq C_\varepsilon \int_V \left\{ |f(\zeta)|\delta(\zeta)^\varepsilon \int_{\partial\Omega} \frac{d\sigma_{2n-1}(z)}{|\widetilde{\Phi}(z,\zeta)|^{n+\varepsilon}} \right\} dV_{n-1}(\zeta)$$

$$\leq C_\varepsilon \int_V |f(\zeta)|\delta(\zeta)^\varepsilon \delta(\zeta)^{-\varepsilon} dV_{n-1}(\zeta)$$

$$\leq C_\varepsilon \int_V |f(\zeta)| dV_{n-1}(\zeta).$$

$\square$

Next we prove C^k, $k = 1, 2, \cdots, \infty$, extensions of holomorphic functions from submanifolds in general position of strictly pseudoconvex domains with C^∞ boundary.

Lemma 3.35 *Let Ω and V be the same notations as in Lemma 3.32.*

(a) For $z \in \Omega$, there exists a constant $C > 0$ such that
$$\int_{\partial V} \frac{1}{|\Phi(z,\zeta)|^{n-1}} d\sigma_{2n-3}(\zeta) \leq C|\log|\rho(z)||.$$

(b) For $z \in \overline{\Omega}$ and $\delta > 0$, there exists a constant $C > 0$ such that
$$\int_{\partial V \cap B(z,\delta)} \frac{|\zeta - z|}{|\Phi(z,\zeta)|^{n-1}} d\sigma_{2n-3}(\zeta) \leq C\delta|\log\delta|.$$

Proof. (a) Suppose $\delta > 0$ is sufficiently small. Then we can choose a local coordinate system $t = (t_1, \cdots, t_{2n-3})$ on $\partial V \cap B(z, \delta)$ such that $t_1 = \operatorname{Im} \Phi(z, \zeta)$. We set $t' = (t_2, \cdots, t_{2n-3})$. Then
$$\int_{\partial V \cap B(z,\delta)} \frac{1}{|\Phi(z,\zeta)|^{n-1}} d\sigma_{2n-3}(\zeta) \leq C \int_{|t| \leq R} \frac{dt_1 \cdots dt_{2n-3}}{(|\rho(z)| + t_1 + |t|^2)^{n-1}}$$
$$\leq C \int_{|t'| \leq R} \frac{dt_2 \cdots dt_{2n-3}}{(|\rho(z)| + |t'|^2)^{n-2}}$$
$$\leq C \int_0^R \frac{r^{2n-5}}{(|\rho(z)| + r^2)^{n-2}} dr$$
$$\leq C|\log|\rho(z)||.$$

This proves (a).

(b) By (a) we have
$$\int_{\partial V \cap B(z,\delta)} \frac{|\zeta - z|}{|\Phi(z,\zeta)|^{n-1}} d\sigma_{2n-3}(\zeta) \leq C\delta \int_{\partial V \cap B(z,\delta)} \frac{1}{|\Phi(z,\zeta)|^{n-1}} d\sigma_{2n-3}(\zeta)$$
$$\leq C\delta|\log\delta|.$$
□

Now we prove that every strictly pseudoconvex domain with C^2 boundary has a peak function by following Range [RAN2].

Lemma 3.36 *Let Ω be a strictly pseudoconvex domain with C^2 boundary and let $\zeta \in \partial\Omega$. Then there exists a function $f \in A(\Omega)$ such that $f(\zeta) = 1$ and $|f(z)| < 1$ for $z \in \overline{\Omega} \setminus \{\zeta\}$.*

Proof. There exists a neighborhood U of $\partial\Omega$ and a C^2 strictly plurisubharmonic function ρ in U such that $\Omega \cap U = \{z \in U \mid \rho(z) < 0\}$. Let $F(z, \zeta)$ be the Levi polynomial. It follows from (3.19) that there exists $\varepsilon > 0$ such that
$$\operatorname{Re} F(z, \zeta) \geq \rho(\zeta) - \rho(z) + C|\zeta - z|^2$$

for $\zeta \in U$, $|z - \zeta| \leq \varepsilon$. Choose $\varphi \in C^\infty(\mathbf{C}^n \times \mathbf{C}^n)$ such that $0 \leq \varphi \leq 1$ and

$$\varphi(z,\zeta) = \begin{cases} 1 & (|\zeta - z| \leq \frac{\varepsilon}{2}) \\ 0 & (|\zeta - z| \geq \varepsilon) \end{cases}.$$

Fix $\zeta \in \partial\Omega$. Define

$$\lambda(z) = \varphi(z,\zeta)F(z,\zeta) + (1 - \varphi(z,\zeta))|\zeta - z|^2.$$

Then

$$\operatorname{Re}\lambda(z) > 0 \quad \text{for} \quad z \in \overline{\Omega}\setminus\{\zeta\}.$$

Then there exists a neighborhood W of $\overline{\Omega}\setminus\{\zeta\}$ such that $\operatorname{Re}\lambda(z) > 0$ for $z \in W$. We set $u(z) = 1/\lambda(z)$ for $z \in W$. Since $\bar{\partial}u = 0$ on $W \cap B(\zeta, \varepsilon/2)$, $\bar{\partial}u$ extends as a C^∞ (0,1)-form to a neighborhood of $\overline{\Omega}$. By Corollary 2.3 there exists a function $v \in C^\infty(\overline{\Omega})$ such that $\bar{\partial}u = \bar{\partial}v$ in a neighborhood of $\overline{\Omega}$. Define $g = (u - v + |v|_{\overline{\Omega}})^{-1}$. Then $\operatorname{Re} g > 0$ on $\overline{\Omega}\setminus\{\zeta\}$ and $\bar{\partial}g = 0$ in W. Hence g is holomorphic in W. Define $h = e^{-g}$. Since $\lim_{z \to \zeta} h(z) = 1$, h is continuous on $\overline{\Omega}$ and $|h(z)| < 1$ for $z \in \overline{\Omega}\setminus\{\zeta\}$. □

Definition 3.23 Let $K(z,\zeta)$ be the $(2n - 3)$ form in Theorem 3.16. We write $K(z,\zeta)$ in the following form

$$K(z,\zeta) = \widetilde{K}(z,\zeta)d\sigma_{2n-3}(\zeta),$$

where $d\sigma_{2n-3}$ is the surface measure on ∂V. Then $\widetilde{K} : \overline{\Omega} \times \partial V \to \mathbf{C}$ is a C^∞ function on $\overline{\Omega} \times \partial V$ and $\widetilde{K}(\cdot,\zeta)$ is holomorphic in Ω.

The following lemma is due to Ahern-Schneider [AHS1].

Lemma 3.37 Let Ω and V be the same notations as in Lemma 3.32. Then $\widetilde{K}(\zeta,\zeta) \neq 0$ for all $\zeta \in \partial V$.

Proof. Assume that $\widetilde{K}(\zeta_0,\zeta_0) = 0$ for some $\zeta_0 \in \partial V$. We show that

$$f(\zeta_0) = \int_{\partial V} f(\zeta)\frac{\widetilde{K}(\zeta_0,\zeta)}{\Phi(\zeta_0,\zeta)^{n-1}}d\sigma_{2n-3}(\zeta) \qquad (3.33)$$

for $f \in A(V)$. Let $z \in V$. By Theorem 3.16 we have

$$f(z) - \int_{\partial V} f(\zeta) \frac{\widetilde{K}(\zeta_0, \zeta)}{\Phi(\zeta_0, \zeta)^{n-1}} d\sigma_{2n-3}(\zeta)$$

$$= \int_{\partial V \setminus B(\zeta_0, \delta)} f(\zeta) \left[\frac{\widetilde{K}(z, \zeta)}{\Phi(z, \zeta)^{n-1}} - \frac{\widetilde{K}(\zeta_0, \zeta)}{\Phi(\zeta_0, \zeta)^{n-1}} \right] d\sigma_{2n-3}(\zeta)$$

$$+ \int_{\partial V \cap B(\zeta_0, \delta)} f(\zeta) \left[\frac{\widetilde{K}(z, \zeta)}{\Phi(z, \zeta)^{n-1}} - \frac{\widetilde{K}(\zeta_0, \zeta)}{\Phi(\zeta_0, \zeta)^{n-1}} \right] d\sigma_{2n-3}(\zeta)$$

$$:= J_1(z) + J_2(z).$$

It follows from Lebesgue's dominated convergence theorem that $\lim_{z \to \zeta_0} J_1(z) = 0$. On the other hand we have

$$J_2(z) \leq \int_{\partial V \cap B(\zeta_0, \delta)} |f(\zeta)| \frac{|\widetilde{K}(\zeta_0, \zeta)|}{|\Phi(\zeta_0, \zeta)|^{n-1}} d\sigma_{2n-3}(\zeta)$$

$$+ \int_{\partial V \cap B(\zeta_0, \delta)} |f(\zeta)| \frac{|\widetilde{K}(z, \zeta)|}{|\Phi(z, \zeta)|^{n-1}} d\sigma_{2n-3}(\zeta)$$

$$:= J_2'(\zeta_0) + J_2'(z).$$

By Lemma 3.35 (b) we have $|J_2'(\zeta_0)| \leq C|f|_{\overline{V}} \delta |\log \delta|$. To estimate $J_2'(z)$, we let z approach ζ_0 along the inward normal to ∂V. Then we have $|z - \zeta_0| \leq C|z - \zeta|$ and $|\zeta - \zeta_0| \leq C|z - \zeta|$. Consequently,

$$|\widetilde{K}(z, \zeta)| \leq |\widetilde{K}(z, \zeta) - \widetilde{K}(z, \zeta_0)| + |\widetilde{K}(z, \zeta_0) - \widetilde{K}(\zeta_0, \zeta_0)|$$

$$\leq C(|\zeta - \zeta_0| + |z - \zeta_0|) \leq C|\zeta - z|.$$

If $|z - \zeta_0| < \delta$, then

$$J_2'(z) \leq C|f|_{\overline{V}} \int_{\partial V \cap B(\zeta_0, \delta)} \frac{|z - \zeta|}{|\Phi(z, \zeta)|^{n-1}} d\sigma_{2n-3}(\zeta)$$

$$\leq C|f|_{\overline{V}} \int_{\partial V \cap B(z, 2\delta)} \frac{|z - \zeta|}{|\Phi(z, \zeta)|^{n-1}} d\sigma_{2n-3}(\zeta)$$

$$\leq C|f|_{\overline{V}} \delta |\log \delta|.$$

By letting z approach ζ_0 along the inward normal, we have (3.33). By Lemma 3.36 there exists $f \in A(\Omega)$ such that $f(\zeta_0) = 1$ and $|f(z)| < 1$ for

$z \in \overline{\Omega} \backslash \{\zeta_0\}$. Then

$$f(\zeta_0)^N = \int_{\partial V} f(\zeta)^N \frac{\widetilde{K}(\zeta_0,\zeta)}{\Phi(\zeta_0,\zeta)^{n-1}} d\sigma_{2n-3}(\zeta).$$

By Lebesgue's dominated convergence theorem, the right side of the above equality tends to 0 as $N \to \infty$. This is a contradiction. $\square$

Theorem 3.21 *Let Ω be a strictly pseudoconvex domain with C^∞ boundary and let $X = \{z_n = 0\}$, $V = X \cap \Omega$. Suppose X intersects $\partial\Omega$ transversally. If $f \in \mathcal{O}(V) \cap C^k(\overline{V})$, $k = 0, 1, \cdots, \infty$, then $E_1 f \in \mathcal{O}(\Omega) \cap C^k(\overline{\Omega})$.*

Proof. We prove by induction on k that

$$G(z) = \int_{\partial V} \frac{f(\zeta)\lambda(z,\zeta)\widetilde{K}(z,\zeta)}{\Phi(z,\zeta)^{n-1}} d\sigma_{2n-3}(\zeta)$$

belongs to $C^k(\overline{\Omega})$ if $f \in \mathcal{O}(V) \cap C^k(\overline{V})$ and $\lambda \in C^{k+1}(\overline{\Omega} \times \partial V)$. Suppose $k = 0$. Then

$$G(z) = \lambda(z,z) \int_{\partial V} \frac{f(\zeta)k(z,\zeta)}{\Phi(z,\zeta)^{n-1}} d\sigma_{2n-3}(\zeta)$$
$$+ \int_{\partial V} \frac{f(\zeta)k(z,\zeta)(\lambda(z,\zeta) - \lambda(z,z))}{\Phi(z,\zeta)^{n-1}} d\sigma_{2n-3}(\zeta)$$
$$:= G_1(z) + G_2(z).$$

G_1 is continuous on $\overline{\Omega}$ by Theorem 3.18. On the other hand, we obtain

$$\frac{\partial G_2}{\partial z_j}(z) = \int_{\partial V} \frac{\lambda_1(z,\zeta)d\sigma_{2n-3}(\zeta)}{\Phi(z,\zeta)^{n-1}} + \int_{\partial V} \frac{\lambda_2(z,\zeta)O(|z-\zeta|)d\sigma_{2n-3}(\zeta)}{\Phi(z,\zeta)^n},$$

where λ_1 and λ_2 are continuous functions on $\overline{\Omega} \times \overline{V}$. There exists a constant $C > 0$ such that

$$\left| \int_{\partial V} \frac{\lambda_2(z,\zeta)O(|z-\zeta|)d\sigma_{2n-3}(\zeta)}{\Phi(z,\zeta)^n} \right| \leq C\sqrt{|\rho(z)|} |\log|\rho(z)||.$$

It follows from Lemma 3.20 that $G_2 \in \Lambda_\alpha(\Omega)$ for any $0 < \alpha < 1/2$. Hence G_2 is continuous on $\overline{\Omega}$. Assume that the assertion has already been proved for $k - 1$. Let $f \in \mathcal{O}(V) \cap C^k(\overline{V})$ and $\lambda \in C^{k+1}(\overline{\Omega} \times \partial V)$.

If $\Omega = \{z \mid \rho(z) < 0\}$, then $d\rho \neq 0$ on $\partial\Omega$. Let $z^0 \in \partial V$. We may assume that there exist constants $\sigma_1 > 0$ and $\gamma_1 > 0$ such that

$$\left| \frac{\partial \rho}{\partial \zeta_1}(\zeta) \right| > \gamma_1 \quad \text{for} \quad \zeta \in \overline{B}(z^0, \sigma_1).$$

In order to prove the assertion it is sufficient to show that

$$\widetilde{G}(z) = \int_{\partial V \cap B(z^0,\sigma_1)} \frac{f(\zeta)\lambda(z,\zeta)\widetilde{K}(z,\zeta)}{\Phi(z,\zeta)^{n-1}} d\sigma_{2n-3}(\zeta)$$

belongs to $C^k(\overline{\Omega})$. We may assume that (see Theorem 3.8)

$$\Phi(z,\zeta) = F(z,\zeta)M(z,\zeta) \quad \text{for} \quad (z,\zeta) \in \overline{B}(z^0,\sigma_1) \times \overline{B}(z^0,\sigma_1),$$

where $F(z,\zeta)$ is the Levi polynomial and

$$M(z,\zeta) \neq 0 \quad \text{for} \quad ((z,\zeta) \in \overline{B}(z^0,\sigma_1) \times \overline{B}(z^0,\sigma_1).$$

Then we obtain

$$\frac{\partial \Phi}{\partial \zeta_1}(z^0, z^0) = \frac{\partial M}{\partial \zeta_1}(z^0, z^0)F(z^0, z^0) + M(z^0, z^0)\frac{\partial F}{\partial \zeta_1}(z^0, z^0)$$
$$= 2M(z^0, z^0)\frac{\partial \rho}{\partial \zeta_1}(z^0) \neq 0,$$

$$\frac{\partial \Phi}{\partial \bar{\zeta}_1}(z^0, z^0) = \frac{\partial M}{\partial \bar{\zeta}_1}(z^0, z^0)F(z^0, z^0) + M(z^0, z^0)\frac{\partial F}{\partial \bar{\zeta}_1}(z^0, z^0) = 0.$$

There exists $\gamma_2 > 0$ such that

$$\left| \frac{\partial \Phi}{\partial \zeta_1} - \frac{\partial \rho}{\partial \zeta_1} \left(\frac{\partial \rho}{\partial \bar{\zeta}_1} \right)^{-1} \frac{\partial \Phi}{\partial \bar{\zeta}_1} \right| > \gamma_2 \quad \text{on} \quad \overline{B}(z^0,\sigma_1) \times \overline{B}(z^0,\sigma_1).$$

We define

$$d\zeta = d\zeta_1 \wedge \cdots \wedge d\zeta_{n-1},$$

$$[d\zeta]_1 = d\zeta_2 \wedge \cdots \wedge d\zeta_{n-1}, \quad [d\bar{\zeta}]_1 = d\bar{\zeta}_2 \wedge \cdots \wedge d\bar{\zeta}_{n-1}.$$

Then we have

$$d\{\Phi^{-n-1}[d\zeta]_1 \wedge [d\bar{\zeta}]_1\} = -m\Phi^{-n} \left\{ \frac{\partial \Phi}{\partial \zeta_1} - \frac{\partial \rho}{\partial \zeta_1} \left(\frac{\partial \rho}{\partial \bar{\zeta}_1} \right)^{-1} \frac{\partial \Phi}{\partial \bar{\zeta}_1} \right\} d\zeta \wedge [d\bar{\zeta}]_1$$

on $\partial V \cap B(z^0, \sigma_1)$. In view of Lemma 3.37 we may assume that $K(z, \zeta) \neq 0$ for $(z, \zeta) \in \overline{B}(z^0, \sigma_1) \times \overline{B}(z^0, \sigma_1)$. Then we obtain a representation

$$\frac{\partial \widetilde{G}}{\partial z_j}(z) = \int_{\partial V \cap B(z^0, \sigma_1)} \frac{f(\zeta)\lambda_1(z,\zeta)\widetilde{K}(z,\zeta)}{\Phi(z,\zeta)^{n-1}} d\sigma_{2n-3}(\zeta)$$
$$+ \int_{\partial V \cap B(z^0, \sigma_1)} \frac{f(\zeta)\lambda_2(z,\zeta)\widetilde{K}(z,\zeta)}{\Phi(z,\zeta)^n} d\sigma_{2n-3}(\zeta)$$
$$:= \widetilde{G}_1(z) + \widetilde{G}_2(z),$$

where $\lambda_1 \in C^k(\overline{\Omega} \times \partial V)$ and $\lambda_2 \in C^{k+1}(\overline{\Omega} \times \partial V)$. It follows from the inductive hypothesis that $\widetilde{G}_1 \in C^{k-1}(\overline{\Omega})$. On the other hand, $\widetilde{G}_2$ is expressed by

$$\widetilde{G}_2(z) = \int_{\partial V \cap B(z^0, \sigma_1)} \frac{f(\zeta)\lambda_3(z,\zeta)\widetilde{K}(z,\zeta)}{\Phi(z,\zeta)^n} d\zeta \wedge [d\bar{\zeta}]_1$$
$$= \int_{\partial V \cap B(z^0, \sigma_1)} f(\zeta)\lambda_4(z,\zeta)\widetilde{K}(z,\zeta) d\{\Phi^{-(n-1)}[d\zeta]_1 \wedge [d\bar{\zeta}]_1\}.$$

Let $\varphi(z, \zeta)$ be a C^∞ function on $\mathbf{C}^n \times \mathbf{C}^n$ satisfying $\varphi = 0$ on $|z-\zeta| > \sigma_1/2$, $\varphi = 1$ on $|z - \zeta| < \sigma_1/4$. Then

$$\widetilde{G}_2(z)$$
$$= \int_{\partial V \cap B(z^0, \sigma_1)} f(\zeta)\lambda_4(z,\zeta)\widetilde{K}(z,\zeta)\varphi(z,\zeta) d\{\Phi^{-(n-1)}[d\zeta]_1 \wedge [d\bar{\zeta}]_1\}$$
$$+ \int_{\partial V \cap B(z^0, \sigma_1)} f(\zeta)\lambda_4(z,\zeta)\widetilde{K}(z,\zeta)(1-\varphi(z,\zeta)) d\{\Phi^{-(n-1)}[d\zeta]_1 \wedge [d\bar{\zeta}]_1\}$$
$$:= \widetilde{G}_3(z) + \widetilde{G}_4(z).$$

Clearly, $\widetilde{G}_4 \in C^{k-1}(\overline{\Omega})$. Using Stokes' theorem, we have

$$\widetilde{G}_3(z) = \int_{\partial V} f(\zeta)\lambda_4(z,\zeta)\widetilde{K}(z,\zeta)\varphi(z,\zeta) d\{\Phi^{-(n-1)}[d\zeta]_1 \wedge [d\bar{\zeta}]_1\}$$
$$= -\int_{\partial V} d\{f(\zeta)\lambda_4(z,\zeta)\widetilde{K}(z,\zeta)\varphi(z,\zeta)\}\Phi^{-(n-1)}[d\zeta]_1 \wedge [d\bar{\zeta}]_1.$$

It follows from the inductive hypothesis that $\widetilde{G}_3 \in C^{k-1}(\overline{\Omega})$. Hence $\frac{\partial \widetilde{G}}{\partial z_j} \in C^{k-1}(\overline{\Omega})$. Similarly, we have $\frac{\partial \widetilde{G}}{\partial \bar{z}_j} \in C^{k-1}(\overline{\Omega})$, which means that $\widetilde{G} \in C^k(\overline{\Omega})$. Hence $G \in C^k(\overline{\Omega})$. Thus we have proved that $E_1 f \in \mathcal{O}(\Omega) \cap C^k(\overline{\Omega})$ if $f \in \mathcal{O}(V) \cap C^k(\overline{V})$. □

Theorem 3.20 was first proved by Cumenge [CUM]. Adachi [ADA1] and Elgueta [ELG] proved Theorem 3.21 in the case when $k = \infty$, independently. Jakobczak [JK1] also proved Theorem 3.21. The proof of Theorem 3.21 given here is an application of the method of Ahern-Schneider [AHS2]. Amar [AMA2] proved C^∞ extensions of holomorphic functions from submanifold of certain weakly pseudoconvex domains. In case $1 \leq p < \infty$, Adachi [ADA4] obtained L^p extensions of L^p holomorphic functions from submanifolds of strictly pseudoconvex domains with non-smooth boundary. Theorem 3.21 is still open when Ω is a strictly pseudoconvex domain with non-smooth boundary.

3.5 The Bergman Kernel

For the preparation of the next section, we study the Bergman kernel. We begin with an orthonormal system in a Hilbert space.

Lemma 3.38 *(Gram-Schmidt orthonormalization process) Suppose H is a Hilbert space. For a sequence $\{x_n\}$ of linearly independent vectors in H, we set*

$$e_1 = \frac{x_1}{\|x_1\|},$$

$$y_2 = x_2 - (x_2, e_1)e_1, \quad e_2 = \frac{y_2}{\|y_2\|},$$

$$\cdots$$

$$y_n = x_n - \sum_{k=1}^{n-1}(x_n, e_k)e_k, \quad e_n = \frac{y_n}{\|y_n\|},$$

$$\cdots .$$

Then $\{e_n\}$ is an orthonormal system.

Proof. We prove Lemma 3.38 by induction on n. When $n = 1$, the proof is trivial. Assume that the assertion is true when $n = m-1$. Let $1 \leq k < m$. Then

$$(e_m, e_k) = \frac{1}{\|y_m\|}(y_m, e_k) = \frac{1}{\|y_m\|}\left\{(x_m, e_k) - \sum_{j=1}^{m-1}(x_m, e_j)(e_j, e_k)\right\}$$

$$= \frac{1}{\|y_m\|}\{(x_m, e_k) - (x_m, e_k)(e_k, e_k)\} = 0.$$

Since $(e_m, e_m) = 1$, $\{e_1, \cdots, e_n\}$ is an orthonormal system. □

Lemma 3.39 *(Bessel's inequality) Let H be a Hilbert space and let $\{x_1, \cdots, x_n\}$ be an orthonormal system in H. Then*

$$\sum_{k=1}^{n} |(x, x_k)|^2 \leq \|x\|^2$$

for all $x \in H$.

Proof. For any complex numbers $\alpha_1, \cdots, \alpha_n$, it follows from Lemma 3.39 that

$$\left\| \sum_{k=1}^{n} \alpha_k x_k \right\|^2 = \sum_{k=1}^{n} \|\alpha_k x_k\|^2 = \sum_{k=1}^{n} |\alpha_k|^2.$$

Consequently,

$$\left\| x - \sum_{k=1}^{n} \alpha_k x_k \right\|^2 = \left(x - \sum_{k=1}^{n} \alpha_k x_k, x - \sum_{k=1}^{n} \alpha_k x_k \right)$$

$$= \|x\|^2 - \left(x, \sum_{k=1}^{n} \alpha_k x_k \right) - \left(\sum_{k=1}^{n} \alpha_k x_k, x \right) + \sum_{k=1}^{n} |\alpha_k|^2$$

$$= \|x\|^2 - \sum_{k=1}^{n} \overline{\alpha_k}(x, x_k) - \sum_{k=1}^{n} \alpha_k \overline{(x, x_k)} + \sum_{k=1}^{n} \alpha_k \overline{\alpha_k}$$

$$= \|x\|^2 - \sum_{k=1}^{n} |(x, x_k)|^2 + \sum_{k=1}^{n} |(x, x_k) - \alpha_k|^2.$$

If we set $\alpha_k = (x, x_k)$, then

$$0 \leq \|x\|^2 - \sum_{k=1}^{n} |(x, x_k)|^2.$$

$\square$

Lemma 3.40 *Let H be a Hilbert space. Suppose $\{x_n\}$ is an orthonormal system in H and $\{\alpha_n\}$ is a sequence of complex numbers. Then $\sum_{n=1}^{\infty} \alpha_n x_n$ converges if and only if*

$$\sum_{n=1}^{\infty} |\alpha_n|^2 < \infty. \tag{3.34}$$

Proof. For positive integers m, k with $m > k > 0$, we have

$$\left\| \sum_{n=k}^{m} \alpha_n x_n \right\|^2 = \sum_{n=k}^{m} |\alpha_n|^2.$$

We set $s_m = \sum_{n=1}^m \alpha_n x_n$. Then

$$\|s_m - s_k\|^2 = \sum_{n=k}^m |\alpha_n|^2. \tag{3.35}$$

If (3.34) holds, then by (3.35) $\{s_m\}$ is a Cauchy sequence, and hence $\{s_n\}$ converges. Conversely, if $\{s_n\}$ converges, then $\{s_n\}$ is a Cauchy sequence, and hence (3.34) follows from (3.35). □

Definition 3.24 Let H be a Hilbert space. An orthonormal system $\{x_n\}$ in H is said to be complete if

$$x = \sum_{n=1}^\infty (x, x_n) x_n$$

for every $x \in H$.

Lemma 3.41 Let H be a Hilbert space and let $\{x_n\}$ be an orthonormal system in H. Then $\{x_n\}$ is complete if and only if the following holds:

$$(x, x_n) = 0 \quad (n = 1, 2, \cdots) \quad \Longrightarrow \quad x = 0. \tag{3.36}$$

Proof. Let $\{x_n\}$ be complete. Then for any $x \in H$ we have

$$x = \sum_{n=1}^\infty (x, x_n) x_n.$$

Hence (3.36) holds. Conversely, assume that (3.36) holds. It follows from the Bessel inequality that

$$\sum_{k=1}^\infty |(x, x_k)|^2 \leq \|x\|^2.$$

By Lemma 3.40, $\sum_{n=1}^\infty (x, x_n) x_n$ converges. We set $y = \sum_{n=1}^\infty (x, x_n) x_n$. Then we have

$$(x - y, x_n) = (x, x_n) - \left(\sum_{k=1}^\infty (x, x_k) x_k, x_n \right)$$

$$= (x, x_n) - \sum_{k=1}^\infty (x, x_k)(x_k, x_n)$$

$$= (x, x_n) - (x, x_n) = 0.$$

Hence, by the assumption we have $x - y = 0$. Therefore we have

$$x = \sum_{n=1}^{\infty} (x, x_n) x_n.$$

Hence $\{x_n\}$ is complete. □

Lemma 3.42 *(Parseval's equality) Let H be a Hilbert space and let $\{x_n\}$ be an orthonormal system in H. Then $\{x_n\}$ is complete if and only if*

$$\|x\|^2 = \sum_{n=1}^{\infty} |(x, x_n)|^2 \qquad (3.37)$$

for all $x \in H$.

Proof. Let $x \in H$. For a positive integer n we have

$$\left\| x - \sum_{k=1}^{n} (x, x_k) x_k \right\|^2 = \|x\|^2 - \sum_{k=1}^{n} |(x, x_k)|^2.$$

Suppose $\{x_n\}$ is complete. Then the left side of the above equality converges to 0 as $n \to \infty$. Hence (3.37) holds. Conversely, assume that (3.37) holds. Then we have

$$x = \sum_{n=1}^{\infty} (x, x_n) x_n,$$

which implies that $\{x_n\}$ is complete. □

Lemma 3.43 *(Riesz-Fischer theorem) Let H be a Hilbert space and let $\{u_j\}$ be a complete orthonormal system in H. Then*

(a) For $x \in H$, we set $\alpha_j = (x, u_j)$. Then $\sum_{j=1}^{N} \alpha_j u_j$ converges to x as $N \to \infty$. Further, we have $\|x\|^2 = \sum_{j=1}^{\infty} |\alpha_j|^2$.

(b) If $\sum_{j=1}^{\infty} |\beta_j|^2 < \infty$, then there exists $x \in H$ such that $(x, u_j) = \beta_j$ for all j, and

$$\|x\|^2 = \sum_{j=1}^{\infty} |\beta_j|^2, \quad x = \sum_{j=1}^{\infty} \beta_j u_j.$$

Proof. We have already proved (a). Suppose $\sum_{j=1}^{\infty} |\beta_j|^2 < \infty$. We set

$$x_n = \sum_{j=1}^{n} \beta_j u_j.$$

For $n \geq m > 0$, we have

$$\|x_n - x_m\|^2 = \left\| \sum_{j=m+1}^{n} \beta_j u_j \right\|^2 = \sum_{j=m+1}^{n} |\beta_j|^2 \to 0 \quad (m, n \to \infty).$$

Hence $\{x_n\}$ is a Cauchy sequence. Since H is a Hilbert space, $\{x_n\}$ converges. Let $\lim_{n \to \infty} x_n = x$. Then we have

$$x = \sum_{j=1}^{\infty} \beta_j u_j.$$

Since $\|x_n\|^2 \to \|x\|^2$, we have

$$\|x\|^2 = \sum_{j=1}^{\infty} |\beta_j|^2.$$

This proves (b). $\square$

Lemma 3.44 *Define*

$$l^2 = \left\{ \alpha = \{a_j\} \;\middle|\; \sum_{j=1}^{\infty} |a_j|^2 < \infty, \; a_j \in \mathbf{C} \right\}.$$

For $\alpha = \{a_j\}, \beta = \{b_j\} \in l^2$, we define an inner product by

$$(\alpha, \beta) = \sum_{j=1}^{\infty} a_j \overline{b_j}.$$

Then l^2 is a Hilbert space. Further we have

$$\|\beta\|_{l^2} = \left(\sum_{j=1}^{\infty} |b_j|^2 \right)^{1/2} = \sup_{\substack{\alpha \in l^2 \\ \|\alpha\| \leq 1}} |(\alpha, \beta)|.$$

Proof. For $\|\alpha\| \leq 1$, we have

$$|(\alpha, \beta)| \leq \|\alpha\| \|\beta\| \leq \|\beta\|.$$

On the other hand, if we set

$$c_j = \frac{b_j}{\|\beta\|}, \quad \gamma = \{c_j\},$$

then $\|\gamma\| = 1$. Moreover we have

$$|(\gamma, \beta)| = \|\beta\|.$$

□

Lemma 3.45 *Let H be a Hilbert space. Then the following statements are equivalent:*

(a) H is separable.
(b) H contains a complete orthonormal system which is at most countable.

Proof. (a) $\Longrightarrow$ (b). Suppose H is separable. Let $E = \{x_n \mid n = 1, 2, \cdots\} \subset H$, $\overline{E} = H$. If x_n is a linear combination of $x_1, \cdots, x_{n-1}$, then we omit x_n from E. Let $\{y_n\}$ be a subsequence of $\{x_n\}$ obtained by this process. Since $\{y_n\}$ is linearly independent, by the Schmidt orthonormalization process we have an orthonormal system $\{e_n\}$. The set of all linear combinations of elements in $\{e_n\}$ is equal to the set of all linear combinations of elements in $\{x_n\}$. Hence $\{e_n\}$ is dense in H. Let $x \in H$. For any $\varepsilon > 0$, there exists positive integer N such that

$$\left\| x - \sum_{n=1}^{N} c_n x_n \right\| < \varepsilon.$$

Since

$$\left\| x - \sum_{n=1}^{N} c_n x_n \right\|^2 \geq \left\| x - \sum_{n=1}^{N} (x, e_n) e_n \right\|^2 = \|x\|^2 - \sum_{n=1}^{N} |(x, e_n)|^2,$$

we have

$$\|x\|^2 \leq \sum_{n=1}^{\infty} |(x, e_n)|^2 + \varepsilon^2.$$

Since $\varepsilon > 0$ is arbitrary, we have

$$\|x\|^2 \leq \sum_{n=1}^{\infty} |(x, e_n)|^2.$$

It follows from the Bessel inequality that

$$\|x\|^2 = \sum_{n=1}^{\infty} |(x, e_n)|^2.$$

Hence $\{e_n\}$ is a complete orthonormal system.

(b) $\Longrightarrow$ (a). Suppose H contains a complete orthonormal system $\{e_n\}$ which is at most countable. We set

$$A = \left\{ \sum_{n=1}^{k} \alpha_n e_n \mid \alpha_n = a_n + ib_n \in \mathbf{Q} + i\mathbf{Q},\ k \in \mathbf{N} \right\}.$$

Then A is a countable set. For any $x \in H$ we have

$$x = \sum_{n=1}^{\infty} (x, e_n) e_n.$$

Hence we have

$$\left\| x - \sum_{n=1}^{N} (x, e_n) e_n \right\| \to 0 \quad (N \to \infty).$$

Then $\overline{A} = H$, and hence H is separable. $\square$

Definition 3.25 Let $\Omega \subset \mathbf{C}^n$ be an open set. We denote by $A^2(\Omega)$ the set of all holomorphic functions f in Ω satisfying

$$\int_{\Omega} |f(\zeta)|^2 dV(\zeta) < \infty.$$

$A^2(\Omega)$ is called the Bergman space.

Lemma 3.46 $A^2(\Omega)$ is a closed subspace of $L^2(\Omega)$.

Proof. For simplicity, we prove Lemma 3.46 in case $n = 1$. The proof of the general case will be left to the reader. Let $K \subset \Omega$ be compact. We choose $r > 0$ such that $\overline{B}(w, r) \subset \Omega$ for every $w \in K$. Let $w \in K$ and $h \in A^2(\Omega)$. It follows from the Cauchy integral formula that

$$h(w) = \frac{1}{2i\pi} \int_{|z-w|=\rho} \frac{h(z)}{z-w} dz = \frac{1}{2\pi} \int_0^{2\pi} h(w + \rho e^{i\theta}) d\theta \qquad (3.38)$$

for $0 < \rho \leq r$. If we multiply by ρ and integrate from 0 to r, then we have

$$\frac{r^2}{2}h(w) = \int_0^r \left(\frac{1}{2\pi}\int_0^{2\pi} h(w+\rho e^{i\theta})d\theta\right)\rho d\rho$$

$$= \frac{1}{2\pi}\int\int_{|z-w|\leq r} h(z)\rho d\rho d\theta$$

$$= \frac{1}{2\pi}\int\int_{|z-w|\leq r} h(z)dxdy.$$

By the Hölder inequality we obtain

$$|h(w)| \leq \frac{1}{\pi r^2}\int_{|z-w|\leq r} |h(z)|dxdy$$

$$\leq \frac{1}{\pi r^2}\left(\int_{|z-w|\leq r} |h(z)|^2 dxdy\right)^{1/2}\left(\int_{|z-w|\leq r} dxdy\right)^{1/2}$$

$$\leq \frac{1}{\pi r^2}\left(\int_\Omega |h(z)|^2 dxdy\right)^{1/2}\sqrt{\pi}r$$

$$= \frac{1}{\sqrt{\pi}r}\|h\|.$$

Suppose $f_j \in A^2(\Omega)$ for $j = 1, 2, \cdots$, $f \in L^2(\Omega)$ and $f_j \to f$. For $\varepsilon > 0$, if we choose N sufficiently large, then for $j, k \geq N$, we have

$$\|f_j - f_k\| < \varepsilon\sqrt{\pi}r.$$

Therefore, we obtain

$$|f_j(w) - f_k(w)| \leq \frac{1}{\sqrt{\pi}r}\|f_j - f_k\| < \varepsilon$$

for $w \in K$, which implies that $\{f_j\}$ converges to a holomorphic function f uniformly on every compact subset of Ω. Now we show that $f \in A^2(\Omega)$. Since $\{f_j\}$ is a Cauchy sequence, there exists positive number M such that $\|f_j\| \leq M$ for all j. On the other hand, for any compact subset $K \subset \Omega$

$$\int_K |f_j(z)|^2 dxdy \leq \int_\Omega |f_j(z)|^2 dxdy = \|f_j\|^2 \leq M^2.$$

Hence we obtain

$$\int_K |f(z)|^2 dxdy \leq M^2.$$

Since K is independent of M, we have

$$\int_\Omega |f(z)|^2 dxdy \le M^2.$$

Thus $f \in A^2(\Omega)$. Hence $A^2(\Omega)$ is a closed subspace of $L^2(\Omega)$. □

Theorem 3.22 $A^2(\Omega)$ *has a countable complete orthonormal system.*

Proof. Since $A^2(\Omega)$ is a separable Hilbert space (see Adams [ADM]), Theorem 3.22 follows from Lemma 3.45. □

Lemma 3.47 *Let $\{\varphi_j\}$ be a complete orthonormal sequence in $A^2(\Omega)$. Then*

$$\sum_{j=1}^\infty |\varphi_j(z)|^2 = \sup_{\substack{f \in A^2(\Omega) \\ \|f\| \le 1}} |f(z)|^2 < \infty$$

for all $z \in \Omega$.

Proof. Let K be a compact subset of Ω and let $z \in K$. By the Riesz-Fischer theorem we have

$$\sum_{j=1}^\infty |\varphi_j(z)|^2 = \sup_{\substack{\{a_j\} \in l^2 \\ \|\{a_j\}\|_{l^2} \le 1}} \left| \sum_{j=1}^\infty a_j \varphi_j(z) \right|^2 = \sup_{\substack{f \in A^2(\Omega) \\ \|f\| \le 1}} |f(z)|^2.$$

Consequently,

$$\sum_{j=1}^\infty |\varphi_j(z)|^2 \le \sup_{\substack{f \in A^2(\Omega) \\ \|f\| \le 1}} c_K \|f\|^2 \le c_K < \infty,$$

where c_K is a constant depending only on K. □

Lemma 3.48 *Let $\Omega \subset \mathbf{C}^n$ be an open set. For $a \in \Omega$, define $\tau_a : A^2(\Omega) \to \mathbf{C}$ by $\tau_a(f) = f(a)$. Then τ_a is a bounded linear functional on $A^2(\Omega)$.*

Proof. It is clear that τ_a is a linear functional. We choose $r_j > 0$ for $j = 1, \cdots, n$ such that

$$\{(z_1, \cdots, z_n) \mid |z_j - a_j| \le r_j, j = 1, \cdots, n\} \subset \Omega.$$

It follows from the proof of Lemma 3.46 that

$$f(a) = \frac{1}{\pi r_1^2} \int_{|z_1-a_1|\leq r_1} f(z_1, a_2, \cdots, a_n) dx_1 dy_1$$

$$= \frac{1}{\pi^n r_1^2 \cdots r_n^2} \int_{|z_1-a_1|\leq r_1} \cdots \int_{|z_n-a_n|\leq r_n} f(z_1, \cdots, z_n) dV.$$

It follows from the Hölder inequality that there exists a constant $C_a > 0$ such that

$$|\tau_a(f)| = |f(a)| \leq C_a \|f\|_{L^2}.$$

Thus τ_a is bounded. □

Definition 3.26 By Lemma 3.48 and the Riesz representation theorem, there exists $g \in A^2(\Omega)$ such that

$$\tau_a(f) = (f, g) \qquad (f \in A^2(\Omega)).$$

We define $g(z) = K_\Omega(z, a)$. We say that $K_\Omega : \Omega \times \Omega \to \mathbf{C}$ is the Bergman kernel for Ω.

By definition we obtain

$$f(z) = \int_\Omega f(\zeta)\overline{K_\Omega(\zeta, z)} dV(\zeta) \qquad (f \in A^2(\Omega)). \tag{3.39}$$

Lemma 3.49 For any $z, \zeta \in \Omega$, $K_\Omega(\zeta, z) = \overline{K_\Omega(z, \zeta)}$.

Proof. For $z \in \Omega$ fixed, we have $K_\Omega(\cdot, z) \in A^2(\Omega)$. If we set $f(\zeta) = K_\Omega(\zeta, z)$, then (3.39) shows that

$$K_\Omega(\zeta, z) = f(\zeta) = \int_\Omega f(w)\overline{K_\Omega(w, \zeta)} dV(w)$$

$$= \int_\Omega K_\Omega(w, z)\overline{K_\Omega(w, \zeta)} dV(w)$$

$$= \overline{\int_\Omega \overline{K_\Omega(w, z)} K_\Omega(w, \zeta) dV(w)}$$

$$= \overline{K_\Omega(z, \zeta)}.$$

□

It follows from Lemma 3.49 and (3.39) that

$$f(z) = \int_\Omega f(\zeta) K_\Omega(z, \zeta) dV(\zeta) \qquad (f \in A^2(\Omega), z \in \Omega). \tag{3.40}$$

Lemma 3.50 There exists a constant $C > 0$ such that

$$\|K_\Omega(\cdot, a)\|_{L^2} \leq C\delta_\Omega^{-n}(a) \qquad (a \in \Omega),$$

where $\delta_\Omega(a) = \mathrm{dist}(a, \partial\Omega)$.

Proof. We choose r such that $r < \delta_\Omega(a)/\sqrt{n}$. Then $\{z \mid |z_i - a_i| \leq r\} \subset \Omega$. Using the same method as the proof of Lemma 3.46, we have

$$f(a) = \frac{1}{\pi^n r^{2n}} \int_{|z_1-a_1|\leq r} \cdots \int_{|z_n-a_n|\leq r} f(z_1, \cdots, z_n) dV.$$

By Hölder's inequality,

$$|f(a)| \leq \frac{\|f\|_{L^2}}{(\sqrt{\pi})^n r^n}.$$

Letting $r \to \delta_\Omega(a)/\sqrt{n}$, we have

$$|f(a)| \leq \left(\sqrt{\frac{n}{\pi}}\right)^n (\delta_\Omega(a))^{-n} \|f\|_{L^2}. \tag{3.41}$$

On the other hand, by the Riesz representation theorem we obtain

$$\|\tau_a\| = \|K_\Omega(\cdot, a)\|_{L^2}.$$

It follows from (3.41) that

$$\|\tau_a\| = \sup_{\|f\|_{L^2}=1} |\tau_a(f)| = \sup_{\|f\|_{L^2}=1} |f(a)| \leq \left(\sqrt{\frac{n}{\pi}}\right)^n (\delta_\Omega(a))^{-n}.$$

We set

$$C = \left(\sqrt{\frac{n}{\pi}}\right)^n.$$

Then we have the desired inequality. □

Lemma 3.51 Let K be a compact subset of Ω, Then there exists a constant $C_K > 0$ such that for every complete orthonormal sequence $\{\varphi_j\}$ in $A^2(\Omega)$,

$$\sup_{z \in K} \sum_{j=1}^{\infty} |\varphi_j(z)|^2 \leq C_K.$$

Proof. For $z \in K$, we have $\delta_\Omega(z) \geq \text{dist}(K, \partial\Omega)$. It follows from Lemma 3.50 that

$$\|K_\Omega(\cdot, z)\|_{L^2} \leq C(\text{dist}(K, \partial\Omega))^{-n} = C_K \qquad (z \in K).$$

Since $K_\Omega(\cdot, z) \in A^2(\Omega)$, we have

$$K_\Omega(\zeta, z) = \sum_{j=1}^\infty (K_\Omega(\cdot, z), \varphi_j)\varphi_j(\zeta). \qquad (3.42)$$

By Lemma 3.42 we obtain

$$\sum_{j=1}^\infty |(K_\Omega(\cdot, z), \varphi_j)|^2 = \|K_\Omega(\cdot, z)\|_{L^2}^2.$$

It follows from (3.40) that

$$\varphi_j(z) = \int_\Omega \varphi_j(\zeta) K_\Omega(z, \zeta) dV(\zeta) = (\varphi_j, K_\Omega(\cdot, z)). \qquad (3.43)$$

Hence we have

$$\sum_{j=1}^\infty |\varphi_j(z)|^2 = \|K_\Omega(\cdot, z)\|_{L^2}^2 \leq C_K.$$

$\square$

Theorem 3.23 *Let $\{\varphi_j\}$ be a complete orthonormal sequence in $A^2(\Omega)$. Then*

$$K_\Omega(\zeta, z) = \sum_{j=1}^\infty \varphi_j(\zeta)\overline{\varphi_j(z)} \qquad ((\zeta, z) \in \Omega \times \Omega). \qquad (3.44)$$

Moreover, the infinite series in the right side of (3.44) converges uniformly on every compact subset of $\Omega \times \Omega$.

Proof. If we substitute (3.43) into (3.42), then we obtain (3.44). Suppose $K \subset \Omega$ is compact. It is sufficient to show that the infinite series in the right side of (3.44) converges uniformly on $K \times K$. It follows from Lemma 3.51 that

$$\sum_{j=1}^\infty |\varphi_j(\zeta)||\varphi_j(z)| \leq \left(\sum_{j=1}^\infty |\varphi_j(\zeta)|^2\right)^{1/2} \left(\sum_{j=1}^\infty |\varphi_j(z)|^2\right)^{1/2} \leq C_K$$

for $\zeta, z \in K$. We set

$$g_n(z,\zeta) = \sum_{j=1}^{n} |\varphi_j(\zeta)||\varphi_j(z)|.$$

Then $\{g_n\}$ converges monotonically on K. In view of Lemma 1.19 $\{g_n\}$ converges uniformly on $K \times K$. Hence the infinite series in the right side of (3.44) converges uniformly on every compact subset of $\Omega \times \Omega$. □

Corollary 3.7 $K_\Omega \in C^\infty(\Omega \times \Omega)$.

Proof. By Theorem 3.23, $K_\Omega(\zeta, z)$ is continuous in $\Omega \times \Omega$. Since $K_\Omega(\zeta, z)$ is holomorphic with respect to $(\zeta, \bar{z})$, $K_\Omega(\zeta, z)$ is expressed by the Cauchy integral. Differentiating under the integral sign, derivatives of $K_\Omega(\zeta, z)$ of any order are continuous in $\Omega \times \Omega$, which completes the proof of Corollary 3.7. □

Lemma 3.52 *Suppose a function $\widetilde{K}_\Omega : \Omega \times \Omega \to \mathbf{C}$ satisfies the following properties:*

(1) $\overline{\widetilde{K}_\Omega(z, \cdot)} \in A^2(\Omega)$ for every fixed $z \in \Omega$.
(2) $f(z) = \int_\Omega f(\zeta) \widetilde{K}_\Omega(z, \zeta) dV(\zeta)$ for every $f \in A^2(\Omega)$.

Then

$$\widetilde{K}_\Omega = K_\Omega.$$

Proof. For $z \in \Omega$, define $k_z(\zeta) = \overline{K_\Omega(z, \zeta)}$. Since $k_z \in A^2(\Omega)$, we obtain

$$K_\Omega(w, z) = \overline{K_\Omega(z, w)} = \overline{k_z(w)} = \overline{\int_\Omega k_z(\zeta) \widetilde{K}_\Omega(w, \zeta) dV(\zeta)}$$
$$= \overline{\int_\Omega k_z(\zeta) \widetilde{K}_\Omega(w, \zeta) dV(\zeta)}$$
$$= \int_\Omega \overline{\widetilde{K}_\Omega(w, \zeta)} K_\Omega(z, \zeta) dV(\zeta)$$
$$= \widetilde{K}_\Omega(w, z).$$

□

Definition 3.27 Let Ω be a domain in $\mathbf{C}^n$. For a C^1 mapping

$$F = (f_1, \cdots, f_n) : \Omega \to \mathbf{C}^n,$$

define

$$F'(z) = \begin{pmatrix} \frac{\partial f_1}{\partial z_1}(z) & \cdots & \frac{\partial f_1}{\partial z_n}(z) \\ \vdots & \vdots & \vdots \\ \frac{\partial f_n}{\partial z_1}(z) & \cdots & \frac{\partial f_n}{\partial z_n}(z) \end{pmatrix}.$$

Theorem 3.24 *Let Ω_j, $j = 1, 2$, be bounded domains in $\mathbf{C}^n$ and let $F : \Omega_1 \to \Omega_2$ be a biholomorphic mapping. Then*

$$K_{\Omega_1}(\zeta, z) = \det F'(\zeta) K_{\Omega_2}(F(\zeta), F(z)) \overline{\det F'(z)} \qquad (\zeta, z \in \Omega_1).$$

Proof. We set

$$H(z, w) = \det F'(\zeta) K_{\Omega_2}(F(\zeta), F(z)) \overline{\det F'(z)}.$$

Then for a fixed point $z \in \Omega_1$, $\overline{H(z, \cdot)}$ is holomorphic in Ω_1. Differentiating $z = F^{-1}(F(z))$, we have

$$1 = \det(F^{-1})'(F(z)) \det F'(z).$$

We set $\tilde{z} = F(z)$. Then using the Cauchy-Riemann equation, the Jacobian of F is equal to $|\det F'|^2$. Hence we have

$$\int_{\Omega_1} |H(z, w)|^2 dV(z)$$
$$= \int_{\Omega_1} |\det F'(z)|^2 |K_{\Omega_2}(F(z), F(w))|^2 |\det F'(w)|^2 dV(z)$$
$$= \int_{\Omega_2} |\det F'(F^{-1}(\tilde{z}))|^2 |K_{\Omega_2}(\tilde{z}, F(w))|^2 |\det F'(w)|^2 |\det(F^{-1})'(\tilde{z})|^2 dV$$
$$= |\det F'(w)|^2 \int_{\Omega_2} |K_{\Omega_2}(\tilde{z}, F(w))|^2 |dV(\tilde{z}) < \infty.$$

Since $H(z, w) = \overline{H(w, z)}$, we have

$$\int_{\Omega_1} |H(z, w)|^2 dV(w) < \infty,$$

which means that $\overline{H(z, \cdot)} \in A^2(\Omega)$. Next we show that $H(z, w)$ is the reproducing kernel for Ω_1. Let $f \in A^2(\Omega_1)$. If we set $\tilde{\zeta} = F(\zeta)$, then we have

$$\int_{\Omega_1} f(\zeta)H(z,\zeta)dV(\zeta)$$
$$= \int_{\Omega_1} f(\zeta)\det F'(z)K_{\Omega_2}(F(z),F(\zeta))\overline{\det F'(\zeta)}dV(\zeta)$$
$$= \det F'(z)\int_{\Omega_2} f(F^{-1}(\tilde{\zeta}))\det(F^{-1})'(\tilde{\zeta})K_{\Omega_2}(F(z),\tilde{\zeta})dV(\tilde{\zeta}).$$

We set

$$g(\tilde{\zeta}) = f(F^{-1}(\tilde{\zeta}))\det(F^{-1})'(\tilde{\zeta}).$$

Then g is holomorphic in Ω_2. Moreover, we have

$$\int_{\Omega_1} |f(\zeta)|^2 dV(\zeta) = \int_{\Omega_2} |g(\tilde{\zeta})|^2 dV(\tilde{\zeta}),$$

which implies that $g \in A^2(\Omega_2)$. Thus we obtain

$$\int_{\Omega_1} f(\zeta)H(z,\zeta)dV(\zeta) = \det F'(z)\int_{\Omega_2} g(\tilde{\zeta})K_{\Omega_2}(F(z),\tilde{\zeta})dV(\tilde{\zeta})$$
$$= \det F'(z)g(F(z)) = f(z).$$

By Lemma 3.52, we have $H(z,w) = K_{\Omega_1}(z,w)$ □

3.6 Fefferman's Mapping Theorem

We prove Fefferman's mapping theorem [FEF] which says that every biholomorphic mapping between two strictly pseudoconvex domains with C^∞ boundary can be extended to a C^∞ mapping up to the boundary. Bell-Ligocka [BEL] gave a simple proof of Fefferman's mapping theorem. In what follows we give the proof of Fefferman's mapping theorem by following the methods of Range [RAN2]. Range obtained C^k extensions up to the boundary under the assumption that $\partial\Omega$ is of class C^{2k+4} ($1 \leq k \leq \infty$). For simplicity, we assume that $\partial\Omega$ is of class C^∞. In order to prove Fefferman's mapping theorem we use the homotopy formula for strictly pseudoconvex domains constructed in 3.2.

Let $\Omega \subset\subset \mathbf{C}^n$ be a strictly pseudoconvex domain with C^∞ boundary. Suppose the neighborhood U of $\partial\Omega$, the functions φ, w and $\tilde{\Phi}$ are as in 3.2. We will adopt the convention of denoting by C any positive constant which does not depend on the relevant parameters in the estimates.

If f is an L^1 holomorphic function in Ω, then it follows from Theorem 3.12 that

$$f(z) = L_\Omega f(z) = \frac{n!}{(2\pi i)^n} \int_\Omega f(\zeta) \omega_\zeta \left(\frac{\chi(\zeta) w(z,\zeta)}{\widetilde{\Phi}(z,\zeta)} \right) \wedge \omega(\zeta)$$

for $z \in \Omega$.

Define

$$dV(\zeta) = dx_1 \wedge dx_2 \wedge \cdots \wedge dx_{2n}$$

for $\zeta_j = x_j + ix_{j+n}$ and $j = 1, \cdots, n$. Then there exists a C^∞ function $G(z,\zeta)$ in $\Omega \times \overline{\Omega}$ which is holomorphic in $z \in \Omega$ for fixed $\zeta \in \overline{\Omega}$ such that

$$\frac{n!}{(2\pi i)^n} \omega_\zeta \left(\frac{\chi(\zeta) w(z,\zeta)}{\widetilde{\Phi}(z,\zeta)} \right) \wedge \omega(\zeta) = G(z,\zeta) dV(\zeta).$$

Then we have

$$f(z) = \int_\Omega f(\zeta) G(z,\zeta) dV(\zeta) \qquad (z \in \Omega). \tag{3.45}$$

For $\zeta \in U$ and $-\varepsilon_0 < \rho(\zeta)$, we have $\chi(\zeta) = 1$. Hence for $\zeta \in U$ and $-\varepsilon_0 < \rho(\zeta)$, we have

$$G(z,\zeta) dV(\zeta) = \frac{n!}{(2\pi i)^n} \bigwedge_{j=1}^n \bar{\partial}_\zeta \left(\frac{w_j(z,\zeta)}{\widetilde{\Phi}(z,\zeta)} \right) \wedge \omega(\zeta).$$

If we choose $\varepsilon > 0$ ($0 < \varepsilon < \varepsilon_0$) sufficiently small, then it follows from Theorem 3.8 and Theorem 3.9 that

$$\Phi(z,\zeta) = F(z,\zeta) M(z,\zeta) = \sum_{j=1}^n w_j(z,\zeta)(z_j - \zeta_j)$$

for $|z - \zeta| \leq \varepsilon$. Differentiating the above equality with respect to ζ_j we have

$$w_j(z,\zeta) = \frac{\partial F}{\partial \zeta_j}(z,\zeta) M(z,\zeta) + O(|\zeta - z|)$$

$$= 2M(z,\zeta) \frac{\partial \rho(\zeta)}{\partial \zeta_j} + O(|\zeta - z|),$$

$$\widetilde{\Phi}(z,\zeta) = (F(z,\zeta) - 2\rho(\zeta)) \widetilde{M}(z,\zeta),$$

$$\bar{\partial}_\zeta \widetilde{\Phi}(z,\zeta) = (\bar{\partial}_\zeta F(z,\zeta) - 2\bar{\partial}\rho(\zeta))\widetilde{M}(z,\zeta) + (F(z,\zeta) - 2\rho(\zeta))\bar{\partial}_\zeta \widetilde{M}(z,\zeta).$$

Let $V_{\partial\Omega}$ be a neighborhood of $\partial\Omega$. If we denote by $N(z,\zeta)$ any C^∞ form in $V_{\partial\Omega} \times V_{\partial\Omega}$, then we have

$$G(z,\zeta)dV(\zeta) = \frac{n!}{(2\pi i)^n} \bigwedge_{j=1}^n \left(\frac{\bar{\partial}_\zeta w_j}{\widetilde{\Phi}} + \frac{w_j \bar{\partial}_\zeta \widetilde{\Phi}}{\widetilde{\Phi}^2} \right) \wedge \omega(\zeta)$$

$$= \frac{n!}{(2\pi i)^n} \sum_{j=1}^n (-1)^{j-1} \frac{w_j \wedge_{k\neq j} \bar{\partial}_\zeta w_k}{\widetilde{\Phi}^{n+1}} \wedge \bar{\partial}_\zeta \widetilde{\Phi} \wedge \omega(\zeta) + \frac{N}{\widetilde{\Phi}^n}$$

$$= \frac{n!}{(2\pi i)^n} \frac{\omega'(w(z,\zeta)) \wedge (\bar{\partial}_\zeta F - 2\bar{\partial}\rho(\zeta))\widetilde{M}}{\widetilde{\Phi}(z,\zeta)^{n+1}} \wedge \omega(\zeta) + \frac{N}{\widetilde{\Phi}^n}.$$

We set

$$P = (P_1, \cdots, P_n) = \left(\frac{\partial \rho}{\partial \zeta_1}, \cdots, \frac{\partial \rho}{\partial \zeta_n} \right).$$

Then we have

$G(z,\zeta)dV(\zeta)$

$$= \frac{n!}{(2\pi i)^n} \frac{\omega'(P(\zeta))(2M)^n \widetilde{M}(\bar{\partial}_\zeta F - 2\bar{\partial}\rho(\zeta))}{\widetilde{\Phi}^{n+1}} \wedge \omega(\zeta) + \frac{O(|z-\zeta|)}{\widetilde{\Phi}^{n+1}} + \frac{N}{\widetilde{\Phi}^n}$$

$$= \frac{n!}{(2\pi i)^n} \frac{\omega'(P(\zeta)) \wedge (-2\bar{\partial}\rho(\zeta))2^n M^n}{(F(z,\zeta) - 2\rho(\zeta))^{n+1}\widetilde{M}^n} \wedge \omega(\zeta) + \frac{O(|z-\zeta|)}{\widetilde{\Phi}^{n+1}} + \frac{N}{\widetilde{\Phi}^n}.$$

On the other hand, for $\zeta_0 \in \partial\Omega$ with $|\zeta - \zeta_0| = \text{dist}(\zeta, \partial\Omega)$, we have $M(z,\zeta_0) = \widetilde{M}(z,\zeta_0)$. Hence we have

$$\frac{M(z,\zeta)^n}{\widetilde{M}(z,\zeta)^n} = \frac{M(\zeta,\zeta)^n}{\widetilde{M}(\zeta,\zeta)^n} + O(|\zeta - z|) = 1 + O(|\zeta - \zeta_0|) + O(|\zeta - z|).$$

Consequently, for $\zeta \in \overline{\Omega}$ we obtain

$$\frac{M(z,\zeta)^n}{\widetilde{M}(z,\zeta)^n \widetilde{\Phi}(z,\zeta)} = \frac{1}{\widetilde{\Phi}(z,\zeta)} + \frac{O(|\zeta - z|)}{\widetilde{\Phi}(z,\zeta)}.$$

Further we have

$$\bar{\partial}\rho(\zeta) \wedge \omega'(P(\zeta)) \wedge \omega(\zeta) = \bar{\partial}\rho(\zeta) \wedge \partial\rho(\zeta) \wedge (\bar{\partial}\partial\rho(\zeta))^{n-1}.$$

We set

$$H(\zeta) = (2\pi i)^{-n} \bar{\partial}\rho(\zeta) \wedge \partial\rho(\zeta) \wedge (\bar{\partial}\partial\rho(\zeta))^{n-1}.$$

Then we have $\overline{H(\zeta)} = H(\zeta)$. Therefore, if we write $H(\zeta) = h(\zeta)dV(\zeta)$, then h is a real-valued function. Hence we have the following lemma.

Lemma 3.53 *For $(z,\zeta) \in \overline{\Omega} \times \overline{\Omega}$ with $-\varepsilon < \rho(\zeta)$ and $|z - \zeta| < \varepsilon$, there exists a real-valued function $a(\zeta)$ such that*

$$G(z,\zeta) = \frac{a(\zeta)}{(F(z,\zeta) - 2\rho(\zeta))^{n+1}} + \frac{O(|\zeta - z|)}{\widetilde{\Phi}(z,\zeta)^{n+1}} + \frac{N(z,\zeta)}{\widetilde{\Phi}(z,\zeta)^n}.$$

Lemma 3.54 *Let $\delta_\Omega(z) = \mathrm{dist}(z, \partial\Omega)$. For $z \in \Omega$, define*

$$I_\alpha := \int_\Omega \frac{dV(\zeta)}{|\widetilde{\Phi}(z,\zeta)|^{n+1+\alpha}}.$$

Then there exists a constant $C > 0$ such that

(a) *If $\alpha < 0$, then $I_\alpha < C$.*
(b) *If $\alpha = 0$, then $I_\alpha \leq C|\log \delta_\Omega(z)|$.*
(c) *If $\alpha > 0$, then $I_\alpha \leq C(\delta_\Omega(z))^{-\alpha}$.*

Proof. We may assume that $\rho(\zeta) > -\varepsilon$ and $|z - \zeta| \leq \varepsilon$. There exists a constant $\beta > 0$ such that

$$\mathrm{Re}\,\widetilde{\Phi}(z,\zeta) \geq -\rho(\zeta) - \rho(z) + \beta|\zeta - z|^2.$$

We choose a coordinate system $t = (t_1, \cdots, t_{2n})$ such that $t_1 = \mathrm{Im}\,\widetilde{\Phi}(z,\zeta)$, $t_2 = \rho(\zeta)$. We set $t' = (t_3, \cdots, t_{2n})$. Then we have

$$I_\alpha \leq C \int_{|t| \leq M} \frac{dt}{(|t_1| + |t_2| + |\rho(z)| + |t|^2)^{n+1+\alpha}}$$

$$\leq C \int_{|t'| \leq M} \frac{dt'}{(|\rho(z)| + |t'|^2)^{n-1+\alpha}}$$

$$\leq C \int_0^M \frac{r^{2n-3}}{(|\rho(z)| + r^2)^{n-1+\alpha}} dr.$$

In case $\alpha < 0$, we have

$$I_\alpha \leq C \int_0^M \frac{1}{r^{1+\alpha}} dr \leq C.$$

In case $\alpha \geq 0$, if we set $r = \lambda\sqrt{|\rho(z)|}$, then

$$I_\alpha \leq C \int_0^{M/\sqrt{|\rho(z)|}} \frac{\lambda^{2n-3}}{|\rho(z)|^\alpha (1+\lambda^2)^{n-1+\alpha}} d\lambda$$

$$\leq \frac{C}{|\rho(z)|^\alpha} \int_1^{M/\sqrt{|\rho(z)|}} \frac{d\lambda}{\lambda^{1+2\alpha}}.$$

In case $\alpha = 0$, we have

$$I_0 \leq C \int_1^{M/\sqrt{|\rho(z)|}} \frac{d\lambda}{\lambda} \leq C|\log|\rho(z)|| \leq C|\log \delta_\Omega(z)|.$$

In case $\alpha > 0$, we have

$$I_\alpha \leq \frac{C}{|\rho(z)|^\alpha} \int_1^\infty \frac{d\lambda}{\lambda^{1+2\alpha}} \leq C(\delta_\Omega(z))^{-\alpha}.$$

□

Definition 3.28 For $z, \zeta \in \overline{\Omega}$, we define

$$\widetilde{F}(z,\zeta) = F(z,\zeta) - 2\rho(\zeta), \quad \widetilde{F}^*(z,\zeta) = \overline{\widetilde{F}(\zeta,z)}.$$

Lemma 3.55 *We have*

(a) $\widetilde{F}(z,\zeta) - \widetilde{F}^*(z,\zeta) = O(|\zeta - z|^3)$.
(b) *If* $\zeta, z \in \overline{\Omega}$ *are sufficiently close to* $\partial\Omega$, *then we have* $|\widetilde{F}^*| \geq C|\widetilde{F}|$.

Proof. (a) It follows from Taylor's formula that

$$\frac{\partial \rho}{\partial \zeta_j}(\zeta) = \frac{\partial \rho}{\partial \zeta_j}(z) + \sum_{k=1}^n \frac{\partial^2 \rho}{\partial \zeta_j \partial \zeta_k}(z)(\zeta_k - z_k)$$

$$+ \sum_{k=1}^n \frac{\partial^2 \rho}{\partial \zeta_j \partial \bar{\zeta}_k}(z)(\bar{\zeta}_k - \bar{z}_k) + O(|\zeta - z|^2), \quad (3.46)$$

$$\frac{\partial^2 \rho}{\partial \zeta_j \partial \zeta_k}(\zeta) = \frac{\partial^2 \rho}{\partial \zeta_j \partial \zeta_k}(z) + O(|\zeta - z|), \quad (3.47)$$

$$\frac{\partial^2 \rho}{\partial \zeta_j \partial \bar{\zeta}_k}(z) = \frac{\partial^2 \rho}{\partial \zeta_j \partial \bar{\zeta}_k}(\zeta) + O(|\zeta - z|). \quad (3.48)$$

By definition, we have

$$F(z,\zeta) = 2\sum_{j=1}^{n}\frac{\partial \rho}{\partial \zeta_j}(\zeta)(\zeta_j - z_j) - \sum_{j,k=1}^{n}\frac{\partial^2 \rho}{\partial \zeta_j \partial \zeta_k}(\zeta)(\zeta_j - z_j)(\zeta_k - z_k). \quad (3.49)$$

Substituting (3,46), (3.47) and (3.48) into (3.49), we obtain

$$F(z,\zeta) = -\overline{F(\zeta,z)} + 2\sum_{j,k=1}^{n}\frac{\partial^2 \rho}{\partial \zeta_j \partial \bar{\zeta}_k}(\zeta)(\zeta_j - z_j)(\bar{\zeta}_k - \bar{z}_k) + O(|\zeta - z|^3). \quad (3.50)$$

On the other hand, it follows from Taylor's formula that

$$\rho(z) = \rho(\zeta) - \frac{1}{2}F(z,\zeta) - \frac{1}{2}\overline{F(z,\zeta)}$$
$$+ \sum_{j,k=1}^{n}\frac{\partial^2 \rho}{\partial \zeta_j \partial \bar{\zeta}_k}(\zeta)(\zeta_j - z_j)(\bar{\zeta}_k - \bar{z}_k) + O(|\zeta - z|^3).$$

Substituting (3.50) into the above equality we obtain

$$\rho(z) = \rho(\zeta) - \frac{1}{2}F(z,\zeta) + \frac{1}{2}\overline{F(\zeta,z)} + O(|\zeta - z|^3).$$

This proves (a).

(b) We may assume that $|z - \zeta|$ is sufficiently small. Then there exist positive constants C_1 and C_2 such that

$$|\widetilde{F}^*| \geq |\widetilde{F}| - |\widetilde{F}^* - \widetilde{F}| \geq \frac{1}{2}|\widetilde{F}| + C_1|\zeta - z|^2 - C_2|\zeta - z|^3 \geq \frac{1}{2}|\widetilde{F}|.$$

This proves (b). □

Definition 3.29 For $(z,\zeta) \in \Omega \times \Omega$, define

$$G^*(z,\zeta) := \overline{G(\zeta,z)}$$

and

$$B(z,\zeta) := G(z,\zeta) - G^*(z,\zeta).$$

By definition, we have $B(\zeta,z) = -\overline{B(z,\zeta)}$.

Theorem 3.25 *Let $s < (2n+2)/(2n+1)$. Then there exists a constant $C > 0$ such that*

$$\int_{\Omega}|B(z,\zeta)|^s dV(\zeta) < C \qquad (z \in \Omega)$$

and
$$\int_\Omega |B(z,\zeta)|^s dV(z) < C \qquad (\zeta \in \Omega).$$

Proof. Define
$$C(z,\zeta) := \frac{a(\zeta)}{\widetilde{F}(z,\zeta)^{n+1}} - \frac{\overline{a(z)}}{\widetilde{F}^*(z,\zeta)^{n+1}}.$$

If we prove the inequality
$$|C(z,\zeta)| \leq C \frac{1}{|\widetilde{\Phi}(z,\zeta)|^{n+\frac{1}{2}}}, \qquad (3.51)$$

then it follows from Lemma 3.53 that $|B(z,\zeta)| \leq |\widetilde{\Phi}(z,\zeta)|^{n+(1/2)}$. Therefore, it is sufficient to show (3.51). Since $a(\zeta)$ is a real-valued C^∞ function, we obtain
$$\frac{a(\zeta)}{\widetilde{F}(z,\zeta)^{n+1}} - \frac{\overline{a(z)}}{\widetilde{F}^*(z,\zeta)^{n+1}} = a(\zeta) \left\{ \frac{1}{\widetilde{F}(z,\zeta)^{n+1}} - \frac{1}{\widetilde{F}^*(z,\zeta)^{n+1}} \right\}$$
$$+ \frac{O(|\zeta - z|)}{\widetilde{F}^*(z,\zeta)^{n+1}}.$$

Since
$$|\widetilde{F}(z,\zeta)| \geq C|\zeta - z|^2 \qquad (\zeta, z \in \overline{\Omega}),$$

it follows from Lemma 3.55 that
$$\left| \frac{1}{\widetilde{F}(z,\zeta)^{n+1}} - \frac{1}{\widetilde{F}^*(z,\zeta)^{n+1}} \right|$$
$$= \left| (\widetilde{F}^* - \widetilde{F}) \sum_{\nu=0}^{n} \frac{1}{\widetilde{F}(z,\zeta)^{n+1-\nu}} \frac{1}{\widetilde{F}^*(z,\zeta)^{\nu+1}} \right|$$
$$\leq C|\zeta - z|^3 \frac{1}{|\widetilde{F}(z,\zeta)|^{n+2}} \leq C \frac{1}{|\widetilde{F}(z,\zeta)|^{n+1/2}}$$
$$\leq C \frac{1}{|\widetilde{\Phi}(z,\zeta)|^{n+1/2}}.$$

This proves (3.51). Thus we obtain
$$|B(z,\zeta)|^s \leq C|\widetilde{\Phi}(z,\zeta)|^{-(2n+1)s/2}.$$

By Lemma 3.55, $|B(z,\zeta)|^s$ is integrable, provided $s < (2n+2)/(2n+1)$. The second inequality follows from the equality $B(z,\zeta) = -\overline{B(\zeta,z)}$. $\square$

Definition 3.30 A C^∞ function $\mathcal{A}(z,\zeta)$ in $\Omega \times \Omega$ is called a simple admissible kernel of order $\lambda = 2n + j - 2t + 2$ if for any $P \in \partial\Omega$, there exist a neighborhood U of P and C^∞ functions $\mathcal{E}_j(z,\zeta)$ in $U \times U$ such that $\mathcal{A}(z,\zeta)$ has a representation

$$\mathcal{A}(z,\zeta) = \frac{\mathcal{E}_j(z,\zeta)}{\widetilde{\Phi}(z,\zeta)^t} \quad \text{or} \quad \mathcal{A}(z,\zeta) = \frac{\mathcal{E}_j(z,\zeta)}{\widetilde{\Phi}^*(z,\zeta)^t},$$

where j and t are positive integers with $t \geq 2$, and $\mathcal{E}_j(z,\zeta)$ satisfy the inequalities

$$|\mathcal{E}_j(z,\zeta)| \leq C|\zeta - z|^j \quad ((z,\zeta) \in U \times U).$$

Definition 3.31 A C^∞ function $\mathcal{A}(z,\zeta)$ in $\Omega \times \Omega$ is called an admissible kernel of order λ if for any positive integer N there exist simple admissible kernels $\mathcal{A}^{(0)}, \cdots, \mathcal{A}^{(N-1)}$ of order $\geq \lambda$ such that

$$\mathcal{A} = \sum_{s=0}^{N-1} \mathcal{A}^{(s)} + \mathcal{R}^{(N)},$$

where $\mathcal{R}^{(N)}$ satisfies that for any nonnegative integer k, if we choose N sufficiently large, then

$$\left| \int_\Omega f(\zeta) \mathcal{R}^{(N)}(\cdot,\zeta) dV(\zeta) \right|_{k,\Omega} \leq C_k \|f\|_{L^2} \quad (f \in L^2(\Omega)).$$

Lemma 3.56 We denote by $\mathcal{A}_\lambda$ a simple admissible kernel of order λ. Then

$$\int_\Omega |\mathcal{A}_\lambda(z,\zeta)| dV(\zeta) \leq C \begin{cases} 1 & (\lambda > 0) \\ |\log \delta_\Omega(z)| & (\lambda = 0) \\ (\delta_\Omega(z))^{\lambda/2} & (\lambda < 0) \end{cases}.$$

Proof. We choose a local coordinate system $u_1, \cdots, u_{2n}$ such that

$$\rho(\zeta) = u_1, \quad \operatorname{Im} \widetilde{\Phi}(z,\zeta) = u_2, \quad |u| \approx |\zeta - z|.$$

We set $u = (u_1, \cdots, u_{2n})$, $u' = (u_3, \cdots, u_{2n})$. Then we have

$$\int_\Omega |\mathcal{A}(z,\zeta)| dV(\zeta) = \int_\Omega \left|\frac{\mathcal{E}_j(z,\zeta)}{\widetilde{\Phi}(z,\zeta)^t}\right| dV(\zeta)$$

$$\leq C \int_{|u|\leq M} \frac{|u|^j}{(|u_1|+|u_2|+|\rho(z)|+|u|^2)^t} du$$

$$\leq C \int_{|u|\leq M} \frac{du}{(|u_1|+|u_2|+|\rho(z)|+|u|^2)^{t-j/2}}$$

$$\leq \int_{|u'|\leq M} \frac{du'}{(|\rho(z)|+|u'|^2)^{t-2-j/2}}$$

$$\leq C \int_0^M \frac{r^{2n-3}}{(|\rho(z)|+r^2)^{t-2-j/2}} dr.$$

In case $\lambda = 2n + j - 2t + 2 > 0$, we have

$$\int_\Omega |\mathcal{A}(z,\zeta)| dV(\zeta) \leq C \int_0^M r^{2n-2t+j+1} dr = C \int_0^M r^{\lambda-1} dr \leq C.$$

In case $\lambda = 2n + j - 2t + 2 \leq 0$, if we set $r = \sqrt{|\rho(z)|}s$, then

$$\int_\Omega |\mathcal{A}(z,\zeta)| dV(\zeta) \leq C|\rho(z)|^{\lambda/2} \int_0^{M/\sqrt{|\rho(z)|}} \frac{s^{2n-3}}{(1+s^2)^{t-2-(j/2)}} ds$$

$$\leq C|\rho(z)|^{\lambda/2} \int_1^{M/\sqrt{|\rho(z)|}} \frac{ds}{s^{1-\lambda}}. \qquad \square$$

Lemma 3.57 Let $\mathcal{E}_j(z,\zeta)$ be a C^∞ function on $\overline{\Omega} \times \overline{\Omega}$ such that $|\mathcal{E}_j(z,\zeta)| \leq |\zeta - z|^j$. For positive integers t_1, t_2, we set

$$\mathcal{A}(z,\zeta) = \frac{\mathcal{E}_j(z,\zeta)}{\widetilde{F}(z,\zeta)^{t_1} \widetilde{F}^*(z,\zeta)^{t_2}}. \tag{3.52}$$

Then $\mathcal{A}(z,\zeta)$ is an admissible kernel of order $\lambda = 2n + j - 2(t_1 + t_2) + 2$.

Proof. We have

$$\frac{1}{(\widetilde{F}^*)^{t_2}} = \frac{1}{(\widetilde{F})^{t_2}} + (\widetilde{F} - \widetilde{F}^*) \sum_{\nu=0}^{t_2-1} \frac{1}{\widetilde{F}^{t_2-\nu}(\widetilde{F}^*)^{\nu+1}}. \tag{3.53}$$

Substituting (3.51) into (3.50), we obtain

$$\mathcal{A} = \frac{\mathcal{E}_j}{\widetilde{F}^{t_1}\widetilde{F}^{t_2}} + \frac{\mathcal{E}_j}{\widetilde{F}^{t_1}}(\widetilde{F} - \widetilde{F}^*) \sum_{\nu=0}^{t_2-1} \frac{1}{\widetilde{F}^{t_2-\nu}(\widetilde{F}^*)^{\nu+1}}. \tag{3.54}$$

If we replace t_2 by $\nu + 1$ in (3.53) and substitute it into the right side of (3.54), we obtain

$$\mathcal{A} = \frac{\mathcal{E}_j}{\widetilde{F}^{t_1}\widetilde{F}^{t_2}} + \frac{\mathcal{E}_j(\widetilde{F} - \widetilde{F}^*)}{\widetilde{F}^{t_1+t_2+1}}$$
$$+ \frac{\mathcal{E}_j(\widetilde{F} - \widetilde{F}^*)^2}{\widetilde{F}^{t_1}} \sum_{\mu=0}^{t_2-1} \frac{1}{\widetilde{F}^{t_2+1-\mu}(\widetilde{F}^*)^{\mu+1}}.$$

Repeating this process, we have

$$\mathcal{A} = \sum_{s=0}^{N-1} \frac{\mathcal{E}_j(\widetilde{F} - \widetilde{F}^*)^s}{\widetilde{F}^{t_1+t_2+s}} + (\widetilde{F} - \widetilde{F}^*)^N \sum_{\nu=0}^{t_2-1} \frac{\mathcal{E}_j}{\widetilde{F}^{t_1+t_2+N-1-\nu}(\widetilde{F}^*)^{\nu+1}}. \quad (3.55)$$

Each term of the first sum in the right side of (3.55) is a simple admissible kernel of degree $\geq \lambda$. We denote the second sum in the right side of (3.55) by $\mathcal{R}^{(N)}$. It follows from Lemma 3.55 that

$$(\widetilde{F} - \widetilde{F}^*)^N \mathcal{E}_j = O(|\zeta - z|^{3N+j}).$$

Hence the absolute values of k-th order derivatives of $\mathcal{R}^{(N)}$ are bounded by the sum of

$$C \frac{|\zeta - z|^{j+3N-\mu}}{|\widetilde{F}(z,\zeta)|^{t_1+t_2+N+k-\mu}} \quad (0 \leq \mu \leq k). \quad (3.56)$$

Since

$$\frac{|\zeta - z|^{j+3N-\mu}}{|\widetilde{F}(z,\zeta)|^{t_1+t_2+N+k-\mu}} \leq C|\zeta - z|^{j+N-2(t_1+t_2+k)}$$

for N with $N \geq 2(t_1+t_2+k)$, derivatives of $\mathcal{R}^{(N)}$ of order $\leq k$ are bounded. Thus $\mathcal{A}(z, \zeta)$ is an admissible kernel of order λ. □

Theorem 3.26 $B(z, \zeta) = G(z, \zeta) - G^*(z, \zeta)$ *is an admissible kernel of order* 1.

Proof. By Lemma 3.53, there exists a real-valued C^∞ function $a(\zeta)$ such that

$$G(z, \zeta) = \frac{a(\zeta)}{\widetilde{F}(z,\zeta)^{n+1}} + \frac{O(|\zeta - z|)}{\widetilde{\Phi}(z,\zeta)^{n+1}} + \frac{N(z,\zeta)}{\widetilde{\Phi}(z,\zeta)^n}.$$

Thus we have a representation

$$B(z, \zeta) = a(\zeta) \left[\frac{1}{\widetilde{F}(z,\zeta)^{n+1}} - \frac{1}{\widetilde{F}^*(z,\zeta)^{n+1}} \right] + \mathcal{A}_1,$$

where $\mathcal{A}_1$ is an admissible kernel of order 1. Since

$$\frac{1}{\widetilde{F}^{n+1}} - \frac{1}{(\widetilde{F}^*)^{n+1}} = (\widetilde{F}^* - \widetilde{F}) \sum_{\nu=0}^{n} \frac{1}{\widetilde{F}^{n+1-\nu}(\widetilde{F}^*)^{\nu+1}},$$

by Lemma 3.57 $1/\widetilde{F}^{n+1} - 1/(\widetilde{F}^*)^{n+1}$ is an admissible kernel of order $2n + 3 - 2(n+2) + 2 = 1$. □

Definition 3.32 A vector field L on $\overline{\Omega}$ is said to be a tangent vector field for $\partial\Omega$ if $L\rho = 0$ on $\partial\Omega$ for any defining function ρ for Ω.

Lemma 3.58 *Let $\Omega \subset \mathbf{R}^n$ be a bounded domain with C^∞ boundary and let L be a tangent vector field of class C^∞ for $\partial\Omega$. Then there exists a first order partial differential operator L^* on $\overline{\Omega}$ of class C^∞ such that*

$$(f, Lg)_\Omega = (L^*f, g)_\Omega$$

for all $f, g \in C^\infty(\overline{\Omega})$.

Proof. Assume that there exists an open set U such that $\operatorname{supp}(L) \subset\subset U$, $U \cap \partial\Omega \neq \phi$ and such that if we set $\rho(\zeta) = x_1$, then $x_1, \cdots, x_n$ form a coordinate system in U. We set $x' = (x_2, \cdots, x_n)$, and

$$L = \sum_{j=1}^{n} a_j \frac{\partial}{\partial x_j} \qquad (a_j \in C_c^\infty(U)).$$

Then we have $Lr(0, x') = a_1(0, x') = 0$. The volume element $dV = \gamma(x)dx_1 \wedge \cdots \wedge dx_n$ satisfies $\gamma(x) > 0$ for $x \in U$. If $f, g \in C^\infty(U)$, then we have

$$(f, Lg)_{\Omega \cap U} = \sum_{j=1}^{n} \int_{x \in U, x_1 \leq 0} f \bar{a}_j \frac{\partial \bar{g}}{\partial x_j} \gamma(x) dx_1 \cdots dx_n$$

$$= -\sum_{j=1}^{n} \int_{x \in U, x_1 \leq 0} \frac{\partial}{\partial x_j}(f \bar{a}_j \gamma) \bar{g} dx_1 \cdots dx_n$$

$$+ \int_{x \in U, x_1 = 0} f \bar{a}_1 \gamma \bar{g} dx_2 \cdots dx_n$$

$$= -\sum_{j=1}^{n} \int_{x \in U, x_1 \leq 0} \frac{\partial}{\partial x_j}(f \bar{a}_j \gamma) \bar{g} dx_1 \cdots dx_n$$

$$= -\sum_{j=1}^{n} \int_{x \in U, x_1 \leq 0} \left[\bar{a}_j \frac{\partial f}{\partial x_j} + \gamma^{-1} \frac{\partial}{\partial x_j}(\bar{a}_j \gamma) f \right] \bar{g} dV(x).$$

Hence if we define

$$L^* := -\sum_{j=1}^n \bar{a}_j \frac{\partial}{\partial x_j} - \gamma^{-1} \sum_{j=1}^n \frac{\partial}{\partial x_j}(\bar{a}_j\gamma),$$

then we have

$$(f, Lg)_\Omega = (f, Lg)_{\Omega \cap U} = \int_{\Omega \cap U} L^* f \bar{g} dV(\zeta) = (L^*f, g)_\Omega.$$

In the general case, we can prove Lemma 3.58 using a partition of unity argument. □

Lemma 3.59 *Let $\Omega \subset\subset \mathbf{C}^n$ be a strictly pseudoconvex domain with C^∞ boundary and let ρ be a defining function for Ω. Then the vector field*

$$Y = \sum_{j=1}^n \frac{\partial \rho}{\partial \bar{\zeta}_j} \frac{\partial}{\partial \zeta_j} - \sum_{j=1}^n \frac{\partial \rho}{\partial \zeta_j} \frac{\partial}{\partial \bar{\zeta}_j}$$

is a tangent vector field for $\partial\Omega$ of class C^∞ and satisfies $(Y\widetilde{\Phi})(\zeta, \zeta) \neq 0$ for $\zeta \in \partial\Omega$.

Proof. Let $\tilde{\rho}$ be a defining function for Ω. Then there exist a C^∞ function $\gamma(z) > 0$ such that $\tilde{\rho}(\zeta) = \gamma(\zeta)\rho(\zeta)$ (see Lemma 1.21). Hence we have $Y\rho(\zeta) = 0$ for $\zeta \in \partial\Omega$. Thus Y is a tangent vector field for $\partial\Omega$. We obtain for $\zeta \in \partial\Omega$

$$(Y\widetilde{\Phi})(\zeta, \zeta) = Y(\widetilde{FM})(\zeta, \zeta) = Y\{(F - 2\rho)\widetilde{M}\}(\zeta, \zeta) = (YF)(\zeta, \zeta)\widetilde{M}(\zeta, \zeta).$$

But we have $\widetilde{M}(\zeta, \zeta) \neq 0$ for $\zeta \in \partial\Omega$ and

$$YF(\zeta, \zeta) = 2 \sum_{j=1}^n \left|\frac{\partial \rho}{\partial \zeta_j}(\zeta)\right|^2 \neq 0 \quad (\zeta \in \partial\Omega),$$

which means that $(Y\widetilde{\Phi})(\zeta, \zeta) \neq 0$ for $\zeta \in \partial\Omega$. □

Lemma 3.60 *Let $\Omega \subset\subset \mathbf{C}^n$ be a strictly pseudoconvex domain with C^∞ boundary and let $\mathcal{A}_\lambda$ denote an arbitrary admissible kernel of order λ, where λ is equal to 0 or 1. Suppose $V^{(z)}$ is a vector field with respect to z of class C^∞ on $\overline{\Omega}$. Then*

$$V^{(z)} \int_\Omega f(\zeta)\mathcal{A}_\lambda(z,\zeta)dV(\zeta) = \int_\Omega (Y^*f)(\zeta)\mathcal{A}_\lambda(z,\zeta)dV(\zeta)$$
$$+ \int_\Omega f(\zeta)\mathcal{A}_\lambda(z,\zeta)dV(\zeta).$$

Proof. We set $\Delta_{\partial\Omega} = \{(\zeta, \zeta) \in \mathbf{C}^{2n} \mid \zeta \in \partial\Omega\}$. Let W be a neighborhood of $\Delta_{\partial\Omega}$ such that $(Y\widetilde{\Phi})(z, \zeta) \neq 0$ for $(z, \zeta) \in W$. We choose $\psi \in C_c^\infty(W)$ with the properties that $0 \leq \psi \leq 1$, and $\psi = 1$ in a neighborhood $W'(\subset\subset W)$ of $\Delta_{\partial\Omega}$. $\psi\mathcal{A}_\lambda$ is expressed by

$$\psi\mathcal{A}_\lambda = \frac{\mathcal{E}_j}{\widetilde{\Phi}^t},$$

where $\lambda = 2n + j - 2t + 2$. Then we have

$$\psi\mathcal{A}_\lambda = \psi\frac{\mathcal{E}_j}{\widetilde{\Phi}^t} = \psi\left[-\frac{1}{t-1}Y\left(\frac{\mathcal{E}_j}{\widetilde{\Phi}^{t-1}}\right)\frac{1}{Y\widetilde{\Phi}}\right] + \psi\frac{1}{t-1}\frac{Y\mathcal{E}_j}{\widetilde{\Phi}^{t-1}}\frac{1}{Y\widetilde{\Phi}}$$
$$-\psi_1 Y\left(\frac{\mathcal{E}_j}{\widetilde{\Phi}^{t-1}}\right) + \frac{\mathcal{E}_{j-1}}{\widetilde{\Phi}^{t-1}}$$
$$= \psi_1 Y\mathcal{A}_{\lambda+2} + \psi_2\mathcal{A}_{\lambda+1},$$

where ψ_1 and ψ_2 are C^∞ functions with compact supports in W. Consequently,

$$V^{(z)}\int_\Omega f(\zeta)\mathcal{A}_\lambda(z,\zeta)dV(\zeta) = \int_\Omega f(\zeta)V^{(z)}\mathcal{A}_\lambda(z,\zeta)dV(\zeta)$$
$$= \int_\Omega f(\zeta)\mathcal{A}_{\lambda-2}(z,\zeta)dV(\zeta)$$
$$= \int_\Omega f\{\psi\mathcal{A}_{\lambda-2} + (1-\psi)\mathcal{A}_{\lambda-2}\}dV(\zeta).$$

$(1-\psi)\mathcal{A}_{\lambda-2}$ is of class C^∞ on $\overline{\Omega}\times\overline{\Omega}$. On the other hand, we have

$$\int_\Omega f\psi\mathcal{A}_{\lambda-2}dV(\zeta) = \int_\Omega f(\psi_1 Y\mathcal{A}_\lambda + \psi_2\mathcal{A}_{\lambda-1})dV(\zeta)$$
$$= \int_\Omega Y^*(f\psi_1)\mathcal{A}_\lambda dV(\zeta)$$
$$+ \int_\Omega f(\psi_3 Y\mathcal{A}_{\lambda+1} + \psi_4\mathcal{A}_\lambda)dV(\zeta)$$
$$= \int_\Omega (Y^*f)\mathcal{A}_\lambda dV(\zeta) + \int_\Omega f\mathcal{A}_\lambda dV(\zeta),$$

which completes the proof of Lemma 3.60. $\square$

Lemma 3.61 *Define*

$$\mathcal{A}_\lambda f(z) = \int_\Omega f(\zeta)\mathcal{A}_\lambda(z,\zeta)dV(\zeta).$$

Then

(a) $\mathcal{A}_0$ is a bounded operator from $\Lambda_\alpha(\Omega)$ to $\Lambda_{\alpha/2}(\Omega)$ for every $0 < \alpha < 1$.
(b) $\mathcal{A}_1$ is a bounded operator from $L^\infty(\Omega)$ to $\Lambda_{1/2}(\Omega)$.

Proof. First we prove (b). It is sufficient to prove the inequality (see Lemma 3.20)

$$|d_z \mathcal{A}_1 f(z)| \leq C|f|_\Omega \delta_\Omega(z)^{-1/2} \quad (z \in \Omega). \tag{3.57}$$

By Lemma 3.56, we have

$$|d_z \mathcal{A}_1 f(z)| = \left| \int_\Omega f(\zeta) \mathcal{A}_{-1}(z, \zeta) dV(\zeta) \right|$$
$$\leq |f|_\Omega \int_\Omega |\mathcal{A}_{-1}(z, \zeta)| dV(\zeta)$$
$$\leq C \delta_\Omega(z)^{-1/2}.$$

This proves (3.57).

Next we prove (a). It is sufficient to show that

$$|d_z(\mathcal{A}_0 f)(z)| \leq C|f|_{\alpha,\Omega} \delta_\Omega(z)^{-1+\alpha/2} \quad (z \in \Omega). \tag{3.58}$$

Let $V^{(z)}$ denote either $\frac{\partial}{\partial z_j}$ or $\frac{\partial}{\partial \bar{z}_j}$. Then we have

$$V^{(z)}(\mathcal{A}_0 f) = \int_\Omega f(\zeta) \mathcal{A}_{-2}(z, \zeta) dV(\zeta)$$
$$= \int_\Omega (f(\zeta) - f(z)) \mathcal{A}_{-2}(z, \zeta) dV(\zeta)$$
$$+ f(z) \int_\Omega \mathcal{A}_{-2}(z, \zeta) dV(\zeta).$$

Since $|\widetilde{\Phi}| \geq C|\zeta - z|^2$, it follows from Lemma 3.54 and the definition of the degree of the admissible kernel that

$$\int_\Omega |(f(\zeta) - f(z)) \mathcal{A}_{-2}(z, \zeta)| dV(\zeta) \leq \int_\Omega C|f|_\alpha \frac{|\zeta - z|^{\alpha+j}}{|\widetilde{\Phi}|^{n+2+(j/2)}} dV(\zeta)$$
$$\leq \int_\Omega C|f|_\alpha |\widetilde{\Phi}|^{-(n+1+1-(\alpha/2))} dV(\zeta)$$
$$\leq C|f|_\alpha \delta_\Omega(z)^{-1+\alpha/2}.$$

On the other hand, using the same method as the proof of Lemma 3.60, we obtain

$$\psi \mathcal{A}_{-2} = \psi_1 Y \mathcal{A}_0 + \psi_1 \mathcal{A}_{-1}.$$

Hence we have

$$\int_\Omega \psi \mathcal{A}_{-2}(z,\zeta)dV(\zeta) = \int_\Omega Y^*\psi_1 \mathcal{A}_0 dV(\zeta) + \int_\Omega \psi_1 \mathcal{A}_{-1} dV(\zeta)$$
$$= \int_\Omega \mathcal{A}_{-1} dV(\zeta).$$

By Lemma 3.56, we obtain

$$\left| \int_\Omega \psi \mathcal{A}_{-2}(z,\zeta)dV(\zeta) \right| \leq \int_\Omega |\mathcal{A}_{-1}| dV(\zeta) \leq \delta_\Omega(z)^{-1/2} \leq C\delta_\Omega(z)^{-1+(\alpha/2)},$$

which completes the proof of Lemma 3.61. $\square$

Definition 3.33 For multi-indices $\alpha = (\alpha_1, \cdots, \alpha_n)$ and $\beta = (\beta_1, \cdots, \beta_n)$, where α_j, β_j are nonnegative integers, define

$$\partial_z^{\alpha \bar{\beta}} = \frac{\partial^{|\alpha|+|\beta|}}{\partial z_1^{\alpha_1} \cdots \partial z_n^{\alpha_n} \partial \bar{z}_1^{\beta_1} \cdots \partial \bar{z}_n^{\beta_n}}.$$

Definition 3.34 For $f \in L^1(\Omega)$, define

$$\mathbf{G}f(z) := \int_\Omega f(\zeta)G(z,\zeta)dV(\zeta) \qquad (z \in \Omega),$$

$$\mathbf{G}^*f(z) := \int_\Omega f(\zeta)G^*(z,\zeta)dV(\zeta) \qquad (z \in \Omega),$$

$$\mathbf{B}f(z) := \int_\Omega f(\zeta)B(z,\zeta)dV(\zeta) \qquad (z \in \Omega).$$

Theorem 3.27 *Let k be a nonnegative integer. Then operators $\mathbf{G}$, $\mathbf{G}^*$ and $\mathbf{B}$ have the following properties:*

(a) $\mathbf{G}$ and $\mathbf{G}^*$ are bounded operators from $C^{k+\alpha}(\overline{\Omega})$ to $C^{k+(\alpha/2)}(\overline{\Omega})$ for every $0 < \alpha < 1$.

(b) $\mathbf{B}$ is a bounded operator from $C^k(\overline{\Omega})$ to $C^{k+(1/2)}(\overline{\Omega})$.

Proof. Let $|\alpha| + |\beta| = j \leq k$. Since $G(z,\zeta)$ is an admissible kernel of order 0, it follows from Lemma 3.60 that

$$\partial_z^{\alpha\bar{\beta}}(\mathbf{G}f)(z) = \int_\Omega f(\zeta)\partial_z^{\alpha\bar{\beta}}G(z,\zeta)dV(\zeta)$$
$$= \int_\Omega f(\zeta)\partial_z^{\alpha\bar{\beta}}\mathcal{A}_0(z,\zeta)dV(\zeta)$$
$$= \sum_{\nu=0}^{j}\int_\Omega ((Y^*)^\nu f)\mathcal{A}_0^{(\nu)}(z,\zeta)dV(\zeta).$$

By Lemma 3.61, we obtain

$$|\partial^{\alpha\bar{\beta}}(\mathbf{G}f|_\alpha \leq C\sum_{\nu=0}^{j}|(Y^*)^\nu f|_{\alpha/2}.$$

Therefore, we have $|\mathbf{G}f|_{k+\alpha} \leq C|f|_{k+(\alpha/2)}$. Similarly, we can prove the desired properties for $\mathbf{G}^*$ and $\mathbf{B}$. □

Lemma 3.62 *Let $p > 0$, $q > 0$ and $r > 0$ be such that*

$$\frac{1}{p} + \frac{1}{q} + \frac{1}{r} = 1.$$

Then

$$fgh \in L^1(\Omega), \quad \|fgh\|_1 \leq \|f\|_p\|g\|_q\|h\|_r$$

for $f \in L^p(\Omega)$, $g \in L^q(\Omega)$ and $h \in L^r(\Omega)$.

Proof. Let $s > 0$ be such that

$$\frac{1}{s} = \frac{1}{p} + \frac{1}{q}.$$

Using the Hölder inequality we have

$$\int_\Omega |fg|^s dV = \int_\Omega (|f|^p)^{s/p}(|g|^q)^{s/q}dV$$
$$\leq \left(\int_\Omega |f|^p dV\right)^{s/p}\left(\int_\Omega |g|^q dV\right)^{s/q}.$$

Hence we have

$$fg \in L^s(\Omega), \quad \|fg\|_s \leq \|f\|_p\|g\|_q.$$

On the other hand, we have

$$\frac{1}{s} + \frac{1}{r} = 1.$$

By applying the Hölder inequality to $fg \in L^s(\Omega)$ and $h \in L^r(\Omega)$, we obtain

$$\|fgh\|_1 \leq \|fg\|_s \|h\|_r \leq \|f\|_p \|g\|_q \|h\|_r.$$

□

Theorem 3.28 *Let $K(z,\zeta)$ be a measurable function on $\Omega \times \Omega$. Suppose there exist constants $M > 0$ and $s \geq 1$ such that*

(a) $\displaystyle\int_\Omega |K(z,\zeta)|^s dV(\zeta) \leq M^s \quad (z \in \Omega).$

(b) $\displaystyle\int_\Omega |K(z,\zeta)|^s dV(z) \leq M^s \quad (\zeta \in \Omega).$

Define

$$\mathbf{K}f(z) = \int_\Omega f(\zeta) K(z,\zeta) dV(\zeta).$$

Then $\mathbf{K}$ is a bounded operator from $L^p(\Omega)$ to $L^q(\Omega)$ with $\|\mathbf{K}\| \leq M$ for all p and q satisfying $1 \leq p, q \leq \infty$ and

$$\frac{1}{q} = \frac{1}{p} + \frac{1}{s} - 1.$$

Proof. We prove the theorem in case $1 \leq q < \infty$, $1 < p, s < \infty$. The case $s = 1$ will be left to the reader. Let $f \in L^p(\Omega)$. Since

$$\frac{1}{q} + \frac{p-1}{p} + \frac{s-1}{s} = 1$$

and

$$|\mathbf{K}f(z)| \leq \int_\Omega (|K|^s |f|^p)^{1/q} |K|^{1-(s/q)} |f|^{1-(p/q)} dV(\zeta),$$

it follows from Lemma 3.62 that

$$|\mathbf{K}f(z)| \leq \left(\int_\Omega |K(z,\zeta)|^s |f(\zeta)|^p dV(\zeta)|\right)^{1/q} \times$$
$$\left(\int_\Omega |K(z,\zeta)|^s dV(\zeta)\right)^{(p-1)/p} \left(\int_\Omega |f|^p dV\right)^{(s-1)/s}.$$

Consequently we have

$$\int_\Omega |\mathbf{K}f(z)|^q dV(z) \leq \int_\Omega \left(\int_\Omega |K(z,\zeta)|^s |f(\zeta)|^p dV(\zeta)\right) dV(z) \times$$
$$M^{sq(p-1)/p} \|f\|^{pq(s-1)/s}.$$

It follows from the condition (b) that

$$\|\mathbf{K}f\|_{L^q}^q \le M^s \|f\|_{L^p}^p M^{sq(p-1)/p} \|f\|_{L^p}^{pq(s-1)/s} = M^q \|f\|_{L^p}^q.$$

□

Theorem 3.29 *Let* $\mathbf{B}^*$ *be the adjoint operator of* $\mathbf{B}$. *Then operators* $\mathbf{B}$ *and* $\mathbf{B}^*$ *have the following properties:*

(a) $\mathbf{B}$ *is a bounded operator from* $L^p(\Omega)$ *to* $L^q(\Omega)$ *for* $1 \le p,q \le \infty$ *and* $\frac{1}{q} > \frac{1}{p} - \frac{1}{2n+2}$.
(b) $\mathbf{B}: L^2(\Omega) \to L^2(\Omega)$ *is a compact operator.*
(c) The kernel of $\mathbf{B}^*$ *is* $B^*(\zeta, z) = \overline{B(z,\zeta)}$.
(d) $\mathbf{B}^* = -\mathbf{B}$.

Proof. (a) follows from Theorem 3.25 and Theorem 3.28.

(b) follows from Theorem 3.25 ($s = 1$) and Proposition A.13 in Appendix A.

(c) Define

$$\mathbf{E}^* f(z) = \int_\Omega f(\zeta) B^*(z,\zeta) dV(\zeta).$$

Using Fubini's theorem we obtain for $f, g \in \mathcal{D}(\Omega)$

$$\begin{aligned}(\mathbf{E}^* f, g) &= \int_\Omega \mathbf{E}^* f(z) \overline{g(z)} dV(z) \\ &= \int_\Omega \left(\int_\Omega f(\zeta) B^*(z,\zeta) dV(\zeta) \right) \overline{g(z)} dV(z) \\ &= \int_\Omega f(\zeta) \left\{ \overline{\int_\Omega g(z) B(\zeta, z) dV(z)} \right\} dV(\zeta) \\ &= \int_\Omega f(\zeta) \overline{(\mathbf{B}g)(\zeta)} dV(\zeta) \\ &= (f, \mathbf{B}g).\end{aligned}$$

Hence $\mathbf{E}^* = \mathbf{B}^*$

(d) Since $B(\zeta, z) = -\overline{B(z,\zeta)}$, we obtain

$$\mathbf{B}^* f(z) = \int_\Omega f(\zeta) B^*(z,\zeta) dV(\zeta) = -\int_\Omega f(\zeta) B(z,\zeta) dV(\zeta) = -\mathbf{B}f(z).$$

Hence $\mathbf{B}^* = -\mathbf{B}$. □

Definition 3.35 *For* $f \in L^2(\Omega)$, *we have a unique decomposition*

$$f = f_1 + f_2 \quad (f_1 \in A^2(\Omega),\ f_2 \in (A^2(\Omega))^\perp).$$

We define $\mathbf{P}_\Omega : L^2(\Omega) \to A^2(\Omega)$ by $\mathbf{P}_\Omega f = f_1$. $\mathbf{P}_\Omega$ is said to be the Bergman projection. By definition we have $\|\mathbf{P}_\Omega f\| = \|f_1\| \leq \|f\|$, $(\mathbf{P}_\Omega f, g) = (f, \mathbf{P}_\Omega g)$.

For $a \in \Omega$, (3.45) shows that

$$\begin{aligned}
\mathbf{P}_\Omega f(a) &= \int_\Omega \mathbf{P}_\Omega f(\zeta) G(a, \zeta) dV(\zeta) \\
&= (\mathbf{P}_\Omega f, \overline{G(a, \cdot)}) \\
&= (\mathbf{P}_\Omega f, \overline{G^*(a, \cdot)} + \overline{B(a, \cdot)}) \\
&= (\mathbf{P}_\Omega f, G(\cdot, a)) + (\mathbf{P}_\Omega f, \overline{B(a, \cdot)}) \\
&= (f, \mathbf{P}_\Omega G(\cdot, a)) + (\mathbf{P}_\Omega f, \overline{B(a, \cdot)}).
\end{aligned}$$

For fixed a, $G(z, a)$ is holomorphic in Ω, and continuous on $\overline{\Omega}$, which implies that $G(\cdot, a) \in A^2(\Omega)$. Hence we have $\mathbf{P}_\Omega G(\cdot, a) = G(\cdot, a)$. Therefore we obtain

$$\mathbf{P}_\Omega f(a) = (f, \overline{G^*(a, \cdot)}) + (\mathbf{P}_\Omega f, \overline{B(a, \cdot)}). \tag{3.59}$$

Theorem 3.30 $\mathbf{G}$, $\mathbf{G}^*$ *are bounded operators from $L^2(\Omega)$ to $L^2(\Omega)$. Moreover, $I - \mathbf{B} : L^2(\Omega) \to L^2(\Omega)$ is an invertible operator and satisfies*

$$\mathbf{P}_\Omega = (I - \mathbf{B})^{-1} \circ \mathbf{G}^*. \tag{3.60}$$

Proof. It follows from (3.59) that $\mathbf{P}_\Omega = \mathbf{G}^* + \mathbf{B} \circ \mathbf{P}_\Omega$. Hence we have

$$\mathbf{G}^* = \mathbf{P}_\Omega - \mathbf{B} \circ \mathbf{P}_\Omega = (I - \mathbf{B}) \circ \mathbf{P}_\Omega,$$

which means that $\mathbf{G}^*$ is bounded. We set $T = I - \mathbf{B}$. If $T(f) = 0$, then by theorem 3.29

$$-(f, f) = (-\mathbf{B}f, f) = (\mathbf{B}^* f, f) = (f, \mathbf{B}f) = (f, f),$$

which implies that $f = 0$. Therefore $\operatorname{Ker} T = \{0\}$, and hence from Proposition A.10 in Appendix A, $T : L^2(\Omega) \to L^2(\Omega)$ is invertible. Hence (3.60) holds. $\square$

In order to prove Theorem 3.31, we need the following two lemmas.

Lemma 3.63 *Let $\Omega \subset\subset \mathbf{R}^n$ be an open set and let $0 < \alpha < \beta < 1$. Then the inclusion mapping $\iota : C^\beta(\overline{\Omega}) \to C^\alpha(\overline{\Omega})$ is a compact operator, where $C^\alpha(\overline{\Omega})$ is the Lipschitz space of order α ($C^\alpha(\overline{\Omega})$ is also denoted by $\Lambda_\alpha(\Omega)$).*

Proof. Let $\{f_n\}$ be a bounded sequence in $C^\beta(\overline{\Omega})$. Then there exists a constant $M > 0$ such that $|f_n|_{\beta,\Omega} \leq M$. Let $\varepsilon > 0$. Then any $x \in \overline{\Omega}$ has a neighborhood V_x which satisfies the following:

$$\frac{|f_n(x) - f_n(y)|}{|x-y|^\alpha} < \varepsilon \quad (y \in V_x).$$

By the Ascoli-Arzela theorem (see Proposition A.1 in Appendix A), there exists a convergent subsequence $\{h_n\}$ of $\{f_n\}$. Let $x, y \in \overline{\Omega}$, $x \neq y$. For a positive integer k, there exist $x', y' \in F$ such that

$$\frac{|h_n(x) - h_n(x')|}{|x - x'|^\alpha} < \frac{1}{k}, \quad |x - x'| \leq |x - y|,$$

$$\frac{|h_n(y) - h_n(y')|}{|y - y'|^\alpha} < \frac{1}{k}, \quad |y - y'| \leq |x - y|.$$

If we choose n, m sufficiently large, then

$$|h_n(x') - h_m(x')| < \frac{|x-y|^\alpha}{k}, \quad |h_n(y') - h_m(y')| < \frac{|x-y|^\alpha}{k}.$$

Consequently we have

$$\frac{|h_n(x) - h_m(x) - (h_n(y) - h_m(y))|}{|x-y|^\alpha} \leq \frac{|h_n(x) - h_n(x')|}{|x-x'|^\alpha}$$
$$+ \frac{|h_n(x') - h_m(x')|}{|x-y|^\alpha} + \frac{|h_m(x') - h_m(x)|}{|x-x'|^\alpha} + \frac{|h_n(y) - h_n(y')|}{|y-y'|^\alpha}$$
$$+ \frac{|h_n(y') - h_m(y')|}{|x-y|^\alpha} + \frac{|h_m(y') - h_m(y)|}{|y-y'|^\alpha}$$
$$\leq \frac{6}{k}.$$

Therefore, $|h_n - h_m|_{\alpha,\Omega} \to 0$ as $n, m \to \infty$, and hence $\{h_n\}$ is a Cauchy sequence. Since $C^\alpha(\overline{\Omega})$ is complete (see Lemma 3.6), $\{h_n\}$ converges, which means that ι is a compact operator. □

Lemma 3.64 *Let E, F and G be Banach spaces. Let $A : E \to F$ be a bounded operator and $B : F \to G$ a compact operator. Then $B \circ A : E \to G$ is a compact operator.*

Proof. Let $\{x_n\} \subset E$ be a bounded sequence. Then $\{A(x_n)\}$ is a bounded sequence in F. Since B is compact, we can choose a convergent subsequence of $\{B(A(x_n))\}$. □

Theorem 3.31 Let $\Omega \subset\subset \mathbf{C}^n$ be a strictly pseudoconvex domain with C^∞ boundary and let k be a nonnegative integer. Then the Bergman projection $\mathbf{P}_\Omega$ is a bounded operator from $C^{k+\alpha}(\overline{\Omega})$ to $C^{k+(\alpha/2)}(\overline{\Omega})$ for every $0 < \alpha < 1$.

Proof. $\mathbf{B}$ is a bounded operator from $C^k(\overline{\Omega})$ to $C^{k+(1/2)}(\overline{\Omega})$ by Theorem 3.27. Hence for $0 < \alpha < 1/2$, $\mathbf{B}$ is a bounded operator from $C^{k+\alpha}(\overline{\Omega})$ to $C^{k+(1/2)}(\overline{\Omega})$. By Lemma 3.63 and Lemma 3.64, $\mathbf{B}$ is a compact operator from $C^{k+\alpha}(\overline{\Omega})$ to $C^{k+\alpha}(\overline{\Omega})$ for $0 < \alpha < 1/2$. By Theorem 3.30, we have Ker$(I - \mathbf{B}) = \{0\}$, which means that $I - \mathbf{B} : C^{k+\alpha}(\overline{\Omega}) \to C^{k+\alpha}(\overline{\Omega})$ is invertible. Since $\mathbf{P}_\Omega = (I - \mathbf{B})^{-1} \circ \mathbf{G}^*$ and by Theorem 3.27 $\mathbf{G}^*$ is a bounded operator from $C^{k+\alpha}(\overline{\Omega})$ to $C^{k+(\alpha/2)}(\overline{\Omega})$, $\mathbf{P}_\Omega$ is a bounded operator from $C^{k+\alpha}(\overline{\Omega})$ to $C^{k+(\alpha/2)}(\overline{\Omega})$. □

Definition 3.36 Let $\Omega \subset\subset \mathbf{C}^n$ be a domain. For a positive integer k, we say that Ω satisfies the condition $(\mathbf{R}_k)$ if there exists a positive integer m_k such that Bergman projection $\mathbf{P}_\Omega : L^2(\Omega) \to A^2(\Omega)$ is a bounded operator from $C^{m_k}(\overline{\Omega})$ to $C^k(\overline{\Omega})$, that is, $\mathbf{P}_\Omega$ satisfies the following properties:

(a) If $f \in C^{m_k}(\overline{\Omega})$, then $\mathbf{P}_\Omega f \in C^k(\overline{\Omega})$.
(b) There exists a constant $c_k > 0$ such that

$$|\mathbf{P}_\Omega f|_{k,\Omega} \le c_k |f|_{m_k,\Omega} \qquad (f \in C^{m_k}(\overline{\Omega})).$$

Definition 3.37 Let $\Omega \subset\subset \mathbf{C}^n$ be a domain. We say that Ω satisfies the condition $(\mathbf{R})$ if Ω satisfies the condition $(\mathbf{R}_k)$ for every positive integer k.

The following theorem follows from Theorem 3.31.

Theorem 3.32 Let $\Omega \subset\subset \mathbf{C}^n$ be a strictly pseudoconvex domain with C^∞ boundary. Then Ω satisfies the condition $(\mathbf{R})$.

Theorem 3.33 Let $\mathbf{P}_\Omega : L^2(\Omega) \to A^2(\Omega)$ be the Bergman projection. Then

$$(\mathbf{P}_\Omega f)(z) = \int_\Omega f(\zeta) K_\Omega(z, \zeta) dV(\zeta)$$

for all $f \in L^2(\Omega)$.

Proof. For $f \in A^2(\Omega)$, we have $\mathbf{P}_\Omega f = f$. Hence we obtain

$$(\mathbf{P}_\Omega f)(z) = (\mathbf{P}_\Omega f, K_\Omega(\cdot, z)) = (f, \mathbf{P}_\Omega K_\Omega(\cdot, z)) = (f, K_\Omega(\cdot, z)).$$

Since $K_\Omega(z, \zeta) = \overline{K_\Omega(\zeta, z)}$, we have the desired equality. □

Corollary 3.8 Let Ω_j, $j = 1, 2$, be bounded domains in $\mathbf{C}^n$ and let $F : \Omega_1 \to \Omega_2$ be a biholomorphic mapping. Then

$$\mathbf{P}_{\Omega_1}((f \circ F) \det F') = (\overline{\det F'}) \mathbf{P}_{\Omega_2}(f) \circ F$$

for all $f \in L^2(\Omega_2)$.

Proof. For $f \in L^2(\Omega_2)$, we define $T_F f = (f \circ F) \det F'$. It follows that

$$\int_{\Omega_2} |f(w)|^2 dV(w) = \int_{\Omega_1} |T_F f(\zeta)|^2 dV(\zeta),$$

which implies that $T_F f \in L^2(\Omega_1)$. Hence it follows from Lemma 3.64 and Theorem 3.24 that

$$\mathbf{P}_{\Omega_1}(T_F f)(z) = \int_{\Omega_1} (T_F f)(\zeta) K_{\Omega_1}(z, \zeta) dV(\zeta)$$
$$= \int_{\Omega_1} (T_F f)(\zeta) \det F'(z) K_{\Omega_2}(F(z), F(\zeta)) \overline{\det F'(\zeta)} dV(\zeta)$$
$$= \det F'(z) \int_{\Omega_1} f \circ F(\zeta) |\det F'(\zeta)|^2 K_{\Omega_2}(F(z), F(\zeta)) dV(\zeta)$$
$$= \det F'(z) \int_{\Omega_2} f(\tilde{\zeta}) K_{\Omega_2}(F(z), \tilde{\zeta}) dV(\tilde{\zeta})$$
$$= \det F'(z) \mathbf{P}_{\Omega_2} f(F(z)).$$

$\square$

Lemma 3.65 Let Ω be a bounded domain in $\mathbf{C}^n$. For $a \in \Omega$, we choose a function $\varphi_a \in \mathcal{D}(\Omega)$ such that φ_a depends only on $|z - a|$ and satisfies $\int \varphi_a dV = 1$. Then

$$K_\Omega(\cdot, a) = \mathbf{P}_\Omega \bar{\varphi}_a.$$

Proof. We may assume that $\mathrm{supp}(\varphi_a) \subset B(a, \varepsilon) \subset\subset \Omega$. If f is holomorphic in Ω, then it follows from the mean value theorem that

$$f(a) \int_{\partial B(a, \rho)} dS = \int_{\partial B(a, \rho)} f dS \qquad (0 < \rho \leq \varepsilon).$$

Since φ_a is constant in $\partial B(a, \rho)$, we have

$$f(a) \int_{\partial B(a, \rho)} \varphi_a dS = \int_{\partial B(a, \rho)} f \varphi_a dS. \qquad (3.61)$$

Integrating from 0 to ε in (3.61), we obtain

$$f(a) \int_{B(a, \rho)} \varphi_a dV = \int_{B(a, \rho)} f \varphi_a dV.$$

Therefore, we have

$$f(a) = (f, \overline{\varphi_a})$$

for any holomorphic function f in Ω. Suppose $f \in A^2(\Omega)$. Since $\mathbf{P}_\Omega f = f$, we have

$$f(a) = (f, \overline{\varphi_a}) = (\mathbf{P}_\Omega f, \overline{\varphi_a}) = (f, \mathbf{P}_\Omega \overline{\varphi_a}).$$

Since $\mathbf{P}_\Omega \bar{\varphi}_a \in A^2(\Omega)$, it follows from Lemma 3.52 that $\overline{\mathbf{P}_\Omega \overline{\varphi_a}} = K_\Omega(a, \cdot)$. Hence we obtain $\mathbf{P}_\Omega \overline{\varphi_a} = \overline{K_\Omega(a, \cdot)} = K_\Omega(\cdot, a)$. □

Theorem 3.34 Let $\Omega \subset\subset \mathbf{C}^n$ satisfy the condition $(\mathbf{R}_k)$. Then for $a \in \Omega$,

$$K_\Omega(\cdot, a) \in C^k(\overline{\Omega}).$$

Proof. Since the function φ_a in Lemma 3.65 belongs to $C^\infty(\overline{\Omega})$, we have $\bar{\varphi}_a \in C^{mk}(\overline{\Omega})$. Hence we have $K_\Omega(\cdot, a) = \mathbf{P}_\Omega \bar{\varphi}_a \in C^k(\overline{\Omega})$. □

Lemma 3.66 Let $\Omega \subset\subset \mathbf{R}^n$ be a domain with C^k boundary, $k \geq 1$. For $a \in \Omega$, we denote by φ_a any function which depends only on $|z - a|$ and satisfies $\varphi_a \in C_c^\infty(\Omega)$, $\int \varphi_a dV = 1$. We set

$$\mathcal{R}(\Omega) = \{\alpha \mid \alpha \text{ is a finite linear combination of } \varphi_a\}.$$

If $f \in C^k(\overline{\Omega})$ satisfies the conditions

$$(\partial^\alpha f)(x) = 0 \qquad (x \in \partial\Omega, \ |\alpha| \leq k),$$

then f is a limit of functions in $\mathcal{R}(\Omega)$ in the $C^k(\overline{\Omega})$ norm.

Proof. First we show that f is a limit of functions in $\mathcal{D}(\Omega)$ in the $C^k(\overline{\Omega})$ norm. Using a partition of unity argument, there exists a neighborhood U of $P \in \partial\Omega$ such that if we denote $\mathbf{n}$ the unit inward normal vector at P, then $\mathrm{supp}(f) \subset U$ and $f(x - \tau\mathbf{n})$ has a compact support in $\Omega \cap U$ for any sufficiently small $\tau > 0$. We define $\tilde{f}$ such that $\tilde{f}(x) = f(x)$ for $x \in \overline{\Omega} \cap U$, $\tilde{f}(x) = 0$ for $x \in U - \overline{\Omega}$. Then by the assumption, $\tilde{f} \in C_c^k(U)$. If we set $\tilde{f}_\tau(x) = \tilde{f}(x - \tau\mathbf{n})$, then $|f - \tilde{f}_\tau|_{k, \Omega \cap U} \to 0$ as $\tau \to 0$. Thus we may assume that $f \in C_c^k(\Omega)$. Suppose $\varphi \in \mathcal{D}(B(0, 1))$ depends only on $|z|$ and satisfies $\int \varphi dV = 1$. We set $\varphi_j(x) = j^n \varphi(jx)$, $j = 1, 2, \cdots$, and

$$f_j(x) = \int_{\mathbf{R}^n} f(y) \varphi_j(x - y) dV(y) = \int_{\mathbf{R}^n} f(x - y) \varphi_j(y) dV(y).$$

Then $f_j \in \mathcal{D}(\Omega)$ for any sufficiently large j. Moreover, we have

$$\partial^\alpha f_j(x) = \int_{\mathbf{R}^n} f(y) \partial_x^\alpha \varphi_j(x-y) dV(y) = \int_{\mathbf{R}^n} (\partial^\alpha f)(y) \varphi_j(x-y) dV(y),$$

which implies that

$$|f - f_j|_{k,\Omega} \to 0 \qquad (j \to \infty).$$

Each $\partial^\alpha f_j(x)$ is a limit of Riemann sums

$$\sum_{\nu=1}^{N} c_\nu f(\eta_\nu) \partial_x^\alpha \varphi_j(x - \eta_\nu), \tag{3.62}$$

where c_ν are positive constants. There exists a constant $M > 0$ such that

$$|c_\nu f(\eta_\nu) \partial_x^\alpha \varphi_j(x - \eta_\nu)| \le M c_\nu |f(\eta_\nu)|.$$

Since $\sum_{\nu=1}^{N} c_\nu |f(\eta_\nu)|$ are Riemann sums of $\int |f| dV$, and hence converge. Hence (3.62) converge to f_j uniformly on $\overline{\Omega}$. Therefore, if we set

$$g_N(x) = \sum_{\nu=1}^{N} c_\nu f(\eta_\nu) \varphi_j(x - \eta_\nu),$$

then $|f_j - g_N|_{k,\Omega} \to 0$ as $N \to \infty$, and hence $g_N \in \mathcal{R}(\Omega)$. □

Now we prove the following lemma (see Bell-Ligocka [BEL]).

Lemma 3.67 *(Bell's density lemma) Let Ω be a bounded domain in C^n with C^∞ boundary. Let Ω satisfy the condition* $(\mathbf{R}_0)$. *Then given $u \in C^{k+1}(\overline{\Omega})$, there exists a function $g \in C^k(\overline{\Omega})$ with $\mathbf{P}_\Omega g = 0$ and such that*

$$\partial^{\alpha \bar{\beta}}(u - g)|_{\partial \Omega} = 0 \qquad (|\alpha| + |\beta| \le k).$$

Proof. Let ρ be a defining function for Ω. There exist a constant $C > 0$, $P_j \in \partial\Omega$, $j = 1, \cdots, N$, and neighborhoods U_j of P_j with the following properties:

(1) $\partial\Omega \subset \bigcup_{i=1}^{N} U_i$

(2) For each i ($1 \le i \le N$), there exists integer j with $1 \le j \le n$ such that

$$\left| \frac{\partial \rho}{\partial z_j}(z) \right| > C \qquad (z \in U_i).$$

Let $\{\alpha_j\}_{j=1}^{N}$ be a partition of unity subordinate to $\{U_j\}_{j=1}^{N}$, that is, $\{\alpha_j\}_{j=1}^{N}$ satisfies the following properties:

(a) $\alpha_j \in C^{\infty}(\mathbf{C}^n)$.
(b) $\operatorname{supp}(\alpha_j) \subset\subset U_j$.
(c) There exists a neighborhood U of $\partial\Omega$ such that $\sum_{j=1}^{N} \alpha_j(z) = 1$ for $z \in U$.

It is sufficient to show that Lemma 3.67 holds for $\alpha_i u$ instead of u. So we rewrite $\alpha_i u$ by u. Thus we have

$$\operatorname{supp}(u) \subset \{w \in \overline{\Omega} \mid \frac{\partial \rho}{\partial z_1}(w) \neq 0\}.$$

Define

$$w_1(z) = \frac{u(z)\rho(z)}{\frac{\partial \rho}{\partial z_1}(z)}, \quad v_1(z) = \frac{\partial w_1}{\partial z_1}(z).$$

Then we have

$$v_1(z) = u(z) + \rho(z)\frac{\partial}{\partial z}\left\{\frac{u(z)}{\frac{\partial \rho}{\partial z_1}(z)}\right\},$$

and hence $v_1 - u = 0$ on $\partial\Omega$. Using the fact that $w_1|_{\partial\Omega} = 0$, $K_\Omega(\cdot, z)$ is holomorphic in Ω and $K_\Omega(\cdot, z) \in C(\overline{\Omega})$, we obtain

$$\mathbf{P}_\Omega v_1(z) = \int_\Omega v_1(\zeta)\overline{K_\Omega(\zeta, z)}dV(\zeta) = -\int_\Omega w_1(\zeta)\overline{\frac{K_\Omega(\zeta, z)}{\partial \bar{\zeta}_j}}dV(\zeta) = 0.$$

Hence Lemma 3.67 holds in case $k = 0$. We assume that w_{k-1} and $v_{k-1} = \frac{\partial w_{k-1}}{\partial z_1}$ have already been constructed and that v_{k-1} is equal to u on $\partial\Omega$ up to derivatives of order $k-2$ and satisfies $\mathbf{P}_\Omega v_{k-1} = 0$. Define a differential operator $\mathcal{D}$ on $\overline{\Omega}$ by

$$\mathcal{D}(\varphi)(z) = \frac{\sum_{\nu=1}^{2n} \frac{\partial \rho}{\partial x_\nu}(z) \frac{\partial \varphi}{\partial x_\nu}(z)}{|\nabla \rho(z)|^2}.$$

Define w_k and v_k by

$$w_k = w_{k-1} + \theta_k \rho^k, \quad v_k = \frac{\partial w_k}{\partial z_1},$$

where θ_k is defined by

$$\theta_k = \frac{\mathcal{D}^{k-1}(u - v_{k-1})}{k!\frac{\partial \rho}{\partial z}}.$$

Since
$$v_k = v_{k-1} + \frac{\partial}{\partial z}(\theta_k \rho^k),$$
we obtain
$$\mathcal{D}^{k-1}(u - v_k) = \mathcal{D}^{k-1}(u - v_{k-1}) - \mathcal{D}^{k-1}\left(\frac{\partial}{\partial z}(\theta_k \rho^k)\right). \tag{3.63}$$

Further, we devide the second term of the right side in (3.63) into a term which involves ρ and a term which does not involve ρ. Then we have a representation
$$\mathcal{D}^{k-1}\left(\frac{\partial}{\partial z}(\theta_k \rho^k)\right) = \theta_k \frac{\partial \rho}{\partial z_1} k! + \text{(the term involving } \rho\text{)}.$$
Consequently,
$$\mathcal{D}^{k-1}(u - v_k) = \mathcal{D}^{k-1}(u - v_{k-1}) - \theta_k \frac{\partial \rho}{\partial z} k! + \text{(the term involving } \rho\text{)}.$$
By the definition of θ_k, we have
$$\mathcal{D}^{k-1}(u - v_k)|_{\partial\Omega} = 0.$$

Next we choose vector fields $\tau_1, \cdots, \tau_{2n-1}$ at $P \in \partial\Omega$ such that $\{\tau_1, \cdots, \tau_{2n-1}, \mathcal{D}\}$ are orthogonal basis at P. Every vector field at $P \in \partial\Omega$ is denoted by a linear combination of $\mathcal{D}$ and τ_i, $i = 1, \cdots, 2n - 1$. For simplicity, we denote all τ_i, $i = 1, \cdots, 2n-1$, by τ. Then, in order that v_k and u coincide on $\partial\Omega$ up to derivatives of order $k - 1$, it is sufficient to show that if $s + t \leq k - 1$, then
$$\tau^s \mathcal{D}^t (u - v_k)|_{\partial\Omega} = 0.$$
In case $s + t < k - 1$, by the inductive hypothesis we have
$$\tau^s \mathcal{D}^t(u - v_k)|_{\partial\Omega} = \tau^s \left\{ \mathcal{D}^t(u - v_{k-1}) + \mathcal{D}^t\left(\frac{\partial}{\partial z}(\theta_k \rho^k)\right) \right\}\bigg|_{\partial\Omega} = 0.$$
In case $s = 0$, $t = k - 1$, we have already proved. In case $s + t = k - 1$, $s \geq 1$, we have
$$\tau^s \mathcal{D}^t(u - v_k)|_{\partial\Omega} = 0.$$
By Lemma 1.21, there exists a C^∞ function h in a neighborhood of $P \in \partial\Omega$ such that
$$\tau^{s-1} \mathcal{D}^t(u - v_k) = \rho h.$$

If τ has a representation in a neighborhood U of P

$$\tau = \sum_{j=1}^{2n} a_j(z)\frac{\partial}{\partial x_j},$$

then we obtain

$$\tau(\rho(z)h(z))|_{\partial\Omega\cap U} = \sum_{j=1}^{2n} a_j(z)\frac{\partial\rho}{\partial x_j}(z)h(z)|_{\partial\Omega\cap U} = 0,$$

which means that v_k and u coincide on $\partial\Omega$ up to derivatives of order $k-1$. By the definition of w_k, we have $w_k = \gamma\rho$ (γ is of class C^∞ on $\overline{\Omega}$), and hence $\mathbf{P}_\Omega v_k = 0$. Lemma 3.67 is proved. □

Theorem 3.35 *Let Ω be a bounded domain in C^n with C^∞ boundary and satisfy the condition $(\mathbf{R}_1)$. Suppose f is a holomorphic function in Ω with $f \in C^\infty(\overline{\Omega})$. Then f is a limit of a sequence of finite linear combinations of the elements in $\{K_\Omega(\cdot, a) \mid a \in \Omega\}$ in $C^1(\overline{\Omega})$ norm.*

Proof. Let f be holomorphic in Ω and $f \in C^\infty(\overline{\Omega})$. Then we have $f = \mathbf{P}_\Omega f$. By Bell's density theorem (Lemma 3.67), there exists $g \in C^{m_1}(\overline{\Omega})$ such that $f - g$ is equal to 0 on $\partial\Omega$ up to derivatives of order m_1. Moreover we have $f = \mathbf{P}_\Omega(f-g)$. By Lemma 3.66 there exist $g_k \in C_c(\Omega)$, $k = 1, 2, \cdots$, such that

$$g_k(z) = \sum_{j=1}^{N_k} \alpha_j^{(k)} \varphi_{a_j^k}^{(k)},$$

and $f - g$ is the limit of $\{g_k\}$ in $C^{m_1}(\overline{\Omega})$ norm, where $\varphi_{a_j^k}^{(k)} \in C_c^\infty(\Omega)$ depend only on $|z - a_{a_j^k}|$ and satisfy $\int \varphi_{a_j^k}^{(k)} dV = 1$. By Lemma 3.65, we obtain

$$\mathbf{P}_\Omega g_k = \sum_{j=1}^{N_k} \alpha_j^{(k)} \mathbf{P}_\Omega \varphi_{a_j^k}^{(k)}(z) = \sum_{j=1}^{N_k} \alpha_j^{(k)} K_\Omega(z, a_j^k).$$

By the condition $(\mathbf{R}_1)$, we have

$$|\mathbf{P}_\Omega(f - g - g_k)|_{1,\Omega} \le c_1 |f - g - g_k|_{m_1,\Omega},$$

which implies that

$$|f - \sum_{j=1}^{N_k} \alpha_j^{(k)} K_\Omega(z, a_j^k)|_{1,\Omega} \le c_1 |f - g - g_k|_{m_1,\Omega}.$$

□

Definition 3.38 Let Ω be a domain in $\mathbf{C}^n$. We say that Ω satisfies the condition $(\mathbf{B}_k)$ if

(a) For any $a \in \Omega$, $K_\Omega(\cdot, a) \in C^k(\overline{\Omega})$.
(b) For any $P \in \overline{\Omega}$, there exist $a_0, a_1, \cdots, a_n \in \Omega$ such that

$$K_\Omega(P, a_0) \neq 0, \qquad (3.64)$$

and

$$\begin{vmatrix} K_\Omega(P,a_0) & \cdots & K_\Omega(P,a_n) \\ \frac{\partial K_\Omega}{\partial z_1}(P,a_0) & \cdots & \frac{\partial K_\Omega}{\partial z_1}(P,a_n) \\ \vdots & \vdots & \vdots \\ \frac{\partial K_\Omega}{\partial z_n}(P,a_0) & \cdots & \frac{\partial K_\Omega}{\partial z_n}(P,a_n) \end{vmatrix} \neq 0, \qquad (3.65)$$

where the derivatives in (3.65) are taken with respect to the first variable in $K_\Omega(\cdot, a_j)$.

Theorem 3.36 Let $\Omega \subset\subset \mathbf{C}^n$ be a domain with C^∞ boundary. If Ω satisfies the condition $(\mathbf{R}_k)$, then Ω satisfies the condition $(\mathbf{B}_k)$.

Proof. The condition $(\mathbf{B}_k)$ (a) follows from Theorem 3.34. We will show $(\mathbf{B}_k)$ (b). We fix $P \in \overline{\Omega}$. Suppose (3.65) is equal to 0 for all $(a_0, a_1, \cdots, a_n) \in \Omega^{n+1}$. For any $g_0, g_1, \cdots, g_n \in \mathcal{O}(\Omega) \cap C^\infty(\overline{\Omega})$ and any $\varepsilon > 0$, by Theorem 3.35, there exist $a_j^k \in \Omega$, $j = 1, \cdots, N_k$, and constants b_j^k, $j = 1, \cdots, N_k$, such that

$$|g_k - \sum_{j=1}^{N_k} b_j^k K_\Omega(\cdot, a_j^k)|_{1,\Omega} < \varepsilon.$$

We set

$$\alpha_k(z) = \sum_{j=1}^{N_k} b_j^k K_\Omega(z, a_j^k).$$

Then by the assumption we have

$$\begin{vmatrix} \alpha_0 & \cdots & \alpha_n \\ \frac{\partial \alpha_0}{\partial z_1} & \cdots & \frac{\partial \alpha_n}{\partial z_1} \\ \vdots & \vdots & \vdots \\ \frac{\partial \alpha_0}{\partial z_n} & \cdots & \frac{\partial \alpha_n}{\partial z_n} \end{vmatrix} = 0.$$

On the other hand we have

$$\left| \begin{array}{ccc} g_0 & \cdots & g_n \\ \frac{\partial g_0}{\partial z_1} & \cdots & \frac{\partial g_n}{\partial z_1} \\ \vdots & \vdots & \vdots \\ \frac{\partial g_0}{\partial z_n} & \cdots & \frac{\partial g_n}{\partial z_n} \end{array} \right| - \left| \begin{array}{ccc} \alpha_0 & \cdots & \alpha_n \\ \frac{\partial \alpha_0}{\partial z_1} & \cdots & \frac{\partial \alpha_n}{\partial z_1} \\ \vdots & \vdots & \vdots \\ \frac{\partial \alpha_0}{\partial z_n} & \cdots & \frac{\partial \alpha_n}{\partial z_n} \end{array} \right| = O(\varepsilon).$$

Since $\varepsilon > 0$ is arbitrary, we have for any $g_0, g_1, \cdots, g_n \in \mathcal{O}(\Omega) \cap C^\infty(\overline{\Omega})$

$$\left| \begin{array}{ccc} g_0 & \cdots & g_n \\ \frac{\partial g_0}{\partial z_1} & \cdots & \frac{\partial g_n}{\partial z_1} \\ \vdots & \vdots & \vdots \\ \frac{\partial g_0}{\partial z_n} & \cdots & \frac{\partial g_n}{\partial z_n} \end{array} \right| = 0.$$

If we set $g_0 = 1$, $g_k(z) = z_k$ for $k = 1, \cdots, n$, then the left side of the above equality is 1, which is a contradiction. This proves $(\mathbf{B}_k)$ (b), which completes the proof of Theorem 3.36. □

Corollary 3.9 *Let $\Omega \subset\subset \mathbf{C}^n$ be a domain with C^∞ boundary and satisfy the condition $(\mathbf{B}_k)$. Then for any $P \in \partial\Omega$ there exist a neighborhood W of P and $a_0, a_1, \cdots, a_n \in \Omega$ such that if we set*

$$u_j(z) = \frac{K_\Omega(z, a_j)}{K_\Omega(z, a_0)}, \quad u = (u_1, \cdots, u_n),$$

then $u_j \in \mathcal{O}(W \cap \Omega) \cap C^k(W \cap \overline{\Omega})$, and

$$\det u'(P) \neq 0. \tag{3.66}$$

Moreover, each component of the inverse mapping $u^{-1} : u(W \cap \overline{\Omega}) \to W \cap \overline{\Omega}$ belongs to $\mathcal{O}(u(W \cap \Omega)) \cap C^k(u(W \cap \overline{\Omega}))$.

Proof. We fix $P \in \partial\Omega$. Since Ω satisfies the condition $(\mathbf{B}_k)$, by Theorem 3.36 there exist $a_0, a_1, \cdots, a_n \in \Omega$ which satisfy (3.64) and (3.65). Hence there exists a neighborhood W of P such that $K_\Omega(z, a_0) \neq 0$ for $z \in W \cap \overline{\Omega}$, which implies that $u_j \in \mathcal{O}(W \cap \Omega) \cap C^k(W \cap \overline{\Omega})$. Since

$$\frac{\partial u_j}{\partial z_k}(P) = K_\Omega(P, a_0)^{-1} \frac{\partial K_\Omega}{\partial z_k}(P, a_j) - \frac{\partial K_\Omega}{\partial z_k}(P, a_0) K_\Omega(P, a_0)^{-2} K_\Omega(P, a_j),$$

we have

$$\begin{vmatrix} K_\Omega(P,a_0) & \cdots & K_\Omega(P,a_n) \\ \frac{\partial K_\Omega}{\partial z_1}(P,a_0) & \cdots & \frac{\partial K_\Omega}{\partial z_1}(P,a_n) \\ \vdots & \vdots & \vdots \\ \frac{\partial K_\Omega}{\partial z_n}(P,a_0) & \cdots & \frac{\partial K_\Omega}{\partial z_n}(P,a_n) \end{vmatrix}$$

$$= K(P,a_0)^n \begin{vmatrix} K_\Omega(P,a_0) & K_\Omega(P,a_0) & K_\Omega(P,a_0) & \cdots & K_\Omega(P,a_n) \\ 0 & \frac{\partial u_1}{\partial z_1} & & \cdots & \frac{\partial u_n}{\partial z_1} \\ \vdots & \vdots & \vdots & & \vdots \\ 0 & \frac{\partial u_1}{\partial z_n} & & \cdots & \frac{\partial u_n}{\partial z_n} \end{vmatrix}$$

$$= K(P,a_0)^{n+1} \begin{vmatrix} \frac{\partial u_1}{\partial z_1} & \cdots & \frac{\partial u_n}{\partial z_1} \\ \vdots & \vdots & \vdots \\ \frac{\partial u_1}{\partial z_n} & \cdots & \frac{\partial u_n}{\partial z_n} \end{vmatrix}.$$

This proves (3.66). Since u_j is holomorphic in $W \cap \Omega$, we have

$$\frac{\partial u_j}{\partial \bar{z}_j}(P) = 0.$$

Thus the Jacobian of u at P $Ju(P) = |\det u'(P)|^2 \neq 0$. Since u_j can be extended to C^k functions in W, by contracting W if necessary, $u : W \to u(W)$ is a C^k diffeomorphism. Since $u : W \cap \Omega \to u(W \cap \Omega)$ is a holomorphic mapping, $u^{-1} : u(W \cap \Omega) \to W \cap \Omega$ is a holomorphic mapping. $\square$

Theorem 3.37 *Let Ω_1 and Ω_2 be bounded domains in $\mathbf{C}^n$ and satisfy the condition $(\mathbf{B}_k)$ for $k \geq 1$. Then every holomorphic mapping $F : \Omega_1 \to \Omega_2$ belongs to $C^k(\overline{\Omega}_1)$.*

Proof. The proof involves three steps.

[1] There exist constants c_1, c_2 such that $0 < c_1 \leq |\det F'(z)| \leq c_2$ for all $z \in \Omega_1$.

(Proof of [1]) We assume that there does not exist c_1 which satisfies $0 < c_1 \leq |\det F'(z)|$ for all $z \in \Omega_1$. Then there exists a sequence $\{p_\nu\} \subset \Omega_1$ such that $\det F'(p_\nu) \to 0$ as $\nu \to 0$. Taking the subsequence of $\{p_\nu\}$, we may assume that $\{p_\nu\}$ converges. We set $\lim_{\nu \to \infty} p_\nu = P$. Then $P \in \partial \Omega_1$. By Theorem 3.24, for $a \in \Omega_1$ we have

$$K_{\Omega_1}(p_\nu, a) = \det F'(p_\nu) K_{\Omega_2}(F(p_\nu), F(a)) \overline{\det F'(a)}. \tag{3.67}$$

From the condition $(\mathbf{B}_k)$, $K_{\Omega_1}(\cdot, a)$ and $K_{\Omega_2}(\cdot, F(a))$ are continuous on $\overline{\Omega}_1$ and $\overline{\Omega}_2$, respectively, and hence $K_{\Omega_1}(P, a) = 0$ for $a \in \Omega_1$. This contradicts the condition ($\mathbf{B}_k$). By adopting the same argument to F^{-1}, there exists $c > 0$ such that $|\det (F^{-1})'(w)| \geq c$ for $w \in \Omega_2$. Hence we have $|\det F'(z)| \leq 1/c$ for $z \in \Omega_1$.

[2] $F \in C(\overline{\Omega}_1)$.

(Proof of [2]) It is sufficient to show that $\frac{\partial f_j}{\partial z_i}$, $i,j = 1, \cdots, n$, are bounded in Ω_1. Suppose there exists $\{p_\nu\} \subset \Omega_1$ such that

$$\max_{j,i} \left| \frac{\partial f_j}{\partial z_i}(p_\nu) \right| \to \infty \quad (\nu \to \infty). \tag{3.68}$$

We set $q_\nu = F(p_\nu)$. Taking subsequences, we may assume that $\{p_\nu\}$ and $\{q_\nu\}$ converge. We set $\lim_{\nu \to \infty} p_\nu = P$ ($P \in \partial\Omega_1$), $\lim_{\nu \to \infty} q_\nu = Q$. Then $Q \in \partial\Omega_2$. For $Q \in \partial\Omega_2$, we choose $b_0, b_1, \cdots, b_n \in \Omega_2$ satisfying the condition ($\mathbf{B}_k$). Then we have $K_{\Omega_2}(Q, b_0) \neq 0$. We define $v = (v_1, \cdots, v_n)$ by

$$v_j(w) = \frac{K_{\Omega_2}(w, b_j)}{K_{\Omega_2}(w, b_0)} \quad (j = 1, \cdots, n).$$

By Corollary 3.9 we have $\det v'(Q) \neq 0$ and there exists a neighborhood W_1 of Q such that each component of v^{-1} belongs to $C^k(W_1 \cap \overline{\Omega}_1)$. If we set $a_j = F^{-1}(b_j)$ for $j = 0, 1, \cdots, n$, then $K_{\Omega_1}(P, a_0) \neq 0$ by substituting $a = a_0$ into (3.67). Thus there exists a neighborhood W_2 of P such that if we set

$$u_j(z) = \frac{K_{\Omega_1}(z, a_j)}{K_{\Omega_1}(z, a_0)},$$

then we have $u_j \in \mathcal{O}(W_1 \cap \Omega_1) \cap C^k(W_1 \cap \overline{\Omega}_1)$. For a sufficiently large ν_0, we have $p_\nu \in W_1 \cap \Omega_1$, $q_\nu \in W_2 \cap \Omega_2$. By Theorem 3.24 and the definition of u_j and v_j, we obtain

$$u_j(z) = v_j(F(z))\lambda_j \quad (z \in W_1 \cap \Omega_1, F(z) \in W_2 \cap \Omega_2), \tag{3.69}$$

where we define

$$\lambda_j = \overline{\left(\frac{\det F'(a_j)}{\det F'(a_0)} \right)}.$$

We set
$$\Lambda = \begin{pmatrix} \lambda_1 & 0 & \cdots & 0 \\ 0 & \lambda_2 & & \vdots \\ \vdots & & \ddots & 0 \\ 0 & & \cdots & \lambda_n \end{pmatrix}.$$

It follows from (3.69) that
$$u(z) = \Lambda v(F(z)). \tag{3.70}$$

By the chain rule we have
$$u'(p_\nu) = \Lambda v'(F(p_\nu))F'(p_\nu),$$
which implies that
$$F'(p_\nu) = v'(q_\nu)^{-1}\Lambda^{-1}u'(p_\nu).$$

Since v' is a coordinate system in a neighborhood of Q, each component of $v'(q_\nu)^{-1}$ is bounded. Further, each component of u' is bounded in a neighborhood of P. Hence each component of $F'(p_\nu)$ remains bounded as $\nu \to \infty$, which contradicts (3.68).

[3] $F \in C^k(\overline{\Omega}_1)$.

(Proof of [3]) It follows from (3.70) that for $z \in W_1 \cap \Omega_1$, we have $F(z) = v^{-1}(\Lambda^{-1}(u(z)))$, which means that $F \in C^k(W_1 \cap \overline{\Omega}_1)$. □

Corollary 3.10 *(Fefferman's mapping theorem) Let Ω_1 and Ω_2 are strictly pseudoconvex domains in $\mathbf{C}^n$ with C^∞ boundary. Then every biholomorphic mapping $F : \Omega_1 \to \Omega_2$ belongs to $C^\infty(\overline{\Omega}_1)$.*

Proof. Since Ω_1 and Ω_2 satisfy the condition $(\mathbf{R}_k)$ for any positive integer k by Theorem 3.33, Ω satisfies the condition $(\mathbf{B}_k)$. Hence it follows from Theorem 3.37 that $F \in C^k(\overline{\Omega}_1)$. □

Exercises

3.1 Prove the following:

Let $\delta > 0$ and
$$\Gamma_\delta = \{(x_1, x') \in \mathbf{R}^N \mid 0 < x_1 < \delta, |x'| < \delta\},$$
and suppose $g \in C^1(\Gamma_\delta)$ satisfies
$$|dg(x)| \leq K x_1^{\alpha-1}$$

for $x \in \Gamma_\delta$. Then there is a constant C depending only on α and δ such that

$$|g(x) - g(y)| \le CK|x-y|^\alpha$$

for $x, y \in \Gamma_{\delta/2}$ with $|x-y| \le \delta/2$.

3.2 Let Ω be a convex domain in $\mathbf{C}^n$ and let $F_1 : \Omega \to \mathbf{C}$ be a C^1 function in Ω. For $w, z \in \Omega$ and $1 < \theta, \lambda \le 1$,

$$\left.\frac{dF_1(z + \lambda\theta(w-z))}{d\lambda}\right|_{\lambda=1} = \theta \frac{dF_1(z + \theta(w-z))}{d\theta}.$$

3.3 Let H be a Hilbert space and let φ be a continuous linear functional on H. Define $M = \{x \in H \mid \varphi(x) = 0\}$. Show that if $M \ne H$, then $M^\perp$ is one dimensional.

3.4 Let $\Omega = \{z \in \mathbf{C} \mid |z| < 1\}$. Define

$$\varphi_n(z) = \frac{\sqrt{\pi}}{\sqrt{n+1}} z^n.$$

Prove that $\{\varphi_n\}$ is a complete orthonormal sequence in $A^2(\Omega)$.

3.5 Let $\Omega = \{z \in \mathbf{C} \mid |z| < 1\}$. Prove that the Bergman kernel $K_\Omega(z, \zeta)$ for Ω is given by

$$K_\Omega(z, \zeta) = \frac{1}{\pi} \frac{1}{(1 - z\bar{\zeta})^2}.$$

3.6 Let $\Omega \subset\subset \mathbf{C}^n$ be a domain. We set $k_\Omega(z) = K_\Omega(z, z)$. Show that

(a) $\left(f(z), \dfrac{\partial K_\Omega(z, \zeta)}{\partial \bar{\zeta}_\nu}\right) = \dfrac{\partial f}{\partial \zeta_\nu}(\zeta)$ ($f \in A^2(\Omega)$, $\zeta \in \Omega$).
(b) $k_\Omega(z) > 0$ ($z \in \Omega$).
(c) $\log k_\Omega(z)$ is strictly plurisubharmonic in Ω.

3.7 (**Bergman metric**) Let $\Omega \subset\subset \mathbf{C}^n$. The Hermitian metric for Ω is defined by

$$g_{ij}(z) = \frac{\partial^2 \log k_\Omega}{\partial z_j \partial \bar{z}_k}(z).$$

Let $\gamma : [0,1] \to \Omega$ be a C^1 curve. The length of γ with respect to the Bergman metric $|\gamma|_{B(\Omega)}$ is defined by

$$|\gamma|_{B(\Omega)} = \int_0^1 \left(\sum_{i,j=1}^n g_{ij}(\gamma(t))\gamma_i'(t)\overline{\gamma_j'(t)} \right)^{1/2} dt.$$

For $z_1, z_2 \in \Omega$, we define the distance of z_1, z_2 with respect to the Bergman metric by

$$\delta_{B(\Omega)}(z_1, z_2) = \inf_\gamma |\gamma|_{B(\Omega)},$$

where the infimum is taken for all C^1 curves in Ω which connect z_1 and z_2.

Prove that if $f : \Omega_1 \to \Omega_2$ is a biholomorphic mapping, then

$$\delta_{B(\Omega_1)}(z_1, z_2) = \delta_{B(\Omega_2)}(f(z_1), f(z_2)).$$

Chapter 4

Integral Formulas with Weight Factors

In this chapter we study the Berndtsson-Andersson formula on bounded domains in $\mathbf{C}^n$ with smooth boundary and the Berndtsson formula on submanifolds in general position of bounded domains in $\mathbf{C}^n$ with smooth boundary. By applying the Berndtsson-Andersson formula, we prove L^p estimates for the $\bar{\partial}$ problem in strictly pseudoconvex domains in $\mathbf{C}^n$ with smooth boundary. Moreover, using the Berndtsson formula we give two counterexamples of L^p ($2 < p \leq \infty$) extensions of bounded holomorphic functions from submanifolds of complex ellipsoids due to Mazzilli [MAZ1] and Diederich-Mazzilli [DIM1]. Finally, we give the alternative proof of the bounded extension of holomorphic functions from affine linear submanifolds of strictly convex domains using the method of Diederich-Mazzilli [DIM2].

4.1 The Berndtsson-Andersson Formula

In this section we study the integral formula obtained by Berndtsson-Andersson [BRA].
Let

$$\mu = <\xi, \eta>^{-n} \omega'(\xi) \wedge \omega(\eta)$$

be a differential form in $\mathbf{C}^n \times \mathbf{C}^n = \{(\xi, \eta) \mid \xi \in \mathbf{C}^n, \eta \in \mathbf{C}^n\}$, where we define

$$\omega'(\xi) := \sum_{j=1}^{n}(-1)^{j-1}\xi_j \bigwedge_{i \neq j} d\xi_i, \quad \omega(\eta) := d\eta_1 \wedge \cdots \wedge d\eta_n,$$

$$<\xi,\eta>:=\sum_{j=1}^{n}\xi_j\eta_j.$$

By Lemma 3.2, if $<\xi,\mu>\neq 0$, then $d\mu = 0$. Let Ω be a bounded domain in $\mathbf{C}^n$ with smooth boundary. Assume that a C^1 mapping $s = (s_1,\cdots,s_n) : \overline{\Omega}\times\overline{\Omega}\to\mathbf{C}^n$ satisfies the following conditions:

(**A**) If $\zeta \neq z$, then $<s(z,\zeta),\zeta - z>\neq 0$.
(**B**) For any compact set $K\subset\Omega$, there exist constants $C_1 = C_1(K) > 0$, $C_2 = C_2(K) > 0$ such that

$$|s(z,\zeta)|\leq C_1|\zeta - z|,\quad |<s(z,\zeta),\zeta - z>|\geq C_2|\zeta - z|^2$$

for $\zeta\in\overline{\Omega}$ and $z\in K$.

In what follows we assume that s satisfies the above conditions (**A**) and (**B**). Define $\psi : \overline{\Omega}\times\overline{\Omega}\backslash\Delta \to E$ to be $\psi(z,\zeta) = (s(z,\zeta),\zeta-z)$, where $\Delta = \{(z,z)\mid z\in\mathbf{C}^n\}$. Let K be the pullback of μ by ψ. Then

$$K = \psi^*\mu = \frac{1}{<s,\zeta-z>^n}\omega'(s)\wedge\omega(\zeta-z)$$

$$= \frac{1}{<s,\zeta-z>^n}\sum_{j=1}^{n}(-1)^{j-1}s_j\underset{i\neq j}{\wedge}d_{z,\zeta}s_i$$

$$\wedge(d\zeta_1 - dz_1)\wedge\cdots\wedge(d\zeta_n - dz_n).$$

We denote by $K_{p,q}$ the component of K which is of degree (p,q) with respect to z and of degree $(n-p, n-q-1)$ with respect to ζ. Then we have the following theorem.

Theorem 4.1 *For $f\in C^1_{(p,q)}(\overline{\Omega})$, one has*
(a) For $q > 0$,

$$f = C\left\{\int_{\partial\Omega}f\wedge K_{p,q} + (-1)^{p+q+1}\left(\int_{\Omega}\bar{\partial}f\wedge K_{p,q} - \bar{\partial}_z\int_{\Omega}f\wedge K_{p,q-1}\right)\right\},$$

where $C = C_{p,q,n}$ is a numerical constant depending only on p, q, n.
(b) For $q = 0$,

$$f = C\left\{\int_{\partial\Omega}f\wedge K_{p,0} + (-1)^{p+1}\int_{\Omega}\bar{\partial}f\wedge K_{p,0}\right\},$$

where $C = C_{p,n}$ is a numerical constant depending only on p, n.

Proof. Let φ be a $C^\infty(n-p, n-q)$ form in Ω with compact support. For $\varepsilon > 0$, we set

$$U_\varepsilon = \{(\zeta, z) \in \Omega \times \Omega \mid |\zeta - z| < \varepsilon\}$$

$$\Omega_\varepsilon = \Omega \times \Omega - U_\varepsilon.$$

If we choose ε sufficiently small in comparison with the distance between $\mathrm{supp}(\varphi)$ and $\partial\Omega$, then

$$\partial\Omega_\varepsilon \cap (\mathbf{C}^n \times \mathrm{supp}(\varphi)) = \{(\partial\Omega \times \Omega) \cup \partial U_\varepsilon\} \cap (\mathbf{C}^n \times \mathrm{supp}(\varphi)).$$

It follows from Stokes' theorem that

$$\int_{\Omega_\varepsilon} d_{z,\zeta}(\varphi(z) \wedge f(\zeta) \wedge K(z,\zeta)) = \int_{\partial\Omega_\varepsilon} \varphi \wedge f \wedge K$$
$$= \int_{\partial\Omega \times \Omega} \varphi \wedge f \wedge K - \int_{\partial U_\varepsilon} \varphi \wedge f \wedge K.$$

Since $dK = \psi^* d\mu = 0$, we obtain

$$\int_{\Omega_\varepsilon} d\varphi \wedge f \wedge K + (-1)^{p+q} \int_{\Omega_\varepsilon} \varphi \wedge df \wedge K = \int_{\partial\Omega \times \Omega} \varphi \wedge f \wedge K - \int_{\partial U_\varepsilon} \varphi \wedge f \wedge K. \tag{4.1}$$

Since

$$K = O\left(\frac{|s|}{|<s, \zeta - z>|^n}\right) = O(|\zeta - z|^{1-2n}),$$

the two integrals in the left side of (4.1) converges as $\varepsilon \to 0$. Next we investigate the second integral in the right side of (4.1). We obtain

$$\lim_{\varepsilon \to 0} \int_{\partial U_\varepsilon} \varphi \wedge f \wedge K = \lim_{\varepsilon \to 0} \int_{\partial U_\varepsilon} \varphi \wedge f \wedge \frac{\omega'(s) \wedge \omega(\zeta - z)}{<s, \zeta - z>^n}. \tag{4.2}$$

Lemma 3.5 implies that $\omega'(s) \wedge \omega(\zeta - z) <s, \zeta - z>^{-n}$ is invariant when we replace s by $s\frac{\overline{<s,\zeta-z>}}{|<s,\zeta-z>|}$. Hence we may assume that $<s, \zeta - z> \gg 0$ for $\zeta \neq z$. Define

$$b = \bar\zeta - \bar z, \quad s_\lambda = \lambda s + (1-\lambda)b \quad (0 < \lambda < 1).$$

Further we define $h : \overline{\Omega} \times \overline{\Omega} \times [0,1] \to \mathbf{C}^n \times \mathbf{C}^n$ by

$$h(z, \zeta, \lambda) = (s_\lambda(z, \zeta), \zeta - z).$$

If we set $H = h^*\mu$, then we obtain

$$H(z,\zeta,\lambda) = <\lambda s + (1-\lambda)b, \zeta - z>^{-n} \omega'(\lambda s + (1-\lambda)b) \wedge \omega(\zeta - z).$$

We set

$$I_\varepsilon = \int_{\partial(\{|\zeta-z|=\varepsilon\}\times[0,1])} \varphi(z) \wedge f(\zeta) \wedge H(z,\zeta,\lambda).$$

Since $dH = h^* d\mu = 0$, it follows from Stokes' theorem that

$$I_\varepsilon = \int_{\{|\zeta-z|=\varepsilon\}\times[0,1]} d(\varphi \wedge f) \wedge H.$$

Since $\mathrm{supp}(\varphi)$ is compact, $\varphi = 0$ on $\partial\{|\zeta - z| = \varepsilon\}$ for any sufficiently small $\varepsilon > 0$. Hence we obtain

$$I_\varepsilon = \int_{|\zeta-z|=\varepsilon} \varphi \wedge f \wedge H(z,\zeta,1) - \int_{|\zeta-z|=\varepsilon} \varphi \wedge f \wedge H(z,\zeta,0). \qquad (4.3)$$

We denote by H' the component of H which involves $d\lambda$. Then

$$|H'| \leq C \left(\frac{|s_\lambda|(|s|+|b|)}{\lambda <s,\zeta-z> +(1-\lambda)|\zeta-z|^2)^n} \right) = O(|\zeta-z|^{2-2n}).$$

Consequently we have $\lim_{\varepsilon \to 0} I_\varepsilon = 0$. It follows from (4.2) and (4.3) that

$$\lim_{\varepsilon \to 0} \int_{\partial U_\varepsilon} \varphi \wedge f \wedge K = \lim_{\varepsilon \to 0} \int_{\partial U_\varepsilon} \varphi(z) \wedge f(\zeta) \wedge H(z,\zeta,0). \qquad (4.4)$$

By the same method as the proof of Theorem 3.3, the right side of (4.4) is equal to $C_{p,q,n} \int_\Omega \varphi \wedge f$. Letting $\varepsilon \to 0$ in (4.1) we obtain

$$\int_{\Omega\times\Omega} d\varphi \wedge f \wedge K + (-1)^{p+q} \int_{\Omega\times\Omega} \varphi \wedge df \wedge K = \int_{\partial\Omega\times\Omega} \varphi \wedge f \wedge K - C_{p,q,n} \int_\Omega \varphi \wedge f \qquad (4.5)$$

and

$$\int_{\Omega\times\Omega} d\varphi(z) \wedge f(\zeta) \wedge K(z,\zeta)$$

$$= \int_{z\in\Omega} d\varphi(z) \wedge \int_{\zeta\in\Omega} f(\zeta) \wedge K(z,\zeta)$$

$$= (-1)^{p+q+1} \int_{z\in\Omega} \varphi(z) \wedge d_z \int_{\zeta\in\Omega} f(\zeta) \wedge K(z,\zeta).$$

Since $\varphi(z) \wedge f(\zeta)$ is of degree n with respect to $d\zeta$ and dz and that $K(z,\zeta)$ is of degree $\geq n$ with respect to $d\zeta$ and dz, we have

$$\int_{\Omega \times \Omega} d\varphi(z) \wedge f(\zeta) \wedge K(z,\zeta) = (-1)^{p+q+1} \int_{z \in \Omega} \varphi(z) \wedge \bar{\partial}_z \int_{\zeta \in \Omega} f(\zeta) \wedge K(z,\zeta).$$

Thus by (4.5), we obtain

$$\int_{\Omega} \varphi \wedge \int_{\partial\Omega} f \wedge K = (-1)^{p+q} \int_{\Omega} \varphi \wedge \left\{ \int_{\Omega} \bar{\partial} f \wedge K - \bar{\partial}_z \int_{\Omega} f \wedge K \right\}$$
$$+ C_{p,q,n} \int_{\Omega} \varphi \wedge f.$$

This proves (a). In case $q = 0$, $\bar{\partial}_z \int_{\Omega} f \wedge K$ is of degree ≥ 1 with respect to $d\bar{z}$, which implies that

$$\int_{\Omega} \varphi \wedge \int_{\partial\Omega} f \wedge K = (-1)^p \int_{\Omega} \varphi \wedge \int_{\Omega} \bar{\partial} f \wedge K + C_{p,n} \int_{\Omega} \varphi \wedge f.$$

This proves (b). □

Corollary 4.1 *Assume that in addition to the conditions (**A**) and (**B**), $s : \overline{\Omega} \times \overline{\Omega} \to \mathbf{C}^n$ satisfies the conditon:*
*(**C**) For $\zeta \in \partial\Omega$, $s(z,\zeta)$ is holomorphic with respect to $z \in \Omega$.*
Then

$$u(z) = (-1)^{p+q} C_{p,q,n} \int_{\Omega} f(\zeta) \wedge K_{p,q-1}(z,\zeta)$$

is a solution of the equation $\bar{\partial} u = f$ for $f \in C^1_{(p,q)}(\overline{\Omega})$ $(q > 0)$ with $\bar{\partial} f = 0$.

Proof. It follows from the condition (**C**) that $K(z,\zeta)$ is of degree 0 with respect to $d\bar{z}$ for $\zeta \in \partial\Omega$, and hence $K_{p,q} = 0$. Therefore, Corollary 4.1 follows from Theorem 4.1 (a). □

Next we study the differential form

$$A = \exp <\xi, \eta> \omega(\xi) \wedge \omega(\eta)$$

in $\mathbf{C}^n \times \mathbf{C}^n$. Suppose a C^1 mapping $Q(z,\zeta) : \overline{\Omega} \times \overline{\Omega} \to \mathbf{C}^n$ is holomorphic in $z \in \Omega$ for ζ fixed. Let $Q = (Q_1, \cdots, Q_n)$. We define $\psi : (\overline{\Omega} \times \overline{\Omega} \backslash \Delta) \times (0, \infty) \to \mathbf{C}^n \times \mathbf{C}^n$ by

$$\psi(z,\zeta,t) = (Q(z,\zeta) + ts(z,\zeta), \zeta - z).$$

We set $N = \psi^* A$. Then N can be written

$$N = \exp <Q+ts, \zeta-z> d(Q_1+ts_1) \wedge \cdots \wedge d(Q_n+ts_n) \wedge \omega(\zeta-z). \quad (4.6)$$

We write $N = N_t + N'$, where N_t is the component of N which contains dt, and N' is the the component of N which does not contain dt. Then

$$N_t = -\exp <Q, \zeta - z> \exp\{t<s, \zeta - z>\} \times$$

$$\{t^{n-1}\omega'(s) \wedge \omega(\zeta - z) \wedge dt + \sum_{k=0}^{n-2} t^k a_k \wedge dt\},$$

where a_k are differential forms which do not contain t. Since $dA = 0$, we have $dN = \psi^* dA = 0$. Consequently,

$$0 = d_{\zeta, z, t} N = d_{\zeta, z} N_t + d_t N' + d_{\zeta, z} N'. \tag{4.7}$$

Since the last term in the right side of (4.7) does not contain dt, we obtain

$$d_{\zeta, z} N_t = -d_t N'. \tag{4.8}$$

In the moment we assume that $\operatorname{Re} <s, \zeta - z> < 0$ for $\zeta \neq z$. Later we show that this assumption is not necessary. It follows from (4.8) that

$$d_{\zeta, z} K = \int_0^\infty d_{\zeta, z} N_t = -\int_0^\infty d_t N' = N'|_{t=0}$$
$$= \exp <Q, \zeta - z> \omega(Q) \wedge \omega(\zeta - z).$$

We set

$$P = \exp <Q, \zeta - z> \omega(Q) \wedge \omega(\zeta - z).$$

Then we have

$$d_{\zeta, z} K = P. \tag{4.9}$$

Since Q is holomorphic in z, dQ does not contain $d\bar{z}_j$, and hence P does not contain $d\bar{z}_j$. By the integration by parts, we obtain

$$\int_0^\infty e^{t<s, \zeta-z>} t^{n-1} dt = \left[\frac{e^{t<s, \zeta-z>}}{<s, \zeta - z>} t^{n-1}\right]_0^\infty$$
$$- \int_0^\infty \frac{e^{t<s, \zeta-z>}}{<s, \zeta - z>}(n-1) t^{n-2} dt$$
$$= \cdots = (-1)^n (n-1)! \frac{1}{<s, \zeta - z>^n}.$$

Hence K is expressed by

$$K = (-1)^{n-1}(n-1)!\exp <Q,\zeta-z> \frac{\omega'(s)\wedge\omega(\zeta-z)}{<s,\zeta-z>^n} \quad (4.10)$$

$$+ \sum_{k=0}^{n-2} O(<s,\zeta-z>^{-(k+1)}).$$

Theorem 4.2 *For $f \in C^1_{(p,q)}(\overline{\Omega})$, one has*
(a) In case $q > 0$,

$$f = C\left\{\int_{\partial\Omega} f \wedge K_{p,q} + (-1)^{p+q+1}\left(\int_\Omega \bar\partial f \wedge K_{p,q} - \bar\partial_z \int_\Omega f \wedge K_{p,q-1}\right)\right\},$$

where $K_{p,q}$ are components of K which are (p,q) forms with respect to z and $(n-p, n-q-1)$ forms with respect to ζ.
(b) In case $q = 0$,

$$f = C\left\{\int_{\partial\Omega} f \wedge K_{p,0} + (-1)^{p+1}\int_\Omega \bar\partial f \wedge K_{p,0} - \int_\Omega f \wedge P_{p,0}\right\}.$$

Proof. Let Ω_ε, U_ε and φ be the same notations as in the proof of Theorem 4.1. It follows from Stokes' theorem that

$$\int_{\Omega_\varepsilon} d_{z,\zeta}(\varphi(z) \wedge f(\zeta) \wedge K(z,\zeta)) = \int_{\partial\Omega_\varepsilon} \varphi \wedge f \wedge K$$
$$= \int_{\partial\Omega\times\Omega} \varphi \wedge f \wedge K - \int_{\partial U_\varepsilon} \varphi \wedge f \wedge K.$$

By (4.9) we obtain

$$\int_{\Omega_\varepsilon} d\varphi \wedge f \wedge K + (-1)^{p+q}\int_{\Omega_\varepsilon} \varphi \wedge df \wedge K + \int_{\Omega_\varepsilon} \varphi \wedge f \wedge P \quad (4.11)$$

$$= \int_{\partial\Omega\times\Omega} \varphi \wedge f \wedge K - \int_{\partial U_\varepsilon} \varphi \wedge f \wedge K.$$

On the other hand, we have

$$K = O\left(\frac{|s|}{|<s,\zeta-z>|^n}\right) = O(|\zeta-z|^{1-2n}),$$

which means that the three integrals in the left side of (4.11) converge as $\varepsilon \to 0$. Next we investigate the second integral in the right side of (4.11).

It follows from (4.10) that

$$K = -(n-1)!(\exp <Q, \zeta-z>)\frac{\omega'(s) \wedge \omega(\zeta-z)}{<s, z-\zeta>^n} + T_1,$$

where $T_1 = O(|\zeta - z|^{2-2n})$. Since

$$\exp <Q, \zeta-z> = 1 + <Q, \zeta-z> + \frac{<Q, \zeta-z>^2}{2!} + \cdots,$$

K can be written

$$K = -(n-1)!\frac{\omega'(s) \wedge \omega(\zeta-z)}{<s, z-\zeta>^n} + T_2,$$

where $T_2 = O(|\zeta - z|^{2-2n})$. Consequently, using the same method as in the proof of Theorem 4.1 we obtain

$$\lim_{\varepsilon \to 0} \int_{\partial U_\varepsilon} \varphi \wedge f \wedge K$$
$$= (-1)^{n-1}(n-1)! \lim_{\varepsilon \to 0} \int_{\partial U_\varepsilon} \varphi \wedge f \wedge \frac{\omega'(s) \wedge \omega(\zeta-z)}{<s, \zeta-z>^n}$$
$$= C_{p,q,n} \int_\Omega \varphi \wedge f,$$

where $C_{p,q,n}$ are numerical constants depending only on p, q, n. Letting $\varepsilon \to 0$ in (4.11) we obtain

$$\int_{\Omega \times \Omega} d\varphi \wedge f \wedge K + (-1)^{p+q} \int_{\Omega \times \Omega} \varphi \wedge df \wedge K + \int_{\Omega \times \Omega} \varphi \wedge f \wedge P \quad (4.12)$$

$$= \int_{\partial \Omega \times \Omega} \varphi \wedge f \wedge K - C_{p,q,n} \int_\Omega \varphi \wedge f.$$

Consequently,

$$\int_{\Omega \times \Omega} d\varphi(z) \wedge f(\zeta) \wedge K(z, \zeta)$$
$$= \int_{z \in \Omega} d\varphi(z) \wedge \int_{\zeta \in \Omega} f(\zeta) \wedge K(z, \zeta)$$
$$= (-1)^{p+q+1} \int_{z \in \Omega} \varphi(z) \wedge d_z \int_{\zeta \in \Omega} f(\zeta) \wedge K(z, \zeta).$$

Using the fact that $\varphi(z) \wedge f(\zeta)$ is of degree n with respect to $d\zeta$ and dz and that $K(z,\zeta)$ is of degree $\geq n$ with respect to $d\zeta$ and dz, we have

$$\int_{\Omega \times \Omega} d\varphi(z) \wedge f(\zeta) \wedge K(z,\zeta) = (-1)^{p+q+1} \int_{z \in \Omega} \varphi(z) \wedge \bar{\partial}_z \int_{\zeta \in \Omega} f(\zeta) \wedge K(z,\zeta).$$

It follows from (4.12) that

$$\int_\Omega \varphi \wedge \int_{\partial\Omega} f \wedge K = (-1)^{p+q} \int_\Omega \varphi \wedge \left\{ \int_\Omega \bar{\partial} f \wedge K - \bar{\partial}_z \int_\Omega f \wedge K \right\}$$
$$+ \int_\Omega \varphi \wedge \int_\Omega f \wedge P + C_{p,q,n} \int_\Omega \varphi \wedge f.$$

Since $Q(z,\zeta)$ is holomorphic in $z \in \Omega$, $\omega(Q)$ is of degree 0 with respect to $d\bar{z}$, which implies that $P_{p,q} = 0$ if $q > 0$. This proves (a). If $q = 0$, then $\varphi(z)$ is of degree n with respect to $d\bar{z}$, which means that

$$\varphi(z) \wedge \bar{\partial}_z \int_\Omega f(\zeta) \wedge K(z,\zeta) = 0.$$

This proves (b). □

Theorem 4.3 *Let $s : \bar{\Omega} \times \bar{\Omega} \to \mathbf{C}^n$ satisfy the conditions (A), (B) and (C), and $q > 0$. If $f \in C^1_{(p,q)}(\bar{\Omega})$ satisfies the equation $\bar{\partial} f = 0$, then*

$$u(z) = (-1)^{p+q} C_{p,q,n} \int_\Omega f(\zeta) \wedge K_{p,q-1}(z,\zeta)$$

is a solution of the equation $\bar{\partial} u = f$.

Proof. The condition (C) implies that $\bar{\partial}_z s_j = 0$ for $j = 1, \cdots, n$ and $\zeta \in \partial\Omega$, and hence $\bar{\partial}_z Q_j = 0$ for $j = 1, \cdots, n$ and $\zeta \in \Omega$. Since Q_j are of class C^1 in $\bar{\Omega} \times \bar{\Omega}$, we have $\bar{\partial}_z Q_j = 0$ for $j = 1, \cdots, n$ and $\zeta \in \partial\Omega$. It follows from the condition (C) that N is of degree 0 with respect to $d\bar{z}$ for $\zeta \in \partial\Omega$, which means that $K_{p,q} = 0$ for $q > 0$ by (4.10). Then Theorem 4.3 follows from Theorem 4.2 (a). □

Definition 4.1 For $a = (a_1, \cdots, a_n) \in \mathbf{C}^n$, define

$$\omega'(a,\xi) = \sum_{j=1}^n (-1)^{j-1} a_j \bigwedge_{i \neq j} d\xi_j.$$

We have the following lemma. We omit the proof.

Lemma 4.1

$$\omega'(a,\xi) \wedge \omega(\eta) = C_n \sum_{k=1}^{n} a_k d\eta_k \wedge \left(\sum_{j=1}^{n} d\xi_j \wedge d\eta_j \right)^{n-1},$$

where $C_n = (-1)^{n(n-1)/2}/(n-1)!$.

Definition 4.2 For $s = (s_1, \cdots, s_n)$, $Q = (Q_1, \cdots, Q_n)$, define

$$s = \sum_{j=1}^{n} s_j(d\zeta_j - dz_j), \quad Q = \sum_{j=1}^{n} Q_j(d\zeta_j - dz_j).$$

Notice that we use notations which have two meanings.
By Lemma 4.1 we have

$$N_t = \exp(<Q, \zeta - z> + t<s, \zeta - z>)dt \wedge \omega'(s, Q+ts) \wedge \omega(\zeta - z)$$
$$= C_n \exp(<Q, \zeta - z> + t<s, \zeta - z>)dt \wedge s \wedge (dQ + tds)^{n-1}$$
$$= C_n \exp(<Q, \zeta - z> + t<s, \zeta - z>)dt \wedge s \wedge$$
$$\sum_{k=0}^{n-1} \binom{n-1}{k} (dQ)^k \wedge (ds)^{n-1-k} t^{n-k-1}.$$

It follows from the definition of K that

$$K = C_n(-1)^n \exp <Q, \zeta - z> \sum_{k=0}^{n-1} (-1)^k \frac{(n-1)!}{k!} \frac{s \wedge (dQ)^k \wedge (ds)^{n-1-k}}{<s, \zeta - z>^{n-k}} \tag{4.13}$$

and

$$P = \frac{(-1)^{n(n-1)/2}}{n!} \exp <Q, \zeta - z> (dQ)^n. \tag{4.14}$$

For a C^1 function $\psi: \overline{\Omega} \times \overline{\Omega} \to \mathbf{C}\backslash\{0\}$, we have

$$\psi s \wedge (d(\psi s))^j = \psi s \wedge (d\psi \wedge s + \psi ds)^j = \psi^{j+1} s \wedge (ds)^j.$$

Hence we may assume in (4.13) that $\text{Re} <s, \zeta - z> < 0$.

Theorem 4.4 *(Berndtsson-Andersson formula)* Assume that s satisfies the conditions **A** and **B**. Let a function G be holomorphic in a simply connected domain which contains $\{<Q(z,\zeta), z - \zeta> + 1 \mid (\zeta, z) \in \overline{\Omega} \times \overline{\Omega}\}$ and $G(1) = 1$. Define

$$\widetilde{K} = C_n(-1)^n \sum_{k=0}^{n-1} \frac{(n-1)!}{k!} G^{(k)}(<Q, z - \zeta> + 1) \frac{s \wedge (dQ)^k \wedge (ds)^{n-1-k}}{<s, \zeta - z>^{n-k}}$$

and
$$\widetilde{P} = \frac{(-1)^{n(n-1)/2}}{n!} G^{(n)}(<Q, z-\zeta>+1)(dQ)^n.$$

Then for $f \in C^1_{(p,q)}(\overline{\Omega})$ one has
(a) In case $q > 0$,
$$f = C\left\{\int_{\partial\Omega} f \wedge \widetilde{K}_{p,q} + (-1)^{p+q+1}\left(\int_{\Omega} \bar{\partial}f \wedge \widetilde{K}_{p,q} - \bar{\partial}_z \int_{\Omega} f \wedge \widetilde{K}_{p,q-1}\right)\right\},$$

where $\widetilde{K}_{p,q}$ are components of $\widetilde{K}$ which are (p,q) forms with respect to z and $(n-p, n-q-1)$ forms with respect to ζ.
(b) In case $q = 0$,
$$f = C\left\{\int_{\partial\Omega} f \wedge \widetilde{K}_{p,0} + (-1)^{p+1}\int_{\Omega} \bar{\partial}f \wedge \widetilde{K}_{p,0} - \int_{\Omega} f \wedge \widetilde{P}_{p,0}\right\}.$$

Proof. First we prove Theorem 4.4 in the case when G is a polynomial. Let
$$G(\alpha) = \sum_{j=0}^N a_j \alpha^j, \quad g = \sum_{j=0}^N a_j \frac{d^j \delta}{d\lambda^j},$$

where δ is the Dirac delta function. We denote by $K^{(\lambda)}$ instead of K when we replace in (4.13) Q by λQ. Similarly, we denote by $P^{(\lambda)}$ instead of P when we replace in (4.14) Q by λQ. After replacing in (4.13) and (4.14), if we multiply by $e^{-\lambda}$ and operate g, then we obtain the desired equalities, where we have used the equations
$$g(e^{-\lambda}) = \sum_{j=0}^N a_j \frac{d^j \delta}{d\lambda^j}(e^{-\lambda}) = \sum_{j=0}^N a_j = G(1) = 1,$$

$$G(\alpha) = g(e^{-\alpha\lambda}), \quad G^{(k)}(<Q, z-\zeta>+1) = g((-\lambda)^k e^{-\lambda(<Q,z-\zeta>+1)}).$$

In the general case, G is approximated uniformly in $\{<Q(z,\zeta), z-\zeta>+1 \mid (\zeta,z) \in \overline{\Omega} \times \overline{\Omega}\}$ by a sequence of polynomials. $\square$

4.2 L^p Estimates for the $\bar{\partial}$ Problem

Let $\Omega \subset\subset \mathbf{C}^n$ be a strictly pseudoconvex domain with C^2 boundary. In this section we prove L^p estimates for the $\bar{\partial}$ problem in Ω. L^p estimates for the $\bar{\partial}$ problem in Ω were first proved by Ovrelid [OV] and Kerzman [KER] using the homotopy formula discussed in Chapter 3. The proof given here is due to Bruna-Cufi-Verdera [BRV] using the Berndtsson-Andersson formula.

Let ρ be a C^2 function in a neighborhood U of $\overline{\Omega}$ such that $\Omega = \{z \in U \mid \rho(z) < 0\}$, $d\rho(z) \neq 0$ for $z \in \partial\Omega$. For ε and $\delta > 0$, define

$$V_\delta = \{z \in U \mid |\rho(z)| < \delta\}, \quad \Omega_\delta = \{z \in U \mid \rho(z) < \delta\},$$

$$U_{\varepsilon,\delta} = \{(z,\zeta) \in \Omega_\delta \times V_\delta \mid |\zeta - z| < \varepsilon\}.$$

We choose $\delta > 0$ sufficiently small such that $V_\delta \subset\subset U$. There exist $\beta > 0$ and $a_{jk} \in C^1(\overline{V}_\delta)$ such that

$$\inf_{\zeta \in \overline{V}_\delta} \sum_{j,k=1}^n \frac{\partial^2 \rho(\zeta)}{\partial \zeta_j \partial \bar{\zeta}_k} \xi_j \bar{\xi}_k \geq 3\beta |\xi|^2 \qquad (0 \neq \xi \in \mathbf{C}^n),$$

$$\sup_{\zeta \in \overline{V}_\delta} \left| \frac{\partial^2 \rho(\zeta)}{\partial \zeta_j \partial \bar{\zeta}_k} - a_{jk}(\zeta) \right| < \frac{\beta}{n^2}.$$

For any sufficiently small $\varepsilon > 0$, if we set $\zeta_j = x_j + ix_{n+j}$ for $j = 1, \cdots, n$, then we have

$$\left| \frac{\partial^2 \rho(\zeta)}{\partial x_j \partial x_k} - \frac{\partial^2 \rho(z)}{\partial x_j \partial x_k} \right| < \frac{\beta}{2n^2} \qquad (\zeta, z \in \overline{V}_\delta, \ |\zeta - z| < 2\varepsilon).$$

Instead of the Levi polynomial, we define $F(z,\zeta)$ by

$$F(z,\zeta) = \sum_{j=1}^n \frac{\partial \rho(\zeta)}{\partial \zeta_j}(\zeta_j - z_j) - \frac{1}{2}\sum_{j,k=1}^n a_{jk}(\zeta)(\zeta_j - z_j)(\zeta_k - z_k).$$

Using Taylor's theorem, we have

$$2\operatorname{Re} F(z,\zeta) \geq \rho(\zeta) - \rho(z) + \beta|\zeta - z|^2 \qquad (4.15)$$

for $\zeta, z \in \overline{V}_\delta$ and $|\zeta - z| < 2\varepsilon$. Moreover, using the same method as the proof of Theorem 3.8, we obtain the following lemma.

Lemma 4.2 *There exist constants ε, δ, $c > 0$ and functions $\Phi \in C^1(\Omega_\delta \times V_\delta)$, $G \in C^1(U_{\varepsilon,\delta})$ with the following properties:*

(a) $\Phi(z,\zeta)$ and $G(z,\zeta)$ are holomorphic in z for fixed ζ.
(b) $\Phi = FG$ in $U_{\varepsilon,\delta}$.
(c) $|G| > c$ in $U_{\varepsilon,\delta}$, $|\Phi| > c$ in $\Omega_\delta \times V_\delta \setminus U_{\varepsilon,\delta}$.

(d) There exist $w_j \in C^1(\Omega_\delta \times V_\delta)$ for $j = 1, \cdots, n$ such that

$$\Phi(z,\zeta) = \sum_{j=1}^n w_j(z,\zeta)(\zeta_j - z_j).$$

Moreover, $w_j(z,\zeta)$, $1 \leq j \leq n$, are holomorphic with respect to z.

We set $w(z,\zeta) = (w_1(z,\zeta), \ldots, w_n(z,\zeta))$. Let $\varphi \in C^\infty(\mathbf{C}^n)$ be a function with the properties that $0 \leq \varphi \leq 1$, $\varphi = 1$ in a neighborhood of $\partial\Omega$, $\varphi = 0$ outside of V_δ. Define

$$u(z,\zeta) = -\rho(\zeta)(\bar{\zeta} - \bar{z}) + w(z,\zeta)\overline{\Phi(z,\zeta)}\varphi(\zeta) \qquad ((z,\zeta) \in \Omega_\delta \times \Omega_\delta).$$

Since $u(z,\zeta) = w(z,\zeta)\overline{\Phi(z,\zeta)}$ for $\zeta \in \partial\Omega$, u is the product of w and a function. Hence u satisfies the condition (**C**) in Corollary 4.1. For $(z,\zeta) \in \overline{\Omega} \times \overline{\Omega}$ we have

$$<u(z,\zeta), \zeta - z> = -\rho(\zeta)|\zeta - z|^2 + |\Phi(z,\zeta)|^2\varphi(\zeta).$$

Hence for $\zeta \in \Omega$ with $\zeta \neq z$, we have

$$<u(z,\zeta), \zeta - z> \geq -\rho(\zeta)|\zeta - z|^2 > 0.$$

For $\zeta \in \partial\Omega$ with $\zeta \neq z$ and $|\zeta - z| < \varepsilon$, (4.15) shows that there exists $c_0 > 0$ such that $|\Phi(z,\zeta)| \geq c_0$. For $\zeta \in \partial\Omega$ with $|\zeta - z| < \varepsilon$, (c) shows that $|\Phi(z,\zeta)| \geq c$. Since $\varphi(\zeta) = 1$ for $\zeta \in \partial\Omega$, u satisfies the condition (**A**) in Theorem 4.1. Let $K \subset \Omega$ be a compact set. For $z \in K$ with $|\zeta - z| \leq 2\varepsilon$, if ζ is contained in a small neighborhood B of $\partial\Omega$, then $\varphi(\zeta) = 1$ and $\rho(\zeta) - \rho(z) > 0$. Then (4.15) shows that $2\operatorname{Re} F(z,\zeta) > \beta|\zeta - z|^2$. Hence there exists $c > 0$ such that $|\Phi(z,\zeta)| > c|\zeta - z|^2$. Therefore, for $\zeta \in \overline{\Omega} \cap B$ and $|\zeta - z| \leq \varepsilon$, there exists $c' > 0$ such that

$$<u(z,\zeta), \zeta - z> \geq -\rho(\zeta)|\zeta - z|^2 + c'|\zeta - z|^2 \geq c'|\zeta - z|^2.$$

For $\zeta \in \overline{\Omega}\setminus B$ and $|\zeta - z| \leq \varepsilon$, there exists $c_1 > 0$ such that $-\rho(\zeta) > c_1 > 0$, and hence $<u(z,\zeta), \zeta - z> \geq c_1|\zeta - z|^2$. If $|\zeta - z| \geq \varepsilon$ and $\zeta \in \overline{\Omega}$, then there exists $c_1' > 0$ such that $|\Phi(z,\zeta)| > c_1'$. Hence for $\zeta \in \overline{\Omega}$ and $z \in K$, there exists $c_1'' > 0$ such that $<u(z,\zeta), \zeta - z> \geq c_1''|\zeta - z|^2$. Thus, u satisfies the condition (**B**) in Theorem 4.1. Next we modify u near the diagonal Δ of $\partial\Omega \times \partial\Omega$. Define

$$V_{\varepsilon,\delta} := \{(\zeta,z) \mid |\rho(z)| < \delta, |\rho(\zeta)| < \delta, |\zeta - z| < \varepsilon\}, \quad W_{\varepsilon,\delta} := V_{\varepsilon,\delta} \cap (\overline{\Omega} \times \overline{\Omega}),$$

$$a(z,\zeta) := -\rho(\zeta) + F(z,\zeta), \quad a^*(z,\zeta) := a(\zeta,z).$$

Further, for $(z, \zeta) \in V_{2\varepsilon,\delta}$ we define

$$v(z,\zeta) := \rho(\zeta)\frac{w(\zeta,z)}{G(\zeta,z)} + a^*(z,\zeta)\frac{w(z,\zeta)}{G(z,\zeta)}.$$

Since $v(\zeta,\zeta) = 0$, we have $|v(z,\zeta)| \leq C|z - \zeta|$. On the other hand, we obtain

$$<v, \zeta - z> = -\rho(\zeta)F(\zeta,z) + +a^*(z,\zeta)F(z,\zeta) = a(z,\zeta)a^*(z,\zeta) - \rho(\zeta)\rho(z).$$

It follows from (4.15) that

$$2\operatorname{Re} a(z,\zeta) \geq -\rho(\zeta) - \rho(z) + \beta|\zeta - z|^2 \qquad (|\zeta - z| < 2\varepsilon). \qquad (4.16)$$

Consequently, for $(z,\zeta) \in W_{2\varepsilon,\delta}$, there exists a constant $c_2 > 0$ such that

$$\begin{aligned}|<v,\zeta-z>| &\geq |a||a^*| - \rho(\zeta)\rho(z) \\ &\geq |\operatorname{Re} a||\operatorname{Re} a^*| + |\operatorname{Im} a||\operatorname{Im} a^*| - \rho(\zeta)\rho(z) \\ &\geq c_2\{(\rho(\zeta) - \rho(z))^2 + |\zeta - z|^4 + (-\rho(\zeta) - \rho(z))|\zeta - z|^2 \\ &\quad + |\operatorname{Re} F(z,\zeta)||\operatorname{Re} F(\zeta,z)|\}.\end{aligned}$$

Hence if $(z,\zeta) \in W_{2\varepsilon,\delta}$, then $|<v, \zeta - z>| \geq (-\rho(\zeta) - \rho(z))|\zeta - z|^2$, which implies that v satisfies (**B**). Since $<v, \zeta - z> \neq 0$ for $\zeta \neq z$, multiplying by $\overline{<v,\zeta-z>}/|<v,\zeta-z>|$, we may assume that $<v,\zeta-z> >> 0$ in $W_{2\varepsilon,\delta}\setminus\Delta$. Let $\lambda \in C^\infty(\mathbf{C}^n \times \mathbf{C}^n)$ be a function such that $0 \leq \lambda \leq 1$, $\lambda = 1$ in $V_{\varepsilon,\delta'}$, where $\delta' < \delta$, $\lambda = 0$ outside of $V_{2\varepsilon,\delta}$. Define

$$s = \lambda v + (1-\lambda)u.$$

Then $s : \overline{\Omega}_\delta \times \overline{\Omega}_\delta \to \mathbf{C}^n$ is of class C^1, $<u,\zeta-z> >> 0$ for $\zeta \neq z$, and hence $<s,\zeta-z> >> 0$ for $\zeta \neq z$. Therefore s satisfies (**A**). For $\zeta \in \partial\Omega$, s is a product of $w(z,\zeta)$ and a function. Thus s satisfies (**C**). Clearly s satisfies (**B**). By Theorem 4.2, if we set $\psi = (s, \zeta - z)$, $K = \psi^*\mu$, then for $f \in C^1_{(p,q)}(\overline{\Omega})$ with $\bar{\partial}f = 0$,

$$Tf(z) = C_{p,q,n}\int_\Omega f(\zeta) \wedge K_{p,q-1}(z,\zeta)$$

satisfies $\bar{\partial}(Tf) = f$. For $\tilde{\psi} = (v, \zeta - z)$, we set $K(v) = \tilde{\psi}^*\mu$. Define

$$F_j(z,\zeta) = \frac{P_j(z,\zeta)}{G(z,\zeta)} \qquad (j = 1,\cdots, n)$$

and
$$\alpha(z,\zeta) = \sum_{j=1}^{n} F_j(z,\zeta)(d\zeta_j - dz_j), \quad \beta(z,\zeta) = \sum_{j=1}^{n} F_j(\zeta,z)(d\zeta_j - dz_j).$$

Then we have
$$v = \sum_{j=1}^{n} v_j(d\zeta_j - dz_j) = \rho(\zeta)\beta(z,\zeta) + a^*(z,\zeta)\alpha(z,\zeta).$$

Consequently,
$$dv = d\rho \wedge \beta + \rho d\beta + da^* \wedge \alpha + a^* d\alpha.$$

Since we can adopt the binomial theorem for 2-forms, it follows from Lemma 4.1 that
$$k(v) = \frac{C_{p,q,n}}{<v,\zeta-z>^n} \sum_{j=1}^{n}(-1)^{j-1} v_j \bigwedge_{i \neq j} dv_i \wedge \omega(\zeta-z)$$
$$= \frac{(-1)^{n(n-1)/2} C_{p,q,n}}{(n-1)!} \frac{v \wedge (dv)^{n-1}}{<v,\zeta-z>^n}$$
$$= C_n <v,\zeta-z>^{-n} (\rho\beta + a^*\alpha) \wedge \{(\rho d\beta + a^* d\alpha)^{n-1}$$
$$+ (n-1)(\rho d\beta + a^* d\alpha)^{n-2} \wedge ((-\rho)da^* + a^* d\rho) \wedge \beta \wedge \alpha\},$$

where C_n is a constant such that
$$C_n = \frac{(-1)^{n(n-1)/2} C_{p,q,n}}{(n-1)!}.$$

It follows from (4.16) that
$$-\rho(\zeta) \leq 2|a^*(z,\zeta)| \qquad ((z,\zeta) \in W_{\varepsilon,\delta}).$$

Since
$$\rho(\zeta)\beta + a^*\alpha = \sum_{j=1}^{n} \{\rho(\zeta)F_j(\zeta,z) - \rho(z)F_j(z,\zeta) + F(\zeta,z)F_j(z,\zeta)\}(d\zeta_j - dz_j)$$

and
$$\beta \wedge \alpha = \sum_{j<k}(F_j(\zeta,z)F_k(z,\zeta) - F_k(\zeta,z)F_j(z,\zeta))(d\zeta_j - dz_j) \wedge (d\zeta_k - dz_k),$$

we have
$$\beta \wedge \alpha = O(|\zeta-z|), \quad \rho(\zeta)\beta + a^*\alpha = O(|\zeta-z|).$$

Consequently,

$$|K_{p,q}(v)(\zeta,z)| \leq C\frac{|a^*(z,\zeta)|^{n-1}|\zeta-z|}{|<v,\zeta-z>|^n} \quad ((z,\zeta) \in W_{\varepsilon,\delta}).$$

For $(z,\zeta) \in V_{\varepsilon,\delta}$, define

$$T(z,\zeta) = |F(z,\zeta)| + |F(\zeta,z)|.$$

Then we have the following lemma.

Lemma 4.3 *Let $(z,\zeta) \in W_{\varepsilon,\delta}$. Then*

(a) $a(z,\zeta) \approx a^*(z,\zeta) \approx -\rho(\zeta) - \rho(z) + |\zeta-z|^2 + |Im\,F(z,\zeta)|$,
(b) $T(z,\zeta) \approx |\rho(\zeta) - \rho(z)| + |\zeta-z|^2 + |Im\,F(z,\zeta)|$,
(c) $|<v,\zeta-z>| \geq C\{T(z,\zeta)^2 + (-\rho(\zeta)-\rho(z))|\zeta-z|^2\}$,
(d) $|\zeta-z|^2|a^*(z,\zeta)| \leq C|<v,\zeta-z>| \leq |a^*(z,\zeta)|^2$,
(e) $|<v,\zeta-z>| \leq C|\zeta-z||a^*(z,\zeta)|$,

where C is a constant which is independent of ζ and z.

Proof. It follows from Taylor's formula that

$$\rho(z) = \rho(\zeta)$$
$$+\mathrm{Re}\left(2\sum_{j=1}^{n}\frac{\partial\rho}{\partial z_j}(\zeta)(z_j-\zeta_j) + \sum_{j,k=1}^{n}\frac{\partial^2\rho}{\partial z_j\partial z_k}(\zeta)(z_j-\zeta_j)(z_k-\zeta_k)\right)$$
$$+\sum_{j,k=1}^{n}\frac{\partial^2\rho}{\partial z_j\partial\bar{z}_k}(\zeta)(z_j-\zeta_j)(\bar{z}_k-\bar{\zeta}_k) + O(|\zeta-z|^3)$$

$$= \rho(\zeta) - 2\mathrm{Re}\,F(z,\zeta)$$
$$+\sum_{j,k=1}^{n}\frac{\partial^2\rho}{\partial z_j\partial\bar{z}_k}(\zeta)(z_j-\zeta_j)(\bar{z}_k-\bar{\zeta}_k) + O(|\zeta-z|^3).$$

Consequently,

$$2\mathrm{Re}\,F(z,\zeta) \leq \rho(\zeta) - \rho(z) + C|\zeta-z|^2.$$

Hence together with (4.16) we obtain

$$a(z,\zeta) \approx -\rho(\zeta) - \rho(z) + |\zeta-z|^2 + |Im\,F(z,\zeta)|.$$

Since

$$|F(z,\zeta) + F(\zeta,z)| = O(|\zeta-z|^2),$$

we have

$$a^*(z,\zeta) \approx -\rho(z) - \rho(\zeta) + |\zeta - z|^2 + |\mathrm{Im} F(\zeta,z)|$$
$$\leq C(-\rho(z) - \rho(\zeta) + |\zeta - z|^2 + |\mathrm{Im} F(z,\zeta)|) \approx a(z,\zeta).$$

This proves (a). If $\rho(\zeta) \geq \rho(z)$, then

$$|F(z,\zeta)| + |F(\zeta,z)|$$
$$\approx |\mathrm{Re}\, F(z,\zeta)| + |\mathrm{Im}\, F(z,\zeta)| + |\mathrm{Re}\, F(\zeta,z)| + |\mathrm{Im}\, F(\zeta,z)|$$
$$\geq |\mathrm{Re}\, F(z,\zeta)| + |\mathrm{Im}\, F(z,\zeta)|$$
$$\geq C(|\rho(\zeta) - \rho(z)| + |\zeta - z|^2 + |\mathrm{Im} F(z,\zeta)|).$$

We can also prove the above inequality in case $\rho(\zeta) \leq \rho(z)$. Since

$$|F(z,\zeta) + F(\zeta,z)| = O(|\zeta - z|^2),$$

we obtain

$$|F(z,\zeta)| + |F(\zeta,z)| \leq 2|F(z,\zeta)| + O(|\zeta - z|^2)$$
$$\leq C(|\rho(\zeta) - \rho(z)| + |\zeta - z|^2 + |\mathrm{Im} F(z,\zeta)|).$$

This proves (b). If $|\mathrm{Im}\, F(\zeta,z)| \geq |\mathrm{Im}\, F(z,\zeta)|$, then

$$|<v, \zeta - z>|$$
$$\geq C\{(\rho(\zeta) - \rho(z))^2$$
$$+|\zeta - z|^4 + |\mathrm{Im}\, F(z,\zeta)|^2 + (-\rho(\zeta) - \rho(z))|\zeta - z|^2\}$$
$$\geq C\{T(z,\zeta)^2 + (-\rho(\zeta) - \rho(z))|\zeta - z|^2\}.$$

We can also prove the above inequality in case $|\mathrm{Im}\, F(\zeta,z)| \leq |\mathrm{Im}\, F(z,\zeta)|$. This proves (c). From the definition of a^* and T we have

$$|a^*| \leq -\rho(z) + |F| \leq -\rho(z) + T.$$

By (b) and (c) we have

$$|\zeta - z|^2 |a^*| \leq -\rho(z)|\zeta - z|^2 + |\zeta - z|^2 T$$
$$\leq |\zeta - z|^2(-\rho(z) - \rho(\zeta)) + CT^2$$
$$\leq C|<v,\zeta-z>|.$$

It follows from (a) that

$$|<v,\zeta-z>| = |aa^* - \rho(\zeta)\rho(z)| \leq C|a^*|^2.$$

This proves (d). We obtain

$$
\begin{aligned}
|<v,\zeta-z>| &= |aa^* - \rho(z)\rho(\zeta)| \\
&= |aa^* + \rho(\zeta)a^* - \rho(\zeta)a^* - \rho(z)\rho(\zeta)| \\
&\leq |a + \rho(\zeta)||a^*| - \rho(\zeta)|a^* + \rho(z)| \\
&\leq C(|a + \rho(\zeta)||a^*| + |a^*||a^* + \rho(z)|) \\
&= C(|F(z,\zeta)||a^*| + |a^*||F(\zeta,z)) \\
&\leq C|\zeta - z||a^*|.
\end{aligned}
$$

This proves (e). □

Lemma 4.4 *There exists a constant $C > 0$ such that*

$$\int_{B(z,r)\cap\overline{\Omega}} |K_{p,q}(z,\zeta)|dV(\zeta) \leq Cr \qquad (z \in \overline{\Omega},\ r > 0).$$

Proof. It is sufficient to prove the lemma under the assumption that $r > 0$ is sufficiently small and z is sufficiently close to $\partial\Omega$. For $(z,\zeta) \in W_{\varepsilon,\delta}$, we obtain $K = K(v)$. By the definition of a^* and T and using (c) we have

$$
\begin{aligned}
|K_{p,q}(z,\zeta)| &\leq C\left\{\frac{|a^*(z,\zeta)|^{n-1}|\zeta - z|}{|<v,\zeta-z>|^n}\right\} \\
&\leq C\left\{\frac{(|\rho(z)|^{n-1} + |F(z,\zeta)|^{n-1})|\zeta - z|}{[T(z,\zeta)^2 + (-\rho(\zeta) - \rho(z))|\zeta - z|^2]^n}\right\} \\
&\leq C\left\{\frac{|\zeta - z|}{T(z,\zeta)^{n+1}} + \frac{|\rho(z)|^{n-1}|\zeta - z|}{[T(z,\zeta)^2 + |\rho(z)||\zeta - z|^2]^n}\right\}.
\end{aligned}
$$

We choose a coordinate system $\eta_1(\zeta), \cdots, \eta_n(\zeta)$ in a neighborhood of z such that

$$\eta_1(\zeta) = \rho(\zeta) - \rho(z) + i\operatorname{Im} F(z,\zeta), \quad \eta(z) = 0, \quad |\eta(\zeta)| \approx |\zeta - z|.$$

We set $\eta_j = t_{2j-1} + it_{2j}$. Then we have

$$|T| \approx |t_1| + |t_2| + |t|^2, \quad |\zeta - z| \approx |t|.$$

In order to prove Lemma 4.4, it is sufficient to show that

$$I_1 = \int_{|t|\leq r} \frac{|t|}{[|t_1| + |t_2| + |t|^2]^{n+1}} dt_1 \cdots dt_{2n} \leq Cr \qquad (4.17)$$

and

$$I_2 = \int_{|t|\leq r} \frac{|\rho(z)|^{n-1}|t|}{[t_1^2 + t_2^2 + |\rho(z)||t|^2]^n} dt_1 \cdots dt_{2n} \leq Cr. \qquad (4.18)$$

We set $t' = (t_3, \cdots, t_{2n})$. Then we obtain

$$I_1 \leq \int_{|t| \leq r} \frac{dt_1 \cdots dt_{2n}}{[|t_1| + |t_2| + |t|^2]^{n+(1/2)}}$$

$$\leq C \int_{|t'| \leq r} \frac{dt'}{|t'|^{2n-3}}$$

$$\leq C \int_0^r ds = Cr.$$

This proves (4.17). We set $m = |\rho(z)|$. Then we have

$$I_2 \leq \int_{|t| \leq r} \frac{m^{n-(3/2)}}{[t_1^2 + t_2^2 + m|t|^2]^{n-(1/2)}} dt_1 \cdots dt_{2n}$$

$$\leq Cm^{n-(3/2)} \int_{|t'| \leq r} \int_0^r \frac{ds}{[s + m|t'|^2]^{n-(1/2)}} dt'$$

$$\leq Cm^{n-(3/2)} \int_0^r \frac{\lambda^{2n-3}}{[m\lambda^2]^{n-(3/2)}} d\lambda$$

$$= C \int_0^r d\lambda = Cr.$$

This proves (4.18). □

Now we are going to prove L^p estimates for the $\bar{\partial}$ problem in a strictly pseudoconvex domain Ω in $\mathbf{C}^n$ with smooth boundary.

Theorem 4.5 *For $f \in C^1_{(p,q)}(\overline{\Omega})$, define*

$$Tf(z) = C_{p,q,n} \int_\Omega f(\zeta) \wedge K_{p,q-1}(z, \zeta).$$

Then T satisfies the following:

(a) If $\bar{\partial} f = 0$, then $\bar{\partial}(Tf) = f$.
(b) If $f \in L^r_{(p,q)}(\Omega)$ and $1 \leq r \leq \infty$, then $Tf \in L^r_{(p,q)}(\Omega)$.

Proof. Since every L^p function in Ω can be approximated uniformly on every compact subset of Ω by functions in $C^1(\overline{\Omega})$, we may assume that $f \in C^1(\overline{\Omega})$. (a) follows from Corollary 4.1. (b) follows from Lemma 4.4 and Theorem 3.26. □

Bruna-Cufi-Verdera [BRV] proved the following theorem. We omit the proof.

Theorem 4.6 *Let $\Omega \subset\subset \mathbf{C}^n$ be a strictly pseudoconvex domain with C^3 boundary. Then each function $f \in C(\overline{D})$ satisfying*

$$\bar{\partial}_i \bar{\partial}_j f = 0 \qquad (1 \leq i, j \leq n) \tag{4.19}$$

in Ω, can be approximated on $\overline{\Omega}$ by functions satisfying (4.19) in a neighborhood of $\overline{\Omega}$.

4.3 The Berndtsson Formula

We study the integral formula on submanifolds of bounded domains with smooth boundary obtained by Berndtsson [BR1].

Let Ω be a bounded domain in $\mathbf{C}^n$ with C^2 boundary. Let $\Omega = \{z \mid \rho(z) < 0\}$, where ρ is a C^2 function in a neighborhood of $\overline{\Omega}$ and $d\rho \neq 0$ on $\partial\Omega$. Let $h_1, \cdots, h_m$ be holomorphic functions in a neighborhood $\widetilde{\Omega}$ of $\overline{\Omega}$ satisfying

$$\partial h_1 \wedge \cdots \wedge \partial h_m \wedge \partial \rho \neq 0 \tag{4.20}$$

on $\partial\Omega$. Suppose there exist holomorphic functions $g_i^j(z, \zeta)$ in $\overline{\Omega} \times \overline{\Omega}$ such that

$$h_j(z) - h_j(\zeta) = \sum_{i=1}^n g_i^j(z, \zeta)(z_i - \zeta_i) \qquad (j = 1, \cdots, m). \tag{4.21}$$

We set

$$X = \{z \in \widetilde{\Omega} \mid h_1(z) = \cdots = h_m(z) = 0\},$$

$$V = X \cap \Omega,$$

$$h = (h_1, \cdots, h_m),$$

$$g^j = \sum_{i=1}^n g_i^j d\zeta_i, \tag{4.22}$$

$$\mu = \frac{g^1 \wedge \cdots \wedge g^m \wedge \overline{\partial h_1} \wedge \cdots \wedge \overline{\partial h_m}}{\|\partial h\|^2} dV_{n-1}, \tag{4.23}$$

where dV_{n-1} is the surface measure on V. Let $s = (s_1, \cdots, s_n)$, $Q = (Q_1, \cdots, Q_n)$ and G denote the same notations as in Theorem 4.4. Moreover, we use the abbreviation

$$s = \sum_{j=1}^{n} s_j d\zeta_j, \quad Q = \sum_{j=1}^{n} Q_j d\zeta_j.$$

We set

$$K = \sum_{k=0}^{n-m-1} \frac{(n-1)!}{m!k!} G^{(k)}(<Q, z-\zeta>+1) \frac{s \wedge (\bar{\partial}s)^{n-m-1-k} \wedge (\bar{\partial}Q)^k \wedge \mu}{<s, \zeta-z>^{n-m-k}},$$

and

$$P = \frac{1}{m!(n-m)!} G^{(n-m)}(<Q, z-\zeta>+1)(\bar{\partial}Q)^{n-m} \wedge \mu.$$

Then we have the following theorem.

Theorem 4.7 *(Berndtsson formula) Let u be a C^1 $(0,q)$ form on $\overline{V}$. For $z \in V$ one has*

(a) In case $q > 0$,

$$u(z) = C \left\{ \int_{\partial V} u \wedge K_q + (-1)^{q+1} \left(\int_V \bar{\partial}u \wedge K_q - \bar{\partial}_z \int_V u \wedge K_q \right) \right\},$$

where K_q is a component of K which is of degree $(0, q)$ with respect to z and of degree $(n - m, n - m - q - 1)$ with respect to ζ and $C = C_{q,n}$ is a constant depending only on q, n.

(b) In case $q = 0$,

$$u(z) = C_n \left(\int_{\partial V} uK_0 - \int_V \bar{\partial}u \wedge K_0 - \int_V uP_0 \right).$$

Proof. We prove Theorem 4.7 in case $m = 1$. Let $h_1 = h$, $g^1 = g$. Suppose $Q^1(z, \zeta)$ and $Q^2(z, \zeta)$ are of class C^1 in $\overline{\Omega} \times \overline{\Omega}$, and holomorphic in $z \in \Omega$. In (4.14) we replace Q by $\lambda_1 Q^1 + \lambda_2 Q^2$ and P by P^λ. Then we have

$$P^\lambda = \frac{(-1)^{n(n-2)/2}}{n!} e^{\lambda_1 <Q^1, \zeta-z>} e^{\lambda_2 <Q^2, \zeta-z>} (\lambda_1 dQ^1 + \lambda_2 dQ^2)^n$$

$$= (-1)^{n(n-1)/2} e^{\lambda_1 <Q^1, \zeta-z>} e^{\lambda_2 <Q^2, \zeta-z>}$$

$$\times \sum_{\alpha_1 + \alpha_2 = n} \frac{\lambda_1^{\alpha_1} \lambda_2^{\alpha_2}}{\alpha_1! \alpha_2!} (dQ^1)^{\alpha_1} (dQ^2)^{\alpha_2}.$$

Suppose ψ_1, ψ_2 are distributions. We set

$$\widetilde{P} = \int_0^\infty \int_0^\infty P^\lambda e^{-\lambda_1} e^{-\lambda_2} \psi_1(\lambda_1)\psi_2(\lambda_2) d\lambda_1 d\lambda_2.$$

Further we set

$$G_1(\alpha) = \int_0^\infty e^{-\alpha\lambda_1} \psi_1(\lambda_1) d\lambda_1, \quad G_2(\alpha) = \int_0^\infty e^{-\alpha\lambda_2} \psi_2(\lambda_2) d\lambda_2.$$

Then we have

$$G_1^{(\alpha_1)}(<Q^1, z-\zeta> +1) = \int_0^\infty (-\lambda_1)^{\alpha_1} e^{-\lambda_1(<Q^1,z-\zeta>+1)} \psi_1(\lambda_1) d\lambda_1$$

and

$$G_2^{(\alpha_2)}(<Q^2, z-\zeta> +1) = \int_0^\infty (-\lambda_2)^{\alpha_2} e^{-\lambda_2(<Q^2,z-\zeta>+1)} \psi_2(\lambda_2) d\lambda_2.$$

Consequently,

$$\widetilde{P} = (-1)^n (-1)^{n(n-1)/2} \sum_{\alpha_1+\alpha_2=n} \frac{1}{\alpha_1!\alpha_2!} G_1^{(\alpha_1)} G_2^{(\alpha_2)} (\bar{\partial}Q^1)^{\alpha_1} \wedge (\bar{\partial}Q^2)^{\alpha_2},$$

where $G_j^{(\alpha_j)} = G_j^{(\alpha_j)}(<Q^j, z-\zeta> +1)$. Since u is a $(0,q)$ form, we have only to consider the terms in dQ^1 and dQ^2 which do not contain dz_j, and hence we may replace dQ^1 and dQ^2 by $\bar{\partial}Q^1$ and $\bar{\partial}Q^2$, respectively. In (4.13) we replace Q by $\lambda_1 Q^1 + \lambda_2 Q^2$ and K by K^λ. We set

$$\widetilde{K} = \int_0^\infty \int_0^\infty K^\lambda e^{-\lambda_1} e^{-\lambda_2} \psi_1(\lambda_1) \psi_2(\lambda_2) d\lambda_1 d\lambda_2.$$

Then we have

$$\widetilde{K} = C_n(-1)^n \times$$
$$\sum_{\alpha_0+\alpha_1+\alpha_2=n-1} \frac{(n-1)!}{\alpha_1!\alpha_2!} G_1^{(\alpha_1)} G_2^{(\alpha_2)} \frac{s \wedge (\bar{\partial}s)^{\alpha_0} \wedge (\bar{\partial}Q^1)^{\alpha_1} \wedge (\bar{\partial}Q^2)^{\alpha_2}}{<s, \zeta-z>^{\alpha_0+1}}.$$

We choose distributions ψ_1, ψ_2 such that $G_1(1) = G_2(1) = 1$. Using the same method as the proof of Theorem 4.4, we may assume that G_j are holomorphic in some simply connected domain containing $\{<Q^j, z-\zeta> +1 \mid (z,\zeta) \in \overline{\Omega} \times \overline{\Omega}\}$. Let $g = (g_1, \cdots, g_n)$. We set

$$Q_\varepsilon^2 = \frac{\overline{h(\zeta)}g}{|h(\zeta)|^2 + \varepsilon}.$$

It follows from (4.21) that

$$< Q_\varepsilon^2, z - \zeta > + 1 = \sum_{j=1}^n \frac{\overline{h(\zeta)} g_j (z_j - \zeta_j)}{|h|^2 + \varepsilon} + 1$$

$$= \frac{\overline{h(\zeta)}(h(z) - h(\zeta))}{|h|^2 + \varepsilon} + 1$$

$$= \frac{\overline{h(\zeta)} h(z) + \varepsilon}{|h|^2 + \varepsilon}.$$

On the other hand we have

$$Q_\varepsilon^2 = \sum_{j=1}^n \frac{\overline{h(\zeta)} g_j(z,\zeta)}{|h|^2 + \varepsilon} d\zeta_j.$$

Consequently,

$$\bar\partial Q_\varepsilon^2 = \sum_{j=1}^n \sum_{k=1}^n \frac{\partial}{\partial \bar\zeta_k}\left(\frac{\overline{h(\zeta)} g_j}{|h|^2 + \varepsilon}\right) d\bar\zeta_k \wedge d\zeta_j = \frac{\varepsilon \overline{\partial h} \wedge g}{(|h|^2 + \varepsilon)^2},$$

where $g = \sum_{j=1}^n g_j d\zeta_j$. Therefore we have

$$(\bar\partial Q_\varepsilon^2)^p = 0 \qquad (p > 1).$$

For simplicity, we assume $h(z) = z_1$. Then we have the following lemma. The proof is the same as the proof of Lemma 2.32. So we omit the proof.

Lemma 4.5 Let $z = (z_1, z')$. For $\varphi \in C^1(\overline{\Omega})$ we have

$$\lim_{\varepsilon \to 0+} \int_\Omega \frac{\varepsilon}{(|z_1|^2 + \varepsilon)^2} \varphi(z) dV(z) = \pi \int_{\{z_1 = 0\} \cap \Omega} \varphi(z) dV_{n-1}(z'), \qquad (4.24)$$

$$\lim_{\varepsilon \to 0+} \int_{\partial\Omega} \frac{\varepsilon}{(|z_1|^2 + \varepsilon)^2} \varphi(z) d\sigma_{2n-1}(z) = \pi \int_{\{z_1 = 0\} \cap \partial\Omega} \varphi(z) d\sigma_{2n-3}(z'),$$

$$(4.25)$$

where $d\sigma_{2n-1}$ and $d\sigma_{2n-3}$ are surface measures on $\partial\Omega$ and $\{z_n = 0\} \cap \partial\Omega$, respectively.

We set $G_2(\alpha) = \alpha$. Then

$$\widetilde{K} = (-1)^n C_n \sum_{\alpha_0 + \alpha_1 + \alpha_2 = n-1} \frac{(n-1)!}{\alpha_1! \alpha_2!} G_1^{(\alpha_1)} \left(\frac{\overline{h(\zeta)} h(z) + \varepsilon}{|h|^2 + \varepsilon}\right)^{1-\alpha_2} \qquad (4.26)$$

$$\times \frac{s \wedge (\bar{\partial} s)^{\alpha_0} \wedge (\bar{\partial} Q^1)^{\alpha_1} \wedge (\bar{\partial} Q_\varepsilon^2)^{\alpha_2}}{<s, \zeta - z>^{\alpha_0+1}}$$

and

$$\widetilde{P} = (-1)^{n(n+1)/2} \sum_{\alpha_1+\alpha_2=n} \frac{1}{\alpha_1! \alpha_2!} G_1^{(\alpha_1)} \left(\frac{\overline{h(\zeta)} h(z) + \varepsilon}{|h|^2 + \varepsilon} \right)^{1-\alpha_2} \quad (4.27)$$

$$\times (\bar{\partial} Q^1)^{\alpha_1} \wedge (\bar{\partial} Q^2)^{\alpha_2}.$$

Let $\alpha_2 = 1$. Then coefficients of $\widetilde{K}$ and $\widetilde{P}$ are bounded by integrable functions which are independent of ε. Let $\alpha_2 = 0$. Then we have

$$\frac{|\overline{h(\zeta)} h(z) + \varepsilon|}{|h|^2 + \varepsilon} \frac{\|s\|}{|<s, \zeta-z>|^n} \leq C \left\{ 1 + \frac{|h(\zeta)||\zeta - z|}{|h|^2 + \varepsilon} \right\} \frac{1}{|\zeta - z|^{2n-1}}. \quad (4.28)$$

In case $|\zeta - z| \leq |h(\zeta)|$, the right side of (4.28) is bounded by $|\zeta - z|^{-2n+1}$. In case $|\zeta - z| \geq |h(\zeta)|$, if $0 < \delta < \frac{1}{2}$, $\zeta' = (\zeta_2, \cdots, \zeta_n)$, then there exist positive constants C_1, C_2 and C_3 such that

$$\int_\Omega \frac{|h(\zeta)||\zeta - z|}{|h|^2 + \varepsilon} \frac{\|s\|}{|<s, \zeta-z>|^n} dV(\zeta)$$

$$\leq C_1 \int_\Omega \frac{|h(\zeta)|^\delta |\zeta - z|^{1+\delta}}{(|h|^2 + \varepsilon)|\zeta - z|^{2n-1}} dV(\zeta)$$

$$\leq C_2 \int_{|\zeta_1|<C_2} \frac{dV(\zeta_1)}{|\zeta_1|^{2-\delta}} \int_{|\zeta'|<C_3} \frac{dV_{n-1}(\zeta')}{|\zeta' - z'|^{2n-2-\delta}}.$$

Hence the right side of (4.28) is bounded by an integrable function which is independent of ε. Next we investigate the integral on $\partial\Omega$. Since $\partial h \wedge \partial \rho \neq 0$ on $V \cap \partial\Omega$, there exist positive constants C_4, C_5 and C_6 such that

$$\int_{\partial\Omega} \frac{|\overline{h(\zeta)} h(z) + \varepsilon|}{|h|^2 + \varepsilon} dV(\zeta) \leq C_4 \int_{|\zeta_1|<C_5} \frac{dV_1(\zeta_1)}{|\zeta_1|} \leq C_6$$

for fixed $z \in \Omega$. Let $z \in V$. Then $h(z) = 0$, which implies that in (4.26) and (4.27) each term in which $\alpha_2 = 0$ converges to 0 as $\varepsilon \to 0$. By Lebesgue's dominated convergence theorem the integral of each term converges to 0. In case $\alpha_2 = 1$, integrals on Ω converge to integrals on V and integrals on $\partial\Omega$ converge to integrals on ∂V as $\varepsilon \to 0$ by Lemma 4.5, which completes the proof of Theorem 4.7. □

Theorem 4.8 Let $\Omega = \{z \mid \rho(z) < 0\}$ be a bounded convex domain with C^2 boundary and let f be holomorphic in V and of class C^1 on $\overline{V}$, N a positive integer. Then for $z \in V$

$$f(z) = C \int_V f(\zeta) \left(\frac{\rho(\zeta)}{<\partial\rho(\zeta), z-\zeta> + \rho(\zeta)} \right)^{N+n-m} \left(\bar{\partial}\left(\frac{Q}{\rho}\right) \right)^{n-m} \wedge \mu, \quad (4.29)$$

where μ is defined by (4.23),

$$Q = \sum_{j=1}^n \frac{\partial \rho}{\partial \zeta_j} d\zeta_j$$

and $C = C_{n,m}$ is a constant depending only on n and m.

Proof. Since the function ρ is convex, we have

$$\sum_{j,k=1}^{2n} \frac{\partial^2 \rho}{\partial x_j \partial x_k}(z) u_j u_k \geq 0 \quad (z \in \overline{\Omega}, \ (u_1, \cdots, u_{2n}) \in \mathbf{R}^{2n}). \quad (4.30)$$

It follows from Taylor's formula that

$$\rho(z) - \rho(\zeta) \geq 2\mathrm{Re} \sum_{j=1}^n \frac{\partial \rho}{\partial \zeta_j}(\zeta)(z_j - \zeta_j). \quad (4.31)$$

For $\varepsilon > 0$, we set

$$Q_j(z,\zeta) = \frac{1}{\rho(\zeta) - \varepsilon} \frac{\partial \rho}{\partial \zeta_j}(\zeta).$$

Then we obtain

$$<Q(z,\zeta), z-\zeta> + 1 = \frac{<\partial \rho(\zeta), z-\zeta> + \rho(\zeta) - \varepsilon}{\rho(\zeta) - \varepsilon},$$

It follows from (4.31) that

$$\mathrm{Re}(<Q, z-\zeta> + 1) \geq \frac{\rho(z) + \rho(\zeta) - 2\varepsilon}{2(\rho(\zeta) - \varepsilon)} > 0$$

for $(z,\zeta) \in \overline{\Omega} \times \overline{\Omega}$. We set $G(\alpha) = \alpha^{-N}$ for $N \geq 1$. Since $G(\alpha)$ is holomorphic in $\mathrm{Re}\,\alpha > 0$, G satisfies the hypothesis in Theorem 4.7. If we let $\varepsilon \downarrow 0$, then by Theorem 4.7 we have for some constant γ_k,

$$K = \sum_{k=0}^{n-m-1} \gamma_k \left(\frac{\rho}{<\partial\rho, z-\zeta> + \rho} \right)^{N+k} \frac{s \wedge (\bar{\partial} s)^{n-m-1-k} \wedge (\bar{\partial} Q)^k \wedge \mu}{<s, \zeta-z>^{n-m-k}}, \quad (4.32)$$

$$P = C_{n,m} \left(\frac{\rho(\zeta)}{<\partial\rho(\zeta), z - \zeta > +\rho(\zeta)} \right)^{N+n-m} (\bar{\partial}Q)^{n-m} \wedge \mu. \qquad (4.33)$$

Since

$$\bar{\partial}Q = \frac{1}{\rho}\bar{\partial}\left(\sum_{j=1}^{n} \frac{\partial\rho}{\partial\zeta_j}(\zeta)d\zeta_j\right) - \frac{1}{\rho^2}\bar{\partial}\rho \wedge \sum_{j=1}^{n} \frac{\partial\rho}{\partial\zeta_j}(\zeta)d\zeta_j,$$

we obtain $(\bar{\partial}Q)^k = O(|\rho|^{-k-1})$. On the other hand we have

$$2(\operatorname{Re} <\partial\rho, z - \zeta > +\rho(\zeta)) < \rho(z) + \rho(\zeta) \leq \rho(z) \quad (z \in \Omega, \zeta \in \overline{\Omega}),$$

which implies that the integral of K on Ω exists. Since $d\rho = 0$ on $\partial\Omega$, we have

$$\bar{\partial}Q = \frac{1}{\rho}\bar{\partial}\left(\sum_{j=1}^{n} \frac{\partial\rho}{\partial\zeta_j}d\zeta_j\right),$$

and hence $(\bar{\partial}Q)^k = O(|\rho|^{-k})$. Since $N \geq 1$, the integral of K on $\partial\Omega$ is equal to 0. Thus by Theorem 4.7 (b) we obtain (4.29). □

Remark 4.1 *In the case when Ω is an analytic polyhedron, L^p and H^p extensions of holomorphic functions from submanifolds of Ω were investigated by Adachi-Andersson-Cho [ADC] using the Berndtsson integral formula.*

4.4 Counterexamples for L^p ($p > 2$) Extensions

We give counterexamples for L^p ($p > 2$) extensions of holomorphic functions from submanifolds in complex ellipsoids due to Mazzilli [MAZ1] and Diederich-Mazzilli [DIM1]. From these examples one can see that the Ohsawa-Takegoshi extension theorem is the best possible.

Let Ω be a complex ellipsoid in $\mathbf{C}^n$. Then there exist positive integers $q_1, \cdots, q_n$ such that

$$\Omega = \{z \in \mathbf{C}^n \mid \rho(z) = \sum_{j=1}^{n} |z_j|^{2q_j} - 1 < 0\}.$$

We set

$$k = \sup_{1 \leq j \leq n}\{q_j\},$$

and
$$Q(z,\zeta) = \left(\frac{\partial \rho}{\partial \zeta_1}(\zeta), \cdots, \frac{\partial \rho}{\partial \zeta_n}(\zeta)\right).$$

The following two lemmas have been proved by Range [RAN1].

Lemma 4.6 *For a positive integer m, we set $g(z) = |z|^{2m}$. Then there exists a constant $C > 0$ such that*
$$g(z+w) - g(z) - 2\mathrm{Re}\left(\frac{\partial g}{\partial z}(z)w\right) \geq C|w|^{2m}.$$

Proof. We set
$$f(z,w) = g(z+w) - g(z) - 2\mathrm{Re}\left(\frac{\partial g}{\partial z}(z)w\right).$$

By Taylor's formula, if we set $z = x + iy$, $w = u + iv$, then there exists θ with $0 < \theta < 1$ such that
$$f(z,w) = \frac{1}{2}\left(\frac{\partial^2 g}{\partial x^2}(x+\theta u, y+\theta v)u^2 + 2\frac{\partial^2 g}{\partial x \partial y}(x+\theta u, y+\theta v)uv \right.$$
$$\left. + \frac{\partial^2 g}{\partial y^2}(x+\theta u, y+\theta v)v^2\right).$$

We set $\psi(t) = t^m$. Then we have $g(z) = \psi(x^2 + y^2)$. We set $X = x + \theta u$, $Y = y + \theta v$. Then
$$f(z,w) = 2\psi''(X^2 + Y^2)(Xu + Yv)^2 + \psi'(X^2 + Y^2)(u^2 + v^2). \quad (4.34)$$

Suppose $(X, Y) = (0, 0)$ for $|w| = 1$. Since $z = -\theta w$, we have $|z| = \theta$. Consequently,
$$f(z,w) = (1-\theta)^{2m} - \theta^{2m} + 2m\theta^{2m-1} = (1-\theta)^{2m} + \theta^{2m-1}(2m - \theta) > 0.$$

From (4.34) we have $f(z,w) = 0$, which is a contradiction. Hence we have $(X, Y) \neq 0$, which means that $f(z,w) > 0$ for $|w| = 1$. In case $|w| = 1$ and $|z| \leq 2$, $f(z,w)$ has a minimum value $c_1 > 0$. In case $|w| = 1$ and $|z| \geq 2$, by (4.34) we have
$$f(z,w) \geq \psi'(X^2 + Y^2) = m(|z + \theta w|^2)^{m-1} \geq m(|z| - 1)^{2m-2} \geq m.$$

Hence if $|w| = 1$, then $f(z,w) \geq \min(c_1, m) := c$. Since
$$f\left(\frac{z}{|w|}, \frac{w}{|w|}\right) = \frac{1}{|w|^{2m}} f(z,w) \geq c,$$

we obtain $f(z,w) \geq c|w|^{2m}$. □

Lemma 4.7 For $(\zeta, z) \in \overline{\Omega} \times \overline{\Omega}$, there exists a constant $C > 0$ such that
$$\mathrm{Re}(<Q, \zeta - z> -\rho(\zeta)) \geq C(-\rho(\zeta) - \rho(z) + \sum_{j=1}^{n} |\zeta_j|^{2q_j-2}|z_j - \zeta_j|^2$$
$$+ |z_j - \zeta_j|^{2q_j}).$$

Proof. We set $g(z) = |z|^{2m}$, $\varphi(t) = g(z + tw)$ for $t \in \mathbf{R}$. Then we have
$$\varphi^{(s)}(t) = \left(w\frac{\partial}{\partial z} + \bar{w}\frac{\partial}{\partial \bar{z}}\right)^s g.$$

In case $2m \geq s$,
$$\varphi^{(s)}(t) = \sum_{j=0}^{n} \frac{s!}{j!(s-j)!}\left(w\frac{\partial}{\partial z}\right)^j \left(\bar{w}\frac{\partial}{\partial \bar{z}}\right)^{s-j} g$$
$$= \sum_{j+k=s} \frac{s!}{j!k!} \frac{\partial^n g}{\partial z^j \partial \bar{z}^k} w^j \bar{w}^k.$$

In case $2m < s$, we have $\varphi^{(s)}(t) = 0$. Hence we have
$$f(z, w) = \sum_{s=2}^{2m} \frac{\varphi^{(s)}(0)}{s!} = \sum_{2 \leq j+k \leq 2m} \frac{1}{j!k!} \frac{\partial^s g}{\partial z^j \partial \bar{z}^k} w^j \bar{w}^k.$$

We obtain
$$\sum_{j+k=2} \frac{1}{j!k!} \frac{\partial^s g}{\partial z^j \partial \bar{z}^k} w^j \bar{w}^k$$
$$= m^2 |z|^{2m-2}|w|^2 + \mathrm{Re}(m(m-1)z^{m-2}\bar{z}^m w^2)$$
$$\geq m^2 |z|^{2m-2}|w|^2 - m(m-1)|z|^{2m-2}|w|^2 = m|z|^{2m-2}|w|^2.$$

On the other hand, if $j + k \geq 3$, then for a with $0 < a < 1$ and $|w| \leq a|z|$, there exists a constant $C > 0$ such that
$$\left|\left(\frac{\partial}{\partial z}\right)^j \left(\frac{\partial}{\partial \bar{z}}\right)^k g(z) w^j \bar{w}^k\right|$$
$$= |m(m-1)\cdots(m-j+1)z^{m-j}m(m-1)\cdots(m-k+1)\bar{z}^{m-k}w^j\bar{w}^k|$$
$$\leq Ca|z|^{2m-2}|w|^2,$$

which means that for any sufficiently small a,
$$f(z, w) \geq m|z|^{2m-2}|w|^2 - Ca|z|^{2m^2}|w|^2 \geq C|z|^{2m-2}|w|^2.$$

If $|w| > a|z|$, then $|w|^{2m} > a^{2m-2}|z|^{2m-2}|w|^2$, and hence by Lemma 4.6 we have
$$f(z,w) \geq C|w|^{2m} \geq C|z|^{2m-2}|w|^2.$$
Consequently, we obtain
$$f(z,w) \geq C(|z|^{2m-2}|w|^2 + |w|^{2m}). \tag{4.35}$$
We set $g_j(z_j) = |z_j|^{2q_j}$. It follows from (4.35) that
$$\sum_{j=1}^n g_j(z_j) - \sum_{j=1}^n g_j(\zeta_j) - 2\text{Re}\left(\sum_{j=1}^n \frac{\partial g_j}{\partial \zeta_j}(\zeta_j)(z_j - \zeta_j)\right)$$
$$\geq C\left(\sum_{j=1}^n |\zeta_j|^{2q_j-2}|z_j - \zeta_j|^2 + \sum_{j=1}^n |z_j - \zeta_j|^{2m_j}\right).$$
Then
$$2\text{Re}<\frac{\partial \rho}{\partial \zeta}(\zeta), \zeta - z> \geq -\rho(z) + \rho(\zeta) + C(\sum_{j=1}^n |\zeta_j|^{2q_j-2}|z_j - \zeta_j|^2$$
$$+ \sum_{j=1}^n |z_j - \zeta_j|^{2m_j}).$$
□

Definition 4.3 For $k > 0$, define
$$B_k(\Omega) = \{f \in C(\Omega) \mid \sup_{z \in \Omega}(|f(z)|\text{dist}(z, \partial\Omega)^k) < \infty\}.$$

Now we give a counterexample for bounded extensions of holomorphic functions from submanifolds of complex ellipsoids due to Mazzilli [MAZ1].

Theorem 4.9 *For $p \geq 1$ and any sufficiently small $\varepsilon > 0$, there exist a complex ellipsoid Ω in $\mathbf{C}^{2p+1}$, a submanifold X in a neighborhood of $\overline{\Omega}$ which intersects $\partial\Omega$ transversally, and a bounded holomorphic function f in $V = X \cap \Omega$ such that if g is a holomorphic function in Ω with $g|_V = f$, then $g \notin B_{\frac{p}{2}-\varepsilon}(\Omega)$. Therefore, f cannot be extended to a bounded holomorphic function in Ω.*

Proof. Define $f_j(z) = z_j^n + z_{p+j}$ for $j = 1, \cdots, p$. Define $\Omega \subset \mathbf{C}^{2p+1}$ and a submanifold M of Ω as follows.
$$\Omega = \{z \in \mathbf{C}^{2p+1} \mid \sum_{j=1}^p |z_j|^{2n+1} + \sum_{j=p+1}^{2p+1} |z_j|^2 - 1 = \rho(z) < 0\},$$

$$V = \{z \in \Omega \mid f_1(z) = \cdots = f_p(z) = 0\}.$$

In Lemma 4.7, we set $\zeta = (0, \cdots, 0, 1)$. Then for $z \in \overline{\Omega}$ we have

$$\operatorname{Re}(1 - z_{2p+1}) \geq C \left(|\rho(z)| + |z_{2p+1} - 1|^2 + \sum_{j=1}^{p} |z_j|^{2^{n+1}} + \sum_{j=p+1}^{2p} |z_j|^2 \right), \tag{4.36}$$

which implies that if $z \in \Omega$, then $\operatorname{Re}(1 - z_{2p+1}) > 0$. Hence if we define for $z \in \Omega$

$$f(z) = \frac{z_1^{n-1} \cdots z_p^{n-1}}{(1 - z_{2p+1})^{\frac{p(n-1)}{2n}}},$$

then f is holomorphic in Ω. It follows from (4.36) that

$$|f(z)| \leq \frac{|z_1^{n-1} \cdots z_p^{n-1}|}{|1 - z_{2p+1}|^{\frac{p(n-1)}{2n}}} \leq \frac{|z_{p+1}|^{\frac{n-1}{n}} \cdots |z_{2p}|^{\frac{n-1}{n}}}{\left(\sum_{j=p+1}^{2p} |z_j|^2\right)^{\frac{p(n-1)}{2n}}} \leq C$$

for $z \in V$, which means that f is bounded on V. It follows from (4.36) that

$$\frac{|z_j|^{2^{n+1}}}{|1 - z_{2p+1}|} \leq C \qquad (j = 1, \cdots, p)$$

for $z \in \Omega$. Consequently,

$$|f(z)| = \left(\frac{|z_1|^{2^{n+1}}}{|1 - z_{2p+1}|}\right)^{\frac{n-1}{2^{n+1}}} \cdots \left(\frac{|z_p|^{2^{n+1}}}{|1 - z_{2p+1}|}\right)^{\frac{n-1}{2^{n+1}}}$$

$$\times |1 - z_{2p+1}|^{\frac{p(n-1)}{2^{n+1}} - \frac{p(n-1)}{2n}}$$

$$\leq C |\rho(z)|^{\frac{p(n-1)}{2^{n+1}} - \frac{p(n-1)}{2n}}.$$

Suppose there exists a holomorphic function g in Ω with the following properties:

(a) For $z \in V$, $g(z) = f(z)$.
(b) There exists $\delta > 0$ such that $|g(z)| \leq C |\rho(z)|^{-\frac{p(n-1)}{2n} + \frac{p(n-1)}{2^{n+1}} + \delta}$.

Since $g - f$ is a holomorphic function in Ω such that $g - f = 0$ on V, it follows from Corollary 5.7 that there exist holomorphic functions a_k for $k = 1, \cdots, p$ in Ω such that

$$g(z) = \frac{1}{(1 - z_{2p+1})^{\frac{p(n-1)}{2n}}} \left(z_1^{n-1} \cdots z_p^{n-1} + \sum_{k=1}^{p} (z_k^n + z_{p+k}) a_k(z) \right). \tag{4.37}$$

For any sufficiently small $\varepsilon > 0$ and $\theta_k \in [0, 2\pi]$ for $k = 1, \cdots, p$, we set

$$\begin{cases} z_k = \varepsilon^{\frac{1}{2n+1}} e^{i\theta_k} & (1 \leq k \leq p) \\ z_k = 0 & (p+1 \leq k \leq 2p) \\ z_{2p+1} = 1 - p\varepsilon \end{cases}.$$

Then we have $\rho(z) = p\varepsilon(p\varepsilon - 1) < 0$, which means that $|\rho(z)| = (1 - p\varepsilon)p\varepsilon$ for $z \in \Omega$. It follows from (b) and (4.37) that

$$\frac{1}{(p\varepsilon)^{\frac{p(n-1)}{2n}}} \left| \frac{1}{z_1 \cdots z_p} + \sum_{k=1}^{p} \frac{a_k(z)}{\prod_{\substack{j=1 \\ j \neq k}}^{p} z_j^n} \right| \leq C_n |\rho(z)|^{-\frac{p(n-1)}{2n} + \frac{p(n-1)}{2n+1} + \delta} \frac{1}{|z_1^n \cdots z_p^n|}.$$

Since $|\rho(z)| \approx \varepsilon$, we obtain

$$\left| \frac{1}{z_1 \cdots z_p} + \sum_{k=1}^{p} \frac{a_k(z)}{\prod_{\substack{j=1 \\ j \neq k}}^{p} z_j^n} \right| \leq C_n \varepsilon^{\delta - \frac{p}{2n+1}}.$$

We set

$$\gamma = \{z \in \mathbf{C} \mid |z| = \varepsilon^{\frac{1}{2n+1}}\},$$

$$\Gamma = \underbrace{\gamma \times \cdots \times \gamma}_{p}.$$

Then we have

$$\int_{\Gamma} \sum_{k=1}^{p} \frac{a_k(z)}{\prod_{\substack{j=1 \\ j \neq k}}^{p} z_j^n} dz_1 \wedge \cdots \wedge dz_p = 0.$$

Consequently,

$$\int_{\Gamma} \frac{1}{z_1 \cdots z_p} dz_1 \wedge \cdots \wedge dz_p = \int_{\Gamma} \left(\frac{1}{z_1 \cdots z_p} + \sum_{k=1}^{p} \frac{a_k(z)}{\prod_{\substack{j=1 \\ j \neq k}}^{p} z_j^n} \right) dz_1 \wedge \cdots \wedge dz_p.$$

The left side of the above equality is equal to $(2\pi i)^p$ and the right side is equal to $O(\varepsilon^\delta)$, which is a contradiction for any sufficiently small ε. Therefore there is no g which satisfies (a) and (b). Suppose for an extension g of f there exists $\varepsilon > 0$ such that

$$|g(z)||\rho(z)|^{\frac{p}{2}-\varepsilon} < C \qquad (z \in \Omega).$$

If we choose n sufficiently large, then we have

$$\frac{p}{2n} + \frac{p(n-1)}{2^{n+1}} < \frac{\varepsilon}{2}.$$

Consequently,

$$|g(z)| \leq C|\rho(z)|^{-\frac{p}{2}+\varepsilon} \leq C|\rho(z)|^{-\frac{p(n-1)}{2n}+\frac{p(n-1)}{2^{n+1}}+\frac{\varepsilon}{2}},$$

which means that g satisfies (a) and (b). This is a contradiction. □

Definition 4.4 Suppose $d(z,\partial\Omega)$ denotes the distance from z to $\partial\Omega$ and that dV_Ω and dV_{n-1} are Lebesgue measures on Ω and V, respectively.

(1) For a measurable function f in Ω, $f \in L^q(d(z,\partial\Omega)^s, dV_\Omega)$ means that

$$\int_\Omega |f(z)|^p d(z,\partial\Omega)^s dV_\Omega < \infty.$$

(2) For a measurable function f in V, $f \in L^q(d(z,\partial\Omega)^s, dV_{n-1}))$ means that

$$\int_V |f(z)|^p d(z,\partial\Omega)^s dV_{n-1} < \infty.$$

Lemma 4.8 *Let n and p be positive integers with $n \geq 2p+1$. For a positive integer $N \geq 2$, define*

$$\Omega = \{z \in \mathbf{C}^n \mid \sum_{j=1}^{p}|z_j|^{2^{N+1}} + \sum_{j=p+1}^{n}|z_j|^2 - 1 = \rho(z) < 0\}.$$

Let $q \geq 2$, $s \geq 0$. Then there exists a constant $C_N > 0$ such that for a holomorphic function f in Ω with $f \in L^q(d(z,\partial\Omega)^s, dV_\Omega)$, $\theta_j \in [0,2\pi]$ and any sufficiently small $\varepsilon > 0$, if we set

$$z = (\varepsilon^{\frac{1}{2^{N+1}}} e^{i\theta_1}, \cdots, \varepsilon^{\frac{1}{2^{N+1}}} e^{i\theta_p}, 0, \cdots, 0, 1-p\varepsilon),$$

then we obtain

$$|f(z)| \leq C_N \|f\|_{L^q(d(z,\partial\Omega)^s, dV_\Omega)} d(z,\partial\Omega)^{-\frac{n-p+1}{q} - \frac{s}{q} - \frac{p}{2^N q}}.$$

Proof. We set

$$P(z,\zeta) = \left(\frac{\rho(\zeta)}{<\partial\rho(\zeta), z-\zeta> + \rho(\zeta)}\right)^{1+\frac{s}{q}+n} \left(\partial\bar\partial \log\left(-\frac{1}{\rho(\zeta)}\right)\right)^n \wedge \mu,$$

where μ is defined by (4.23). It follows from Theorem 4.4 that for $f \in \mathcal{O}(\Omega) \cap L^q(d(z,\partial\Omega)^s, dV_\Omega)$

$$f(z) = \int_\Omega f(\zeta) P(z,\zeta).$$

By the Hölder inequality we obtain

$$|f(z)|$$
$$\leq \left(\int_\Omega |f(\zeta)|^q |\rho(\zeta)|^s dV_\Omega(\zeta)\right)^{\frac{1}{q}} \left(\int_\Omega (|\rho(\zeta)|^{-\frac{s}{q}} |P(z,\zeta)|)^{\frac{q}{q-1}} dV_\Omega(\zeta)\right)^{\frac{q-1}{q}}.$$

Further we have

$$\left(\bar\partial\left(\frac{\partial\rho}{\rho}\right)\right)^n = \frac{(\bar\partial\partial\rho)^n}{\rho^n} - n\frac{(\bar\partial\partial\rho)^{n-1} \wedge \partial\rho \wedge \bar\partial\rho}{\rho^{n+1}}.$$

Consequently,

$$|\rho(\zeta)|^{-\frac{s}{q}} |P(z,\zeta)| \leq C\frac{\|(\bar\partial\partial\rho(\zeta))^{n-1} \wedge \partial\rho(\zeta) \wedge \bar\partial\rho(\zeta)\|}{|<\partial\rho(\zeta), z-\zeta>+\rho(\zeta)|^{1+\frac{s}{q}+n}}.$$

It follows from Lemma 4.7 that

$$\operatorname{Re} <\partial\rho(\zeta), \zeta-z> -\rho(\zeta)$$
$$\geq C(|\rho(\zeta)| + |\rho(z)| + \sum_{j=1}^p |z_j - \zeta_j|^{2^{N+1}} + \sum_{j=p+1}^n |z_j - \zeta_j|^2).$$

We set

$$P_1(z,\zeta) = \rho(\zeta)^{-\frac{s}{q}} P(z,\zeta)$$

and

$$u_n = \rho(\zeta) + i\operatorname{Im} <\partial\rho(\zeta), z-\zeta>.$$

Then we have

$$|P_1(z,\zeta)| \leq C\frac{\prod_{j=1}^p |\zeta_j|^{2^{N+1}-2}}{(|\rho(z)| + \sum_{j=1}^p |\zeta_j - z_j|^{2^{N+1}} + \sum_{j=p+1}^{n-1} |\zeta_j|^2 + |u_n|)^{n+1+\frac{s}{q}}}.$$

Since

$$z = (\varepsilon^{\frac{1}{2^{N+1}}} e^{i\theta_1}, \cdots, \varepsilon^{\frac{1}{2^{N+1}}} e^{i\theta_p}, 0, \cdots, 0, 1-p\varepsilon),$$

we obtain

$$|\rho(z)| = 1 - \sum_{j=1}^{p} |z_j|^{2^{N+1}} - |z_n|^2 = p\varepsilon(1-p\varepsilon),$$

which implies that $|\rho(z)| \approx |z_j|^{2^{N+1}}$ for $j = 1, \cdots, p$. On the other hand we have

$$\prod_{j=1}^{p} |\zeta_j|^{2^{N+1}-2} \leq C \prod_{j=1}^{p} (|\zeta_j - z_j|^{2^{N+1}-2} + |z_j|^{2^{N+1}-2})$$

and

$$|\zeta_j - z_j|^{2^{N+1}-2} = (|\zeta_j - z_j|^{2^{N+1}})^{1-\frac{1}{2^N}},$$

which means that

$$|P_1(z,\zeta)|$$
$$\leq C \frac{1}{(|\rho(z)| + \sum_{j=1}^{p} |\zeta_j - z_j|^{2^{N+1}} + \sum_{j=p+1}^{n-1} |\zeta_j|^2 + |u_n|)^{n+1-p+\frac{s}{q}+\frac{p}{2^N}}}.$$

We set

$$\alpha = -\frac{q}{q-1}(n-p+1+\frac{s}{q}+\frac{p}{2^N})$$

and

$$u = (u_1, \cdots, u_n), \quad u' = (u_1, \cdots, u_{n-1}), \quad u'' = (u_1, \cdots, u_p).$$

Define $u_j = \zeta_j - z_j$ for $j = 1, \cdots, n-1$. Then we have

$$\int_\Omega |P_1(z,\zeta)|^{\frac{q}{q-1}} d\lambda_\Omega(\zeta)$$
$$\leq C \int \frac{du'}{(|\rho(z)| + \sum_{j=1}^{p} |u_j|^{2^{N+1}} + \sum_{j=p+1}^{n-1} |u_j|^2)^{-\alpha-2}}$$
$$\leq C \int \frac{du''}{(|\rho(z)|^{\frac{1}{2}} + \sum_{j=1}^{p} |u_j|^{2^N})^{-2\alpha-4-2(n-p-1)}}$$
$$\leq C \int \frac{du''}{(|\rho(z)|^{\frac{1}{2^N}} + \sum_{j=1}^{p} |u_j|^2)^{-2^N(\alpha+n-p+1)}}$$
$$\leq C \int_0^R \frac{r^{2p-1} dr}{(|\rho(z)|^{\frac{1}{2^N}} + r^2)^{-2^N(\alpha+n-p+1)}}$$
$$\leq C |\rho(z)|^{\frac{p}{2^N}+\alpha+n-p+1}.$$

Consequently,

$$\left(\int_\Omega |P_1(z,\zeta)|^{\frac{q}{q-1}} d\lambda_\Omega(\zeta)\right)^{\frac{q-1}{q}} \leq C(|\rho(z)|^{\frac{p}{2^N}+\alpha+n-p+1})^{\frac{q-1}{q}}$$

$$= C|\rho(z)|^{-\frac{p}{2^N q}-\frac{n-p+1}{q}-\frac{s}{q}}.$$

$\square$

Lemma 4.9 *Let p and N be positive integers with $N \geq 2$. Define*

$$\Omega = \{z \in \mathbf{C}^n \mid \sum_{j=1}^{p} |z_j|^{2^{N+1}} + \sum_{j=p+1}^{2p+1} |z_j|^2 - 1 = \rho(z) < 0\}.$$

Then there exist a constant $C_N > 0$ such that for any holomorphic function f in Ω with $f \in L^q(\Omega)$ ($q \geq 2$), any sufficiently small $\varepsilon > 0$ and $\theta_j \in [0, 2\pi]$, if we define

$$z = (\varepsilon^{\frac{1}{2^{N+1}}} e^{i\theta_1}, \cdots, \varepsilon^{\frac{1}{2^{N+1}}} e^{i\theta_p}, 0, \cdots, 0, 1-p\varepsilon),$$

then we obtain

$$|f(z)| \leq C_N \|f\|_{L^q(\Omega)} d(z, \partial\Omega)^{-\frac{n+2}{q}-\frac{p}{2^N q}}.$$

Proof. In Lemma 4.8 we set $n = 2p+1$, $s = 0$. Then we have the desired inequality. $\square$

Diederich-Mazzilli [DIM1] obtained a counterexample for the L^p ($p > 2$) extension of holomorphic functions from submanifolds of complex ellipsoids.

Theorem 4.10 *Given $\varepsilon > 0$, there exist a positive integer p, a complex ellipsoid $\Omega \subset \mathbf{C}^{2p+1}$, a submanifold X of $\mathbf{C}^{2p+1}$ which intersects $\partial\Omega$ transversally and a bounded holomorphic function f in $V = X \cap \Omega$ such that if g is a holomorphic function in Ω which satisfies $g = f$ in V, then $g \notin L^{2+\varepsilon}(\Omega)$.*

Proof. Let N be a positive integer. Define

$$V = \{z \in \Omega \mid z_1^N + z_{p+1} = \cdots = z_p^N + z_{2p} = 0\}$$

and

$$f(z) = \frac{z_1^{N-1} \cdots z_p^{N-1}}{(1-z_n)^{\frac{p(N-1)}{2N}+\frac{2}{q}}}.$$

It follows from the proof of Theorem 4.9 that f is bounded on V. We set

$$z = (\varepsilon^{\frac{1}{2^{N+1}}} e^{i\theta_1}, \cdots, \varepsilon^{\frac{1}{2^{N+1}}} e^{i\theta_p}, 0, \cdots, 0, 1-p\varepsilon).$$

Using the same method as the proof of Theorem 4.9, for $\delta > 0$ any holomorphic extension g of f to Ω cannot satisfy the inequality

$$|g(z)| \leq C|\rho(z)|^{\frac{p(N-1)}{2^{N+1}} - \frac{p(N-1)}{2^N} + \delta}.$$

On the other hand, by Lemma 4.9 if $g \in L^q(\Omega)$, then we have

$$|g(z)| \leq C\|g\|_{L^q(\Omega)} d(z, \partial\Omega)^{-\frac{p+2}{q} - \frac{p}{2^N q}},$$

which means that

$$-\frac{p+2}{q} - \frac{p}{2^N q} \leq \frac{p(N-1)}{2^{N+1}} - \frac{p(N-1)}{2N}.$$

Then q must satisfy the inequality

$$q \leq \frac{2 + \frac{4}{p} + \frac{1}{2^{N-1}}}{1 - \frac{1}{N} - \frac{N-1}{2^N}}.$$

Consequently, there is no holomorphic extension g of f which satisfies conditions

$$g \in L^q(\Omega), \quad q > \frac{2 + \frac{4}{p} + \frac{1}{2^{N-1}}}{1 - \frac{1}{N} - \frac{N-1}{2^N}}.$$

If we choose p and N sufficiently large, then

$$2 + \varepsilon > \frac{2 + \frac{4}{p} + \frac{1}{2^{N-1}}}{1 - \frac{1}{N} - \frac{N-1}{2^N}},$$

which implies that there is no holomorphic function g in Ω which satisfies $g \in L^{2+\varepsilon}(\Omega)$ and $g|_V = f$. □

Mazzilli [MAZ2] investigated L^p extensions of holomorphic functions from submanifolds of complex ellipsoids. Cho [CHO] also obtained a counterexample for the L^p ($p > 2$) extension of holomorphic functions from submanifolds of some pseudoconvex domain. Tsuji [TSU] gave a counterexample for the bounded extension of holomorphic functions from submanifolds of certain unbounded pseudoconvex domain in $\mathbf{C}^2$.

4.5 Bounded Extensions by Means of the Berndtsson Formula

In this section we study the bounded extension of holomorphic functions from complex affine linear hypersurfaces in strictly convex domains. The result has already been proved in 3.3. The aim of the proof is to introduce the method of Diederich-Mazzilli [DIM2] which was used to prove the bounded extension of holomorphic functions from the intersection of a complex affine linear hypersurface with a convex domain of finite type. It is also interesting to compare the method of Henkin-Leiterer (Lemma 3.22) with the method of Diederich-Mazzilli (Theorem 4.12) concerning the integral representation on submanifolds.

Let $\Omega \subset\subset \mathbf{C}^n$ be a convex domain with C^∞ boundary. Then there exists a C^∞ function ρ in $\mathbf{C}^n$ such that $\Omega = \{z \in \mathbf{C}^n \mid \rho(z) < 0\}$. Let h be a holomorphic function in a neighborhood $\widetilde{\Omega}$ of $\overline{\Omega}$. We set $X = \{z \in \widetilde{\Omega} \mid h(z) = 0\}$, $V = \Omega \cap X$. Since Ω is convex, there exist holomorphic functions g_j, $j = 1, \cdots, n$, in $\overline{\Omega} \times \overline{\Omega}$ such that

$$h(z) - h(\zeta) = \sum_{j=1}^n g_j(z,\zeta)(z_j - \zeta_j).$$

Suppose $dh \neq 0$ on X. We set

$$Q^1(z,\zeta) = \frac{1}{\rho(\zeta)} \sum_{i=1}^n \frac{\partial \rho}{\partial \zeta_i}(\zeta) d\zeta_i.$$

Then by (4.29) we have the following theorem.

Theorem 4.11 *Let f be a bounded holomorphic function in V. We define*

$$Ef(z) := C_n \int_V f(\zeta) \left(\frac{\rho(\zeta)}{<\partial\rho(\zeta), z - \zeta> + \rho(\zeta)} \right)^{N+n-1} (\bar{\partial} Q^1)^{n-1}$$

$$\wedge \frac{\overline{\partial h(\zeta)} \wedge (\sum_{j=1}^n g_j(z,\zeta) d\zeta_j)}{\|\partial h\|^2} dV_{n-1}(\zeta)$$

for $z \in \Omega$, where C_n is a numerical constant depending only on n, and dV_{n-1} is the Lebesgue measure on V. Then Ef is holomorphic in Ω and satisfies $Ef|_V = f$.

Remark 4.2 *The integral in the right side of the above equality means to integrate coefficients of forms of degree (n,n) with respect to ζ on V.*

From now on we assume $h(z) = z_n$. For $a > 0$, we set

$$L_a = \{z \mid |z_n| \leq a|\rho(z)|, \ (z_1, \cdots, z_{n-1}, 0) \in V\}.$$

Since Ω is convex, we have $L_a \subset \Omega$ for any sufficiently small a. Let f be a bounded holomorphic function in V. Now we extend f to a C^∞ function in Ω as follows. Let $(\pi_\gamma)_{\gamma \geq 0}$ be a family of C^∞ functions in $\mathbf{R}$ such that $\pi_\gamma \equiv 1$ on $\{x \leq \frac{\gamma}{2}\}$, $\pi_\gamma \equiv 0$ on $\{x \geq \gamma\}$. For $z \in L_a$, we set $f(z_1, \cdots, z_n) = f(z_1, \cdots, z_{n-1}, 0)$. Then f is holomorphic in L_a. Define

$$\psi_\gamma(f)(z) = \pi_\gamma\left(\frac{|z_n|^2}{\rho(z)^2}\right) f(z).$$

Then for any sufficiently small γ, we have $\psi_\gamma(f) \in L^\infty(\Omega) \cap C^\infty(\Omega)$. Since

$$\bar{\partial}(\psi_\gamma(f)) = \pi'_\gamma\left(\frac{|z_n|^2}{\rho(z)^2}\right) f(z) \bar{\partial}\left(\frac{|z_n|^2}{\rho(z)^2}\right),$$

we obtain

$$\|\bar{\partial}(\psi_\gamma(f))\| \leq \frac{C_\gamma}{|\rho|},$$

where C_γ is a constant depending on γ such that $C_\gamma \to \infty$ as $\gamma \to 0$.

Lemma 4.10 Let $X = \{z \mid h(z) = 0\}$, $h(z) = z_n$. Let N be an integer such that $N \geq 2$. Suppose $s(z, \zeta)$ satisfies the conditions (A) and (B) in Theorem 4.1. Define

$$Q_\varepsilon^2 = \sum_{j=1}^n \frac{\overline{h(\zeta)}g_j(z,\zeta)}{|h|^2 + \varepsilon} d\zeta_j = \frac{\bar{\zeta}_n}{|\zeta_n|^2 + \varepsilon} d\zeta_n$$

$$P(z,\zeta) = A_n^1 (<Q^1, z - \zeta> +1)^{-N-n} \frac{z_n \bar{\zeta}_n + \varepsilon}{|\zeta_n|^2 + \varepsilon} (\bar{\partial} Q^1)^n$$
$$+ A_n^2 (<Q^1, z - \zeta> +1)^{-N-n+1} (\bar{\partial} Q^1)^{n-1} \wedge \bar{\partial} Q_\varepsilon^2$$
$$=: P_0(z,\zeta) + P_1(z,\zeta)$$

$$K(z,\zeta)$$
$$= B_n^1 \sum_{k=0}^{n-1}(<Q^1, z-\zeta>+1)^{-N-k}\frac{z_n\bar{\zeta}_n + \varepsilon}{|\zeta_n|^2 + \varepsilon}(\bar{\partial}Q^1)^k \wedge \frac{s \wedge (\bar{\partial}s)^{n-1-k}}{<s,\zeta-z>^{n-k}}$$
$$+ B_n^2 \sum_{k=0}^{n-2}(<Q^1, z-\zeta>+1)^{-N-k}(\bar{\partial}Q^1)^k$$
$$\wedge \frac{s \wedge (\bar{\partial}s)^{n-2-k}}{<s,\zeta-z>^{n-k-1}} \wedge \bar{\partial}Q_\varepsilon^2$$
$$=: \sum_{k=0}^{n-1} K_0^k(z,\zeta) + \sum_{k=0}^{n-2} K_1^k(z,\zeta).$$

Then we can choose constants A_n^i, B_n^i for $i = 1,2$ such that

$$\psi_\gamma(f)(z) = \int_\Omega \psi_\gamma(f)(\zeta)P(z,\zeta) + \int_\Omega \bar{\partial}(\psi_\gamma(f))(\zeta) \wedge K(z,\zeta) \qquad (z \in \Omega).$$

Proof. In the proof of Theorem 4.7, we set

$$G_1(\alpha) = \alpha^{-N} \ (N \geq 2), \quad G_2(\alpha) = \alpha.$$

Then we have

$$\widetilde{K} = A_n \sum_{\alpha_0+\alpha_1+\alpha_2=n-1} \frac{1}{(<Q^1,z-\zeta>+1)^{N+\alpha_1}} \left(\frac{\overline{h(\zeta)}h(z)+\varepsilon}{|h(\zeta)|^2+\varepsilon}\right)^{1-\alpha_2} \times$$

$$\times \frac{s \wedge (\bar{\partial}s)^{\alpha_0} \wedge (\bar{\partial}Q^1)^{\alpha_1} \wedge (\bar{\partial}Q_\varepsilon^2)^{\alpha_2}}{<s,\zeta-z>^{\alpha_0+1}}, \tag{4.38}$$

$$\widetilde{P} = B_n \sum_{\alpha_1+\alpha_2=n} \frac{1}{(<Q^1,z-\zeta>+1)^{N+\alpha_1}} \left(\frac{\overline{h(\zeta)}h(z)+\varepsilon}{|h(\zeta)|^2+\varepsilon}\right)^{1-\alpha_2} \times$$

$$\times (\bar{\partial}Q^1)^{\alpha_1} \wedge (\bar{\partial}Q_\varepsilon^2)^{\alpha_2}. \tag{4.39}$$

α_2 takes only 0 or 1. By the definition of Q_1, the integral on $\partial\Omega$ is equal to 0. □

Next we assume that $\Omega \subset\subset \mathbf{C}^n$ is a strictly convex domain with C^∞ boundary. Since ρ is strictly convex, it follows from Taylor's theorem that

$$2\mathrm{Re}\sum_{j=1}^n \frac{\partial \rho}{\partial \zeta_j}(\zeta)(\zeta_j - z_j) \geq \rho(\zeta) - \rho(z) + C|\zeta-z|^2 \qquad (z,\zeta \in \overline{\Omega}).$$

Define

$$\Phi(z,\zeta) = \sum_{j=1}^{n} \frac{\partial \rho}{\partial \zeta_j}(\zeta)(\zeta_j - z_j).$$

Then we have

$$2\mathrm{Re}\Phi(z,\zeta) \geq \rho(\zeta) - \rho(z) + C|\zeta - z|^2 \quad (z,\zeta \in \overline{\Omega}).$$

Consequently,

$$2|\Phi(z,\zeta) - \rho(\zeta)| \geq |\rho(\zeta)| + |\rho(z)| + |\mathrm{Im}\Phi(z,\zeta)| + C|\zeta - z|^2 \quad (4.40)$$

for $z,\zeta \in \overline{\Omega}$. We take $s(z,\zeta)$ as follows:

$$s(z,\zeta) = -\rho(z)\sum_{i=1}^{n}(\bar{\zeta}_i - \bar{z}_i)d\zeta_i - \overline{\Phi(\zeta,z)}\sum_{i=1}^{n}\frac{\partial \rho}{\partial \zeta_i}(z)d\zeta_i.$$

For $z \in \partial\Omega\setminus X$, $\zeta \in \Omega$, we have

$$s(z,\zeta) = -\overline{\Phi(\zeta,z)}\sum_{j=1}^{n}\frac{\partial \rho}{\partial \zeta_j}(z)d\zeta_i,$$

$$2\mathrm{Re}\Phi(\zeta,z) \geq -\rho(\zeta) + C|\zeta - z|^2$$

and

$$s \wedge \bar{\partial}_\zeta s = 0.$$

Hence for $z \in \partial\Omega\setminus X$, if we choose γ sufficiently small, then $\psi_\gamma(f)(z) = 0$, and hence there is no singular point except $\zeta = z$. For $z \in \partial\Omega\setminus X$, we obtain

$$\psi_\gamma(f)(z) = \int_\Omega \psi_\gamma(f)(\zeta) P(z,\zeta) \quad (4.41)$$

$$+ \int_\Omega \bar{\partial}(\psi_\gamma(f))(\zeta) \wedge (K_0^{n-1}(z,\zeta) + K_1^{n-2}(z,\zeta)).$$

Since $\bar{\partial}(\psi_\gamma(f)) = 0$ on V, by Lemma 4.5 we have

$$\lim_{\varepsilon \to 0} \int_\Omega \bar{\partial}(\psi_\gamma(f))(\zeta) \wedge K_1^{n-2}(z,\zeta)$$
$$= \int_V \bar{\partial}(\psi_\gamma(f)(\zeta))\pi B_n^2(<Q^1, z-\zeta>+1)^{-N-n+2}(\bar{\partial}Q^1)^{n-2}$$
$$\wedge \frac{s}{<s,\zeta-z>}dV_{n-1} = 0.$$

On the other hand, by Lebesgue's dominated convergence theorem we have

$$\lim_{\gamma \to 0} \int_\Omega \psi_\gamma(f) \wedge P(z,\zeta) = 0.$$

By Lemma 4.5 we have

$$\lim_{\varepsilon \to 0} \int_\Omega \psi_\gamma(f) P_1(z,\zeta)$$
$$= \int_V \psi_\gamma(f) \pi A_n^2 (<Q^1, z-\zeta>+1)^{-N-n+1}(\bar{\partial}Q^1)^{n-1} dV_{n-1}$$
$$= C_n E f(z).$$

Thus in (4.41) after letting $\varepsilon \to 0$, we let $\gamma \to 0$. Then we obtain

$$0 = C_n E f(z) + \lim_{\gamma \to 0} \int_\Omega \bar{\partial}(\psi_\gamma(f))(\zeta) \widetilde{K}_0^{n-1}(z,\zeta),$$

where

$$\widetilde{K}_0^{n-1}(z,\zeta) = B_n^1(<Q^1, z-\zeta>+1)^{-N-n+1}\frac{z_n}{\zeta_n}(\bar{\partial}Q^1)^{n-1} \wedge \frac{s}{<s,\zeta-z>}. \tag{4.42}$$

In this setting, we have the following theorem.

Theorem 4.12 *For $z \in \partial\Omega \backslash X$ one has*

$$Ef(z) = C_n \int_V z_n f(\zeta) d\bar{\zeta}_n \wedge \frac{1}{(<Q^1, z-\zeta>+1)^{N+n-1}}(\bar{\partial}Q^1)^{n-1}$$
$$\times \frac{s}{<s,\zeta-z>}dV_{n-1}.$$

Proof. By (4.42), $\widetilde{K}_0^{n-1}(z,\zeta)$ is expressed by

$$\widetilde{K}_0^{n-1}(z,\zeta) = \frac{z_n}{\zeta_n} T^{n-1}(z,\zeta) = \lim_{\varepsilon \to 0} \frac{z_n \bar{\zeta}_n}{|\zeta_n|^2 + \varepsilon} T^{n-1}(z,\zeta).$$

It follows from Stokes' theorem that

$$0 = \int_{\partial\Omega} \frac{z_n \bar{\zeta}_n}{|\zeta_n|^2 + \varepsilon} \psi_\gamma(f)(\zeta) T^{n-1}(z,\zeta)$$

$$= \int_\Omega \psi_\gamma(f)(\zeta) \bar{\partial}\left(\frac{z_n \bar{\zeta}_n}{|\zeta_n|^2 + \varepsilon}\right) \wedge T^{n-1}(z,\zeta)$$

$$+ \int_\Omega \frac{z_n \bar{\zeta}_n}{|\zeta_n|^2 + \varepsilon} \bar{\partial}(\psi_\gamma(f))(\zeta) \wedge T^{n-1}(z,\zeta)$$

$$+ \int_\Omega \frac{z_n \bar{\zeta}_n}{|\zeta_n|^2 + \varepsilon} \psi_\gamma(f)(\zeta) \bar{\partial}(T^{n-1}(z,\zeta)).$$

If γ is sufficiently small, then there exists $\delta > 0$ such that $\psi_\gamma(f) = 0$ in $B(z,\delta)$. Since $\lim_{\gamma \to 0} \psi_\gamma(f)(\zeta) = 0$ for $\zeta \notin V$, by Lebesgue's dominated convergence theorem we obtain

$$\lim_{\gamma \to 0} \int_\Omega \frac{z_n \bar{\zeta}_n}{|\zeta_n|^2 + \varepsilon} \psi_\gamma(f)(\zeta) \bar{\partial}(T^{n-1}(z,\zeta))$$

$$= \lim_{\gamma \to 0} \int_{\Omega \setminus B(z,\delta)} \frac{z_n \bar{\zeta}_n}{|\zeta_n|^2 + \varepsilon} \psi_\gamma(f)(\zeta) \bar{\partial}(T^{n-1}(z,\zeta))$$

$$= 0.$$

Consequently,

$$\lim_{\gamma \to 0} \int_\Omega \bar{\partial}(\psi_\gamma(f))(\zeta) \widetilde{K}_0^{n-1}(z,\zeta)$$

$$= C_n \int_V z_n f(\zeta) d\bar{\zeta}_n \wedge T^{n-1}(z,\zeta) dV_{n-1}$$

$$= C_n \int_V z_n f(\zeta) d\bar{\zeta}_n \wedge \frac{1}{(<Q^1, z-\zeta> +1)^{N+n-1}} (\bar{\partial} Q^1)^{n-1}$$

$$\times \frac{s}{<s, \zeta - z>} dV_{n-1}.$$

□

The following theorem was proved in 3.3 in more general situation. However, we prove theorem 4.13 using the integral formula in Theorem 4.12.

Theorem 4.13 *Let $\Omega \subset\subset \mathbf{C}^n$ be a strictly convex domain with C^∞ boundary and let $X = \{z \in \mathbf{C}^n \mid z_n = 0\}$, $V = \Omega \cap X$. Then every bounded holomorphic function in V can be extended to a bounded holomorphic function in Ω.*

Proof. Let f be a bounded holomorphic function in V. It is sufficient to show that $\sup_{z \in \partial \Omega \setminus X} |Ef(z)| \leq C \|f\|_\infty$ (see the proof of Theorem 3.15). By Theorem 4.12, if $z \in \partial \Omega \setminus X$, then we have

$$Ef(z) = C_n \int_V z_n f(\zeta) d\bar\zeta_n \wedge \frac{\rho(\zeta)^{N+n-1}}{(-\Phi(z,\zeta) + \rho(\zeta))^{N+n-1}} (\bar\partial Q^1)^{n-1}$$
$$\times \frac{\sum_{j=1}^n \frac{\partial \rho}{\partial \zeta_j}(z) d\zeta_j}{\Phi(\zeta,z)} dV_{n-1}.$$

Taking into account that

$$(\bar\partial Q^1)^{n-1} = \left(\bar\partial \left(\frac{\partial \rho}{\rho}\right)\right)^{n-1} = \frac{(\bar\partial \partial \rho)^{n-1}}{\rho^{n-1}} - (n-1)\frac{(\bar\partial \partial \rho)^{n-2} \wedge \partial \rho \wedge \bar\partial \rho}{\rho^n}$$

and

$$\partial \rho(\zeta) \wedge \bar\partial \rho(\zeta) \wedge \partial \rho(z) = (\partial \rho(\zeta) - \partial \rho(z)) \wedge \bar\partial \rho(\zeta) \wedge \partial \rho(z) = O(|\zeta - z|),$$

it is sufficient to estimate the following two integrals:

$$I_1 = \int_V \frac{|\zeta - z|}{(|\rho(\zeta)| + |\Phi(z,\zeta)|)^n |\Phi(\zeta,z)|} dV_{n-1}(\zeta)$$

and

$$I_2 = \int_V \frac{1}{(|\rho(\zeta)| + |\Phi(z,\zeta)|)^{n-1} |\Phi(\zeta,z)|} dV_{n-1}(\zeta).$$

By Lemma 3.43 we have

$$\Phi(\zeta, z) = \Phi(z, \zeta) - \rho(\zeta) + O(|\zeta - z|^3).$$

It follows from (4.40) that

$$I_1 \leq C \int_V \frac{dV_{n-1}}{(|\rho(\zeta)| + |\operatorname{Im}\Phi(z,\zeta)| + |\zeta - z|^2)^3 |\zeta - z|^{2n-5}}.$$

We choose a coordinate system $t_1, t_2, \cdots, t_{2n-2}$ such that $t_1 = \rho(\zeta) - \rho(z)$, $t_2 = \operatorname{Im}\Phi(z,\zeta)$ and $|t| \approx |\zeta - z|$. We set $t' = (t_3, \cdots, t_{2n-2})$. Then we have

$$I_1 \leq C \int_{|t| \leq R} \frac{dt}{(|t_1| + |t_2| + |z_n|^2 + |t'|^2)^3 |t'|^{2n-5}}$$
$$\leq C \int_{|t'| < R} \frac{dt_3 \cdots dt_{2n-2}}{(|z_n|^2 + |t'|^2) |t'|^{2n-5}}$$
$$\leq C \int_0^R \frac{dr}{|z_n|^2 + r^2} \leq \frac{C}{|z_n|}.$$

Similarly we have $I_2 \leq C/|z_n|$. □

The following theorem was proved by Diederich-Mazzilli [DIM2]. We omit the proof.

Theorem 4.14 *Let Ω be a smooth convex domain of finite type m and X a complex affine linear subspace of $\mathbf{C}^n$ with $V = \Omega \cap X$. Then there is a bounded linear extension operator $E : H^\infty(V) \to H^\infty(\Omega)$, where $H^\infty(\cdot)$ denotes the Banach space of bounded holomorphic functions in the corresponding domain.*

Exercises

4.1 We define the n dimensional polar coordinate transformation

$$\Phi_n : (r, \theta_1, \cdots, \theta_{n-1}) \to (x_1, \cdots, x_n)$$

by

$$x_1 = r \sin\theta_1 \sin\theta_2 \cdots \sin\theta_{n-3} \sin\theta_{n-2} \sin\theta_{n-1}$$
$$x_2 = r \sin\theta_1 \sin\theta_2 \cdots \sin\theta_{n-3} \sin\theta_{n-2} \cos\theta_{n-1}$$
$$\cdots$$
$$x_{n-1} = r \sin\theta_1 \cos\theta_2$$
$$x_n = r \cos\theta_1$$

$(r \geq 0,\ 0 \leq \theta_k \leq \pi\ (1 \leq k \leq n-2),\ 0 \leq \theta_{n-1} \leq 2\pi)$.

Show that the Jacobian J_n of Φ_n is given by

$$J_n = \pm r^{n-1} \sin^{n-2}\theta_1 \sin^{n-3}\theta_2 \cdots \sin^2\theta_{n-3} \sin\theta_{n-2}.$$

4.2 Let R be a positive constant and j a nonnegative integer. For $A > 0$, $q \geq 1$ and $z = x + iy$, prove that

$$\int_{|z|<R} \frac{|z+w|^j dx dy}{(A+|z+w|^j |z|^2)^q} = \begin{cases} O(A^{1-q}) & (q > 1) \\ O(\log A) & (q = 1). \end{cases}$$

4.3 Let Ω be a complex ellipsoid in $\mathbf{C}^n$, that is, Ω is given by

$$\Omega = \{z \in \mathbf{C}^n \mid \rho(z) < 0\}, \qquad \rho(z) = \sum_{j=1}^n |z_i|^{2m_j} - 1.$$

We set
$$M = \max\{2m_j\}, \quad \alpha = \frac{1}{M}.$$
Show that for $f \in C^1_{(0,q)}(\Omega)$, $1 \leq q \leq n$, with $\bar{\partial}f = 0$, there exists $u \in \Lambda_{\alpha,(0,q-1)}(\Omega)$ such that $\bar{\partial}u = f$.

4.4 Let p be a positive integer.
(1) Prove that for every $t, \tau \in \mathbf{R}$
$$2p\tau^{2p-1}(t-\tau) + \tau^{2p} \leq t^{2p}.$$
(2) Prove that there exists $\delta > 0$ such that for $t, \tau \in \mathbf{R}$
$$t^{2p} - \tau^{2p} - 2p\tau^{2p-1}(t-\tau) \geq \delta\{\tau^{2p-2}(t-\tau)^2 + (t-\tau)^{2p}\}$$

4.5 Let m be a positive integer. For $\sigma > 0$, define $\Gamma_\sigma = \{z = x+iy \mid |y| < \sigma|x|\}$. Prove that there exist $\sigma > 0$ and $\varepsilon > 0$ such that
$$\mathrm{Re}(z^{2m}) \geq \varepsilon|z|^{2m}$$
on Γ_σ.

4.6 Prove the following:
(a) For $q \geq 1$, $l = 0$ or 1, $j = l, l+1, \cdots$, and A positive close to 0
$$\int_{|z|<R} \frac{|t+x|^{j-l}|x|^l dx dy}{(A+|t+x|^j(x^2+y^2))^q} = \begin{cases} O(A^{1-q}) & (q > 1) \\ O(\log A) & (q = 1) \end{cases},$$
independent of $t \in (-R, R)$.
(b) For $q \geq 1$, $j \geq 1$, and A positive, close to 0
$$\int_{|z|<R} \frac{|t+x|^{j-1}|y| dx dy}{(A+|t+x|^j r^2 + r^{j+2})^q} = \begin{cases} O(A^{1-q}) & (q > 1) \\ O(\log A) & (q = 1) \end{cases},$$
independent of $t \in (-R, R)$.

4.7 For $z_k = x_k + iy_k$ $(k = 1, \cdots, N)$, define
$$\rho(z) = \sum_{k=1}^N \{x_k^{2n_k} + y_k^{2m_k}\} - 1, \quad \Omega = \{z \in \mathbf{C}^n \mid \rho(z) < 0\},$$
where n_k and m_k are positive integers with $m_k \leq n_k$. For $\zeta = \xi + i\eta \in \partial\Omega$, $z \in \bar{\Omega}$ and $\gamma > 0$, define
$$P_j(z,\zeta) = \frac{\partial \rho}{\partial z_j}(\zeta) - \gamma[(\eta_j^{2m_j-2} - \xi_j^{2n_j-2})(z_j - \zeta_j) + (z_j - \zeta_j)^{2m_j-1}]$$

and
$$\Phi(z,\zeta) = \sum_{j=1}^{N} P_j(z,\zeta)(z_j - \zeta_j).$$

Prove the following (1) and (2).

(1) If we choose $\gamma > 0$ sufficiently small, then there exists $\varepsilon > 0$ such that
$$2\text{Re}\Phi(z,\zeta) \leq \rho(z) - \varepsilon \sum_{k=1}^{N} \{(\xi_k^{2n_k-2} + \eta_k^{2m_k-2})|z_k - \zeta_k|^2$$
$$+ |z_k - \zeta_k|^{2m_k}\}$$
for $(z,\zeta) \in \overline{\Omega} \times \partial\Omega$.

(2) Let $q = \max_j \min\{2n_j, 2m_j\}$. Then there exists a constant $C > 0$ such that for every bounded, $\bar{\partial}$ closed $(0,1)$ form f on Ω, there exists a $1/q$-Hölder continuous function u in Ω such that $\bar{\partial} u = f$ (in the sense of distributions) and $\|u\|_{1/q} \leq C\|f\|_{0,\Omega}$. (See Diederich-Fornaess-Wiegerinck [DIK]).

Chapter 5

The Classical Theory in Several Complex Variables

In this chapter we first prove the Poincaré theorem, and then we investigate the Weierstrass preparation theorem, the properties of the coherent analytic sheaf and the Cousin problem. Some of theorems in Chapter 5 were used to prove the theorems in the previous chapters.

5.1 The Poincaré Theorem

In this section we study the Poincaré theorem which says that there is no biholomorphic mapping from a ball to a polydisc in $\mathbf{C}^n$ ($n \geq 2$). Here we give the proof due to Krantz [KR3].

Definition 5.1 Let B be the unit disc in the complex plane. Let Ω be a domain in $\mathbf{C}$ or $\mathbf{C}^2$. For $P \in \Omega$, define

$$(B, \Omega)_P := \{f : \Omega \to B \mid f \text{ is holomorphic}, f(P) = 0\}$$

and

$$(\Omega, B)_P := \{f : B \to \Omega \mid f \text{ is holomorphic}, f(0) = P\}.$$

Moreover, we define for $\Omega \subset \mathbf{C}^2$ and $f \in (B, \Omega)_P$

$$\operatorname{Jac}_{\mathbf{C}} f(P) = \left(\frac{\partial f}{\partial z_1}(P), \frac{\partial f}{\partial z_2}(P) \right).$$

Definition 5.2 (a) For $P \in \Omega \subset \mathbf{C}^2$, $\xi \in \mathbf{C}^2$, we define the Carathéodory metric $F_C^\Omega(P, \xi)$ of ξ at P by

$$F_C^\Omega(P, \xi) = \sup\{|\operatorname{Jac}_{\mathbf{C}} f(P)\xi| \mid f \in (B, \Omega)_P\}.$$

(b) For $P \in \Omega \subset \mathbf{C}$, $\xi \in \mathbf{C}$, we define the Carathéodory metric $F_C^\Omega(P,\xi)$ of ξ at P by

$$F_C^\Omega(P,\xi) = \sup\{|f'(P)\xi| \mid f \in (B,\Omega)_P\}.$$

Definition 5.3 (a) Let $\Omega \subset \mathbf{C}^2$ be a domain. For $P \in \Omega$, $\xi \in \mathbf{C}^2$, we define the Kobayashi metric $F_K^\Omega(P,\xi)$ of ξ at P by

$$F_K^\Omega(P,\xi) = \inf\left\{\frac{|\xi|}{|g'(0)|} \mid g \in (\Omega,B)_P, \text{ there exists } \lambda \text{ such that } g'(0) = \lambda\right\}.$$

(b) Let $\Omega \subset \mathbf{C}$ be a domain. For $P \in \Omega$, $\xi \in \mathbf{C}$, we define the Kobayashi metric $F_K^\Omega(P,\xi)$ of ξ at P by

$$F_K^\Omega(P,\xi) = \inf\left\{\frac{|\xi|}{|g'(0)|} \mid g \in (\Omega,B)_P\right\}.$$

Theorem 5.1 *Let $\Omega_1 \subset \mathbf{C}^2$ and $\Omega_2 \subset \mathbf{C}^n$, $1 \leq n \leq 2$, be domains and let $f : \Omega_1 \to \Omega_2$ be a holomorphic mapping. For $P \in \Omega_1$ and $\xi \in \mathbf{C}^2$, define*

$$f_*(P)\xi = \mathrm{Jac}_{\mathbf{C}} f(P)\xi.$$

Then we have

$$F_C^{\Omega_1}(P,\xi) \geq F_C^{\Omega_2}(f(P), f_*(P)\xi) \tag{5.1}$$

and

$$F_K^{\Omega_1}(P,\xi) \geq F_K^{\Omega_2}(f(P), f_*(P)\xi). \tag{5.2}$$

Proof. Let $n = 2$. Let $\varphi \in (B,\Omega_2)_{f(P)}$. Then $\varphi \circ f \in (B,\Omega_1)_P$. We obtain

$$F_C^{\Omega_1}(P,\xi) \geq \mathrm{Jac}_{\mathbf{C}}(\varphi \circ f)(P)\xi$$
$$= \left|\left(\frac{\partial(\varphi \circ f)}{\partial z_1}(P), \frac{\partial(\varphi \circ f)}{\partial z_2}(P)\right)\begin{pmatrix}\xi_1\\\xi_2\end{pmatrix}\right|$$
$$= \left|\left(\frac{\partial\varphi}{\partial w_1}(f(P)), \frac{\partial\varphi}{\partial w_2}(f(P))\right) f_*(P)\xi\right|$$
$$= |\mathrm{Jac}_{\mathbf{C}}\varphi(f(P))f_*(P)\xi|.$$

Since $\varphi \in (B,\Omega_2)_{f(P)}$ is arbitrary, we have (5.1). When $n = 1$ we can prove (5.1) similarly. Next we prove (5.2). Let $n = 2$. Let $g \in (\Omega_1,B)_P$

and $g'(0) = \lambda\xi$. Then $f \circ g \in (\Omega_2, B)_{f(P)}$. We set $g = (g_1, g_2)$. Then we have

$$(f \circ g)'(0)$$
$$= \left(\frac{\partial f_1}{\partial z_1}(P)g_1'(0) + \frac{\partial f_1}{\partial z_2}(P)g_2'(0), \frac{\partial f_2}{\partial z_1}(P)g_1'(0) + \frac{\partial f_2}{\partial z_2}(P)g_2'(0)\right)$$
$$= \lambda f_*(P)\xi.$$

Consequently,

$$F_K^{\Omega_2}(f(P), f_*(P)\xi) = \left\{\frac{|f_*(P)\xi|}{|h'(0)|} \mid h \in (\Omega_2, B)_{f(P)}, \ h'(0) = \mu f_*(P)\xi\right\}$$
$$\leq \frac{|f_*(P)\xi|}{|(f \circ g)'(0)|} = \frac{1}{|\lambda|} = \frac{|\xi|}{|g'(0)|}.$$

Since g is arbitrary, we have (5.2). When $n = 1$, we can prove (5.2) in the same way. □

Corollary 5.1 *Let Ω_1 and Ω_2 be domains in $\mathbf{C}^2$ and let $f : \Omega_1 \to \Omega_2$ be a biholomorphic mapping. Then*

$$F_C^{\Omega_1}(P, \xi) = F_C^{\Omega_2}(f(P), f_*(P)\xi)$$

and

$$F_K^{\Omega_1}(P, \xi) = F_K^{\Omega_2}(f(P), f_*(P)\xi)$$

for $P \in \Omega_1$ and $\xi \in \mathbf{C}^2$.

Proof. For $f^{-1} : \Omega_2 \to \Omega_1$, $f(P) \in \Omega_2$ and $\eta \in \mathbf{C}^2$, we apply Theorem 5.1. Then we have

$$F_C^{\Omega_2}(f(P), \eta) \geq F_C^{\Omega_1}(f^{-1}(f(P)), (f^{-1})_*(f(P))\eta)$$

and

$$F_K^{\Omega_2}(f(P), \eta) \geq F_K^{\Omega_1}(f^{-1}(f(P)), (f^{-1})_*(f(P))\eta).$$

If we set $\eta = f_*(P)\xi$, then we have

$$F_C^{\Omega_2}(f(P), f_*(P)\xi) \geq F_C^{\Omega_1}(P, \xi)$$

and

$$F_K^{\Omega_2}(f(P), f_*(P)\xi) \geq F_K^{\Omega_1}(P, \xi).$$

Together with (5.1) and (5.2) we have the desired equalities. □

Definition 5.4 Let $\Omega \subset \mathbf{C}^2$ be a domain. We define the length of a C^1 curve $\gamma : [a,b] \to \Omega$ with respect to the Carathéodory metric and the Kobayashi metric by

$$l_C(\gamma) = \int_a^b F_C^\Omega(\gamma(t), \gamma'(t)) dt$$

and

$$l_K(\gamma) = \int_a^b F_K^\Omega(\gamma(t), \gamma'(t)) dt,$$

respectively.

Corollary 5.2 Let $f : \Omega_1 \to \Omega_2$ be a holomorphic mapping. Then

$$l_C(f \circ \gamma) \leq l_C(\gamma), \quad l_K(f \circ \gamma) \leq l_K(\gamma)$$

for every C^1 curve $\gamma : [a,b] \to \Omega_1$.

Proof. Let $f = (f_1, f_2)$. Then we have

$$(f \circ \gamma)'(t) = ((f_1 \circ \gamma)'(t), (f_2 \circ \gamma)'(t)) = f_*(\gamma(t))\gamma'(t).$$

By Theorem 5.1 we obtain

$$l_K(f \circ \gamma) = \int_a^b F_K^{\Omega_2}(f \circ \gamma(t), (f \circ \gamma)'(t)) dt$$

$$= \int_a^b F_K^{\Omega_2}(f(\gamma(t)), f_*(\gamma(t))\gamma'(t)) dt$$

$$\leq \int_a^b F_K^{\Omega_1}(\gamma(t), \gamma'(t)) dt = l_K(\gamma).$$

We can prove $l_C(f \circ \gamma) \leq l_C(\gamma)$ similarly. □

Definition 5.5 Let $\Omega \subset \mathbf{C}^2$ be a domain. For $P \in \Omega$, define

$$\mathbf{i}_P^C(\Omega) = \{\xi \in \mathbf{C}^2 \mid F_C^\Omega(P, \xi) < 1\}$$

and

$$\mathbf{i}_P^K(\Omega) = \{\eta \in \mathbf{C}^2 \mid F_K^\Omega(P, \eta) < 1\}.$$

Theorem 5.2 Let Ω_1 and Ω_2 be domains in $\mathbf{C}^2$ and let $f : \Omega_1 \to \Omega_2$ be a biholomorphic mapping. For $P \in \Omega_1$, we set $Q = f(P)$. Then linear mappings

$$Jac_\mathbf{C} f(P) : \mathbf{i}_P^C(\Omega_1) \to \mathbf{i}_Q^C(\Omega_2)$$

and

$$Jac_{\mathbf{C}}f(P) : \mathbf{i}_P^K(\Omega_1) \to \mathbf{i}_Q^K(\Omega_2)$$

are bijective.

Proof. We set $g(\xi) = Jac_{\mathbf{C}}f(P)\xi$. It follows from Corollary 5.1 that for $\xi \in \mathbf{i}_P^C(\Omega_1)$

$$F_C^{\Omega_2}(Q, g(\xi)) = F_C^{\Omega_2}(f(P), f_*(P)\xi) = F_C^{\Omega_1}(P, \xi) < 1,$$

which implies that $g(\xi) \in \mathbf{i}_Q^C(\Omega_2)$. Clearly g is linear. For $\eta \in \mathbf{i}_Q^C(\Omega_2)$ we set $h(\eta) = Jac_{\mathbf{C}}f^{-1}(Q)\eta$. Then we have

$$F_C^{\Omega_1}(P, h(\eta)) = F_C^{\Omega_1}(f^{-1}(Q), Jac_{\mathbf{C}}f^{-1}(Q)\eta) = F_C^{\Omega_2}(Q, \eta) < 1,$$

which means that $h(\eta) \in \mathbf{i}_P^C(\Omega_1)$. Differentiating $f \circ f^{-1}(w) = w$ with respect to w_1 and w_2, we have $Jac_{\mathbf{C}}f(P)Jac_{\mathbf{C}}f^{-1}(Q) = E$ (E is the unit matrix). Similarly, we have $Jac_{\mathbf{C}}f^{-1}(Q)Jac_{\mathbf{C}}f(P) = E$. Consequently, we have $g \circ h(\eta) = \eta$, $h \circ g(\xi) = \xi$. Hence $g : \mathbf{i}_P^C(\Omega_1) \to \mathbf{i}_Q^C(\Omega_2)$ is bijective. Similarly, $Jac_{\mathbf{C}}f(P) : \mathbf{i}_P^K(\Omega_1) \to \mathbf{i}_Q^K(\Omega_2)$ is bijective. □

Lemma 5.1 *Let a and b be complex numbers such that*

$$|az_1 + bz_2| \leq 1$$

for any complex numbers z_1, z_2 with $|z_1|^2 + |z_2|^2 = 1$. Then

$$|a|^2 + |b|^2 \leq 1.$$

Proof. Let $a = r_1 e^{i\theta_1}$, $b = r_2 e^{i\theta_2}$. For $z_1 = t_1 e^{-i\theta_1}$, $z_2 = t_2 e^{-i\theta_2}$ with $t_1^2 + t_2^2 = 1$, we set $t_1 = \cos\theta$, $t_2 = \sin\theta$. Then we have

$$1 \geq |az_1 + bz_2| = t_1 r_1 + t_2 r_2 = r_1 \cos\theta + r_2 \sin\theta$$
$$= (r_1^2 + r_2^2)\left(\frac{r_1}{r_1^2 + r_2^2}\cos\theta + \frac{r_2}{r_1^2 + r_2^2}\sin\theta\right)$$
$$= (r_1^2 + r_2^2)\sin(\theta + \alpha).$$

We can choose t_1 and t_2 in such a way that $\theta + \alpha = \frac{\pi}{2}$, which means that $r_1^2 + r_2^2 \leq 1$. □

Theorem 5.3 *Let $B(0,1) = \{z \in \mathbf{C}^2 \mid |z| < 1\}$. Then $\mathbf{i}_0^K(B(0,1)) = B(0,1)$.*

Proof. Let $B = \{z \in \mathbf{C} \mid |z| < 1\}$ and let $\varphi \in (B(0,1), B)_0$. For $\eta \in \mathbf{C}^2$ with $|\eta| = 1$, define

$$h(\zeta) = \varphi(\zeta) \cdot \eta \qquad (\zeta \in B).$$

Then $h : B \to B$ satisfies $\varphi(0) = 0$. It follows from the Schwarz lemma that

$$|h'(0)| \leq 1.$$

Since η is arbitrary so far as $|\eta| = 1$, it follows from Lemma 5.1 that $|\varphi'(0)| \leq 1$. Consequently,

$$F_K^{B(0,1)}(0, \xi) = \inf\left\{ \frac{|\xi|}{|\varphi'(0)|} \mid \varphi \in (B(0,1), B)_0 \right\} \geq |\xi|.$$

for every $\xi \in \mathbf{C}^2$. On the other hand, for $\xi \neq 0$ we set

$$\varphi_0(\zeta) = \frac{\zeta}{|\xi|}\xi \qquad (\zeta \in B).$$

Then $\varphi_0 \in (B(0,1), B)_0$, which implies that

$$F_K^{B(0,1)}(0, \xi) \leq \frac{|\xi|}{|\varphi_0'(0)|} = |\xi|.$$

Hence we have $F_K^{B(0,1)}(0, \xi) = |\xi|$, and hence $\mathbf{i}_0^K(B(0,1)) = B(0,1)$. □

Theorem 5.4 *Define $P(0,1) = \{z \in \mathbf{C}^2 \mid |z_1| < 1, |z_2| < 1\}$. Then*

$$\mathbf{i}_0^K(P(0,1)) = P(0,1).$$

Proof. Define mappings $\pi_1 : P(0,1) \to B$ and $\pi_2 : P(0,1) \to B$ by

$$\pi_1(z_1, z_2) = z_1, \quad \pi_2(z_1, z_2) = z_2.$$

For $\eta = (\eta_1, \eta_2) \in \mathbf{C}^2$, it follows from Theorem 5.1 that

$$F_K^{P(0,1)}(0, \eta) \geq F_K^B(\pi_1(0), (\pi_1)_*\eta) = F_K^B(0, \eta_1).$$

By the Schwarz lemma, for a holomorphic mapping $\varphi : B \to B$ with $\varphi(0) = 0$, we have $|\varphi'(0)| \leq 1$. Moreover, if we define $\varphi_0 : B \to B$ by $\varphi_0(\zeta) = \zeta$, then we have $\varphi'(0) = 1$. Hence we have

$$F_K^B(0, \eta_1) = \left\{ \frac{|\eta_1|}{|\varphi'(0)|} \mid \varphi \in (B, B)_0 \right\} = |\eta_1|.$$

Consequently,
$$F_K^{P(0,1)}(0,\eta) \geq |\eta_1|$$
and
$$F_K^{P(0,1)}(0,\eta) \geq |\eta_2|.$$

Therefore we obtain
$$F_K^{P(0,1)}(0,\eta) \geq \max\{|\eta_1|,|\eta_2|\},$$
which means that $\mathbf{i}_0^K(P(0,1)) \subset P(0,1)$. Next, for $0 \neq \eta \in \mathbf{C}^2$, we set
$$\psi(\zeta) = \left(\frac{\zeta\eta_1}{\max\{|\eta_1|,|\eta_2|\}}, \frac{\zeta\eta_2}{\max\{|\eta_1|,|\eta_2|\}}\right).$$
Then $\psi \in (P(0,1), B)_0$ and we have $\psi'(0) = \mu\eta$ ($\mu > 0$). Consequently,
$$F_K^{P(0,1)}(0,\eta) \leq \frac{|\eta|}{|\psi'(0)|} = \max\{|\eta_1|,|\eta_2|\},$$
which means that $\mathbf{i}_0^K(P(0,1)) \supset P(0,1)$. □

Theorem 5.5 *(Poincaré theorem) There is no biholomorphic mapping from the unit polydisc $P(0,1)$ in $\mathbf{C}^2$ onto the unit ball $B(0,1)$ in $\mathbf{C}^2$.*

Proof. Suppose there is a biholomorphic mapping $\Phi : P(0,1) \to B(0,1)$. We set $\Phi^{-1}(0) = \alpha$. Then $\alpha \in P(0,1)$. Let $\alpha = (\alpha_1, \alpha_2)$. Then $\alpha_1 \in B$, $\alpha_2 \in B$. For $z \in B$, we set
$$\varphi_1(z) = \frac{z-\alpha_1}{1-\bar{\alpha}_1 z}, \quad \varphi_2(z) = \frac{z-\alpha_2}{1-\bar{\alpha}_2 z}$$
and
$$\varphi(\zeta) = (\varphi_1(\zeta_1), \varphi_2(\zeta_2)).$$
Then $\varphi : P(0,1) \to P(0,1)$ is holomorphic and bijective. Further we have $\varphi(\alpha) = 0$. Define $g = \Phi \circ \varphi^{-1}$. Then $g : P(0,1) \to B(0,1)$ is biholomorphic and $g(0) = 0$. Next we show that g does not exist. By Theorem 5.2, $\mathrm{Jac}_{\mathbf{C}} g(0)$ is a bijective linear mapping from $\mathbf{i}_0^K(P(0,1))$ onto $\mathbf{i}_0^K(P(0,1))$. By Theorem 5.3 and Theorem 5.4, $\mathrm{Jac}_{\mathbf{C}} g(0)$ is a bijective linear mapping from $P(0,1)$ onto $B(0,1)$. We set $h = \mathrm{Jac}_{\mathbf{C}} g(0)$. Since $h : P(0,1) \to B(0,1)$ is biholomorphic, h maps $\partial P(0,1)$ to $\partial B(0,1)$. Therefore a segment $A = \{(t,1) \mid 0 \leq t \leq 1\} \subset \partial P(0,1)$ is mapped by h to $\partial B(0,1)$. Since h is

linear, $h(A)$ is also a segment. Since B is strictly convex, $\partial B(0,1)$ cannot contain a segment (see Lemma 3.12), which is a contradiction. Hence a biholomorphic mapping Φ does not exist. □

5.2 The Weierstrass Preparation Theorem

We prove the Weierstrass preparation theorem using the Cauchy integral formula. Further we prove the implicit function theorem for holomorphic functions.

Definition 5.6 Let f be a holomorphic function in a neighborhood of $a \in \mathbf{C}^n$, and let $f(a) = 0$. We set $a = (a', a_n)$. We say that f is regular of order k in z_n at the point a if $f(a', z_n)$, considered as a holomorphic function of the single variable z_n, has a zero of order k at the point $z_n = a_n$. Equivalently, the condition can be stated as follows:

$$g(a_n) = g'(a_n) = \cdots = g^{(k-1)}(a_n) = 0, \quad g^{(k)}(a_n) \neq 0,$$

where $g(z_n) = f(a', z_n)$.

Lemma 5.2 *Let f be a holomorphic function in $B(a, \varepsilon)$ and let $f(a) = 0$, $f(z) \not\equiv 0$. Then after a suitable complex linear change of coordinates in $\mathbf{C}^n$, the function will be regular of order k, $k \geq 1$, in z_n at the point a.*

Proof. There exists $p \in B(a, \varepsilon)$, $p \neq a$, such that $f(p) \neq 0$. There exist constants b_{ij} such that the linear change of coordinates

$$z_i = (p_i - a_i)(\zeta_n - a_n) + \sum_{j=1}^{n-1} b_{ij}(\zeta_j - a_j) + a_i \qquad (i = 1, \cdots, n)$$

is nonsingular. We set $g(\zeta) = f(z(\zeta))$. Then

$$g(a_1, \cdots, a_{n-1}, 1 + a_n) = f(p_1, \cdots, p_n) \neq 0,$$

$$g(a_1, \cdots, a_{n-1}, a_n) = f(a_1, \cdots, a_n) = 0.$$

we set $h(\zeta_n) = g(a_1, \cdots, a_{n-1}, \zeta_n)$. Then by the identity theorem, $h(\zeta_n)$ has a zero of order k at the point $\zeta_n = a_n$ for some positive integer k. Hence g is regular of order k in ζ_n at a. □

Lemma 5.3 *Let f be a holomorphic function in a neighborhood of 0 and let $f(0) = 0$. Let f be regular of order k in z_n at 0, $k \geq 1$. Then for each sufficiently small $\delta_n > 0$ there is $\delta' = (\delta_1, \cdots, \delta_{n-1})$, such that for* each fixed $z' \in P(0', \delta')$, the equation $f(z', z_n) = 0$ has precisely k solutions (counted with multiplicities) in the disc $\{|z_n| < \delta_n\}$.

Proof. We set $g(z_n) = f(0', z_n)$. By the assumption $g(z_n)$ has a zero of order k at the point 0. Since the zero set of any non-constant holomorphic function in one variable is discrete, there exists $\delta_n > 0$ such that $g(z_n)$ is holomorphic and nowhere vanishing in $\{0 < |z_n| \leq \delta_n\}$. We set $m = \min_{|z_n|=\delta_n} |g(z_n)|$. If we choose $\delta' = (\delta_1, \cdots, \delta_{n-1})$ sufficiently small, then for $z' \in P(0', \delta')$, $|z_n| \leq \delta_n$, using the uniform continuity we have

$$|f(0', z_n) - f(z', z_n)| < m.$$

Hence for $|z_n| = \delta_n$ we obtain

$$|g(z_n)| \geq m > |g(z_n) - f(z', z_n)|.$$

By the Rouché theorem, the number of zeros of $g(z_n)$ in $\{|z_n| < \delta_n\}$ counting multiplicities equals the number of zeros of $f(z', z_n)$ in $\{|z_n| < \delta_n\}$ counting multiplicities. Since $g(z_n)$ has a zero of order k in 0 and does not vanish except 0, $f(z', z_n)$ has k zeros in $\{|z_n| < \delta_n\}$. □

Let f be a holomorphic function in a neighborhood of $0 \in \mathbf{C}^n$ and let $f(0) = 0$. Suppose f is regular of order k in z_n ($k \geq 1$) at 0. By Lemma 5.3, there exists $\delta = (\delta_1, \cdots, \delta_n)$ such that for any $z' \in P(0', \delta')$ $f(z', \cdot)$ has k zeros $\varphi_1(z'), \cdots, \varphi_k(z')$ in $\{|z_n| < \delta_n\}$ counting multiplicities. We set

$$\omega(z', z_n) = (z_n - \varphi_1(z')) \cdots (z_n - \varphi_k(z')) = z_n^k + a_{k-1}(z')z_n^{k-1} + \cdots + a_0(z').$$

Then the number of zeros of $f(z', \cdot)$ in $\{|z_n| < \delta_n\}$ counting multiplicities equals the number of zeros of $\omega(z', \cdot)$ in $\{|z_n| < \delta_n\}$ counting multiplicities. Hence there exists a nonvanishing function u in $P(0, \delta)$ such that $f = \omega u$. The Weierstrass preparation theorem says that we can choose ω and u to be holomorphic. Moreover, since $|\varphi_j(z')| < \delta_n$, we have $\omega(z', z_n) \neq 0$ for $z' \in P(0', \delta')$, $|z_n| = \delta_n$.

Definition 5.7 A function

$$\omega(z'.z_n) = z_n^k + a_{k-1}(z')z_n^{k-1} + \cdots + a_0(z')$$

is called a Weierstrass polynomial at 0 if $a_j(z')$ for $j = 0, \cdots, k-1$ are holomorphic in a neighborhood of $0'$ and satisfies $a_j(0') = 0$ for $j = 0, \cdots, k-1$.

Theorem 5.6 *(Weierstrass preparation theorem) Let f be a holomorphic function in a neighborhood of $0 \in \mathbf{C}^n$ and let $f(0) = 0$. Suppose f*

is regular of order k, $k \geq 1$, in z_n at 0. Then there exists $\delta = (\delta_1, \cdots, \delta_n)$ such that for $z \in P(0, \delta)$ we have a unique factorization

$$f(z) = \omega(z)u(z),$$

where $\omega(z) = z_n^k + a_{k-1}(z')z_n^{k-1} + \cdots + a_0(z')$ is a Weierstrass polynomial at 0 and u is a nowhere vanishing holomorphic function in $P(0, \delta)$.

Proof. The uniqueness is obvious. By Lemma 5.3 there exist $\delta = (\delta', \delta_n)$ such that for any fixed $z' \in P(0, \delta')$, $f(z', \cdot)$ has k zeros $\varphi_1(z'), \cdots, \varphi_k(z')$ in $\{|z_n| < \delta_n\}$. By the argument principle, we have for $z' \in P(0, \delta')$

$$S_m(z') \equiv \sum_{j=1}^{k} \varphi_j^m(z') = \frac{1}{2\pi i} \int_{|\zeta|=\delta_n} \frac{\zeta^m \frac{\partial f}{\partial \zeta}(z', \zeta)}{f(z', \zeta)} d\zeta.$$

Hence $S_m(z')$ is holomorphic as a function in z'. Moreover, $a_0, \cdots, a_k$ are holomorphic since they are polynomials of $S_0, \cdots, S_{k-1}$. Since $f(0', z_n) = z_n^k g(z_n)$ and $g(0) \neq 0$, we have $\varphi_1(0') = \cdots = \varphi_k(0') = 0$. Hence $a_0(0') = \cdots = a_{k-1}(0') = 0$, which means that $\omega(z) = z_n^k + a_{k-1}(z')z_n^{k-1} + \cdots + a_0(z')$ is a Weierstrass polynomial at 0. We set $u = f/\omega$. Since $u(z', \cdot)$ is holomorphic in $\{|z_n| \leq \delta_n\}$ for fixed z', we obtain

$$u(z', z_n) = \frac{1}{2\pi i} \int_{|\zeta|=\delta_n} \frac{(f/\omega)(z', \zeta)}{\zeta - z_n} d\zeta.$$

Since $\omega(z', z_n) \neq 0$ for $|z_n| = \delta_n$, u is holomorphic in $P(0, \delta)$. □

Theorem 5.7 *(Weierstrass division theorem) Let ω be a Weierstrass polynomial of degree k at 0, and let f be holomorphic in a neighborhood of $0 \in \mathbf{C}^n$. Then there is a unique factorization in some sufficiently small neighborhood of 0*

$$f = \omega q + r,$$

where q and r are holomorphic in a neighborhood of 0 and r is a polynomial in z_n of degree less than k with coefficients that are holomorphic functions of $z_1, \cdots, z_{n-1}$.

Proof. As we mentioned in the remark before Theorem 5.6, there exists $\delta = (\delta_1, \cdots, \delta_n) = (\delta', \delta_n)$ such that if $z' \in P(0', \delta')$, $|z_n| = \delta_n$, then $\omega(z', z_n) \neq 0$. We set

$$q(z', z_n) = \frac{1}{2\pi i} \int_{|\zeta|=\delta_n} \frac{f(z', \zeta)}{\omega(z', \zeta)(\zeta - z_n)} d\zeta.$$

Then q is holomorphic in $P(0,\delta)$. Define
$$r = f - q\omega.$$
Then r is holomorphic in $P(0,\delta)$. Consequently, we have
$$r(z) = \frac{1}{2\pi i} \int_{|\zeta|=\delta_n} \left\{ f(z',\zeta) - \omega(z',z_n) \frac{f(z',\zeta)}{\omega(z',\zeta)} \right\} \frac{d\zeta}{\zeta - z_n}$$
$$= \frac{1}{2\pi i} \int_{|\zeta|=\delta_n} \frac{f(z',\zeta)}{\omega(z',\zeta)} \left(\frac{\omega(z',\zeta) - \omega(z',z_n)}{\zeta - z_n} \right) d\zeta.$$
Taking into account that
$$\frac{\omega(z',\zeta) - \omega(z',z_n)}{\zeta - z_n} = \frac{(\zeta^k - z_n^k) + \sum_{j=0}^{k-1} a_j(z')(\zeta^j - z_n^j)}{\zeta - z_n},$$
r is a polynomial in z_n of degree less than k. Next we show the uniqueness. Suppose we have two factorizations
$$q_1\omega + r_1 = q_2\omega + r_2 = f.$$
In the equation
$$r_1 - r_2 = (q_2 - q_1)\omega,$$
the left side is a polynomial in z_n of degree less than k and the right side has k zeros in z_n, which means that $r_1 \equiv r_2$. Hence we have $q_1 = q_2$. $\square$

Theorem 5.8 *Let f be a holomorphic function in a polydisc $P(w,r) \subset \mathbf{C}^n$ with the following properties:*

(a) $f(w) = 0$.
(b) $\frac{\partial f}{\partial z_n}(w) = 1$.

Let $w = (w', w_n)$. Then there exist $\delta = (\delta_1, \cdots, \delta_n) = (\delta', \delta_n)$ and a holomorphic function $\varphi(z_1, \cdots, z_{n-1})$ in a polydisc $P(w', \delta')$ such that in $P(w,\delta)$, $f(z_1, \cdots, z_n) = 0$ is equivalent to $z_n = \varphi(z_1, \cdots, z_{n-1})$.

Proof. We set
$$f = u + iv, \quad z_j = x_j + ix_{n+j} \quad (j = 1, \cdots, n),$$
$$x = (x_1, \cdots, x_{2n}), \quad x' = (x_1, \cdots, x_{n-1}, x_{n+1}, \cdots, x_{2n}).$$

Then by the Cauchy-Riemann equations we have

$$\left|\frac{\partial f}{\partial z_n}(w)\right|^2 = \left|\frac{\partial(u,v)}{\partial(x_n, x_{2n})}(w)\right|.$$

By the implicit function theorem in real variables, there exist C^∞ functions $g(x')$, $h(x')$ in a neighborhood of w' such that $u(x) = 0$, $v(x) = 0$ are equivalent to $x_n = g(x')$, $x_{2n} = h(x')$. We set $\varphi(z') = g(x') + ih(x')$. Then in a neighborhood of w, $f(z) = 0$ is equivalent to $z_n = \varphi(z')$, which implies that $f(z_1, \cdots, z_{n-1}, \varphi(z')) = 0$. By the condition (b) and Lemma 5.3, there exists $\delta = (\delta_1, \cdots, \delta_n)$ such that for $z' = (z_1, \cdots, z_{n-1}) \in P(w', \delta')$, $f(z', \cdot)$ has only zero $\varphi(z')$ of order 1 in $|w_n - z_n| < \delta_n$. By the argument principle, we have

$$\varphi(z') = \frac{1}{2\pi i} \int_{|\zeta_n - w_n| = \delta_n} \frac{\zeta_n \frac{\partial f(z', \zeta_n)}{\partial \zeta_n}}{f(z', \zeta_n)} d\zeta_n.$$

Therefore $\varphi(z')$ is holomorphic in $P(w', \delta')$. $\square$

Next we prove the implicit function theorem for holomorphic functions.

Theorem 5.9 *(Implicit function theorem) Let $1 \leq k \leq n$. Let $f_{k+1}, \cdots, f_n$ be holomorphic functions in a polydisc $P(w, r) \subset \mathbf{C}^n$ satisfying the following properties:*

(a) $f_j(w) = 0 \quad (j = k+1, \cdots, n)$.
(b) $\frac{\partial f_j}{\partial z_i}(w) = \delta_i^j \quad (i, j = k+1, \cdots, n)$.

Let $w = (w'', w_{k+1}, \cdots, w_n)$. Then there exist $\delta = (\delta'', \delta_{k+1}, \cdots, \delta_n)$ and holomorphic functions $\varphi_j(z_1, \cdots, z_k)$ for $j = k+1, \cdots, n$ in a polydisc $P(w'', \delta'')$ such that in a neighborhood of w, equations $f_j(z_1, \cdots, z_n) = 0$ for $j = k+1, \cdots, n$ are equivalent to equations $z_j = \varphi_j(z_1, \cdots, z_k)$ for $j = k+1, \cdots, n$.

Proof. We prove Theorem 5.9 by induction on $m = n - k$. In case $m = 1$, Theorem 5.9 follows from Theorem 5.8. Assume that Theorem 5.9 has already been proved for $m - 1$. Suppose $f_{k+1}, \cdots, f_n$ are holomorphic in $P(w, r)$ and satisfy conditions (a) and (b). We set $z = (z', z_n)$, $w = (w', w_n)$. We apply Theorem 5.8 to $f_n(z)$. Then there exist $\eta = (\eta', \eta_n)$ and a holomorphic function $\varphi(z')$ in $P(w', \eta')$ such that in $P(w, \eta)$, an equation $f(z', z_n) = 0$ is equivalent to $z_n = \varphi(z')$. We set

$$g_j(z') = f_j(z', \varphi(z')) \quad (j = k+1, \cdots, n-1).$$

Then $f_j(z) = 0$, $j = k+1, \cdots, n$, are equivalent to

$$g_j(z') = 0 \qquad (j = k+1, \cdots, n-1),$$

$$z_n = \varphi(z').$$

Consequently, we have $g_j(w') = 0$, $j = k+1, \cdots, n-1$. For $i = k+1, \cdots, n-1$ we have

$$\frac{\partial g_j}{\partial z_i}(w') = \frac{\partial f_j}{\partial z_i}(w) + \frac{\partial f_j}{\partial z_n}(w)\frac{\partial \varphi}{\partial z_i}(w') = \delta_i^j.$$

By the inductive hypothesis, there are $\delta' = (\delta'', \delta_{k+1}, \cdots, \delta_{n-1})$ and holomorphic functions $\varphi_{k+1}(z''), \cdots, \varphi_{n-1}(z'')$ in $P(w'', \delta'')$ such that equations $g_j(z') = 0$, $j = k+1, \cdots, n-1$, are equivalent to equations $z_j = \varphi_j(z'')$, $j = k+1, \cdots, n-1$ in $P(w', \delta')$. We set

$$\varphi_n(z'') = \varphi(z'', \varphi_{k+1}(z''), \cdots, \varphi_{n-1}(z'')).$$

Then equations $f_j(z) = 0$, $j = k+1, \cdots, n$, are equivalent to equations $z_j = \varphi_j(z'')$, $j = k+1, \cdots, n$. $\square$

Definition 5.8 Let Ω be a domain in $\mathbf{C}^n$, and let $F = (f_1, \cdots, f_m) : \Omega \to \mathbf{C}^m$ be a holomorphic mapping. Define

$$F'(z) = \left(\frac{\partial f_i}{\partial z_j}(z)\right).$$

$F'(z)$ is called the Jacobian matrix of F at $z \in \Omega$. We say that F is nonsingular at z if the rank of $F'(z)$ equals $\min(m, n)$.

Theorem 5.10 Let $\Omega \subset \mathbf{C}^n$ be a domain with $0 \in \Omega$. Let $n \geq m$. Suppose $F : \Omega \to \mathbf{C}^m$ is a nonsingular holomorphic mapping and $F(0) = 0$. Then there exist a linear change of variables $w_i = \sum_{j=1}^n a_{ij} z_j$ for $i = 1, \ldots, n$, $\delta = (\delta', \delta_{n-m+1}, \cdots, \delta_n)$ where $\delta' = (\delta_1, \cdots, \delta_{n-m})$ and holomorphic functions $\varphi_j(w_1, \cdots, w_{n-m})$ for $j = n-m+1, \cdots, n$ in $P(0.\delta')$ such that in $P(0, \delta)$, equations $F(w_1, \cdots, w_n) = 0$ are equivalent to equations $w_j = \varphi_j(w_1, \cdots, w_{n-m})$, $j = n-m+1, \cdots, n$.

Proof. Let $F = (f_{n-m+1}, \cdots, f_n)$. By the hypothesis there exist an $(m \times m)$ matrix B and an $(n \times n)$ matrix A such that

$$BF'(0)A^{-1} = (0, I),$$

where I is an $(m \times m)$ unit matrix. We set

$$B = (b_{ij}) \qquad (n - m + 1 \leq i, j \leq n),$$

$$A = (a_{ij}), \quad A^{-1} = (a'_{ij}) \qquad (1 \leq i, j \leq n),$$

$$g_i = \sum_{j=n-m+1}^{n} b_{ij} f_j \qquad (i = n - m + 1, \cdots, n),$$

$$w_i = \sum_{j=1}^{n} a_{ij} z_j \qquad (i = 1, \cdots, n).$$

Further we set $G = (g_{n-m+1}, \cdots, g_n)$. Then $F(z) = 0$ is equivalent to $G(z) = 0$. Taking into account that

$$\frac{\partial g_i}{\partial w_j} = \sum_{k=n-m+1}^{n} b_{ik} \frac{\partial f_k}{\partial w_j} = \sum_{k=n-m+1}^{n} b_{ik} \sum_{s=1}^{n} \frac{\partial f_k}{\partial z_s} a'_{sj},$$

we obtain

$$\frac{\partial g_i}{\partial w_j}(0) = \delta_j^i.$$

Theorem 5.10 follows from Theorem 5.9. □

Corollary 5.3 *(Inverse mapping theorem) Let F be a nonsingular holomorphic mapping from a neighborhood of $z \in \mathbf{C}^n$ into $\mathbf{C}^n$ and let $F(z) = w$. Then there exist a neighborhood U ($U \subset W$) of z and a neighborhood V of w such that $F : U \to V$ has the holomorphic inverse mapping $F^{-1} : V \to U$.*

Proof. We assume that $z = w = 0$. We denote by $J(0)$ the Jacobian matrix of F at 0. Without loss of generality we may assume that $J(0)$ is the unit matrix. We set $H(z, w) = w - F(z)$. By Theorem 5.10, there exist $\varepsilon > 0$, a polydisc $P(0, \varepsilon)$ in $\mathbf{C}^n$ and a holomorphic mapping G in $P(0, \varepsilon)$ such that the equation $H(z, w) = 0$ is equivalent to the equation $z = G(w)$. Therefore we have $w = F \circ G(w)$, $z = G \circ F(z)$, which implies that $G = F^{-1}$. □

Theorem 5.11 *Let $\Omega \subset \mathbf{C}^n$ be an open set. If a holomorphic mapping $F : \Omega \to \mathbf{C}^n$ is injective, then $\det F'(z) \neq 0$ for $z \in \Omega$.*

Proof. We prove Theorem 5.11 by induction on n. When $n = 1$, Theorem 5.11 is true. Assume that Theorem 5.11 has already been proved for $n - 1$. Now we prove the following lemma:

Lemma 5.4 *Let $\Omega \subset \mathbf{C}^n$ be an open set. Suppose a holomorphic mapping $F : \Omega \to \mathbf{C}^n$ is injective. If $F'(a) \neq 0$ for $a \in \Omega$, then $\det F'(a) \neq 0$.*

Proof of Lemma 5.4 Without loss of generality, we may assume that $F = (f_1, \cdots, f_n)$, $\frac{\partial f_n}{\partial z_n}(a) \neq 0$. We set $w(z) = (z_1, \cdots, z_{n-1}, f_n(z))$. Then $\det\left(\frac{\partial w_k}{\partial z_j}\right)(a) \neq 0$. By Corollary 5.3, w^{-1} is a holomorphic mapping in a neighborhood of a. We set $\widetilde{F} = F \circ w^{-1}$. Then we have

$$\widetilde{F}(w) = (g_1(w), \cdots, g_{n-1}(w), w_n),$$

where $g_1, \cdots, g_{n-1}$ are holomorphic at $b = w(a)$. Set $w' = (w_1, \cdots, w_{n-1})$ and $G(w') = (g_1(w', b_n), \cdots, g_{n-1}(w', b_n))$. Then G is injective in a neighborhood of $b' = (b_1, \cdots, b_{n-1})$. By the inductive hypothesis, we have $\det G'(b') \neq 0$, which implies that $\det \widetilde{F}'(b) \neq 0$. Hence we have $\det F'(a) \neq 0$, which completes the proof of Lemma 5.4.

We continue the proof of Theorem 5.11. We set $h = \det F' \in \mathcal{O}(\Omega)$. Suppose $Z(h) = \{z \in \Omega \mid h(z) = 0\} \neq \phi$. Then $Z(h)$ contains a $n - 1$ dimensional submanifold M. By Lemma 5.4, we have $F'(z) = 0$ for $z \in Z(h)$, and hence $F'(z) = 0$ for $z \in M$. Consequently, F is locally constant in M. Since $\dim_{\mathbf{C}} M = n - 1 > 0$, this contradicts that F is injective. □

Corollary 5.4 *Let $\Omega \subset \mathbf{C}^n$ be an open set. If a holomorphic mapping $F : \Omega \to \mathbf{C}^n$ is injective, then $F(\Omega)$ is an open set and $F^{-1} : F(\Omega) \to \Omega$ is holomorphic.*

Proof. By Theorem 5.11, we have $\det F'(z) \neq 0$ for $z \in \Omega$. Hence $F(\Omega)$ is open. By Corollary 5.3, $F^{-1} : F(\Omega) \to \Omega$ is holomorphic. □

Theorem 5.12 *Let Ω be a pseudoconvex domain in $\mathbf{C}^n$ and let m be a positive integer with $m \leq n$. Suppose that $f_1, \cdots, f_m$ are holomorphic functions in Ω and that $F = (f_1, \cdots, f_m)$ is nonsingular in Ω. We set*

$$M = \{z \in \Omega \mid f_1(z) = \cdots = f_m(z) = 0\}.$$

Let $a \in \Omega$. If f is holomorphic in Ω and vanishes everywhere on M, then there exist a neighborhood U of a and holomorphic functions $g_1, \cdots, g_m$ in U such that

$$f(z) = \sum_{j=1}^{m} f_j(z) g_j(z) \qquad (z \in U).$$

Proof. By Theorem 5.10 using a complex linear change of variables there exist a neighborhood U of $a = (a', a_{n-m+1}, \cdots, a_n)$ and holomorphic functions $\varphi_j(w_1, \cdots, w_{n-m})$, $n - m + 1 \leq j \leq n$, in a neighborhood U' of a' such that

$$M \cap U = \{w \in U \mid w_j = \varphi_j(w_1, \cdots, w_{n-m}),\ n - m + 1 \leq j \leq n\}.$$

If we set

$$\zeta_1 = w_{n-m+1} - \varphi_{n-m+1}(w_1, \cdots, w_{n-m})$$
$$\cdots$$
$$\zeta_m = w_n - \varphi_n(w_1, \cdots, w_{n-m})$$
$$\zeta_{m+1} = w_1$$
$$\cdots$$
$$\zeta_n = w_{n-m},$$

then $U \cap M$ is expressed by

$$U \cap M = \{\zeta \in U \mid \zeta_1 = \cdots = \zeta_m = 0\}.$$

In case $m = 1$, we set

$$\frac{f(\zeta)}{\zeta_1} = \psi_1(\zeta).$$

Then ψ_1 is holomorphic in a neighborhood of 0 and satisfies

$$f(\zeta) = \zeta_1 \psi_1(\zeta).$$

This proves Theorem 5.12 in case $m = 1$. Assume that we have already proved Theorem 5.12 for $m-1$. Since f is holomorphic in $\Omega_1 = \{\zeta_1 = 0\} \cap \Omega$ and vanishes in $\Omega_1 \cap \{\zeta_2 = \cdots \zeta_m = 0\}$, by the inductive hypothesis, there exist holomorphic functions $\tilde{g}_j(\zeta_2, \cdots, \zeta_n)$ $(2 \leq j \leq m)$ in a neighborhood of $0 \in \mathbf{C}^{n-1}$ such that

$$f(\zeta) = \sum_{j=2}^{m} \zeta_j \tilde{g}_j(\zeta).$$

By Theorem 2.14 there exist holomorphic functions g_j in a neighborhood W of 0 such that $g_j = \tilde{g}_j$ in $\{\zeta_1 = 0\} \cap W$. We set

$$g_1(\zeta) = \frac{1}{\zeta_1}\left(f(\zeta) - \sum_{j=2}^{m}\zeta_j g_j(\zeta)\right).$$

Then we have the desired equality. □

5.3 Oka's Fundamental Theorem

We give a proof of Oka's fundamental theorem [OkA2] which is the prototype of the sheaf theory. Moreover, we state the Cartan theorems A and B without giving proofs.

Definition 5.9 Let $a \in \mathbf{C}^n$. For a neighborhood U of a and $f : U \to \mathbf{C}$, we say that (f, U) is a function element at a. We say that two function elements (f, U) and (g, V) at a are equivalent if there exists a neighborhood $W \subset U \cap V$ of a such that $f|_W = g|_W$. The equivalence class of a function element (f, U) at a is called a germ of functions at a and is denoted by $\mathbf{f}_a$ or $\gamma_a(f)$. Further, we denote by $\mathcal{F}_a$ the set of all germs at a. The set of all germs $\mathbf{f}_a$ such that $\mathbf{f}_a$ has a representative (f, U) with $f \in C(U)$ ($f \in C^k(U)$, $f \in \mathcal{O}(U)$) is denoted by $\mathcal{C}_a$ ($\mathcal{C}_a^k$, $\mathcal{O}_a$).

By definition we have

$$\mathcal{O}_a \subset \mathcal{C}_a^\infty \subset \mathcal{C}_a^k \subset \mathcal{C}_a \subset \mathcal{F}_a,$$

where k is an integer with $1 \le k < \infty$. Let (f, U) and (g, V) be representatives of $\mathbf{f}_a$ and $\mathbf{g}_a$, respectively. We define $\mathbf{f}_a + \mathbf{g}_a$ and $\mathbf{f}_a \mathbf{g}_a$ by an equivalent class of $(f + g, U \cap V)$ and $(fg, U \cap V)$, respectively. Then $\mathcal{F}_a$, $\mathcal{C}_a^k$ and $\mathcal{O}_a$ become commutative rings.

Definition 5.10 For the set $\mathcal{O}_a$ of all germs of holomorphic functions at $a \in \mathbf{C}^n$, define

$$\mathcal{O} = \mathcal{O}_{\mathbf{C}^n} = \bigcup_{a \in \mathbf{C}^n} \mathcal{O}_a.$$

We introduce the basis of all open sets in $\mathcal{O}$ as follows:
For an open set U in $\mathbf{C}^n$ and a holomorphic function f in U, define

$$U_f = \{\mathbf{f}_z \mid z \in U\}.$$

We define the basis in $\mathcal{O}$ to be the set of all U_f. Define $\pi : \mathcal{O} \to \mathbf{C}^n$ by $\pi(\mathbf{f}_a) = a$. Let $\mathbf{f}_a \in \mathcal{O}_a$ and let (f, U) be a representative of $\mathbf{f}_a$. Then U_f is a neighborhood of $\mathbf{f}_a$ and $\pi : U_f \to U$ is bijective, continuous and open mapping, which means that π is a local homeomorphism.

Definition 5.11 Let X be a topological space. We say that a topological space $\mathcal{S}$ is a sheaf over X if there is a surjective local homeomorphism $\pi : \mathcal{S} \to X$. Hence π is an open mapping. For $x \in X$, $\mathcal{S}_x = \pi^{-1}(x)$ is called a stalk.

Definition 5.12 Let $\mathcal{S}$ be a sheaf over X and $Y \subset X$. We say that a continuous mapping $s : Y \to \mathcal{S}$ is a section of $\mathcal{S}$ over Y if $\pi \circ s(x) = x$ for all $x \in Y$. We denote by $\Gamma(Y, \mathcal{S})$ the collection of all sections of $\mathcal{S}$ over Y.

Definition 5.13 We say that a sheaf $\mathcal{S}$ over X is a sheaf of Abelian groups over X if each $\mathcal{S}_x$ ($x \in X$) carries the structure of an Abelian group, so that if $Y \subset X$, $s_1, s_2 \in \Gamma(Y, \mathcal{S})$, then $s_1 - s_2 : Y \to \mathcal{S}$, being defined by

$$(s_1 - s_2)(x) = s_1(x) - s_2(x) \qquad (x \in Y),$$

is continuous. The sheaf of rings over X is defined similarly.

Definition 5.14 Let $\mathcal{R}$ be a sheaf of rings over X and let $\mathcal{S}$ be a sheaf of Abelian groups over X. We say that $\mathcal{S}$ is a sheaf of modules over $\mathcal{R}$ (or a sheaf of $\mathcal{R}$-modules), if $\mathcal{S}_x$ is a $\mathcal{R}_x$-module, and the product of a section of $\mathcal{R}$ and a section of $\mathcal{S}$ is a section of $\mathcal{S}$. We say that $\mathcal{S}$ is an analytic sheaf if X is a complex manifold and $\mathcal{R}$ is a sheaf $\mathcal{O}$ of germs of holomorphic functions.

Definition 5.15 We say that $\mathcal{S}' \subset \mathcal{S}$ is a subsheaf of $\mathcal{S}$ if $\pi|_{\mathcal{S}'} : \mathcal{S}' \to X$ is a sheaf. Hence $\mathcal{S}'$ is a subsheaf of $\mathcal{S}$ if and only if $\mathcal{S}'$ is an open subset of $\mathcal{S}$ and $\pi(\mathcal{S}') = X$. If $\mathcal{S}$ is a sheaf of Abelian groups, we assume that $\mathcal{S}'_x$ is a subgroup of $\mathcal{S}_x$. Suppose $\mathcal{S}'$ and $\mathcal{S}$ are sheaves of Aberian groups over X. We say that a continuous mapping $\varphi : \mathcal{S}' \to \mathcal{S}$ is a sheaf homomorphism if $\varphi(\mathcal{S}'_x) \subset \mathcal{S}_x$ and $\varphi_x = \varphi|_{\mathcal{S}'_x} : \mathcal{S}'_x \to \mathcal{S}_x$ is a group homomorphism for each $x \in X$.

Definition 5.16 Let $\varphi : \mathcal{S}' \to \mathcal{S}$ be a sheaf homomorphism. Define

$$\operatorname{Ker} \varphi = \bigcup_{x \in X} \operatorname{Ker} \varphi_x, \qquad \operatorname{Im} \varphi = \bigcup_{x \in X} \operatorname{Im} \varphi_x.$$

Then $\operatorname{Ker}\varphi$ is a subsheaf of $\mathcal{S}'$ and $\operatorname{Im}\varphi$ is a subsheaf of $\mathcal{S}$. We have the exact sequence

$$0 \to \operatorname{Ker}\varphi \xrightarrow{\iota} \mathcal{S}' \xrightarrow{\varphi} \operatorname{Im}\varphi \to 0.$$

Definition 5.17 Let $\mathcal{S}$ be a sheaf of Abelian groups and let $\mathcal{S}'$ be a subsheaf of $\mathcal{S}$. Define

$$\mathcal{S}/\mathcal{S}' = \bigcup_{x \in X} \mathcal{S}_x/\mathcal{S}'_x.$$

We define the quotient sheaf $\mathcal{S}/\mathcal{S}'$ as the union of all the quotient groups $\mathcal{S}_x/\mathcal{S}'_x$ for $x \in X$, equipped with the quotient topology, that is, the finest topology which makes the stalkwise defined quotient mapping $q : \mathcal{S} \to \mathcal{S}/\mathcal{S}'$ continuous. Then q is a sheaf homomorphism and we have the exact sequence

$$0 \to \mathcal{S}' \xrightarrow{\iota} \mathcal{S} \xrightarrow{q} \mathcal{S}/\mathcal{S}' \to 0.$$

Lemma 5.5 *If $s_1, s_2 \in \Gamma(Y, \mathcal{S})$ satisfy $s_1(x_0) = s_2(x_0)$, then there exists a neighborhood W of x_0 such that $s_1(x) = s_2(x)$ for all $x \in W$.*

Proof. We set $s_1(x_0) = s_2(x_0) = z_0$. Then there exists a neighborhood U of z_0 such that $\pi : U \to \pi(U)$ is a homeomorphism. We set $W = s_1^{-1}(U) \cap s_2^{-1}(U)$. Then W is a neighborhood of x_0 and $\pi \circ s_1|_W = \pi \circ s_2|_W = I_W$, which implies that $s_1 = s_2 = \pi^{-1}|_W$. □

Lemma 5.6 *Let $\Omega \subset \mathbf{C}^n$ be an open set. For a holomorphic function f in Ω, define $s_f : \Omega \to \mathcal{O}$ by $s_f(a) = \mathbf{f}_a$ for $a \in \Omega$. Then s_f is continuous. Moreover, $f \in \mathcal{O}(\Omega) \to s_f \in \Gamma(\Omega, \mathcal{O})$ is bijective.*

Proof. Let $a \in \Omega$. For a neighborhood of a, we set $U_f = \{\mathbf{f}_z \mid z \in U\}$. Then we have $s_f^{-1}(U_f) = U$, and hence s_f is continuous. Since $\pi \circ s|_\Omega = I_\Omega$, we have $s_f \in \Gamma(\Omega, \mathcal{O})$. If $f_1 \neq f_2$, then there exist $z \in \Omega$ such that $s_{f_1}(z) \neq s_{f_2}(z)$, which means that $s_{f_1} \neq s_{f_2}$. Next, assume that $s \in \Gamma(\Omega, \mathcal{O})$. For $a \in \Omega$, we have $s(a) = \mathbf{f}_a$. Let (f, U) be a representative of $\mathbf{f}_a$. Since s is continuous, there exists a neighborhood W of a such that $s(z) = \mathbf{f}_z$ for $z \in W$. Hence there exists $f \in \mathcal{O}(\Omega)$ such that $s = s_f$. □

Definition 5.18 We say that a commutative ring A with unit is Noetherian if every ideal $I \subset A$ is finitely generated, that is, if there exist elements $f_1, \cdots, f_j \in I$ so that every $f \in I$ can be written

$$f = \sum_{i=1}^{j} a_i f_i$$

for some $a_1, \cdots, a_j \in A$.

Theorem 5.13 $\mathcal{O}_0$ *is a Noetherian ring.*

Proof. In case $n = 1$, Theorem 5.13 is trivial since every ideal in $\mathcal{O}_0$ is generated by a power of z using the Taylor expansion. Assume that Theorem 5.13 has already been proved for the ring $\mathcal{O}_0(\mathbf{C}^{n-1})$. Suppose I is an ideal in $\mathcal{O}_0$ which contains some non-zero element. Let $f \in I$ be a non-zero element. Then by a change of coordinates we may assume that f is regular of order k in z_n. For $g \in I$, by the Weierstrass division theorem we have a representation

$$g = qf + r,$$

where r is a polynomial in z_n of degree less than k. Let M be a set $g \in I$ such that g is a polynomial in z_n of degree less than k. Then M is regarded a submodule in $\mathcal{O}_0(\mathbf{C}^{n-1})^p$. By the inductive hypothesis, $\mathcal{O}_0(\mathbf{C}^{n-1})$ is a Noetherian ring, and hence M is finitely generated. If $f_1, \cdots, f_r$ is the generators for M, then $f_1, \cdots, f_r, f$ generate I. Consequently, $\mathcal{O}_0$ is a Noetherian ring. $\square$

Lemma 5.7 *Let f, g and ω be holomorphic functions in a neighborhood of $0 \in \mathbf{C}^n$. Suppose ω is a Weierstrass polynomial in z_n and f is a polynomial in z_n. If*

$$f = g\omega,$$

then g is a polynomial in z_n.

Proof. Since the coefficient of the term of ω of the maximal degree in z_n equals 1, f is expressed by

$$f = q\omega + r,$$

where q and r are polynomials in z_n and the degree of r is less than the degree of ω. By the uniqueness of the Weierstrass division theorem we have $r = 0$, $q = g$. $\square$

In order to prove Oka's fundamental theorem, we need the following lemma.

Lemma 5.8 *Let $\{f_\lambda\}$ be a sequence of at most countable non-zero holomorphic functions in a neighborhood U of $0 \in \mathbf{C}^n$. Then there exists an* *invertible linear change of variables*

$$z_j = \sum_{k=1}^{n} \alpha_{jk} z_k^* \qquad (j = 1, \cdots, n)$$

such that for all λ, $f_\lambda^*(z^*) = f_\lambda(z)$ *satisfy the following properties:*

$$f_\lambda^*(z_1^*, 0, \cdots, 0) \not\equiv 0, \quad f_\lambda^*(0, z_2^*, 0, \cdots, 0) \not\equiv 0, \quad \cdots, \quad f_\lambda^*(0, \cdots, 0, z_n^*) \not\equiv 0. \tag{5.3}$$

Proof. In the power series expansion of f_λ

$$f_\lambda(z_1, \cdots, z_n) = \sum_{\nu_1 \cdots \nu_n = 0}^{\infty} a_{\nu_1 \cdots \nu_n}^{(\lambda)} z_1^{\nu_1} \cdots z_n^{\nu_n},$$

we rewrite the right side of the above equality by the series of homogeneous polynomials so that s_λ is the least homogeneous degree. Since

$$f_\lambda^*(0, \cdots, 0, z_k^*, 0, \cdots, 0) = \sum_{\nu_1 \cdots \nu_n = 0}^{\infty} a_{\nu_1, \cdots, \nu_n}^{(\lambda)} (\alpha_{1k} z_k^*)^{\nu_1} \cdots (\alpha_{nk} z_k^*)^{\nu_n},$$

it is sufficient to choose α_{jk} $(j, k = 1, \cdots, n)$ with the following properties:

$$\sum_{\nu_1 + \cdots + \nu_n = s_\lambda} a_{\nu_1, \cdots, \nu_n}^{(\lambda)} (\alpha_{1k})^{\nu_1} \cdots (\alpha_{nk})^{\nu_n} \neq 0 \qquad (k = 1, \cdots, n),$$

$$\det(\alpha_{jk}) \neq 0.$$

By the Baire theorem (Lemma 1.4), there exist α_{jk} $(j, k = 1, \cdots, n)$ which satisfy the above properties. $\square$

Theorem 5.14 *(Oka's fundamental theorem) Suppose* Ω *is an open set in* $\mathbf{C}^n$. *For* $z \in \Omega$ *and* $F_1, \cdots, F_q \in \mathcal{O}(\Omega)^p$, *define a submodule* $R_z(F_1, \cdots, F_q)$ *of* $\mathcal{O}_z^q$ *as follows:*

$$R_z(F_1, \cdots, F_q) = \{G = (g^1, \cdots, g^q) \in \mathcal{O}_z^q \mid \sum_{j=1}^{q} g^j \gamma_z(F_j) = 0\}.$$

Given $z_0 \in \Omega$, *one can then find a neighborhood* $\omega \subset \Omega$ *of* z_0 *and finitely many elements* $G_1, \cdots, G_r \in \mathcal{O}(\omega)^q$ *such that for any* $z \in \omega$, $R_z(F_1, \cdots, F_q)$ *is generated by* $\gamma_z(G_1), \cdots, \gamma_z(G_r)$.

Proof. We assume $z_0 = 0$.

(a) Suppose that $p > 1$. Assume that the theorem has already been proved for $p-1$. Let $F_j = (F_j^1, \cdots, F_j^p)$. Evidently we have

$$R_z(F_1, \cdots, F_q) \subset R_z(F_1^1, \cdots, F_q^1).$$

By the inductive hypothesis, there exist a neighborhood ω' of 0 and $H_1, \cdots, H_r \in \mathcal{O}(\omega')^q$ such that for any $z \in \omega'$, $R_z(F_1^1, \cdots, F_q^1)$ is generated by $\gamma_z(H_1), \cdots, \gamma_z(H_r)$. Consequently, we have

$$R_z(F_1, \cdots, F_q) \subset \left\{ \sum_{j=1}^r c^j \gamma_z(H_j) \mid c^j \in \mathcal{O}_z \right\} \qquad (z \in \omega').$$

Let $H_j = (H_j^1, \cdots, H_j^q)$. Then $\sum_{j=1}^r c^j \gamma_z(H_j) \in R_z(F_1, \cdots, F_q)$ if and only if

$$\sum_{k=1}^q \sum_{j=1}^r c^j \gamma_z(H_j^k) \gamma_z(F_k) = 0. \tag{5.4}$$

(5.4) is equivalent to the following equations.

$$\sum_{k=1}^q \sum_{j=1}^r c^j \gamma_z(H_j^k F_k^i) = 0 \qquad (i = 1, \cdots, p). \tag{5.5}$$

Since $\gamma_z(H_1), \cdots, \gamma_z(H_r) \in R_z(F_1^1, \cdots, F_q^1)$, we obtain

$$\sum_{k=1}^q \gamma_z(H_j^k) \gamma_z(F_k^1) = \sum_{k=1}^q \gamma_z(H_j^k F_k^1) = 0 \qquad (j = 1, \cdots, r).$$

Hence (5.5) holds when $i = 1$. We set

$$L_j = \left(\sum_{k=1}^q H_j^k F_k^2, \cdots, \sum_{k=1}^q H_j^k F_k^p \right).$$

It follows from (5.5) that $\sum_{j=1}^r c^j \gamma_z(L_j) = 0$. By the inductive hypothesis, there exist a neighborhood ω of 0 and $K_1, \cdots, K_s \in \mathcal{O}(\omega)^r$ such that for $z \in \omega$, $(c_1, \cdots, c_r)$ is generated by $\gamma_z(K_1), \cdots, \gamma_z(K_s)$. Hence there exist $\alpha_k \in \mathcal{O}_z$ for $k = 1, \cdots, s$ such that

$$c^j = \sum_{k=1}^s \alpha_s \gamma_z(K_k^j).$$

Therefore every element of $R_z(F_1,\cdots,F_q)$ has a representation

$$\sum_{j=1}^{r} c^j \gamma_z(H_j) = \sum_{j=1}^{r}\sum_{k=1}^{s} \alpha_k \gamma_z(K_k^j)\gamma_z(H_j),$$

which implies that if we set

$$G_k = \sum_{j=1}^{r} K_k^j H_j \qquad (k=1,\cdots,s),$$

then for $z \in \omega$, $R_z(F_1,\cdots,F_q)$ is generated by $\gamma_z(G_1),\cdots,\gamma_z(G_s)$.

(b) When $n = 0$, the theorem holds for every p. Assume that the theorem has already been proved for $n-1$ dimension and for all p. We will prove the theorem for n dimension and for all p. In (a) we have proved that if the theorem is true for $p-1$, then the theorem is true for p when $p > 1$. Hence it is sufficient to prove the theorem when $p = 1$. By Lemma 5.8, without loss of generality we may assume that $F_1,\cdots,F_q$ are Weierstrass polynomials in z_n at 0. We denote by μ the maximum of degrees in z_n of $F_1,\cdots,F_q$. We assume that μ is the degree in z_n of F_q. We prove the following lemma.

Lemma 5.9 *Let $\zeta = (\zeta',\zeta_n) \in \mathbf{C}^n$. Then $R_\zeta(F_1,\cdots,F_q)$ is generated by finitely many elements whose components are gems of functions in ${}_{n-1}\mathcal{O}_{\zeta'}[z_n]$ with a degree in z_n which does not exceed μ.*

Proof of Lemma 5.9 By the Weierstrass preparation theorem we have

$$\gamma_\zeta(F_q) = F'F'',$$

where F' is a germ of a Weierstrass polynomial in $z - \zeta$ and F'' is a germ of holomorphic functions which do not vanish at ζ. Since

$$F_q = z_n^\mu + a_{\mu-1}(z')z_n^{\mu-1} + \cdots + a_0(z'),$$

F_q is a polynomial in $z_n - \zeta_n$. By Lemma 5.7, F'' is a polynomial in $z_n - \zeta_n$, and hence a polynomial in z_n with leading coefficient equal to 1. We denote degrees of F' and F'' in z_n by μ' and μ'', respectively. Let $(c_1,\cdots,c^q) \in R_\zeta(F_1,\cdots,F_q)$. Then by the Weierstrass division theorem we have

$$c^i = t'^i F' + r^i = \gamma_\zeta(F_q) t^i + r^i \qquad (i=1,\cdots,q-1),$$

where $t'^i, t^i, r^i \in \mathcal{O}_\zeta$ and each r_i is a polynomial in z_n with the degree less than μ'. We set

$$r^q = c^q - \sum_{i=1}^{q-1} \gamma_\zeta(F_i) t^i.$$

Then we obtain

$$(c^1, \cdots, c^q) = \gamma_\zeta(F_q, 0, \cdots, 0, -F_1) t^1 + \gamma_\zeta(0, F_q, 0, \cdots, 0, -F_2) t^2 \quad (5.6)$$

$$+ \cdots + \gamma_\zeta(0, \cdots, 0, F_q, -F_{q-1}) t^{q-1} + (r^1, \cdots, r^q).$$

Consequently we have $(r^1, \cdots, r^q) \in R_\zeta(F_1, \cdots, F_q)$. Hence we have

$$\sum_{i=1}^{q} r^i \gamma_\zeta(F_i) = \sum_{i=1}^{q-1} r^i \gamma_\zeta(F_i) + (r^q F'') F' = 0.$$

Since the degree of $\sum_{i=1}^{q-1} r^i \gamma_\zeta(F_i)$ in z_n is less than $\mu + \mu'$, by Lemma 5.7 the degree of $r^q F''$ in z_n is less than μ. In the equality

$$(r^1, \cdots, r^q) = \frac{1}{F''}(F'' r^1, \cdots, F'' r^q),$$

the degrees of $F'' r^j$ ($j = 1, \cdots, q$) in z_n are less than μ. Hence Lemma 5.9 follows from (5.6), which completes the proof of Lemma 5.9.

End of the proof of (b) Let $(c^1, \cdots, c^q)$ be one of the elements in $R_\zeta(F_1, \cdots, F_q)$ described in Lemma 5.9. Then we have

$$c^j = \sum_{k=0}^{\mu} c^{jk} \gamma_\zeta(z_n^k) \quad c^{jk} \in \mathcal{O}_{\zeta'}.$$

Since $(c^1, \cdots, c^q) \in R_\zeta(F_1, \cdots, F_q)$, we have

$$\sum_{j=1}^{q} \sum_{k=0}^{\mu} c^{jk} \gamma_\zeta(z_n^k) \gamma_\zeta(F_j) = 0.$$

Consequently, we have

$$\sum_{k=0}^{\mu} (c^{1k} \gamma_\zeta(F_1) + \cdots + c^{qk} \gamma_\zeta(F_q)) \gamma_\zeta(z_n^k) = 0. \quad (5.7)$$

Let

$$F_j(z) = a_{j\mu}(z') z^\mu + a_{j\mu-1}(z') z^{\mu-1} + \cdots + a_{j0}(z').$$

Since coefficients in z_n^k for $k = 0, \cdots, 2\mu$ in (5.7) are 0, the coefficient of z_n^μ in (5.7) is equal to 0. Hence we have

$$c^{10} a_{1\mu} + \cdots + c^{q0} a_{q\mu} + c^{11} a_{1\mu-1} + \cdots$$

$$+ c^{q1} a_{q\mu-1} + \cdots + c^{1\mu} a_{1,0} + \cdots + c^{q\mu} a_{q0} = 0.$$

By the inductive hypothesis, there exist a neighborhood ω' of $0 \in \mathbf{C}^{n-1}$ and $C_{1k}, \cdots, C_{r_k k} \in \mathcal{O}(\omega')^q$, $k = 0, \cdots, \mu$, such that $(c^{1k}, \cdots, c^{qk})$ is generated by $C_{1k}, \cdots, C_{r_k k}$. Since

$$(c^1, \cdots, c^q) = \sum_{k=0}^{\mu} (c^{1k}, \cdots, c^{qk}) \gamma_\zeta(z_n^k),$$

$R_\zeta(F_1, \cdots, F_q)$ is generated by germs of $C_{1k} z_n^k, \cdots, C_{r_k k} z_n^k$ ($k = 0, \cdots, \mu$) for $\zeta = (\zeta', \zeta_n)$ with $\zeta' \in \omega'$. This proves (b). □

Definition 5.19 An analytic sheaf $\mathcal{S}$ on the complex manifold Ω is said to be locally finitely generated if for every $z \in \Omega$ there exists a neighborhood ω of z and a finite number of sections $f_1, \cdots, f_q \in \Gamma(\omega, \mathcal{S})$ such that $\mathcal{S}_\zeta$ is generated by $\gamma_\zeta(f_1), \cdots, \gamma_\zeta(f_q)$ as an $\mathcal{O}_\zeta$ module for every $\zeta \in \omega$.

Lemma 5.10 *Suppose that an analytic sheaf $\mathcal{S}$ is locally finitely generated. Let $f_1, \cdots, f_q$ be sections of $\mathcal{S}$ in a neighborhood of z such that $\gamma_z(f_1), \cdots, \gamma_z(f_q)$ generate $\mathcal{S}_z$. Then $\gamma_\zeta(f_1), \cdots, \gamma_\zeta(f_q)$ generate $\mathcal{S}_\zeta$ for every ζ in a neighborhood of z.*

Proof. Since $\mathcal{S}$ is locally finitely generated, there exist a neighborhood ω of z and $g_1, \cdots, g_r \in \Gamma(\omega, \mathcal{S})$ such that for any $\zeta \in \omega$, $\gamma_\zeta(g_1), \cdots, \gamma_\zeta(g_r)$ generate $\mathcal{S}_\zeta$. On the other hand, by the hypothesis we have

$$\gamma_z(g_i) = \sum_{j=1}^{q} \gamma_z(c_{ij}) \gamma_z(f_j) \qquad (i = 1, \cdots, r).$$

By Lemma 5.5 there exists a neighborhood W of z such that for $\zeta \in W$,

$$\gamma_\zeta(g_i) = \sum_{j=1}^{q} \gamma_\zeta(c_{ij}) \gamma_\zeta(f_j) \qquad (i = 1, \cdots, r).$$

□

Definition 5.20 Let $\mathcal{S}$ be an analytic sheaf on the complex manifold Ω and let ω be an open subset of Ω. For $f_1, \cdots, f_q \in \Gamma(\omega, \mathcal{S})$, define the sheaf homomorphism $h : \mathcal{O}^q \to \mathcal{S}$ by

$$\mathcal{O}^q \supset \mathcal{O}^q_z \ni (g^1, \cdots, g^q) \xrightarrow{h} \sum_{j=1}^{q} g^j \gamma_z(f_j) \in \mathcal{S}_z \subset \mathcal{S}.$$

The subsheaf $\mathcal{R}(f_1, \cdots, f_q)$ of $\mathcal{O}^q$ is defined by

$$\mathcal{R}(f_1, \cdots, f_q) = \bigcup_{z \in \omega} \{(g^1, \cdots, g^q) \in \mathcal{O}_z \mid h(g^1, \cdots, g^q) = 0\},$$

and is called the sheaf of relations between $f_1, \cdots, f_q$.

Definition 5.21 Let $\mathcal{S}$ be an analytic sheaf on the complex manifold Ω. $\mathcal{S}$ is called coherent if

(1) $\mathcal{S}$ is locally finitely generated.
(2) If ω is an open subset of Ω and $f_1, \cdots, f_q \in \Gamma(\omega, \mathcal{S})$, then the sheaf of relations $\mathcal{R}(f_1, \cdots, f_q)$ is locally finitely generated.

Theorem 5.15 *Every locally finitely generated subsheaf of $\mathcal{O}^p$ is coherent.*

Proof. We show (2) in the definition of the coherent sheaf. Since $f_1, \cdots, f_q \in \mathcal{O}(\omega)^p$, by Oka's fundamental theorem (Theorem 5.14) the sheaf of relations $\mathcal{R}(f_1, \cdots, f_q)$ is locally finitely generated. $\square$

Theorem 5.16 *Let $\mathcal{S}$ be a coherent sheaf on the complex manifold Ω and let ω be an open subset of Ω. If $f_1, \cdots, f_q \in \Gamma(\omega, \mathcal{S})$, then the sheaf of relations $\mathcal{R}(f_1, \cdots, f_q)$ is also coherent.*

Proof. Since $\mathcal{S}$ is coherent, $\mathcal{R}(f_1, \cdots, f_q)$ is locally finitely generated. Theorem 5.16 follows from Theorem 5.15 and the fact that $\mathcal{R}(f_1, \cdots, f_q)$ is a subsheaf of $\mathcal{O}^q$. $\square$

Example 5.1 There is a subsheaf of $\mathcal{O}$ which is not coherent.

Proof. Suppose that ω and Ω are open sets in $\mathbf{C}$ with $\phi \neq \omega \subset \Omega$, $\omega \neq \Omega$. Define

$$\mathcal{S}_z = \begin{cases} \mathcal{O}_z & (z \in \omega) \\ 0 & (z \in \Omega \setminus \omega) \end{cases}.$$

Then $\mathcal{S}$ is a subsheaf of $\mathcal{O}$. Every section of $\mathcal{S}$ over a connected open set which intersects $\Omega \setminus \omega$ must be 0 (see Exercise 1.5). Hence if $\mathcal{S}$ is finitely generated in some connected neighborhood of a boundary point of ω, then we have $\mathcal{S}_z = 0$ in the neighborhood, which is a contradiction. $\square$

Definition 5.22 Let X be a paracompact Hausdorff space and let $\mathcal{S}$ be a sheaf of Abelian groups in X. Let $\mathcal{U} = \{U_j \mid j \in J\}$ be an open cover of X and let q be a nonnegative integer. We say that c is a q-cochain for $\mathcal{U}$ with coefficients in $\mathcal{S}$ if c is a mapping which assigns to each $(q+1)$ tuple $(j_0, j_1, \cdots, j_q) \in J^{q+1}$ with $U_{j_0} \cap \cdots \cap U_{j_q} \neq \phi$ a section $c_{j_0, j_1, \cdots, j_q} \in \Gamma(U_{j_0} \cap \cdots \cap U_{j_q}, \mathcal{S})$. Define $c_{j_0, \cdots, j_q} = \varepsilon c_{i_0, \cdots, i_q}$, where $\varepsilon = \pm 1$ is a sign of the permutation

$$\begin{pmatrix} j_0, j_1, \cdots, j_q \\ i_0, i_1, \cdots, i_q \end{pmatrix}.$$

We denote by $C^q(\mathcal{U}, \mathcal{S})$ the set of all q-cochains. A coboundary mapping $\delta_q : C^q(\mathcal{U}, \mathcal{S}) \to C^{q+1}(\mathcal{U}, \mathcal{S})$ is defined as follows:

$$(\delta_q c)_{j_0 \cdots j_{q+1}} = \sum_{k=0}^{q+1} (-1)^k c_{j_0 \cdots \hat{j}_k \cdots j_{q+1}},$$

where $j_0 \cdots \hat{j}_k \cdots j_{q+1}$ means that j_k is omitted. By definition we have

$$\delta_{q+1} \circ \delta_q = 0 \quad (q \geq 0). \tag{5.8}$$

Define

$$Z^q(\mathcal{U}, \mathcal{S}) = \{c \in C^q(\mathcal{U}, \mathcal{S}) \mid \delta_q c = 0\} \quad (q \geq 0)$$

and

$$B^q(\mathcal{U}, \mathcal{S}) = \{\delta_{q-1} c \mid c \in C^{q-1}(\mathcal{U}, \mathcal{S})\} \quad (q \geq 1).$$

An element $Z^q(\mathcal{U}, \mathcal{S})$ is called a q-cocycle and an element of $B^q(\mathcal{U}, \mathcal{S})$ is called a q-coboundary. Define $B^0(\mathcal{U}, \mathcal{S}) = 0$. It follows from (5.8) that $B^q(\mathcal{U}, \mathcal{S}) \subset Z^q(\mathcal{U}, \mathcal{S})$. Define

$$H^q(\mathcal{U}, \mathcal{S}) := Z^q(\mathcal{U}, \mathcal{S}) / B^q(\mathcal{U}, \mathcal{S}).$$

$H^q(\mathcal{U}, \mathcal{S})$ is called a q-th Čech cohomology group of $\mathcal{U}$ with coefficients in $\mathcal{S}$.

Definition 5.23 Let $\mathcal{V} = \{V_i \mid i \in I\}$ be a refinement of $\mathcal{U}$, that is, there exists a mapping $\tau : I \to J$ such that $V_i \subset U_{\tau(i)}$ for $i \in I$. Define

$$\tau_q^* : C^q(\mathcal{U}, \mathcal{S}) \to C^q(\mathcal{V}, \mathcal{S}) \quad (q \geq 0)$$

by

$$(\tau_q^*(c))_{i_0 \cdots i_q} = c_{\tau(i_0) \cdots \tau(i_q)} |_{V_{i_0} \cap \cdots \cap V_{i_q}}.$$

Since $\tau_{q+1}^* \circ \delta_q = \delta_q \circ \tau_q^*$, we define

$$\rho_q^{\mathcal{U}\mathcal{V}} : H^q(\mathcal{U}, \mathcal{S}) \to H_q(\mathcal{V}, \mathcal{S})$$

by

$$\rho_q^{\mathcal{U}\mathcal{V}}([c]) = [\tau_q^* c].$$

We have the following lemma. We omit the proof.

Lemma 5.11 $\rho_q^{\mathcal{U}\mathcal{V}}$ *is independent of the choice of* τ.

Definition 5.24 For two open covers $\mathcal{U}$, $\mathcal{W}$ of X, we say that $[c] \in H^q(\mathcal{U}, \mathcal{S})$ and $[d] \in H^q(\mathcal{W}, \mathcal{S})$ are equivalent if there exists a refinement $\mathcal{V}$ of $\mathcal{U}$ and $\mathcal{W}$ such that

$$\rho_q^{\mathcal{U}\mathcal{V}}([c]) = \rho_q^{\mathcal{W}\mathcal{V}}([d]).$$

We denote by $H^q(X, \mathcal{S})$ the set of all equivalent classes by this equivalent relation. $H^q(X, \mathcal{S})$ is an Abelian group. $H^q(X, \mathcal{S})$ is called the q-th Čech cohomology group of X with coefficients in $\mathcal{S}$.

By definition we have the following lemma.

Lemma 5.12 *For an open cover* $\mathcal{U}$ *of* X, *we have*

$$H^0(X, \mathcal{S}) = H^0(\mathcal{U}, \mathcal{S}) = Z^0(\mathcal{U}, \mathcal{S}) = \Gamma(X, \mathcal{S}).$$

Definition 5.25 Let $\varphi : \mathcal{S}' \to \mathcal{S}$ be a sheaf homomorphism and $\mathcal{U}$ an open cover of X. Define $\varphi : C^q(\mathcal{U}, \mathcal{S}') \to C^q(\mathcal{U}, \mathcal{S}')$ by

$$\varphi(c) = \varphi \circ c.$$

Moreover, we define $\varphi_\mathcal{U}^q$ and φ^q using φ such that

$$\varphi_\mathcal{U}^q : H^q(\mathcal{U}, \mathcal{S}') \to H^q(\mathcal{U}, \mathcal{S})$$

and

$$\varphi^q : H^q(X, \mathcal{S}') \to H^q(X, \mathcal{S}).$$

Then we have the following theorem.

Theorem 5.17 *Suppose*

$$0 \to \mathcal{S}' \xrightarrow{\varphi} \mathcal{S} \xrightarrow{\psi} \mathcal{S}'' \to 0$$

is an exact sequence of sheaf homomorphisms over X and $H^1(X, \mathcal{S}') = 0$. Then

$$\psi^0 : \Gamma(X, \mathcal{S}) \to \Gamma(X, \mathcal{S}'')$$

is surjective.

Proof. Let $s'' \in \Gamma(X, \mathcal{S}'')$. For $x \in X$, we have $s''(x) \in \mathcal{S}''_x$. Since ψ is surjective, there exists $s_x \in \mathcal{S}_x$ such that $\psi(s_x) = s''(x)$. There exist a neighborhood W of x and a section $\hat{s}_x \in \Gamma(W, \mathcal{S})$ such that $\hat{s}_x(x) = s_x$ (see Exercise 5.2). Consequently, we have $\psi \circ \hat{s}_x(x) = s''(x)$. By Lemma 5.5 there exists a neighborhood $U_x \subset W$ of x such that $\psi \circ \hat{s}_x = s''$ in U_x. We set $\mathcal{U} = \{U_x \mid x \in X\}$. Then we have $\hat{s} = \{s_x\} \in C^0(\mathcal{U}, \mathcal{S})$. Since $\psi \circ (\delta_0 \hat{s}) = \delta_0(\psi \circ \hat{s}) = \delta_0 s'' = 0$, we have $\delta_0 \hat{s} \in B(\mathcal{U}, \operatorname{Ker}\psi)$. Since $\operatorname{Im} \varphi = \operatorname{Ker} \psi$, there exist a refinement $\mathcal{V} = \{V_x \mid x \in X\}$ ($V_x \subset U_x$) of $\mathcal{U}$ and $s' \in C^1(\mathcal{V}, \mathcal{S}')$ such that

$$\varphi \circ s' = \delta_0 \hat{s}.$$

Consequently,

$$\varphi \circ (\delta_1 s') = \delta_1(\varphi \circ s') = \delta_1(\delta_0 \hat{s}) = 0.$$

Since φ is injective, we have $\delta_1 s' = 0$, which implies that $s' \in Z^1(\mathcal{V}, \mathcal{S}')$. Since $H^1(X, \mathcal{S}') = 0$, taking a refinement of $\mathcal{V}$ if necessary, we may assume that $s' \in B^1(\mathcal{V}, \mathcal{S}')$. Hence there exists $g \in C^0(\mathcal{V}, \mathcal{S}')$ such that $s' = \delta_0 g$. Thus we have

$$\delta_0 \hat{s} = \varphi \circ s' = \varphi \circ (\delta_0 g) = \delta_0(\varphi \circ g).$$

If we set $s = \hat{s} - \varphi \circ g$, then we have $\delta_0 s = 0$, and hence $s \in \Gamma(X, \mathcal{S})$. Moreover, we have

$$\psi \circ s = \psi \circ \hat{s} - \psi \circ \varphi \circ g = s'',$$

which means that $\psi^0(s) = s''$. Hence ψ^0 is surjective. □

Definition 5.26 A σ compact complex manifold Ω is said to be a Stein manifold if

(a) Ω is holomorphically convex, that is, for any compact subset K of Ω,

$$\hat{K}_\Omega^\mathcal{O} = \{z \in \Omega \mid |f(z)| \leq \sup_K |f| \text{ for all } f \in \mathcal{O}(\Omega)\}$$

is compact.

(b) For $z_1, z_2 \in \Omega$ $z_1 \neq z_2$, there exists $f \in \mathcal{O}(\Omega)$ such that $f(z_1) \neq f(z_2)$.
(c) For every $z \in \Omega$, one can find n functions $f_1, \cdots, f_n \in \mathcal{O}(\Omega)$ which form a coordinate system at z.

Remark 5.1 *Every pseudoconvex domain in $\mathbf{C}^n$ satisfies (a), (b) and (c), and hence it is a Stein manifold.*

Theorem 5.18 *Every submanifold of a Stein manifold is a Stein manifold.*

Proof. Let V be a submanifold of a Stein manifold Ω. Since $\mathcal{O}(\Omega) \subset \mathcal{O}(V)$, we have $\hat{K}_V^\mathcal{O} \subset \hat{K}_\Omega^\mathcal{O}$. Hence V is holomorphically convex. This proves (a). (b) is trivial. Let $v \in V$. Then there exist a neighborhood ω of v and a local coordinate system $z_1, \cdots, z_n$ in ω such that

$$\omega \cap V = \{w \in \omega \mid z_{m+1}(w) = \cdots = z_n(w) = 0\}.$$

Let $f_1, \cdots, f_n \in \mathcal{O}(\Omega)$ be a coordinate system at v. Then at $z(v)$ we have

$$\det\left(\frac{\partial f_i}{\partial z_j}\right) \neq 0 \quad (i, j = 1, \cdots, n).$$

Hence we can choose $i_1, \cdots, i_m$ such that

$$\det\left(\frac{\partial f_{i_\mu}}{\partial z_j}\right) \neq 0 \quad (\mu, j = 1, \cdots, m).$$

Thus the restrictions of $f_{i_1}, \cdots, f_{i_m}$ to V form a local coordinate system at v. □

Definition 5.27 We say that a subset A of a Stein manifold Ω is an analytic subset if A is a closed subset of Ω and for any $p \in A$ there exist a neighborhood U_p of p and holomorphic functions $h_1, \cdots, h_{k_p}$ in U_p such that

$$U_p \cap A = \{z \in U_p \mid h_1(z) = \cdots = h_{k_p}(z) = 0\}.$$

Definition 5.28 Let A be an analytic subset of a Stein manifold Ω. A continuous function $f : A \to \mathbf{C}$ is said to be holomorphic in A if for any $a \in A$ there exist a neighborhood U_a of a in Ω and a holomorphic function h_a in U_a such that $f(z) = h_a(z)$ for all $z \in A \cap U_a$.

Definition 5.29 Let A be an analytic subset of a Stein manifold Ω. We define a subsheaf $\mathcal{F}_A$ of $\mathcal{O}$ in such a way that $(\mathcal{F}_A)_z = \mathcal{O}_z$ for $z \notin A$, and $(\mathcal{F}_A)_z = \{\mathbf{f}_z \in \mathcal{O}_z \mid f|_A = 0\}$ for $z \in A$. $\mathcal{F}_A$ is called the sheaf of ideals of the analytic subset A.

We have the following theorem. The proof is omitted (see Gunning-Rossi [GUR]).

Theorem 5.19 *Every sheaf of ideals $\mathcal{I}$ of an analytic subset of a Stein manifold is coherent.*

We omit the proof of the following lemma (see Gunning-Rossi [GUR]).

Lemma 5.13 *If in an exact sequence of sheaves*
$$0 \to \mathcal{S}' \to \mathcal{S} \to \mathcal{S}'' \to 0$$
any two of the sheaves $\mathcal{S}'$, $\mathcal{S}$, $\mathcal{S}''$ are coherent sheaves, then the third is also coherent.

The following theorem is known as Theorem A and Theorem B of Cartan. We omit the proof (see Hörmander [HR2], Gunning-Rossi [GUR]).

Theorem 5.20 *Let Ω be a Stein manifold and $\mathcal{A}$ a coherent sheaf over Ω. Then*

(a) (**Cartan theorem A**) *Let $z \in \Omega$. For any $s \in \mathcal{A}_z$ there exist $f_1, \cdots, f_k \in \Gamma(\Omega, \mathcal{A})$ and $s_1, \cdots, s_k \in \mathcal{O}_z$ such that*
$$s = \sum_{j=1}^{k} s_j (f_j)_z.$$

(b) (**Cartan theorem B**) *$H^q(\Omega, \mathcal{A}) = 0 \quad (q \geq 1)$.*

Corollary 5.5 *Let Ω be a Stein manifold and let A be an analytic subset of Ω. Then for any $z \notin A$ there exists $f \in \mathcal{O}(\Omega)$ such that $f(z) \neq 0$, $f|_A = 0$.*

Proof. Let $z \notin A$. Then there exist a neighborhood U of z and a holomorphic function s in U such that $s(z) \neq 0$, $s|_A = 0$. Let $\mathcal{F}_A$ be the sheaf of ideals of A. By the Cartan theorem A, there exist holomorphic functions $s_1, \cdots, s_k$ in a neighborhood of z and $f_1, \cdots, f_k \in \Gamma(\Omega, \mathcal{F}_A)$ such that
$$s = \sum_{j=1}^{k} s_j (f_j)_z.$$
Hence there exists j_0 with $1 \leq j_0 \leq k$ such that $f_{j_0}(z) \neq 0$. □

Corollary 5.6 *Let Ω be a Stein manifold and let A be an analytic subset of Ω. Then every holomorphic function in A can be extended to a holomorphic function in Ω.*

Proof. We denote by $\mathcal{F}_A$ the sheaf of ideals of A. $\Gamma(\Omega, \mathcal{O}(\Omega)/\mathcal{F}_A)$ can be regarded as the set of all holomorphic functions in A. Since $\mathcal{F}_A$ is coherent, we have $H^1(\Omega, \mathcal{F}_A) = 0$. By the exact sequence of sheaves

$$0 \to \mathcal{F}_A \to \mathcal{O}(\Omega) \to \mathcal{O}(\Omega)/\mathcal{F}_A \to 0,$$

$\Gamma(\Omega, \mathcal{O}(\Omega)) \to \Gamma(\Omega, \mathcal{O}(\Omega)/\mathcal{F}_A)$ is surjective, which means that for a holomorphic function f in A there exists $F \in \Gamma(\Omega, \mathcal{O}(\Omega))$ with $F|_A = f$. □

Definition 5.30 Let X be a complex manifold. An open cover $\mathcal{U} = \{U_i\}_{i \in I}$ of X is said to be a Stein cover if $\mathcal{U}$ is a locally finite cover and each U_i is a Stein open set.

The following theorem holds. We omit the proof (see Grauert and Remmert [GRR]).

Theorem 5.21 Let X be a complex manifold and let $\mathcal{U}$ be a Stein cover of X, $\mathcal{S}$ a coherent sheaf over X. Then

$$H^q(\mathcal{U}, \mathcal{S}) = H^q(X, \mathcal{S}) \qquad (q \geq 0).$$

The following theorem follows from Theorem 5.19, Theorem 5.20 and Theorem 5.21.

Theorem 5.22 Let $\Omega \subset \mathbf{C}^n$ be a pseudoconvex domain. Let A be an analytic subset of Ω and let $\mathcal{F}_A$ be the sheaf of ideals of A, $\{U_j\}_{j \in I}$ a Stein cover of Ω. Suppose $f_{ij} \in \Gamma(U_i \cap U_j, \mathcal{F}_A)$ satisfy the equalities

$$f_{ij}(z) + f_{jk}(z) + f_{ki}(z) = 0 \qquad (z \in U_i \cap U_j \cap U_k,\ i,j,k \in I).$$

Then there exist $f_j \in \Gamma(U_j, \mathcal{F}_A)$ such that

$$f_{ij}(z) = f_i(z) - f_j(z) \qquad (z \in U_i \cap U_j,\ i,j \in I).$$

Theorem 5.23 Let $\mathcal{A}$ be a coherent analytic sheaf over a Stein manifold Ω and let $\mathcal{S}$ be a subsheaf of $\mathcal{A}$. Let $s_1, \cdots, s_k \in \Gamma(\Omega, \mathcal{A})$. Suppose that $s_1, \cdots, s_k$ generate $\mathcal{S}_z$ over $\mathcal{O}_z$ for each $z \in \Omega$. Then for $s \in \Gamma(\Omega, \mathcal{S})$, there exist $f_1, \cdots, f_k \in \Gamma(\Omega, \mathcal{O})$ such that

$$s = \sum_{j=1}^{k} f_j s_j.$$

Proof. Define $\varphi : \mathcal{O}(\Omega)^k \to \mathcal{A}$ by

$$\varphi(b_1, \cdots, b_k) = \sum_{j=1}^{k} b_j s_j \qquad ((b_1, \cdots, b_k) \in \mathcal{O}_z^k, \, z \in \Omega).$$

By Theorem 5.16 $\operatorname{Ker} \varphi$ is coherent. By the Cartan theorem B, we have

$$H^1(\Omega, \operatorname{Ker} \varphi) = 0.$$

By applying Theorem 5.18 to the exact sequence of sheaves

$$0 \to \operatorname{Ker} \varphi \to \mathcal{O}(\Omega)^k \xrightarrow{\varphi} \mathcal{S} \to 0$$

we have that $\varphi^0 : \Gamma(\Omega, \mathcal{O}(\Omega)^k) \to \Gamma(\Omega, \mathcal{S})$ is surjective. Consequently, for $s \in \Gamma(\Omega, \mathcal{S})$, there exist $(f_1, \cdots, f_k) \in \Gamma(\Omega, \mathcal{O}(\Omega)^k)$ such that

$$s = \sum_{j=1}^{k} f_j s_j.$$

□

Corollary 5.7 *Let Ω be a pseudoconvex domain in $\mathbf{C}^n$ and let A be an analytic subset of Ω. Suppose that there exist holomorphic functions $s_1(z)$, $\cdots$, $s_k(z)$, $k \leq n$, in Ω such that*

$$A = \{z \in \Omega \mid s_1(z) = \cdots = s_k(z) = 0\}$$

and $F = (s_1, \cdots, s_k)$ is nonsingular in Ω. If g is a holomorphic function in Ω with $g|_A = 0$, then there exist holomorphic functions $f_1, \cdots, f_k$ in Ω such that

$$g(z) = \sum_{j=1}^{k} f_j(z) s_j(z) \qquad (z \in \Omega).$$

Proof. Let $\mathcal{I}$ be the sheaf of ideals of A. We apply Theorem 5.23 by setting $\mathcal{A} = \mathcal{O}(\Omega)$, $\mathcal{S} = \mathcal{I}$. We have $g \in \Gamma(\Omega, \mathcal{I})$. By Theorem 5.12 $(s_1, \cdots, s_k)$ satisfies the hypothesis of Theorem 5.23. □

Theorem 5.24 *Let Ω be a Stein manifold and let K be a compact subset of Ω, ω a neighborhood of $\hat{K}$. Then there exists $\varphi \in C^\infty(\Omega)$ with the following properties:*

(a) *φ is a strictly plurisubharmonic function in Ω.*
(b) *$\varphi < 0$ in K and $\varphi > 0$ in $\Omega \backslash \omega$.*
(c) *For every $c \in \mathbf{R}$, $\{z \in \Omega \mid \varphi(z) < c\} \subset\subset \Omega$.*

Proof. For simplicity, we adopt the notation $\hat{K}$ instead of $\hat{K}_\Omega^\mathcal{O}$. Since Ω is σ compact, there exists a sequence $\{K_j\}$ of compact sets such that $\Omega = \cup_{j=1}^\infty K_j$, $K_j \subset K_{j+1}$. Hence we have $\Omega = \cup_{j=1}^\infty \hat{K}_j$, $\hat{K}_j \subset \hat{K}_{j+1}$. By replacing K_j by $\hat{K}_j$, we can obtain a sequence $\{K_j\}$ of compact subsets of Ω such that

$$K_1 = \hat{K}, \ K_j \subset K_{j+1}^\circ, \ \hat{K}_j = K_j, \ \Omega = \bigcup_{j=1}^\infty K_j.$$

We choose open sets ω_j with the properties that $K_j \subset \omega_j \subset K_{j+1}$, $\omega_1 \subset \omega$. Let $a \in K_{j+2} - \omega_j$. Since $a \notin K_j$, there exists $f_{ja} \in \mathcal{O}(\Omega)$ such that

$$|f_{ja}(a)| > \sup_{K_j} |f_{ja}|.$$

We choose α_{ja} such that

$$|f_{ja}(a)| > \alpha_{ja} > \sup_{K_j} |f_{ja}|.$$

We set $g_{ja} = f_{ja}/\alpha_{ja}$. Then we have

$$|g_{ja}(a)| > 1, \quad \sup_{K_j} |g_{ja}| < 1.$$

Since $K_{j+2} - \omega_j$ is a compact set, there exist an open set W_{jk} and functions $g_{jk} \in \mathcal{O}(\Omega)$, $k = 1, \cdots, k_j$, such that

$$K_{j+2} - \omega_j \subset \bigcup_{k=1}^{k_j} W_{jk}, \quad \sup_{K_j} |g_{jk}| < 1, \quad |g_{jk}(z)| > 1 \ (z \in W_{jk}).$$

Consequently, we have

$$\sup_{K_j} |g_{jk}| < 1 \ (k = 1, \cdots, k_j), \quad \max_k |g_{jk}(z)| > 1 \ (z \in K_{j+2} - \omega_j).$$

Replacing g_{jk} by g_{jk}^m (m is sufficiently large), we obtain

$$\sum_{k=1}^{k_j} |g_{jk}(z)|^2 < \frac{1}{2^j} \quad (z \in K_j) \tag{5.9}$$

and

$$\sum_{k=1}^{k_j} |g_{jk}(z)|^2 > j \quad (z \in K_{j+2} - \omega_j). \tag{5.10}$$

Further we may assume that g_{jk}, $k = 1, \cdots, k_j$, contains n functions which form the coordinate system at any point in K_j. Define

$$\varphi(z) = \sum_{j=1}^{\infty} \sum_{k=1}^{k_j} |f_{jk}(z)|^2 - 1. \tag{5.11}$$

By (5.11) the series in the right side of (5.11) converges. By (5.10) we have $\varphi > j - 1$ in ω_j^c. Therefore we have $\varphi > 0$ in ω^c. By (5.9) we have $\varphi < \sum_{j=1}^{\infty} 2^{-j} = 1$ in K. On the other hand

$$\sum_{j=1}^{\infty} \sum_{k=1}^{k_j} f_{jk}(z)\overline{f_{jk}(\zeta)}$$

converges uniformly on every compact subset of $\Omega \times \Omega$ and is holomorphic in $(z, \bar{\zeta})$, and hence can be expanded to a power series. Hence we obtain

$$\sum_{s,t=1}^{n} \frac{\partial^2 \varphi}{\partial z_s \partial \bar{z}_t}(z) w_s \bar{w}_t = \sum_{j=1}^{\infty} \sum_{k=1}^{k_j} \left| \sum_{s=1}^{n} \frac{\partial f_{jk}}{\partial z_s}(z) w_s \right|^2.$$

Assume that for all j, k

$$\sum_{s=1}^{n} \frac{\partial f_{jk}}{\partial z_s}(z) w_s = 0.$$

Since f_{jk} ($k = 1, \cdots, k_j$) contain n functions which form a coordinate system at z, we have $w = 0$. Hence φ is strictly plurisubharmonic. (c) is trivial, which completes the proof of Theorem 5.24. □

Lemma 5.14 *Let Ω be a Stein manifold and let K be a compact subset of Ω with $K = \hat{K}_{\Omega}^{\mathcal{O}}$. Let ω be a neighborhood of K. Then there exists an analytic polyhedron P such that $K \subset P \subset\subset \omega$.*

Proof. We may assume that $\omega \subset\subset \Omega$. Let $z \in \partial \omega$. Since $z \notin \hat{K}_{\Omega}^{\mathcal{O}}$, there exists $f \in \mathcal{O}(\Omega)$ such that

$$|f(z)| > \sup_K |f|.$$

We choose α such that $|f(z)| > \alpha > \sup_K |f|$. We set $g = f/\alpha$. Then $|g| < 1$ in K, $|g(z)| > 1$. By the Heine-Borel theorem, there exist $f_1, \cdots, f_N \in \mathcal{O}(\Omega)$ such that if we set

$$P' = \{z \in \Omega \mid |g_j(z)| < 1 \ (j = 1, \cdots, N)\},$$

then $K \subset P'$, $\partial \omega \cap P' = \phi$. Let $P = \omega \cap P'$. Then P is an analytic polyhedron we seek. □

5.4 The Cousin Problem

We study the Cousin problem using the L^2 estimate for solutions of the $\bar{\partial}$ problem on Stein manifolds due to Hörmander [HR2].

Hörmander [HR2] proved the following theorem. We omit the proof.

Theorem 5.25 *Let Ω be a Stein manifold. Then for $f \in C^\infty_{(p,q+1)}(\Omega)$ with $\bar{\partial} f = 0$, there exists $u \in C^\infty_{(p,q)}(\Omega)$ such that $\bar{\partial} u = f$.*

Theorem 5.26 *(First Cousin problem) Let Ω be a Stein manifold and let $\{\omega_j\}$ be a sequence of open subsets of Ω with $\Omega = \cup_{j=1}^\infty \omega_j$. Suppose that $g_{jk} \in \mathcal{O}(\omega_j \cap \omega_k)$, $j, k = 1, 2, \cdots$, satisfy the following conditions:*

(a) $g_{jk} = -g_{kj}$.
(b) $g_{ij} + g_{jk} + g_{ki} = 0$ in $\omega_i \cap \omega_j \cap \omega_k$.

Then there exist $g_j \in \mathcal{O}(\omega_j)$ such that

$$g_{jk} = g_k - g_j$$

in $\omega_j \cap \omega_k$

Proof. Let $\{\varphi_\nu\}$ be a partition of unity subordinate to $\{\omega_j\}$. Then we have $\varphi_\nu \in C_c^\infty(\omega_{i_\nu})$. Further, for any compact subset K of Ω, φ_ν equals identically zero on K except for a finite number of ν and

$$\sum_{\nu=1}^\infty \varphi_\nu = 1.$$

Suppose that g_{jk} is expressed by $g_{jk} = g_k - g_j$ in $\omega_j \cap \omega_k$. Let $j = i_\nu$. Then we have

$$g_{i_\nu k} = g_k - g_{i_\nu}.$$

If we multiply by φ_ν and add with respect to ν, then we obtain

$$\sum_{\nu=1}^\infty \varphi_\nu g_{i_\nu k} = g_k - \sum_{\nu=1}^\infty \varphi_\nu g_{i_\nu}.$$

Define
$$h_k = \sum_{\nu=1}^{\infty} \varphi_\nu g_{i_\nu k}.$$

Then we have $h_k \in C^\infty(\omega_k)$. Moreover we have
$$h_k - h_j = \sum_{\nu=1}^{\infty} \varphi_\nu(g_{i_\nu k} - g_{i_\nu j}) = \sum_{\nu=1}^{\infty} \varphi_\nu g_{jk} = g_{jk}$$

in $\omega_j \cap \omega_k$. If we set $\psi = \bar{\partial} h_k$ in ω_k, then $\psi \in C^\infty_{(0,1)}(\Omega)$ and $\bar{\partial}\psi = 0$. By Theorem 5.25, there exist $u \in C^\infty(\Omega)$ such that $\bar{\partial} u = -\psi$. We set $g_k = h_k + u$. Then g_k are solutions we seek. □

Lemma 5.15 *Let $\Omega \subset \mathbf{R}^N$ be a simply connected domain. Suppose $f : \Omega \to \mathbf{C}$ is continuous and nowhere vanishing. Then there exists a continuous function g in Ω such that $f = e^g$.*

Proof. Let $P \in \Omega$. Without loss of generality we may assume that $\operatorname{Re} f(P) > 0$. Then there exists a neighborhood U_P of P such that $f(U_P) \subset \{z \mid \operatorname{Re} z > 0\}$. Hence we can define a continuous function $\log f$ in U_P. Fix $P_0 \in \Omega$. Let $\gamma : [0,1] \to \Omega$ be a smooth Jordan closed curve such that $\gamma(0) = \gamma(1) = P_0$. For each P on γ we can choose a neighborhood U_p of P having the property mentioned above. Then we can define a function $\log f(\gamma(t))$ for $0 \le t < 1$. Assume that $\log f \circ \gamma(0) \ne \lim_{t \to 1-} \log f \circ \gamma(t)$. Since Ω is simply connected, there exists a continuous function $u(s,t)$ on $[0,1] \times [0,1]$ such that
$$u(0,t) = \gamma(t) \qquad (0 \le t \le 1),$$
$$u(s,0) = u(s,1) = P_0 \qquad (0 \le s \le 1),$$
$$u(1,t) = P_0 \qquad (0 \le t \le 1).$$

If we set
$$\rho(s) = \frac{1}{2\pi i}\{\lim_{t \to 1-} \log f(u(s,t)) - \log f(u(s,0))\},$$

then $\rho(s)$ is an integer valued continuous function and equals 0 when s is close to 1. Then $\rho(0) \ne 0$, which is a contradiction. Hence we can define $\log f(\gamma(t))$ for $0 \le t \le 1$. Since γ is an arbitrary closed Jordan curve, we can define $\log f$ in Ω. We set $g = \log f$. Then $f = e^g$. □

Lemma 5.16 Let $\Omega \subset \mathbf{C}^n$ be a simply connected domain and let $f : \Omega \to \mathbf{C}$ be holomorphic and nowhere vanishing. Then there exists a holomorphic function g in Ω such that $f = e^g$.

Proof. By Lemma 5.15 there exists a continuous function g in Ω such that $f = e^g$. Then we have in the sense of distributions

$$0 = \frac{\partial f}{\partial \bar{z}_j} = e^g \frac{\partial g}{\partial \bar{z}_j},$$

and hence

$$\frac{\partial g}{\partial \bar{z}_j} = 0.$$

Hence g is holomorphic. $\square$

Definition 5.31 We denote by $\mathcal{O}^*(\Omega)$ the set of all nowhere vanishing holomorphic functions in a complex manifold Ω. We also denote by $C^*(\Omega)$ the set of all nowhere vanishing continuous functions on a complex manifold Ω.

Theorem 5.27 *(Second Cousin problem)* Let Ω be a Stein manifold and let $\{\omega_j\}$ be a sequence of open subsets of Ω with $\Omega = \cup_{j=1}^{\infty} \omega_j$. Suppose that $g_{jk} \in \mathcal{O}^*(\omega_j \cap \omega_k)$, $j, k = 1, 2, \cdots$, satisfy the following properties:

(a) $g_{jk} g_{kj} = 1$.
(b) $g_{ij} g_{jk} g_{ki} = 1$ in $\omega_i \cap \omega_j \cap \omega_k$.

Moreover, suppose there exist $c_j \in C^*(\omega_j)$ with the properties

$$g_{jk} = c_k c_j^{-1}$$

in $\omega_j \cap \omega_k$. Then there exist $g_j \in \mathcal{O}^*(\omega_j)$ such that

$$g_{jk} = g_k g_j^{-1}$$

in $\omega_j \cap \omega_k$.

Proof. By the assumption there exist $c_j \in C^*(\omega_j)$ such that

$$g_{jk} = c_k c_j^{-1}$$

in $\omega_j \cap \omega_k$. First we assume that ω_j is simply connected. By Lemma 5.16 there exist $b_j \in C(\Omega)$ such that $c_j = e^{b_j}$. We set $h_{jk} = b_k - b_j$. Then we have

$$g_{jk} = c_k c_j^{-1} = e^{h_{jk}}.$$

Using the same method as in the proof of Lemma 5.16, h_{jk} is holomorphic in $\omega_j \cap \omega_k$. Evidently we have

$$h_{ij} = -h_{ji}, \quad h_{ij} + h_{jk} + h_{ki} = 0.$$

By Theorem 5.26 there exist $h_k \in \mathcal{O}(\omega_k)$ such that

$$h_{jk} = h_k - h_j$$

in $\omega_j \cap \omega_k$. We set $g_k = e^{h_k}$. Then

$$g_k g_j^{-1} = g_{jk}.$$

Next we prove the general case. Let $\{\omega'_\nu\}$ be a refinement of $\{\omega_j\}$ whose elements are simply connected open subsets of Ω. Then for any ν, there exists i_ν such that $\omega'_\nu \subset \omega_{i_\nu}$. Define

$$g'_{\nu\mu} = g_{i_\nu i_\mu}$$

in $\omega'_\nu \cap \omega'_\mu$. Then from the proof of the first half, there exist $g'_\mu \in \mathcal{O}^*(\omega'_\mu)$ such that

$$g'_{\nu\mu} = g'_\mu {g'_\nu}^{-1}.$$

in $\omega'_\nu \cap \omega'_\mu$. Consequently, we obtain

$$g'_\mu {g'_\nu}^{-1} g_{i_\mu i} g_{i i_\nu} = 1$$

in $\omega_i \cap \omega'_\nu \cap \omega'_\mu \subset \omega_i \cap \omega_{i_\nu} \cap \omega_{i_\mu}$. Hence we have $g'_\mu g_{i_\mu i} = g'_\nu g_{i_\nu i}$ in $\omega_i \cap \omega'_\nu \cap \omega'_\mu$. If we define $g_i = g'_\nu g_{i_\nu i}$ in $\omega_i \cap \omega'_\nu$, then $g_i \in \mathcal{O}^*(\omega_i)$. Therefore we obtain

$$g_k g_j^{-1} = g'_\nu g_{i_\nu k} (g'_\nu g_{i_\nu j})^{-1} = g_{i_\nu k} g_{j i_\nu} = g_{jk}$$

in $\omega'_\nu \cap \omega_j \cap \omega_k$. □

Exercises

5.1 (Poincaré theorem) Define

$$\Delta = \{z \in \mathbf{C} \mid |z| < 1\}, \quad B = \{w \in \mathbf{C}^2 \mid |w| < 1\}.$$

Show that there is no biholomorphic mapping $F = (f_1, f_2) : \Delta \times \Delta \to B$ by proving the following:

(a) For $w \in \Delta$, define a holomorphic mapping $F_w : \Delta \to B$ by
$$F_w(z) = \left(\frac{\partial f_1}{\partial w}(z, w), \frac{\partial f_2}{\partial w}(z, w)\right).$$
Then for any $z_0 \in \partial \Delta$ we have
$$\lim_{z \to z_0} F_w(z) = 0.$$

(b) $F(z, w)$ is constant with respect to w.

5.2 Let $(\mathcal{S}, \pi, X)$ be a sheaf over X. Show that if $s_x \in \mathcal{S}_x$, then there exists a neighborhood U of x and $s \in \Gamma(U, \mathcal{S})$ such that $s(x) = s_x$.

5.3 Suppose C^1 curve $\varphi : [0, 2\pi] \to \mathbf{C} \setminus \{0\}$ satisfies $\varphi(0) = \varphi(2\pi)$. Show that
$$N(\varphi) = \frac{1}{2\pi i} \int_0^{2\pi} \frac{\varphi'(\theta)}{\varphi(\theta)} d\theta$$
is an integer.

5.4 Suppose g is a C^1 function in $\{z \in \mathbf{C} \mid |z| \leq 1\}$ and nowhere vanishing. Prove that if we set $\varphi(\theta) = g(e^{i\theta})$, then $N(\varphi) = 0$.

5.5 (Oka's counterexample) Define
$$\Omega = \{(z_1, z_2) \in \mathbf{C}^2 \mid \tfrac{3}{4} < |z_j| < \tfrac{5}{4}, j = 1, 2\}.$$
Then Ω is a domain of holomorphy. Define
$$A = \{z \in \Omega \mid z_2 - z_1 + 1 = 0\},$$
$$\omega_1 = A \cap \{z \in \Omega \mid \operatorname{Im} z_1 < 0\}, \quad \omega_2 = A \cap \{z \in \Omega \mid \operatorname{Im} z_1 > 0\}$$
and
$$U_1 = \Omega - \omega_1, \quad U_2 = \Omega - \omega_2.$$
Show that

(a) $A \cap \{z \in \Omega \mid \operatorname{Im} z_1 = 0\} = \phi$, $A = \omega_1 \cup \omega_2$, $\Omega = U_1 \cup U_2$.

(b) Define $f_1 = z_2 - z_1 + 1$ in U_1, $f_2 = 1$ in U_2. Then $f_2 f_1^{-1} \in \mathcal{O}^*(U_1 \cap U_2)$.

(c) There is no $f \in \mathcal{O}(\Omega)$ which satisfies $f/f_2 \in \mathcal{O}^*(U_2)$, $f/f_1 \in \mathcal{O}^*(U_1)$.

Appendix A

Compact Operators

In Appendix A we prove Proposition A.10 and Proposition A.13 concerning compact operators which are needed to prove Theorem 3.30 and Theorem 3.29, respectively. For the proofs we refer to Berezansky-Sheftel-Us [BES].

Let E_1 and E_2 be normed spaces and let $A : E_1 \to E_2$ be a bounded operator. Define $A^* : E_2' \to E_1'$ by

$$(A^*f)(x) = f(Ax) \qquad (f \in E_2', \ x \in E_1). \tag{A.1}$$

A^* is called a conjugate operator of A.

For a bounded linear operator $A : E_1 \to E_2$, $A^* : E_2' \to E_1'$ is a bounded linear operator. Moreover, we have $\|A^*\| = \|A\|$.

Let X and Y be normed spaces.

(1) We denote by $\mathcal{B}(X,Y)$ the set of all bounded linear operators $T : X \to Y$. Moreover, we denote $\mathcal{B}(X,X)$ by $\mathcal{B}(X)$.

(2) A linear operator $T : X \to Y$ is called a compact operator if for any bounded sequence $\{x_n\}$ of X, there exists a subsequence $\{x_{n_i}\}$ of $\{x_n\}$ such that $\{T(x_{n_i})\}$ converges to a point in Y.

(3) We say that $T \in \mathcal{B}(X,Y)$ is invertible if there exists $S \in \mathcal{B}(Y,X)$ such that

$$ST = I_X, \quad TS = I_Y,$$

where I_X is the identity mapping from X onto X and I_Y is the identity mapping from Y onto Y. In this case we write $S = T^{-1}$.

Proposition A.1 *(Ascoli-Arzela theorem) Let X be a compact topological space and let $C(X)$ be a Banach space consisting of all continuous functions on X. That is, if we define the metric for $f, g \in C(X)$ by*

$$d(f,g) = \|f - g\| = \sup\{|f(x) - g(x)| \mid x \in X\},$$

then $C(X)$ is a complete metric space). Suppose $\Phi \subset C(X)$ satisfies the following properties:

(a) $\sup\{|f(x)| \mid x \in X, f \in \Phi\} = M < \infty$.
(b) For any $\varepsilon > 0$ and any $x \in X$ there exist a neighborhood V such that
$$|f(y) - f(x)| < \varepsilon \quad (y \in V, f \in \Phi).$$
Then every sequence $\{f_n\}$ in Φ contains a convergent subsequence.

Proof. Since X is compact, for any positive integer k, it follows from (b) that there exist a finite subset F_k of X and a neighborhood V_y^k of $y \in F_k$ such that
$$X = \bigcup_{y \in F_k} V_y^k$$
and
$$|f(x) - f(y)| < \frac{1}{k} \quad (f \in \Phi, x \in V_y^k).$$

We set $F = \bigcup_{k=1}^{\infty} F_k$. Then F is at most countable. Suppose F is countable, say $F = \{x_1, x_2, \cdots\}$. Since $|f_n(x_1)| \leq M$, there exists a subsequence $\{g_n^1\}$ of $\{f_n\}$ such that $\{g_n^1(x_1)\}$ converges. Since $|g_n^1(x_2)| \leq M$, there exists a subsequence $\{g_n^2\}$ of $\{g_n^1\}$ such that $\{g_n^2(x_2)\}$. Repeating this process, there exist $\{g_n^i\}$, $i = 1, 2, \cdots$, satisfying the following properties:

(a) $\{g_n^1\}$ is a subsequence of $\{f_n\}$.
(b) Each $\{g_n^{i+1}\}$ $(i = 1, 2, \cdots)$ is a subsequence of $\{g_n^i\}$.
(c) $\lim_{n \to \infty} g_n^i(x_j)$ $(j = 1, \cdots, i)$ exist.

We set $h_n = g_n^n$. Then $\{h_n\}$ is a subsequence of $\{f_n\}$ and $\lim_{n \to \infty} h_n(x_i)$, $i = 1, 2, \cdots$, exist. Next we show that $\{h_n\}$ is a Cauchy sequence in $C(X)$. We fix k. For $y \in F$, there exists a positive integer n_0 such that
$$|h_n(y) - h_m(y)| < \frac{1}{k}$$
for $n, m \geq n_0$. For $x \in X$ there exists $y \in F_k$ such that $x \in V_y$. Hence
$$|h_n(x) - h_m(x)| \leq |h_n(x) - h_n(y)| + |h_n(y) - h_m(y)|$$
$$+ |h_m(y) - h_m(x)|$$
$$< \frac{1}{k} + \frac{1}{k} + \frac{1}{k} = \frac{3}{k}$$

for $n, m \geq n_0$. Consequently,

$$\|h_n - h_m\| \leq \frac{3}{k} \quad (n, m \geq n_0),$$

which means that $\{h_n\}$ is a Cauchy sequence, and hence $\{h_n\}$ converges. □

Proposition A.2 *Let E be a Banach space. If a bounded operator $A: E \to E$ is compact, then $A^*: E' \to E'$ is compact.*

Proof. Let A be a compact operator. Suppose $f_n \in E'$ and $\{f_n\}$ is bounded. We set

$$S_1(0) = \{y \in E \mid \|y\| = 1\}, \quad Q = A(S_1(0)).$$

Then

$$\begin{aligned}\|A^*(f_n)\| &= \sup\{|(A^*(f_n))(y)| \mid \|y\| = 1\} \\ &= \sup\{|f_n(A(y))| \mid \|y\| = 1\} \\ &= \sup\{|f_n(z)| \mid z \in A(S_1(0))\}.\end{aligned}$$

Hence Q is a compact set. We set

$$c = \sup\{\|f_n\| \mid n = 1, 2, \cdots\}, \quad c_1 = \sup\{\|z\| \mid z \in Q\}.$$

Then we have $|f_n(z)| \leq \|f_n\| \|z\| \leq cc_1$, which implies that $\{f_n\}$ is uniformly bounded on Q. Moreover we have

$$|f_n(z_1) - f_n(z_2)| \leq c\|z_1 - z_2\| \quad (z_1, z_2 \in Q),$$

which means that $\{f_n\}$ is equicontinuous on Q. By the Ascoli-Arzela theorem, there exists a convergent subsequence $\{f_{n_k}\}$ of $\{f_n\}$. Taking into account that

$$\|f_{n_k} - f_{n_m}\| = \max\{|f_{n_k}(z) - f_{n_m}(z)| \mid z \in Q\} \to 0 \quad (k, m \to \infty),$$

we have $\|A^*(f_{n_k}) - A^*(f_{n_m})\| \to 0$. Since E' is complete, $\{A^*(f_{n_k})\}$ converges, which means that A^* is a compact operator. □

Proposition A.3 *Let E be a normed space and let V be a closed subspace of E. For $y \in E$, $y \notin V$, define*

$$V^* = \{x + \lambda y \mid x \in V, \lambda \in F\}.$$

Then V^* is a closed subspace of E, where F is the set of all scalars.

Proof. Suppose $z \in E$, $z_n \in V^*$, $z = \lim_{n \to \infty} z_n$. Then we have a representation $z_n = x_n + \lambda_n y$ with $x_n \in V, \lambda_n \in F$. Since $\{x_n + \lambda_n y\}$ is a bounded sequence, there exists $M > 0$ such that $\|x_n + \lambda_n y\| < M$ for all n. Assume that $|\lambda_n| \to \infty$. Then we have

$$\left\|\frac{x_n}{\lambda_n} + y\right\| < \frac{M}{|\lambda_n|} \to 0,$$

which means that $\lim_{n \to \infty} \lambda_n^{-1} x_n = -y$. Since V is a closed set, we have $-y \in V$, which contradicts that $y \notin V$. Therefore there exists $N > 0$ such that there are infinitely many n with $|\lambda_n| \leq N$. Hence we can choose a convergent subsequence $\{\lambda_{k_n}\}$ of $\{\lambda_n\}$. We set $\lim_{n \to \infty} \lambda_{k_n} = \lambda$. Taking into account that

$$\|x_{k_n} - (z - \lambda y)\| \leq \|x_{k_n} + \lambda_{k_n} y - z\| + \|\lambda_{k_n} y - \lambda y\| \to 0,$$

we have $\lim_{n \to \infty} x_{k_n} = z - \lambda y$. If we set $\lim_{n \to \infty} x_{k_n} = x$, then $x \in V$ and $x = z - \lambda y$, and hence $z \in V^*$. Hence V^* is closed. □

Proposition A.4 *Let E be a normed space and let G be a closed subspace of E with $E \neq G$. Then for any $\varepsilon > 0$ there exists $y_\varepsilon \notin G$ such that*

$$\|y_\varepsilon\| = 1, \quad \|y_\varepsilon - x\| > 1 - \varepsilon \quad (x \in G).$$

Proof. Let $z \notin G$. Since G is closed, $\delta = \rho(z, G) = \inf\{\|z - x\| \mid x \in G\} > 0$. By the definition of the infimum, for any $\eta > 0$ there exists $x_\eta \in G$ such that

$$\delta \leq \|z - x_\eta\| < \delta + \eta.$$

We choose η such that $\varepsilon = \eta(\delta + \eta)^{-1}$ and set $y_\varepsilon = \|z - x_\eta\|^{-1}(z - x_\eta)$. We show that y_ε satisfies the desired properties. Clearly we have $y_\varepsilon \notin G$, $\|y_\varepsilon\| = 1$. Let $x \in G$. Then we have

$$\|y_\varepsilon - x\| = \|z - x_\eta\|^{-1}\|z - (x_\eta + x\|z - x_\eta\|)\|.$$

Taking into account that $x_\eta + x\|z - x_\eta\| \in G$, we obtain

$$\|y_\varepsilon - x\| \geq \|z - x_\eta\|^{-1} \delta > \frac{\delta}{\delta + \eta} = 1 - \frac{\eta}{\delta + \eta} = 1 - \varepsilon.$$
□

Proposition A.5 *Let E be a normed space. If every bounded sequence in E contains a convergent subsequence, then E is a finite dimensional space.*

Proof. Suppose E is an infinite dimensional space. Let $x_1 \in E$ be such that $\|x_1\| = 1$. We set $G_1 = \{\lambda x_1 \mid \lambda \in F\}$, where F is the set of all scalars. It follows from Theorem A3 that there exists $x_2 \notin G_1$ such that

$$\|x_1\| = 1, \quad \|x_2 - x\| > \frac{1}{2} \quad (x \in G_1).$$

In particular, we have $\|x_2 - x_1\| > \frac{1}{2}$. Let G_2 be a vector space generated by x_1, x_2. By Theorem A3 there exists $x_3 \notin G_2$ such that

$$\|x_3\| = 1 \quad \|x_3 - x\| > \frac{1}{2} \quad (x \in G_2).$$

In particular, we have

$$\|x_3 - x_2\| > \frac{1}{2}, \quad \|x_3 - x_1\| > \frac{1}{2}.$$

Repeating this process, there exists a sequence $\{x_n\}$ such that $\|x_n\| = 1$, $\|x_n - x_m\| > \frac{1}{2}$ for $m \neq n$. Then $\{x_n\}$ does not contain any convergent subsequence, which contradicts the hypothesis. Hence E is a finite dimensional space. □

Proposition A.6 *Let E be a Banach space and let $A : E \to E$ be a compact operator, $T = A - I$, where $I: E \to E$ is the identity mapping. Then $\operatorname{Ker} T = \{x \in E \mid Tx = 0\}$ is a finite dimensional space.*

Proof. Let $\{x_n\}$ be a bounded sequence in $\operatorname{Ker} T$. Since $A(x_n) = x_n$, $\{x_n\}$ contains a convergent subsequence. By Theorem A4, $\operatorname{Ker} T$ is a finite dimensional space. □

Proposition A.7 *Let E be a Banach space and let $A : E \to E$ be a compact operator, $T = A - I$. Then $T(E)$ is a closed subspace of E.*

Proof. $T(E)$ is a vector space. We show that $T(E)$ is a closed subset of E. First we show that there is a constant $c > 0$ depending only on T such that for $y \in T(E)$ there exists x with

$$Tx = y, \quad \|x\| \leq c\|y\|. \tag{A.2}$$

Suppose x_0 satisfies $Tx_0 = y$. Then any solution x of the equation $Tx = y$ can be written $x = x_0 + z$, where z is a solution of $Tz = 0$. Hence we have

$$d := \inf\{\|x\| \mid Tx = y\} = \inf\{\|x_0 + z\| \mid z \in \operatorname{Ker} T\}.$$

Then there exists a sequence $\{z_n\} \subset \operatorname{Ker} T$ such that $\|x_0 + z_n\| \to d$. Consequently, $\{z_n\}$ is bounded. Since $\operatorname{Ker} T$ is a finite dimensional space in view of Proposition A.6, we have a representation $z_n = a_1^n x_1 + \cdots + a_k^n x_k$, where $\{x_1, \cdots, x_k\}$ is a basis of $\operatorname{Ker} T$. Suppose $\{a_1^n\}$ is not bounded. Then there exists a subsequence $\{a_1^{n_i}\}$ of $\{a_1^n\}$ such that $\lim_{i \to \infty} a_1^{n_i} = \infty$. We set

$$\alpha_j^n = \frac{a_j^n}{\sqrt{\sum_{i=1}^k (a_i^n)^2}}.$$

Then $|\alpha_j^n| \leq 1$. We can choose a convergent subsequence $\{\alpha_j^{m_i}\}$ of $\{\alpha_j^{n_i}\}$. Since we have

$$z_{m_i} = \left\{ \sum_{i=1}^k (a_i^{m_i})^2 \right\}^{1/2} (\alpha_1^{m_i} x_1 + \cdots + \alpha_k^{m_i} x_k),$$

which implies that $\lim_{i \to \infty} \|z_{m_i}\| = \infty$. This contradicts the hypothesis. Hence $\{a_1^n\}$ is bounded, and hence $\{a_1^n\}$ contains a convergent subsequence, which means that $\{z_n\}$ contains a convergent subsequence $\{z_{n_k}\}$. We set

$$z_0 = \lim_{k \to \infty} z_{n_k}.$$

Then we have

$$\|x_0 + z_0\| = \lim_{k \to \infty} \|x_0 + z_{n_k}\| = d.$$

We set $\hat{x} = x_0 + z_0$. Then $T(\hat{x}) = y$. Now we show that $\hat{x}$ satisfies (A.2). Suppose (A.2) does not hold. For any positive integer n there exists $y_n \in T(E)$ such that

$$\|\hat{x}_n\| > n\|y_n\|, \quad T(\hat{x}_n) = y_n. \tag{A.3}$$

We set

$$\hat{\xi}_n = \|\hat{x}_n\|^{-1} \hat{x}_n, \quad \eta_n = \|\hat{x}_n\|^{-1} y_n.$$

If $T\xi = \eta_n$, then $\|\xi\| \geq 1$. Hence $\hat{\xi}_n$ has the smallest norm among solutions of the equation $T\xi = \eta_n$. Since $\{\hat{\xi}_n\}$ is bounded, $\{A\hat{\xi}_n\}$ contains a convergent subsequence $\{A(\hat{\xi}_{n_k})\}$. We set $\xi_0 = \lim_{k \to \infty} A(\hat{\xi}_{n_k})$. By (A.3) we have $\lim_{n \to \infty} \eta_n = 0$. Taking into account that $A(\hat{\xi}_{n_k}) - \hat{\xi}_{n_k} = \eta_{n_k}$, we have

$$\xi_0 = \lim_{k \to \infty} \hat{\xi}_{n_k}.$$

Since A is continuous, we have

$$A(\xi_0) = \lim_{k \to \infty} A(\hat{\xi}_{n_k}).$$

We obtain $A(\xi_0) = \xi_0$, and hence $\xi_0 \in \text{Ker}\, T$. Thus we have $T(\hat{\xi}_{n_k} - \xi_0) = \eta_{n_k}$, which implies that $\|\hat{\xi}_{n_k} - \xi_0\| \geq 1$. This is a contradiction. This proves (A.2). Suppose $y_n \in T(E)$, $y \in E$, $y_n \to y$. Taking a subsequence, if necessary, we may assume that

$$\|y_n - y\| < 2^{-n-1}, \quad \|y_{n+1} - y_n\| < 2^{-n}.$$

Choose $\hat{x}_0$ such that

$$T(\hat{x}_0) = y_1, \quad \|\hat{x}_0\| \leq c\|y_1\|.$$

Choose $\hat{x}_n$, $n \geq 1$, such that

$$T(\hat{x}_n) = y_{n+1} - y_n, \quad \|\hat{x}_n\| \leq c\|y_{n+1} - y_n\|.$$

We set

$$\hat{x} = \sum_{k=0}^{\infty} \hat{x}_k.$$

Then we have

$$T(\hat{x}) = T\left(\lim_{n \to \infty} \sum_{k=0}^{n} \hat{x}_k\right) = \lim_{n \to \infty} \sum_{k=0}^{n} T(\hat{x}_k)$$

$$= \lim_{n \to \infty} \left[y_1 + \sum_{k=1}^{n}(y_{k+1} - y_k)\right] = \lim_{n \to \infty} y_{n+1} = y,$$

which means that $y \in T(E)$. Hence $T(E)$ is closed. □

Proposition A.8 *Let E be a Banach space and let $A : E \to E$ be a compact operator, $T = A - I$. Then the equation $Tx = y$ has solutions if and only if for any solution $f \in E'$ of the equation $T^*(f) = 0$, one has $f(y) = 0$. That is, $T(E) = E$ if and only if $\text{Ker}\, T^* = \{0\}$.*

Proof. (Necessity) Let $x \in E$ be a solution of $Tx = y$. Then $f(y) = f(Tx) = (T^*f)(x) = 0$.

(Sufficiency) Suppose $y \in E$ satisfies $f(y) = 0$ for any solution f of the equation $T^*(f) = 0$. If $Tx = y$ does not have solutions, then $y \notin T(E)$. By Proposition A7, $T(E)$ is a closed subspace. By the Hahn-Banach theorem, there exists $h \in E'$ such that $h = 0$ in $T(E)$, $h(y) \neq 0$. Thus we have $(T^*h)(x) = h(Tx) = 0$, and hence $T^*h = 0$. This contradicts the assumption. □

Proposition A.9 *Let E be a Banach and let $A : E \to E$ be a compact operator, $T = A - I$. Then the equation $T^*(f) = g$ has solutions if and only if $g(x) = 0$ for any $x \in \operatorname{Ker} T$. That is, $T^*(E') = E'$ if and only if $\operatorname{Ker} T = \{0\}$.*

Proof. (Necessity) Let $f \in E'$ satisfy $T^*(f) = g$ and let $x \in \operatorname{Ker} T$. Then we have $g(x) = (T^*(f))(x) = f(Tx) = f(0) = 0$.

(Sufficiency) Suppose $g \in E'$ satisfies $g(x) = 0$ for any $x \in \operatorname{Ker} T$. Define a linear functional f_0 on $T(E)$ by $f_0(y) = g(x)$ for $y \in T(E)$, where x is one of the solutions of the equation $T(x) = y$. If $T(x_1) = T(x_2) = y$, then $T(x_1 - x_2) = 0$, which means that $g(x_1) = g(x_2 + (x_1 - x_2)) = g(x_2)$. Hence f_0 is well defined. f_0 is linear since g is linear. Now we show that f_0 is bounded. By (A.2) there exists a solution $\hat{x}$ of the equation $Tx = y$ such that $\|\hat{x}\| \leq c\|y\|$. Hence we have

$$|f_0(y)| = |g(\hat{x})| \leq \|g\|\,\|\hat{x}\| \leq c\|g\|\,\|y\|.$$

Hence f_0 is bounded. By the Hahn-Banach theorem, f_0 is extended to a bounded linear functional F on E. Then for $x \in E$ we have

$$(T^*(F))(x) = F(Tx) = f_0(Tx) = g(x).$$

Hence we have $T^*(F) = g$. □

Proposition A.10 *Let E be a Banach space and let $A : E \to E$ be a compact operator, $T = A - I$. Then $T(E) = E$ if and only if $\operatorname{Ker} T = \{0\}$. In this case $T : E \to E$ is surjective and invertible.*

Proof. (Necessity) Let $G_n = \operatorname{Ker} T^n$. Then G_n is a closed subspace of E and $G_n \subset G_{n+1}$. Suppose $T(E) = E$. Assume that there exist $x_1 \neq 0$ such that $T(x_1) = 0$. We set $T(x_2) = x_1$. Repeating this process, we have $T^{k-1}(x_k) = x_1 \neq 0$. Since $T^k(x_k) = T(x_1) = 0$, we have $x_k \in G_k \setminus G_{k-1}$. By Proposition A4 there exists $y_k \in G_k$ such that $\|y_k\| = 1$, $\|y_k - x\| \geq \frac{1}{2}$ ($x \in G_{k-1}$). Since $\{y_k\}$ is bounded, $\{A(y_k)\}$ contains a convergent subsequence. On the other hand, if $n > m$, then we have

$$T^{n-1}(y_m + T(y_n) - T(y_m)) = T^{n-1}(y_m) + T^n(y_n) - T^n(y_m) = 0,$$

which implies that $y_m + T(y_n) - T(y_m) \in G_{n-1}$. Consequently we have

$$\|A(y_n) - A(y_m)\| = \|y_n - (y_m + T(y_n) - T(y_m))\| \geq \frac{1}{2},$$

which contradicts $\{A(y_k)\}$ contains a convergent subsequence. Hence we have $\operatorname{Ker} T = \{0\}$.

(Sufficiency) Suppose $\operatorname{Ker} T = \{0\}$. By Proposition A.9, we have $T^*(E') = E'$. A^* is a compact operator and $T^* = A^* - I$. Using the same method as the proof of the first half, we have $\operatorname{Ker} T^* = \{0\}$. By Proposition A8, we obtain $T(E) = E$. Finally we show that T is invertible. Since $T: E \to E$ is surjective, there is an inverse mapping $T^{-1} : E \to E$. For $Tx = y$, x satisfies (A.2), $\|T^{-1}y\| \leq c\|y\|$, which means that T^{-1} is bounded. Hence T is invertible. □

Proposition A.11 *Let X and Y be Banach spaces. Then*

(a) Every compact operator $T : X \to Y$ is bounded.
(b) Let $T \in \mathcal{B}(X,Y)$ and let $T(X)$ be a finite dimensional subspace of Y. Then T is a compact operator.
(c) The set of all compact operators from X to Y is a closed subset of $\mathcal{B}(X,Y)$.

Proof. (a) Suppose the compact operator $T : X \to Y$ is not bounded. Then there exists $\{x_n\}$ such that $\|x_n\| = 1$, $\|T(x_n)\| \to \infty$. Since $\{T(x_n)\}$ does not contain any convergent subsequence, which contradicts that T is compact.

(b) We denote the basis of $T(X)$ by $\{e_1, \cdots, e_k\}$. Let $\{x_n\}$ be a bounded sequence in X. Then we have a representation

$$T(x_n) = a_n^1 e_1 + \cdots + a_n^k e_k,$$

where $\{a_n^j\}$, $j = 1, \cdots, k$, are bounded sequences. Then $\{a_n^1\}$ contains a convergent subsequence $\{a_{j_n}^1\}$. Similarly, $\{a_{j_n}^2\}$ contains a convergent subsequence $\{a_{s_n}^2\}$. Repeating this process, $\{T(x_n)\}$ contains a convergent subsequence $\{T(x_{t_n})\}$. Hence T is compact.

(c) Let $T_n : X \to Y$, $n = 1, 2, \cdots$, be a compact operators and let $T \in \mathcal{B}(X,Y)$, $\|T_n - T\| \to 0$. Suppose $\{x_n\}$ is a bounded sequence in X. Then there exists $c > 0$ such that $\|x_n\| \leq c$. Since T_1 is a compact operator, $\{T_1(x_n)\}$ contains a convergent subsequence $\{T_1(x_{n1})\}$. Similarly, $\{T_2(x_{n1})\}$ contains a convergent subsequence $\{T_2(x_{n2})\}$. Repeating this process, $\{T_k(x_{nn})\}$ converges for any k. On the other hand, we have

$$\|T(x_{mm}) - T(x_{nn})\|$$
$$\leq \|T(x_{mm}) - T_k(x_{mm})\|$$
$$+ \|T_k(x_{mm}) - T_k(x_{nn})\| + \|T(x_{nn}) - T_k(x_{nn})\|$$
$$\leq \|T - T_k\|(\|x_{mm}\| + \|x_{nn}\|) + \|T_k(x_{nn}) - T_k(x_{mm})\|$$
$$\leq 2c\|T - T_k\| + \|T_k(x_{nn}) - T_k(x_{mm})\|,$$

which means that $\{T(x_{nn})\}$ is a Cauchy sequence, and hence $\{T(x_{nn})\}$ converges. Hence T is a compact operator. □

Proposition A.12 *Let $\{K_n(x,y)\}$ be a sequence of measurable functions in $\Omega \times \Omega$ which satisfies the following properties:*

(1) There exists $M > 0$ such that $|K_n(x,y)| \leq M$ $(x, y \in \Omega)$.
(2) $\lim_{n\to\infty} K_n(x,y) = 0$ $(x, y \in \Omega)$.

For $1 \leq p < \infty$, define a linear operator $\mathbf{K}_n : L^p(\Omega) \to L^p(\Omega)$ by

$$\mathbf{K}_n f(y) = \int_\Omega K_n(x,y) f(x) dV(x).$$

Then

$$\lim_{n\to\infty} \|\mathbf{K}_n\|_p = 0.$$

Proof. By the Hölder inequality we have

$$|\mathbf{K}_n f(y)| \leq \left[\int_\Omega |K_n(x,y)| dV(x)\right]^{1/q} \left[\int_\Omega |K_n(x,y)||f(x)|^p dV(x)\right]^{1/p}.$$

Consequently we have

$$\|\mathbf{K}_n f\|_{L^p}^p \leq M \|f\|_{L^p}^p \int_\Omega \left[\int_\Omega |K_n(x,y)| dV(x)\right]^{p/q} dV(y).$$

Therefore we have $\lim_{n\to\infty} \|\mathbf{K}_n\|_p = 0$. □

Proposition A.13 *Let $\Omega \subset \mathbf{R}^n$ be a bounded open set and let $K(x,y)$ be a measurable function in $\Omega \times \Omega$. Suppose there exists $C > 0$ with the properties that*

(1) $\int_\Omega |K(x,y)| dV(x) \leq C$ $(y \in \Omega)$.

(2) $\int_\Omega |K(x,y)| dV(y) \leq C$ $(x \in \Omega)$.

For $1 \leq p < \infty$, we define a linear operator $\mathbf{K} : L^p(\Omega) \to L^p(\Omega)$ by

$$\mathbf{K} f(y) = \int_\Omega K(x,y) f(x) dV(x).$$

Then $\mathbf{K}$ is a compact operator.

Proof. First we assume that $K(x,y)$ is bounded. Then there exists $C > 0$ such that $|K(x,y)| \leq C$. Since K is expressed by

$$K(x,y) = K_1(x,y)^+ + K_1(x,y)^- + i(K_2(x,y)^+ + K_2(x,y)^-),$$

where $K_i^{\pm}(x,y) \geq 0$, there exists a sequence $\{K_n(x,y)\}$ of simple functions which are finite linear combinations of characteristic funtions of product sets in $\Omega \times \Omega$ such that $|K_n(x,y)| \leq 2C$ and $K_n(x,y) \to K(k,y)$ in $\Omega \times \Omega$ almost everywhere. Since

$$\int_\Omega \chi_{A \times B}(x,y) f(x) dV(x) = \int_\Omega \chi_A(x) f(x) dV(x) \chi_B(y),$$

the range of $\mathbf{K}_n : L^p(\Omega) \to L^p(\Omega)$ is a finite dimensional space. Hence By Proposition A.11 (b), $\mathbf{K}_n$ is compact. By Proposition A.12, $\|\mathbf{K} - \mathbf{K}_n\|_p \to 0$. By Proposition A.11 (c), $\mathbf{K}$ is compact. In the general case, we set

$$K^{(j)}(x,y) = \begin{cases} K(x,y) & (|K(x,y)| \leq j) \\ 0 & (|K(x,y)| > j) \end{cases}.$$

Then $K^{(j)}(x,y)$ is bounded, and hence compact on $L^p(D)$ by the first part of the proof. It follows from (2) that

$$\int_\Omega \left[\int_\Omega |K(x,y) - K^{(j)}(x,y)| |f(x)|^p dV(x) \right] dV(y)$$
$$\leq 2 \int_\Omega |K(x,y)| dV(y) \int_\Omega |f(x)|^p dV(x)$$
$$\leq 2C \|f\|^p,$$

which implies that

$$\|(\mathbf{K} - \mathbf{K}^{(j)})f\|_{L^p}^p \leq 2C \|f\|_{L^p}^p \int_\Omega \left[\int_\Omega |K - K^{(j)}| dV(x) \right]^{p/q} dV(y).$$

We set

$$g_j(y) = \int_\Omega |K(x,y) - K^{(j)}(x,y)| dV(x).$$

It follows from (1) that $|g_j(y)| \leq 2C$ and $g_j(y) \to 0$ (pointwise). By the Lebesgue dominated convergence theorem

$$\int_\Omega \left[\int_\Omega |K - K^{(j)}| dV(x) \right]^{p/q} dV(y) \to 0 \quad (j \to \infty).$$

By Proposition A.11 (c), $\mathbf{K}$ is compact. $\square$

Appendix B

Solutions to the Exercises

1.2 Suppose u is upper semicontinuous in Ω, that is, for any real number c, $\{z \mid u(z) < c\}$ is an open set. For $\varepsilon > 0$, $\{z \mid u(z) < u(a) + \varepsilon\}$ is an open set containing a. Hence for sufficiently small $\delta > 0$, if $|z - a| < \delta$, then $u(z) < u(a) + \varepsilon$. Consequently we have $\sup_{|z-a|<\delta} u(z) \leq u(a) + \varepsilon$, and hence

$$\limsup_{z \to a} u(z) = \lim_{\delta \to 0} \left(\sup_{|z-a|<\delta} u(z) \right) \leq u(a) + \varepsilon.$$

Since $\varepsilon > 0$ is arbitrary, we obtain $\limsup_{z \to a} u(z) \leq u(a)$.

1.3 Suppose $|f|$ attains the maximum at $a \in \Omega$. We choose $r = (r_1, \cdots, r_n)$ with $r_j > 0$ such that $\overline{P(a,r)} \subset \Omega$. It follows from Theorem 1.7 that

$$f(a) = \frac{1}{(2\pi i)^n} \int_{\partial P_1} \cdots \int_{\partial P_n} \frac{f(\zeta) d\zeta_1 \cdots d\zeta_n}{(\zeta_1 - a_1) \cdots (\zeta_n - a_n)}.$$

Consequently,

$$|f(a)| \leq \frac{1}{(2\pi)^n} \int_0^{2\pi} \cdots \int_0^{2\pi} |f(a_1 + r_1 e^{i\theta_1}, \cdots, a_n + r_n e^{i\theta_n})| d\theta_1 \cdots d\theta_n$$
$$\leq |f(a)|.$$

Hence we have

$$\int_0^{2\pi} \cdots \int_0^{2\pi} \{|f(a)| - |f(a_1 + r_1 e^{i\theta_1}, \cdots, a_n + r_n e^{i\theta_n})|\} d\theta_1 \cdots d\theta_n = 0.$$

Thus we have

$$|f(a)| = |f(a_1 + r_1 e^{i\theta_1}, \cdots, a_n + r_n e^{i\theta_n})| \quad (0 \leq \theta_j \leq 2\pi, j = 1, \cdots, n),$$

which means that $|f(z)| = |f(a)|$ for $z \in P(a,r)$. Hence $|f|$ is constant. Since f is holomorphic in each variable, f is constant.

1.4 Since f is the limit of continuous functions which converges uniformly on every compact subset of Ω, f is continuous in Ω. Let $a \in \Omega$. We choose $r > 0$ such that $\overline{P(a,r)} \subset \Omega$. It follows from Theorem 1.7 that for $z \in P(a,r)$

$$f_j(z) = \frac{1}{(2\pi i)^n} \int_{\partial P_1} \cdots \int_{\partial P_n} \frac{f_j(\zeta)d\zeta_1 \cdots d\zeta_n}{(\zeta_1 - z_1)\cdots(\zeta_n - z_n)}. \qquad \text{(B.1)}$$

Letting $j \to \infty$ in (B.1) we have

$$f(z) = \frac{1}{(2\pi i)^n} \int_{\partial P_1} \cdots \int_{\partial P_n} \frac{f(\zeta)d\zeta_1 \cdots d\zeta_n}{(\zeta_1 - z_1)\cdots(\zeta_n - z_n)}. \qquad \text{(B.2)}$$

The right side of (B.2) can be expanded to a power series with center a (or from (B.2) one can prove $\partial f/\partial \bar{z}_j = 0$), which implies that f is holomorphic in $P(a,r)$.

1.5 We choose $r = (r_1, \cdots, r_n)$ such that $P(\xi,r) \subset \Omega$. It follows from Theorem 1.7 that for $z \in P(\xi,r)$ f is expressed by

$$f(z) = \sum_{k_1,\cdots,k_n} a_{k_1,\cdots,k_n}(z_1 - \xi_1)^{k_1} \cdots (z_n - \xi_n)^{k_n},$$

where

$$a_{k_1,\cdots,k_n} = \frac{\partial^\alpha f(\xi)}{\alpha!} \quad (\alpha = (k_1, \cdots, k_n)).$$

Hence we have $f(z) = 0$ for $z \in P(\xi,r)$. Since Ω is connected, we have $f = 0$.

1.6 Let $0 < r < 1$. Define for t with $-\infty < t < \infty$

$$g(t) = \begin{cases} e^{-(t-r)^{-1}}e^{-(1-t)^{-1}} & (r < t < 1) \\ 0 & (\text{otherwise}) \end{cases}.$$

Then g is a C^∞ function in $\mathbf{R}$. We set

$$A = \int_{\mathbf{C}^n} g\left(\sqrt{|z_1|^2 + \cdots + |z_n|^2}\right) dV(z),$$

$$\lambda(z) = A^{-1} g\left(\sqrt{|z_1|^2 + \cdots + |z_n|^2}\right).$$

Then λ satisfies the following properties:

(a) $\lambda \in C^\infty(\mathbf{C}^n)$.

(b) $\lambda(z) = 0$ for $|z| \geq 1$.
(c) $\int_{\mathbf{C}^n} \lambda(z) dV(z) = 1$.
(d) λ depends only on $|z_1|, \cdots, |z_n|$.

1.7 Define g such that $g(z) = f(z)/z$ for $z \neq 0$ and $g(0) = f'(0)$ for $z = 0$. Then g is holomorphic in $B(0,1)$. By the maximum principle, we have $|g(z)| \leq 1$.

1.8 Define $\Phi(z) = (z + z_1)/(1 + \bar{z}_1 z)$ and $\Psi(z) = (z - w_1)/(1 - \bar{w}_1 z)$. Then the mapping $\Psi \circ f \circ \Phi : B(0,1) \to B(0,1)$ is one-to-one and onto and satisfies $\Psi \circ f \circ \Phi(0) = 0$. Apply Schwarz's lemma.

1.9 By Taylor's formula, there exist holomorphic functions φ and ψ at a such that $f(z) = (z-a)\varphi(z)$, $g(z) = (z-a)\psi(z)$, $\psi(a) \neq 0$ Use $\varphi(a) = f'(a)$ and $\psi(a) = g'(a)$.

1.10 Since $f(a) = 0$, there exists a holomorphic function g in an open neighborhood W of a such that $f(z) = (z-a)g(z)$ ($z \in W$) and $g(a) \neq 0$. By the continuity, there exists an open neighborhood $U \subset W$ of a such that $g(z) \neq 0$ for all $z \in U$. On the other hand, there exists a natural number N such that $z_n \in U$ whenever $n \geq N$. Then $0 = f(z_n) = (z_n - a)g(z_n)$ whenever $n \geq N$, which is a contradiction.

1.11 Let $w_0 \in f(\Omega)$. It is sufficient to show that $f(\Omega)$ contains an open neighborhood of w_0. There exists $z_0 \in \Omega$ such that $w_0 = f(z_0)$. We may assume that $w_0 = z_0 = 0$. By the uniqueness theorem (Exercise 1.10), we can choose $\delta > 0$ such that $\{z \mid |z| \leq \delta\} \subset \Omega$ and $f(z) \neq 0$ for $|z| = \delta$. We set $d = \min_{|z|=\delta} |f(z)|$. Then $d > 0$. Suppose there exists w such that $w \notin f(\Omega)$ and $|w| < d$. Since $\varphi(z) = (f(z) - w)^{-1}$ is holomorphic in Ω, it follows from the maximum principle that $1/|w| \leq 1/((d - |w|))$, which implies that $\{w \mid |w| < d/2\} \subset f(\Omega)$.

1.12 $a \in \Omega$. Since f'/f is holomorphic in a simply connected domain Ω, we can define a holomorphic function φ in Ω such that

$$\varphi(z) = \int_a^z \frac{f'(\zeta)}{f(\zeta)} d\zeta.$$

We set $\psi = e^\varphi$. Then a simple calculation yields

$$\frac{d}{dz}\left(\frac{f(z)}{\psi(z)}\right) = 0.$$

There is a constant C such that $f(z) = Ce^{\varphi(z)}$. Let α be an n-th root of

C. Then $g(z) = \alpha e^{(\psi(z)/m)}$ satisfies $f = g^m$. Let β satisfy $C = e^\beta$. Then $h = \psi + \beta$ satisfies $f = e^h$.

1.13 Suppose there exists $a \in \Omega$ such that $f'(a) = 0$. By the uniqueness theorem, there exists a positive integer m ($m \geq 2$) such that

$$0 = f'(a) = \cdots f^{(m-1)}(a) = 0, \quad f^{(m)}(a) \neq 0.$$

Using Taylor expansion, there exists a holomorphic function g in a neighborhood of a such that

$$f(z) = f(a) + (z-a)^m g(z), \quad g(a) \neq 0.$$

By continuity, there exists $\delta > 0$ such that $g(z) \neq 0$ for $z \in B(a, \delta)$. By Exercise 1.12, there exists a holomorphic function h in $B(a, \delta)$ such that $g(z) = h(z)^m$ for $z \in B(a, \delta)$. Define $\varphi(z) = (z-a)h(z)$. Then

$$f(z) = f(a) + \varphi(z)^m \quad (z \in B(a, \delta)), \quad \varphi(a) = 0.$$

Since $\varphi(B(a, \delta))$ is an open set containing 0 by Exercise 1.11, there is $\varepsilon > 0$ such that $\{w \mid |w| = \varepsilon\} \subset \varphi(B(a, \delta))$. Let $\varphi(z_0) = w_0$ and $|w_0| = \varepsilon$. We denote by $w_0, w_1, \cdots, w_{m-1}$, the m-th roots of w_0^m. Then there exist $z_0, z_1, \cdots, z_{m-1} \in B(a, \delta)$ such that $\varphi(z_i) = w_i$ ($0 \leq i \leq m-1$). Therefore, $f(z_i) = f(a) + w_0^m$ for $i = 0, 1, \cdots, m-1$, which contradicts f is a one-to-one mapping.

1.14 First we show that f^{-1} is continuous. Suppose $w_n, w_0 \in f(\Omega)$ and $w_n \to w_0$. We set $f^{-1}(w_n) = z_n$ and $f^{-1}(w_0) = z_0$. Since Ω is open, there exists $r > 0$ such that $\overline{B}(z_0, r) \subset \Omega$. For any $\varepsilon > 0$ with $0 < \varepsilon < r$, $F(B(z_0, \varepsilon))$ is an open set containing w_0 by Exercise 1.11. There exists an integer N such that $w_n \in f(B(z_0, \varepsilon))$ whenever $n \geq N$, and hence $z_n \in B(z_0, \varepsilon)$ whenever $n \geq N$. Thus we have $\lim_{n \to \infty} f^{-1}(w_n) = f^{-1}(w_0)$. Hence f^{-1} is continuous. We set $f(z) = w$, $f(z_0) = w_0$. Then if $w \to w_0$, then $z \to z_0$, Consequently,

$$\lim_{w \to w_0} \frac{f^{-1}(w) - f^{-1}(w_0)}{w - w_0} = \lim_{z \to z_0} \frac{z - z_0}{f(z) - f(z_0)} = \frac{1}{f'(z_0)}.$$

By Exercise 1.13, we have $f'(z_0) \neq 0$. Hence f^{-1} is holomorphic.

2.2 Let K be a compact subset of Ω. For $z \in K$ there exists $\varepsilon(z) > 0$ such that

$$B(z, \varepsilon(z)) = \{w \in \mathbf{C}^n \mid |w - z| < \varepsilon(z)\} \subset \Omega.$$

Since K is compact, by the Heine-Borel theorem there exist $z_i \in K$ ($i = 1, \cdots, p$) such that

$$K \subset \bigcup_{i=1}^{p} B(z_i, \varepsilon(z_i)).$$

We set $L = \cup_{i=1}^{p} B(z_i, \varepsilon(z_i))$. Let d be the distance between K and the boundary of L. Choose ρ such that $0 < \rho < d/(3n)$. For $z', z'' \in K$, $|z' - z''| < \rho$, we set $\Gamma = \{w \mid |w_j - z'_j| = 2\rho\}$. Then by the Cauchy integral formula we have

$$f_\lambda(z') - f_\lambda(z'') = \frac{1}{(2\pi i)^n} \int_\Gamma \frac{f_\lambda(\zeta_1, \cdots, \zeta_n)}{(\zeta_1 - z'_1) \cdots (\zeta_n - z'_n)} d\zeta_1 \cdots d\zeta_n$$
$$- \frac{1}{(2\pi i)^n} \int_\Gamma \frac{f_\lambda(\zeta_1, \cdots, \zeta_n)}{(\zeta_1 - z''_1) \cdots (\zeta_n - z''_n)} d\zeta_1 \cdots d\zeta_n.$$

Since $\mathcal{F}$ is uniformly bounded, there exists a constant $M > 0$ such that

$$|f_\lambda(\zeta)| < M \qquad (\lambda \in \Lambda, \zeta \in \Omega).$$

Hence there exists a constant $C > 0$ such that

$$|f_\lambda(z') - f_\lambda(z'')| \leq \frac{CM|z' - z''|}{\rho^{2n}}.$$

Thus for any $\varepsilon > 0$, if we set $\delta = \rho^{2n}\varepsilon(CM)^{-1}$, then

$$z', z'' \in K, |z' - z''| < \delta \Rightarrow |f_\lambda(z') - f_\lambda(z'')| < \varepsilon,$$

which means that $\mathcal{F}$ is equicontinuous on K.

2.3 Let $\{K_j\}$ be a sequence of compact subsets of Ω which satisfies

$$K_j \subset (K_{j+1})^\circ, \quad \bigcup_{j=1}^{\infty} K_j = \Omega.$$

We choose a countable set $E \subset \Omega$ such that each $E \cap K_j$ is dense in K_j. Let $E = \{w_i\}$. Since $\{u_m(w_1)\}$ is a bounded sequence, $\{u_m\}$ contains a subsequence $\{u_{m,1}\}$ which converges at w_1. By the same reason, $\{u_{m,1}\}$ contains a subsequence $\{u_{m,2}\}$ which converges at w_2. Repeating this process, there exists a subsequence $\{u_{m,m}\}$ of $\{u_m\}$ which converges pointwise in E. It follows from Exercise 2.2 that $\{u_m\}$ is equicontinuous in K_j, which means that for $\varepsilon > 0$, there exists δ_j such that

$$z', z'' \in K_j, |z' - z''| < \delta_j \Rightarrow |u_{m,m}(z') - u_{m,m}(z'')| < \varepsilon \qquad (j = 1, 2, \cdots).$$

Let $K_j \cap E = \{a_i\}$. Since $K_j \cap E$ is dense in K_j, we have
$$K_j \subset \bigcup_{i=1}^{\infty} B(a_i, \delta_j).$$
Since K_j is compact, there exists a positive integer p such that
$$K_j \subset \bigcup_{i=1}^{p} B(a_i, \delta_j). \tag{B.3}$$
Since $\{u_{m,m}\}$ converges in $K_j \cap E$, there exists a positive integer n_0 such that if $r, s > n_0$, then
$$|u_{r,r}(a_i) - u_{s,s}(a_i)| < \varepsilon \quad (i = 1, \cdots, p).$$
Suppose $z \in K_j$. It follows from (B.3) that there exists i ($1 \le i \le p$) such that $|z - a_i| < \delta_j$. Hence, if $r, s > n_0$, then
$$|u_{r,r}(z) - u_{s,s}(z)| \le |u_{r,r}(z) - u_{r,r}(a_i)| + |u_{r,r}(a_i) - u_{s,s}(a_i)|$$
$$+ |u_{s,s}(a_i) - u_{s,s}(z)| < 3\varepsilon,$$
which implies that $\{u_{m,m}(z)\}$ converges uniformly on K_j. Let K be an arbitrary compact subsets of Ω. Then there exists K_j such that $K \subset K_j$. Hence $\{u_{m,m}\}$ converges uniformly on every compact subset of Ω.

2.4 Let $b > 0$. Define for x with $-\infty < x < \infty$
$$g_b(x) = \begin{cases} e^{-\frac{b}{x}} e^{-\frac{b}{a-x}} & (0 < x < a) \\ 0 & (\text{otherwise}) \end{cases}.$$
Then $g_b \in C^{\infty}(\mathbf{R})$. We set
$$f_b(x) = \frac{\int_x^a g_b(t) dt}{\int_0^a g_b(t) dt}.$$
Then we have $f_b(x) = 1$ ($x \le 0$), $f_b(x) = 0$ ($x \ge a$), $f_b \in C^{\infty}(\mathbf{R})$, $0 \le f_b(x) \le 1$. Since $\lim_{b \to 0} g_b(x) = 1$ ($0 < x < a$),
$$\int_0^a g_b(x) dx \to a \text{ as } b \to 0.$$
Hence if we choose $b > 0$ sufficiently small, then
$$|f_b'(x)| = \frac{|g_b(x)|}{\int_0^a g_b(t) dt} < \frac{c}{a}.$$
f_b satisfies (a), (b) and (c).

3.1 Let $x, y \in \Gamma_{\delta/2}$ and $d = |x - y| \leq \delta/2$. Then
$$|g(x_1, x') - g(x_1 + d, x')| \leq \int_{x_1}^{x_1+d} \left|\frac{\partial g}{\partial x_1}(t, x')\right| dt \leq C_1 K d^\alpha.$$
By the mean value theorem, there exists θ such that
$$|g(x_1 + d, x') - g(y_1 + d, y')| \leq K\theta^{\alpha-1} d,$$
where θ is a point between $x_1 + d$ and $y_1 + d$. Since $\theta > d$, we have
$$|g(x_1 + d, x') - g(y_1 + d, y')| \leq K d^\alpha.$$
Then
$$\begin{aligned}|g(x) - g(y)| &\leq |g(x_1, x') - g(x_1 + d, x')| \\ &+ |g(x_1 + d, x') - g(y_1 + d, y')| \\ &+ |g(y_1 + d, y') - g(y_1, y')| \\ &\leq C_2 K d^\alpha.\end{aligned}$$

3.2
$$\begin{aligned}&\frac{dF_1(z + \lambda\theta(w - z))}{d\lambda}\bigg|_{\lambda=1} \\ &= \sum_{j=1}^{n} \left\{\frac{\partial F_1}{\partial z_j}(z + \lambda\theta w)\theta w_j + \frac{\partial F_1}{\partial \bar{z}_j}(z + \lambda\theta w)\theta \bar{w}_j\right\}\bigg|_{\lambda=1} \\ &= \theta \frac{dF_1(z + \theta(w - z))}{d\theta}.\end{aligned}$$

3.3 By the Riesz representation theorem, there exists $y \in H$ such that
$$\varphi(x) = (x, y) \qquad (x \in H).$$
For $x \in M$, we have
$$0 = \varphi(x) = (x, y),$$
which implies that $y \in M^\perp$. Suppose there exists $x \in M^\perp$ such that x, y are linearly independent. We set
$$e_1 = \frac{y}{\|y\|}, \quad y_1 = x - (x, e_1)e_1, \quad e_2 = \frac{y_1}{\|y_1\|}.$$
Since $\{e_1, e_2\}$ is an orthonormal system, $(e_1, e_2) = 0$. Hence $(y_1, y) = 0$, and hence $\varphi(y_1) = 0$. Thus $y_1 \in M$. Since y_1 is a linear combination of

x and y, $y_1 \in M^\perp$. Since $M \cap M^\perp = \{0\}$, we have $y_1 = 0$. Hence x and y are linearly dependent, which contradicts our assumption. Hence $M^\perp$ is one dimensional.

3.4 When $j \neq k$, we have

$$(z^j, z^k) = \int_\Omega z^j \bar{z}^k dxdy = \int_0^1 \int_0^{2\pi} r^{j+k} e^{i\theta(j-k)} r\,dr\,d\theta = 0,$$

$$\|z^n\| = \frac{\sqrt{\pi}}{\sqrt{n+1}}.$$

Hence $\{\varphi_n(z)\}$ is an orthonormal sequence in $A^2(\Omega)$. We define $\psi_j : A^2(\Omega) \to \mathbf{C}$ by

$$\psi_j(f) = \left(\frac{\partial}{\partial z}\right)^j f(0) (= f^{(j)}(0)).$$

Let $0 < r_1 \leq \rho \leq r_2 < 1$. By the Cauchy integral formula

$$f^{(j)}(0) = \frac{j!}{2\pi i} \int_{|z|=\rho} \frac{f(z)}{z^{j+1}} dz = \frac{j!}{2\pi} \int_0^{2\pi} \frac{f(\rho e^{i\theta})}{(\rho e^{i\theta})^j} d\theta.$$

If we multiply by ρ and integrate from r_1 to r_2, then we obtain

$$\frac{(r_2-r_1)^2}{2} f^{(j)}(0) = \frac{j!}{2\pi} \int_{r_1 \leq |z| \leq r_2} \frac{f(z)}{z^j} dxdy.$$

Consequently, we have for some constant $C > 0$

$$|f^{(j)}(0)| \leq \frac{j!}{\pi(r_2-r_1)^2 r_1^j} \int_{r_1 \leq |z| \leq r_2} |f(z)| dxdy$$
$$\leq C\|f\|.$$

Hence ψ_j is a continuous linear functional. We set

$$M_j = \{f \in A^2(\Omega) \mid \psi_j(f) = 0\}.$$

For $f \in M_j$, we have a representation

$$f(z) = \sum_{k=0}^\infty a_k z^k \quad (a_k = \frac{f^{(k)}(0)}{k!}, |z| \leq R < 1).$$

Since $a_j = 0$, we have

$$\int_{B(0,R)} \varphi_j(z)\overline{f(z)}dxdy = \int_0^R \int_0^{2\pi} \varphi_j(z)\overline{f(re^{i\theta})}rdrd\theta$$

$$= \sum_{k=0}^\infty \int_0^R \int_0^{2\pi} r^j e^{i(j-k)\theta} \overline{a_k} r^{k+1} dr d\theta = 0.$$

Since $R < 1$ is arbitrary, $(f, \varphi_j) = 0$, and hence $\varphi_j \in M_j^\perp$. Since $M_j^\perp$ is one dimensional, $M_j^\perp = \{c\varphi_j | c \in \mathbf{C}\}$. By the Riesz representation theorem, there exists $x_j \in A^2(\Omega)$ such that

$$\psi_j(f) = (f, x_j) \quad (f \in A^2(\Omega)).$$

We set

$$x_j = x_j' + x_j'', \quad (x_j' \in M_j, x_j'' \in M_j^\perp).$$

Then we have $x_j'' = c_j \varphi_j$. If we set $f = f_1 + f_2$ ($f_1 \in M_j, f_2 \in M_j^\perp$), then

$$\psi_j(f) = (f_2, x_j'') = (f, c_j \varphi_j),$$

which means that

$$(f, \varphi_j) = 0 \text{ for all } j \implies a_j = 0 \text{ for all } j \implies f = 0.$$

Hence $\{\varphi_n\}$ is complete.

3.5 It follows from Exercise 3.4 that

$$K_\Omega(z, \zeta) = \sum_{j=0}^\infty \frac{\sqrt{j+1}}{\sqrt{\pi}} z^j \frac{\sqrt{j+1}}{\sqrt{\pi}} \bar{\zeta}^j = \frac{1}{\pi} \sum_{j=0}^\infty (j+1)(z\bar{\zeta})^j$$

$$= \frac{1}{\pi} \frac{1}{(1-z\bar{\zeta})^2}.$$

3.6 Assume that $n = 1$. (a) Let $\zeta \in \Omega$. For $f \in A^2(\Omega)$, define

$$\psi(f) = \frac{\partial f}{\partial z}(\zeta).$$

Then ψ is a continuous linear functional on $A^2(\Omega)$. By the Riesz representation theorem, there exists a function $u(z, \zeta) \in A^2(\Omega)$ such that

$$\psi(f) = (f, u(z, \zeta)) \quad (f \in A^2(\Omega)).$$

Then
$$u(z_0, \zeta) = (u(z,\zeta), K_\Omega(z, z_0)) = \overline{(K_\Omega(z,z_0), u(z,\zeta))}$$
$$= \overline{\psi(K_\Omega(\cdot, z_0))} = \overline{\frac{\partial K_\Omega(\zeta, z_0)}{\partial \zeta}} = \frac{\partial K_\Omega(z_0, \zeta)}{\partial \bar\zeta},$$

which implies that
$$\frac{\partial f}{\partial z}(\zeta) = \left(f(z), \frac{\partial K_\Omega(z, \zeta)}{\partial \bar\zeta} \right).$$

(b) We have a representation
$$K_\Omega(z, z) = \sum_{n=1}^\infty |\varphi_n(z)|^2,$$

where $\{\varphi_j(z)\}$ is a complete orthonormal system in $A^2(\Omega)$. If $K(z,z)=0$, then for any $f \in A^2(\Omega)$ we have $f(z) = 0$. Thus $K_\Omega(z,z) > 0$.

(c) We have
$$K_\Omega(\zeta, \zeta) \frac{\partial^2 \log K_\Omega(\zeta, \zeta)}{\partial \zeta \partial \bar\zeta} = \frac{\partial^2 K_\Omega(\zeta, \zeta)}{\partial \zeta \partial \bar\zeta}$$
$$- \frac{1}{K_\Omega(\zeta,\zeta)} \frac{\partial K_\Omega(\zeta,\zeta)}{\partial \zeta} \frac{\partial K_\Omega(\zeta,\zeta)}{\partial \bar\zeta}.$$

We fix $\zeta \in \Omega$. We set
$$L(z) = \frac{\partial K_\Omega(z,\zeta)}{\partial \bar\zeta}.$$

Then $L \in A^2(\Omega)$. We set
$$H_0 = \{ f \in A^2(\Omega) \mid (f, K_\Omega(\cdot, \zeta)) = 0 \}.$$

Then by the property of the Bergman kernel, $H_0 = \{f \in A^2(\Omega) \mid f(\zeta) = 0\}$. Moreover we have
$$\|K_\Omega(\cdot, \zeta)\|^2 = (K_\Omega(\cdot, \zeta), K_\Omega(\cdot, \zeta)) = \int_\Omega K_\Omega(z, \zeta) \overline{K_\Omega(z, \zeta)} dx dy$$
$$= \int_\Omega K_\Omega(z, \zeta) K_\Omega(\zeta, z) dx dy = K_\Omega(\zeta, \zeta).$$

Now we have
$$L(\cdot) - \alpha K_\Omega(\cdot, \zeta) \in H_0$$

$\iff$
$$0 = (L - \alpha K_\Omega(\cdot,\zeta), K_\Omega(\cdot,\zeta)) = (L, K_\Omega(\cdot,\zeta)) - \alpha K_\Omega(\zeta,\zeta)$$

$\iff$
$$\alpha = \frac{(L, K_\Omega(\cdot,\zeta))}{K_\Omega(\zeta,\zeta)}.$$

We choose $\alpha = (L, K_\Omega(\cdot,\zeta))/K_\Omega(\zeta,\zeta)$. Then we have

$$\|L - \alpha K_\Omega(\cdot,\zeta)\|^2 = (L - \alpha K_\Omega(\cdot,\zeta), L - \alpha K_\Omega(\cdot,\zeta))$$

$$= \|L\|^2 - \alpha(K_\Omega(\cdot,\zeta), L) - \bar{\alpha}(L, K_\Omega(\cdot,\zeta)) + |\alpha|^2\|K_\Omega(\cdot,\zeta)\|^2$$

$$= \left(\frac{\partial K_\Omega(\cdot,\zeta)}{\partial \bar{\zeta}}, \frac{\partial K_\Omega(\cdot,\zeta)}{\partial \bar{\zeta}}\right) - \frac{|(L, K_\Omega(\cdot,\zeta))|^2}{K_\Omega(\zeta,\zeta)}$$

$$= \frac{\partial^2 K_\Omega(\zeta,\zeta)}{\partial\zeta\partial\bar{\zeta}} - \frac{1}{K_\Omega(\zeta,\zeta)} \frac{\partial K_\Omega(\zeta,\zeta)}{\partial\zeta} \frac{\partial K_\Omega(\zeta,\zeta)}{\partial\bar{\zeta}}.$$

Hence we obtain
$$\frac{\partial^2 \log K_\Omega(\zeta,\zeta)}{\partial\zeta\partial\bar{\zeta}} \geq 0.$$

Suppose there exists $\zeta \in \Omega$ such that $\partial^2 \log K_\Omega(\zeta,\zeta)/\partial\zeta\partial\bar{\zeta} = 0$. Then $L - \alpha K_\Omega(\cdot,\zeta) = 0$. For $f \in H_0$ we have

$$0 = (f, L - \alpha K_\Omega(\cdot,\zeta)) = (f, L) = \left(f, \frac{\partial K_\Omega(\cdot,\zeta)}{\partial\bar{\zeta}}\right) = \frac{\partial f}{\partial\zeta}(\zeta).$$

We set $f(z) = z - \zeta$. Since Ω is bounded, $f \in A^2(\Omega)$ and $f(\zeta) = 0$ which implies that $f \in H_0$. Further we have $\frac{\partial f}{\partial\zeta}(\zeta) = 1$, which is a contradiction. Hence we have $\partial^2 \log K_\Omega(\zeta,\zeta)/\partial\zeta\partial\bar{\zeta} > 0$.

3.7 It follows from Theorem 3.24 that

$$g_{ij}^{\Omega_1}(z) = \frac{\partial^2}{\partial z_i \partial \bar{z}_j} \log K_{\Omega_1}(z,z)$$

$$= \frac{\partial^2}{\partial z_i \partial \bar{z}_j} \log\{|\det f'(z)|^2 K_{\Omega_2}(f(z), f(z))\}$$

$$= \frac{\partial^2}{\partial z_i \partial \bar{z}_j} \log K_{\Omega_2}(f(z), f(z)).$$

4.1 By the expansion formula of the determinant we have

$$J_n = \begin{vmatrix} \frac{\partial x_1}{\partial r} & \frac{\partial x_1}{\partial \theta_1} & \frac{\partial x_1}{\partial \theta_2} & \cdots & \frac{\partial x_1}{\partial \theta_{n-1}} \\ \vdots & \vdots & \vdots & \vdots & \vdots \\ \frac{\partial x_{n-1}}{\partial r} & \frac{\partial x_{n-1}}{\partial \theta_1} & \frac{\partial x_{n-1}}{\partial \theta_2} & \cdots & \frac{\partial x_{n-1}}{\partial \theta_{n-1}} \\ \cos\theta_1 & -r\sin\theta_1 & 0 & \cdots & 0 \end{vmatrix}$$

$$= (-1)^{n+1} \cos\theta_1 \begin{vmatrix} \frac{\partial x_1}{\partial \theta_1} & \cdots & \frac{\partial x_1}{\partial \theta_{n-1}} \\ \vdots & \vdots & \vdots \\ \frac{\partial x_{n-1}}{\partial \theta_1} & \cdots & \frac{\partial x_{n-1}}{\partial \theta_{n-1}} \end{vmatrix}$$

$$+ (-1)^{n+2}(-r\sin\theta_1) \begin{vmatrix} \frac{\partial x_1}{\partial r} & \frac{\partial x_1}{\partial \theta_2} & \cdots & \frac{\partial x_1}{\partial \theta_{n-1}} \\ \vdots & \vdots & \vdots & \vdots \\ \frac{\partial x_{n-1}}{\partial r} & \frac{\partial x_{n-1}}{\partial \theta_2} & \cdots & \frac{\partial x_{n-1}}{\partial \theta_{n-1}} \end{vmatrix}.$$

We set

$$y_1 = \sin\theta_2 \cdots \sin\theta_{n-3} \sin\theta_{n-2} \sin\theta_{n-1}$$
$$y_2 = \sin\theta_2 \cdots \sin\theta_{n-3} \sin\theta_{n-2} \cos\theta_{n-1}$$
$$y_3 = \sin\theta_2 \cdots \sin\theta_{n-3} \sin\theta_{n-3} \cos\theta_{n-2}$$
$$\cdots$$
$$y_{n-1} = \cos\theta_2.$$

Then we have

$$= (-1)^{n+1} r \begin{vmatrix} y_1 & \frac{\partial x_1}{\partial \theta_2} & \cdots & \frac{\partial x_1}{\partial \theta_{n-1}} \\ \vdots & \vdots & \vdots & \vdots \\ y_{n-1} & \frac{\partial x_{n-1}}{\partial \theta_2} & \cdots & \frac{\partial x_{n-1}}{\partial \theta_{n-1}} \end{vmatrix}$$

$$= \cdots$$

$$= (-1)^{n+1} r^{n-1} \sin^{n-2}\theta_1 (-1)^n \sin^{n-3}\theta_2 \cdots (-1)^5 \sin^2\theta_{n-3}$$
$$\times \begin{vmatrix} \sin\theta_{n-1} & \sin\theta_{n-2}\cos\theta_{n-1} \\ \cos\theta_{n-1} & -\sin\theta_{n-2}\sin\theta_{n-1} \end{vmatrix}$$
$$= \pm r^{n-1} \sin^{n-2}\theta_1 \sin^{n-3}\theta_2 \cdots \sin^2\theta_{n-3} \sin\theta_{n-2}.$$

4.2 Divide the domain of integration into three parts

$$\{z \mid |z| < R\} = \{z \mid |z| < R, |z| < |w|/2\}$$
$$\cup \{z \mid |z| < R, |z| \geq |w|/2, |z+w| < |w|/2\}$$
$$\cup \{z \mid |z| < R, |z| \geq |w|/2, |z+w| \geq |w|/2\},$$

and use the polar coordinate system.

4.3 We set

$$\rho_j(z) = \frac{\partial \rho}{\partial z_j}(z), \quad \Phi(z,\zeta) = \sum_{j=1}^{m} \rho_j(\zeta)(z_j - \zeta_j)$$

Then by Lemma 4.6, we have

$$|\Phi(z,\zeta)| \geq C(|\operatorname{Im} \Phi(z,\zeta)| + |\rho(z)| + \sum_{j=1}^{n} |\zeta_j|^{2m_j-2}|z_j - \zeta_j|^2 + |z-\zeta|^M).$$

For z with $\rho_n(z) \neq 0$, define

$$t(\zeta) = \rho(\zeta) + |\rho(z)|, \quad y(\zeta) = \operatorname{Im} \Phi(z,\zeta),$$

$$x_{2j-1}(\zeta) = \operatorname{Re}(z_j - \zeta_j), \quad x_{2j}(\zeta) = \operatorname{Im}(z_j - \zeta_j), \quad j = 1, \cdots, n-1.$$

Then $t, y, x_1, \cdots, x_{2n-2}$ form a coordinate system in a neighborhood of z. Then apply the method in the proof of Theorem 3.11 and Exercise 4.2.

4.4 (1) Use the strict convexity of t^{2p}. (2) Apply (1).

4.5 By the binomial theorem, there exist positive integers $\alpha_1, \cdots, \alpha_m$ such that

$$\operatorname{Re}(z^{2m}) = x^{2m} + \alpha_1 x^{2m-2}(iy)^2 + \cdots + \alpha_m (iy)^{2m}$$

Then on Γ_σ,

$$\operatorname{Re}(z^{2m}) \geq x^{2m} - \alpha_1 x^{2m-2} y^2 - \alpha \cdots \alpha_- m y^{2m}$$
$$\geq x^{2m} - \alpha_1 \sigma^2 x^{2m} - \cdots - \alpha_m \sigma^{2m} x^{2m} \geq x^{2m}/2$$

for sufficiently small $\sigma > 0$. On the other hand if we choose $\varepsilon > 0$ sufficiently small, then $\varepsilon |z|^{2m} \leq \operatorname{Re}(z^{2m})$ on Γ_σ.

4.6 Divide the domain of integration into 3 parts:

$$\{z \mid |z| < R\} = \{z \mid |z| < R, |x| < |t|/2\}$$
$$\cup \{z \mid |z| < R, |x+t| < |t|/2, |x| > |t|/2\}$$
$$\cup \{z \mid |z| < R, |x| > |t|/2, |x+t| > |t|/2\}.$$

4.7 (1) Apply Exercise 4.4 (2) to the equation

$$2\operatorname{Re}\Phi(z,\zeta) = \sum_{k=1}^{N}\{2n_k\xi_k^{2n_k-1}(x_k-\xi^k) + 2m_k\eta_k^{2m_k-1}(y_k-\eta_k)\}$$

$$-\gamma\sum_{k=1}^{N}\{(\eta_k^{2m_k-2}-\xi_k^{2n_k-2})((x_k-\xi_k)^2-(y_k-\eta_k)^2) + \operatorname{Re}((z_k-\zeta_k)^{2m_k})\}$$

$$= \rho(z) + \sum_{k=1}^{N}\{\xi_k^{2n_k}-x_k^{2n_k}+2n_k\xi_k^{2n_k-1}(x_k-\xi_k)$$

$$+\eta_k^{2m_k}-y_k^{2m_k}+2m_k\eta_k^{2m_k-1}(y_k-\eta_k)\}$$

$$-\gamma\sum_{k=1}^{N}\{(\eta_k^{2m_k-2}-\xi_k^{2n_k-2})((x_k-\xi_k)^2-(y_k-\eta_k)^2) + \operatorname{Re}((z_k-\zeta_k)^{2m_k})\}.$$

Then there exists $\delta > 0$ such that

$$2\operatorname{Re}\Phi(z,\zeta) \le \rho(z) - \sum_{k=1}^{N}\{\xi_k^{2n_k-2}((\delta-\gamma)(x_k-\xi_k)^2 + \gamma(y_k-\eta_k)^2)$$

$$+\eta_k^{2m_k-2}((\delta-\gamma)(y_k-\eta_k)^2 + \gamma(x_k-\xi_k)^2)\}$$

$$-\sum_{k=1}^{N}\{\delta(y_k-\eta_k)^{2m_k} + \gamma\operatorname{Re}((z_k-\zeta_k)^{2m_k})\}$$

If $0 < \gamma < \delta$, $\alpha = \min\{\gamma, \delta-\gamma\}$, then

$$2\operatorname{Re}\Phi(z,\zeta) \le \rho(z) - \alpha\sum_{k=1}^{N}(\xi_k^{2n_k-2}-\eta_k^{2m_k-2})|z_k-\zeta_k|^2$$

$$-\sum_{k=1}^{N}\{\delta(y_k-\eta_k)^{2m_k} + \gamma\operatorname{Re}((z_k-\zeta_k)^{2m_k})\}.$$

By Exercise 4.5, if we choose $\gamma > 0$ small enough, then there exists $\beta > 0$

such that
$$\sum_{k=1}^{N}\{\delta(y_k - \eta_k)^{2m_k} + \gamma \mathrm{Re}\,((z_k - \zeta_k)^{2m_k})\} \geq \beta |z_k - \zeta_k|^{2m_k}.$$

(2) Use (1) and Exercise 4.6.

5.1 Let $\{z_\nu\} \subset \Delta$ be a sequence such that $z_\nu \to z_0$. If we set $\varphi_\nu^j(w) = f_j(z_\nu, w)$ for $j = 1, 2$, then $\varphi_\nu^j : \Delta \to \Delta$ are holomorphic. By the Montel theorem, $\{\varphi_\nu^j\}$ contains a subsequence $\{\varphi_{\nu_k}^j\}$ which converges uniformly on every compact subset of Ω. Let $\lim_{k\to\infty} \varphi_{\nu_k}^j = \varphi_j$. Since $F(z, w)$ is a biholomorphic mapping, $F(z_{\nu_k}, w) = (\varphi_{\nu_k}^1(w), \varphi_{\nu_k}^2(w))$ converges to a point in ∂B. Hence $(\varphi_1(w), \varphi_2(w)) \in \partial B$ which means that $|\varphi_1(w)|^2 + |\varphi_2(w)|^2 = 1$. Operating $\partial^2/\partial \bar{w} \partial w$ we have $|\varphi_1'(w)|^2 + |\varphi_2'(w)|^2 = 0$, Hence we have $\varphi_1' = \varphi_2' = 0$ in Δ. Consequently we have $\lim_{k\to\infty} F_w(z_{\nu_k}, w) = (\varphi_1'(w), \varphi_2'(w)) = 0$. Suppose that $\lim_{z\to z_0} F_w(z) = 0$ does not hold. Then there exists a sequence $\{z_n\}$ and $\delta > 0$ such that $z_n \to z_0$, $|F_w(z_n)| \geq \delta$, which is a contradiction. This proves (a). For fixed $w \in \Delta$, define $F_w(z) = 0$ ($z \in \partial \Delta$). Then by (a), F_w is continuous on $\overline{\Delta}$, holomorphic in Δ and equals 0 on $\partial \Delta$. By the maximum principle, $F_w = 0$, and hence $f_1(z, w)$ and $f_2(z, w)$ are constant with respect to w. This proves (b). It follows from (b) that F is not one-to-one, which is a contradiction.

5.2 By the definition of the sheaf, $\pi : \mathcal{S} \to X$ is a local homeomorphism. Hence there exists a neighborhood W of $s_x \in \pi^{-1}(x)$ such that $\pi : W \to \pi(W) = U$ is a homeomorphism. Hence we have $\pi^{-1}(x) \cap W = \{s_x\}$. Define $s = (\pi|_W)^{-1}$. Then we have $\pi \circ s(y) = y$ ($y \in U$), which implies that s is a section over U. Since $s(x) \in \pi^{-1}(x) \cap W$, we have $s(x) = s_x$.

5.3 We set
$$h(s) = \varphi(s) \exp\left[-\int_0^s \frac{\varphi'(\theta)}{\varphi(\theta)} d\theta\right].$$

Then $h'(s) = 0$, and hence h is constant. Thus $h(2\pi) = h(0) = \varphi(0)$. Since
$$h(2\pi) = \varphi(2\pi)\exp(-2\pi i N(\varphi)) = \varphi(0)\exp(-2\pi i N(\varphi)),$$
$\exp(-2\pi i N(\varphi)) = 1$. Therefore $N(\varphi)$ is an integer.

5.4 Let $I = [0, 1]$. For $t \in I$, we set $\varphi_t(\theta) = g(te^{i\theta})$. Then $\varphi_t : [0, 2\pi] \to \mathbb{C}\backslash\{0\}$ is a C^1 curve. By exercise 5.3 $N(\varphi_t)$ is an integer. Moreover $\varphi_t(\theta)$ and $\varphi_t'(\theta)$ are continuous on $[0, 2\pi] \times I$. Hence $N(\varphi_t)$

is continuous with respect to t, which means that $N(\varphi_1) = N(\varphi_0) = 0$.

5.5 (a) Since $\operatorname{Im} z_1 = \operatorname{Im} z_2$ in A, we have for $z \in A \cap \{z \in \Omega \mid \operatorname{Im} z_1 = 0\}$, $x_2 - x_1 + 1 = 0$, $3/4 < |x_1| < 5/4$, $3/4 < |x_2| < 5/4$, which is a contradiction. (b) follows from $U_1 \cap U_2 = \Omega \setminus A$. (c) Since $f_1(1, e^{i\theta}) = e^{i\theta}$, $f_1(-1, e^{i\theta}) = e^{i\theta} + 2$, we have

$$N(f_1(1, e^{i\theta})) = \frac{1}{2\pi i} \int_0^{2\pi} \frac{ie^{i\theta}}{e^{i\theta}} d\theta = 1$$

and

$$N(f_1(-1, e^{i\theta})) = \frac{1}{2\pi i} \int_0^{2\pi} \frac{ie^{i\theta}}{e^{i\theta} + 2} d\theta = -\frac{1}{2\pi} \int_{|z|=1} \frac{dz}{z+2} = 0.$$

Consequently,

$$N(f_1(-1, e^{i\theta})) = 0 \neq 1 = N(f_1(1, e^{i\theta})). \tag{B.4}$$

Suppose that $f \in \mathcal{O}(\Omega)$ satisfies $f/f_2 \in \mathcal{O}^*(U_2)$, $h = f/f_1 \in \mathcal{O}^*(U_1)$. We set $\varphi_t(\theta) = f(e^{it}, e^{i\theta})$ $(0 \leq \theta \leq 2\pi, -\pi \leq t \leq 0)$. Then $f = f/f_2$ does not vanish in U_2, by the continuity of $N(\varphi_t)$ with respect to t

$$N(f(-1, e^{i\theta})) = N(\varphi_{-\pi}) = N(\varphi_0) = N(f(1, e^{i\theta})).$$

Similarly, taking into account that $h(e^{it}, e^{i\theta}) \neq 0$ $(0 \leq \theta \leq 2\pi, 0 \leq t \leq \pi)$, we have

$$N(h(-1, e^{i\theta})) = N(h(1, e^{i\theta})).$$

Since $f = hf_1$ in U_1, we obtain

$$N(f(\zeta, e^{i\theta})) = N(h(\zeta, e^{i\theta})) + N(f_1(\zeta, e^{i\theta})) \qquad (\zeta = \pm 1).$$

Hence we have $N(f_1(-1, e^{i\theta})) = N(f_1(1, e^{i\theta}))$, which contradicts (B.4).

Bibliography

[ADA1] Adachi, K. (1980). Continuation of A^∞-functions from submanifolds to strictly pseudoconvex domains, J. Math. Soc. Japan **32**, pp. 331–341.
[ADA2] Adachi,K. (1984). Extending bounded holomorphic functions from certain subvarieties of a weakly pseudoconvex domain, Pacific J. Math. **110**, 1, pp. 9–19.
[ADA3] Adachi, K. (1987). Continuation of bounded holomorphic functions from certain subvarieties to weakly psudoconvex domains, Pacific J. Math. **130**, 1, pp. 1–8.
[ADA4] Adachi, K. (2003). L^p extension of holomorphic functions from submanifolds to strictly pseudoconvex domains with nonsmooth boundary, Nagoya Math. J. **172**, pp. 103–110.
[ADC] Adachi, K., Andersson, M and Cho, H.R. (1999). L^p and H^p extensions of holomorphic functions from subvarieties of analytic polyhedra, Pacific J. Math. **189**, 2, pp. 201–210.
[ADK] Adachi, K. and Kajimoto, H. (1993). On the extension of Lipschitz functions from boundaries of subvarieties to strongly psudoconvex domains, Pacific. J. Math. **158**, 1, pp. 201–222.
[ADM] Adams, R.A. (1975). Sobolev spaces, Academic Press.
[AHS1] Ahern, P. and Schneider, R. (1976). The boundary behavior of Henkin's kernel, Pacific J. Math. **66**, 1, pp. 9–14.
[AHS2] Ahern, P. and Schneider, R. (1979). Holomorphic Lipschitz functions in pseudoconvex domains, Amer. J. Math. **101**, 3, pp. 543–565.
[AMA1] Amar, E. (1983). Extension de fonctions holomorphes et courants, Bull. Sci. Math. **107**, pp. 25–48.
[AMA2] Amar, E. (1984). Cohomologie complexe et applications, J. London Math. Soc. **29**, pp. 127–140.
[BEA] Beatrous, F. (1985). L^p estimates for extensions of holomorphic functions, Michigan Math. J. **32**, pp. 361–380.
[BEL] Bell, S.R. and Ligocka, E. (1980). A simplification and extension of Fefferman's theorem on biholomorphic mappings, Invent. Math. **57**, pp. 283–289.
[BES] Berezansky, Y.M., Sheftel, Z.G. and Us, G.F. (1996). Functional Analysis, Birkhäuser.

[BR1] Berndtsson, B. (1983). A formula for interpolation and division in $\mathbf{C}^n$, *Math. Ann.* **263**, pp. 399–418.

[BR2] Berndtsson, B. (1996). The extension theorem of Ohsawa-Takegoshi and the theorem of Donnelly-Fefferman, *Ann. Inst. Fourier* **46**, pp. 1083-1094.

[BRA] Berndtsson, B. and Andersson, M. (1982). Henkin-Ramirez formulas with weight factors, *Ann. Inst. Fourier* **32**, pp. 91–110.

[BRE] Bremermann, H.J. (1954). Über die Äquivalenz der pseudokonvexen Gebiete und der Holomorphiegebiete im Raum von n komplexen Veränderlichen, *Math. Ann.* **128**, pp. 63–91.

[BRG] Bruna,J. and Burgués, M. (1987), Holomorphic approximation and estimates for the $\bar{\partial}$ equation on strictly pseudoconvex nonsmooth domains, *Duke Math. J.* **55**, 3, pp. 539–596.

[BRJ] Bruna, J. and Castillo, J. (1984). Hölder and L^p-estimates for the $\bar{\partial}$ equation in some convex domains with real-analytic boundary, *Math. Ann.* **269**, pp. 527–539.

[BRV] Bruna, J., Cufi, J. and Verdera, J. (1985). Cauchy kernels in strictly pseudoconvex domains and an application to a Mergelyan type approximation problem, *Math. Z.* **189**, pp. 41–53.

[CH] Chen, S. C. (1990). Real analytic boundary regularity of the Cauchy transform on convex domains, *Proc. Amer. Math. Soc.* **108**, pp. 423–432.

[CHE] Chen, Z. (1990). Local real analytic boundary regularity of an integral solution operator of the $\bar{\partial}$-equation on convex domains, *Pacific J. Math.* **156**, pp. 97–105.

[CHK] Chen, Z, Krantz, S.G. and Ma, D. (1993). Optimal L^p estimates for the $\bar{\partial}$ estimates for the $\bar{\partial}$-equation on complex ellipsoids in $\mathbf{C}^n$, *Manuscripta Math.* **80**, pp. 131–149.

[CHO] Cho, H.R. (1998). A counterexample to the L^p extension of holomorphic functions from subvarieties to pseudoconvex domains, *Complex Variables* **35**, pp. 89–91.

[CUM] Cumenge, A. (1983). Extension dans des classes de Hardy de fonctions holomorphes et estimation de type "measures de Carleson" pour l'equation ∂, *Ann. Inst. Fourier* **33**, pp. 59–97.

[DA] D'Angelo, J. (1993). Several complex variables and the geometry of real hypersurfaces, CRC Press, Inc.

[DIK] Diederich, K., Fornaess, J.E. and Wiegerinck, J. (1986). Sharp Hölder estimates for $\bar{\partial}$ on ellipsoids, *Manuscripta Math.* **56**, pp. 399–417.

[DIM1] Diederich, K. and Mazzilli, E. (1997). Extension and restriction of holomorphic functions, *Ann. Inst. Fourier* **47**, pp. 1079-1099.

[DIM2] Diederich, K. and Mazzilli, E. (2001). Extension of bounded holomorphic functions in convex domains, *Manuscripta Math.* **105**, pp. 1–12.

[ELG] Elgueta, M. (1980). Extensions to strictly pseudoconvex domains of functions holomorphic in a submanifold in general position and $C^i nfty$ up to the boundary, *Ill. J. Math.* **24**, pp. 1–17.

[FEF] Fefferman, C. (1974). The Bergmann kernel and biholomorphic mappings of pseudoconvex domains, *Invent. Math.* **26**, pp. 1–65.

[FEK] Fefferman, C. and Kohn, J.J. H (1988). ölder estimates on domains in two complex dimensions and on three dimensional CR manifolds,*Adv. Math.* **69**, pp. 233–303.

[FIL] Fischer, W. and Lieb, I. (1974). Lokale Kerne und beschränkte Lösungen für den $\bar{\partial}$-Operator auf q-konvexen Gebieten, *Math. Ann.* **208**, pp. 249–265.

[FLE] Fleron, J.F. (1996). Sharp Hölder estimates for $\bar{\partial}$ on ellipsoids and their compliments via order of contact, *Proc. Amer. Math. Soc.* **124**, pp. 3193–3202.

[FOR] Fornaess, J.E. (1976). Embedding strictly pseudoconvex domains in convex domains, *Amer. J. Math.* **98**, pp. 529–569.

[GRL] Grauert, H. and Lieb, I. (1970). Das Ramirezsche integral und die Lösung der Gleichung $\bar{\partial}f = \alpha$ im Bereich der beschränkten Formen, *Proc. Conf. Complex Analysis, Rice University*, **56**, pp. 29–50.

[GRR] Grauert, H. and Remmert, R. (1979). Theory of Stein spaces, Springer-Verlag.

[GUR] Gunning, R. and Rossi, H. (1965). Analytic functions of several complex variables, Prentice Hall, Englewood Cliffs, N.J.

[HAT1] Hatziafratis, T.E. (1986). Integral Representation formulas on analytic varieties, *Pacific J. Math.* **123**, 1, pp. 71–91.

[HAT2] Hatziafratis, T. (1987). An explicit extension formula of bounded holomorphic functions from analytic varieties to strictly convex domains, *J. Func. Anal.* **70**, pp. 289–303.

[HEN1] Henkin, G.M. (1969). Integral representations of functions holomorphic in strictly pseudoconvex domains and some applications, *Math. USSR Sb.* **7**, pp. 597–616.

[HEN2] Henkin, G.M. (1970). Integral representations of functions in strictly pseudoconvex domains and applications to the $\bar{\partial}$-problem, *Math. USSR Sb.* **11**, pp. 273–281.

[HEN3] Henkin, G.M. (1972). Continuation of bounded holomorphic functions from submanifolds in general position in a strictly pseudoconvex domain, *Math. USSR Izv.* **6**, pp. 536–563.

[HER] Henkin, G.M. and Leiterer, J. (1984). Theory of functions on complex manifolds, Birkhäuser.

[HEV] Henkin, G.M. and Romanov, A.V. (1971). Exact Hölder estimates for the solutions of the $\bar{\partial}$-equation, Math. USSR Izvestija, **5**, pp. 1180–1192.

[HO1] Ho, L.H. (1991). $\bar{\partial}$-problem on weakly q-convex domains, *Math. Ann.* **290**, pp. 3–18.

[HO2] Ho, L.H. (1993). Hölder estimates for local solutions for $\bar{\partial}$ on a class of nonpseudoconvex domains, *Rocky Mount. J. Math.* **23**, pp. 593–607.

[HR1] Hörmander, L. (1965). L^2 estimates and existence theorems for the $\bar{\partial}$-operator, *Acta Math.* **113**, pp. 89–152.

[HR2]
Hörmander, L. (1990). An introduction to complex analysis in several variables, Third edition, North Holland.

[JK1] Jakóbczak, P. (1983). On the regularity of extension to strictly pseudoconvex domains of functions holomorphic in a submanifold in general position, *Ann. Polon. Math.* **42**, pp. 115–124.

[JK2] Jakóbczak, P. (1985). Extension and decomposition with local boundary regularity properties in pseudoconvex domains, *Math. Z.* **188**, pp. 513–533.
[JP] Jarnicki, M. and Pflug, R.P. (2000). Extension of holomorphic functions, De Gruyter expositions in Mathematics **34**.
[KER] Kerzman, N. (1971). Hölder and L^p estimates for solutions of $\bar{\partial}u = f$ in strongly pseudoconvex domains, *Comm. Pure Appl. Math.* **24**, pp. 301–379.
[KES] Kerzman, N. and Stein, E.M. (1978).The Szegö kernel in terms of Cauchy-Fantappiè kernels, *Duke Math. J.* **45**, pp. 197–224.
[KON] Kohn, J.J. (1977). Methods of partial differential equations in complex analysis, *Proc. symp. Pure Math.* **30**, pp. 215-237.
[KR1] Krantz, S.G. (1976). Optimal Lipschitz and L^p regularity for the equation $\bar{\partial}u = f$ on strongly pseudo-convex domains, *Math. Ann.* **219**, pp. 233–260.
[KR2] Krantz, S.G. (1982). Function Theory of Several Complex Variables, John Wiley & Sons, New York.
[KR3] Krantz, S.G. (1990). Complex Analysis: The geometric viewpoint, The Carus Mathematical Monographs **23**.
[KRP] Krantz, S.G. and Parks, H. (1999). The geometry of domains in space, Birkhäuser.
[LI1] Lieb, I. (1969). Ein Approximationssatz auf streng pseudokonvex Gebieten, *Math. Ann.* **184**, pp. 56–60.
[LI2] Lieb, I. (1970). Die Cauchy-Riemannschen Differentialgleichungen auf streng pseudokonvexen Gebieten, *Math. Ann.* **190**, pp. 6–44.
[LI3] Lieb, I. (1972). Die Cauchy-Riemannschen Differentialgleichungen auf streng pseudokonvexen Gebieten: Stetige Randwerte, *Math. Ann.* **199**, pp. 241–256.
[LIM] Lieb, I. and Michel, J. (2002). The Cauchy-Riemann Complex, Vieweg.
[LIR] Lieb, I. and Range, R.M. (1980). Ein Lösungsoperator für den Cauchy-Riemann Komplex mit C^k-Abschätzungen, *Math. Ann.* **253**, pp. 145–164.
[MA] Ma, L. (1992). Hölder and L^p estimates for the ∂-equation on non-smooth strictly q-convex domains, *Manuscripta Math.* **74**, pp. 177–193.
[MAZ1] Mazzilli, E. (1995). Extension des fonctions holomorphes, *C. R. Acad. Sci. Paris* **321**, pp. 831–836.
[MAZ2] Mazzilli, E. (1998). Extension des fonctions holomorphes dans les pseudo ellipsoïdes, *Math. Z.* **227**, pp. 607–622.
[MEN] Menini, C. (1997). Estimations pour la resolution du $\bar{\partial}$ sur une intersection d'ouverts strictement pseudoconvexes, *Math. Z.* **225**, pp. 87–93.
[MIC] Michel, J. (1988). Randregularität des $\bar{\partial}$-Problems für stückweise streng pseudokonvexe Gebiete in $\mathbf{C}^n$, *Math. Ann.* **280**, pp. 45–68.
[MIP] Michel, J. and Perotti, A. (1990). C^k-Regularity for the $\bar{\partial}$-equation on strictly pseudoconvex domains with piecewise smooth boundaries, *Math. Z.* **203**, pp. 415–427.
[NOR] Norguet, F. (1954). Sur les domains d'holomorphie des fonctions uniformes de plusieurs variable complexes(passage du local au global), *Bull. Soc. Math. France* **82**, pp. 137–159.
[OHT] Ohsawa, T. and Takegoshi, K. (1987). On the extension of L^2 holomorphic functions, *Math. Z.* **195**, pp. 197–204.

[OKA1] Oka, K. (1942). Sur les fonctions de plusieurs variables, VI. Domaines pseudoconvexes, *Tohoku Math. J.* **49**, pp. 15–52.

[OKA2] Oka, K. (1950). Sur les fonctions de plusieurs variables, VII. Sur quelques notions arithmétiques, *Bull. Soc. Math. France* **78**, pp. 1–27.

[OKA3] Oka, K. (1953). Sur les fonctions de plusieurs variables, Domaines finis sans point critique intéerieur, *Jap. J. Math.* **23**, pp. 97–155.

[OV] Ovrelid, N. (1971). Integral representation formulas and L^p estimates for the $bar\partial$-equation, *Math. Scand.* **29**, pp. 137–160.

[POL] Polking, J. (1991). The Cauchy Riemann equations on convex sets, *Proc. Symp. Pure Math.* **52**, pp. 309–322.

[RAM] Ramirez, E. (1970). Ein Divisionsproblem und Randintegraldarstellungen in der komplexen Analysis, *Math. Ann.* **184**, pp. 172–187.

[RAN1] Range, R.M. (1976). On Hölder estimates for $\bar\partial u = f$ on weakly pseudoconvex domains, *Int. Conf. Cortona*.

[RAN2] Range, R.M. (1986). Holomorphic functions and integral representations in several complex variables, Springer.

[RAN3] Range, R.M. (1990). Integral kernels and Hölder estimates for $\bar\partial$ on pseudo-convex domains of finite type in $\mathbf{C}^2$, *Math. Ann.* **288**, pp. 63–74.

[RAN4] Range, R.M. (1992). On Hölder and BMO estimates for $\bar\partial$ on convex domains in $\mathbf{C}^2$, *J. Geom. Anal.* **2**, pp. 575–583.

[RAS] Range, R.M. and Siu. Y.T. (1973). Uniform estimates for the $\bar\partial$-equation on domains with piecewise smooth strictly pseudoconvex boundaries, *Math. Ann.* **206**, pp. 325–354.

[SCH] Schmalz, G. (1989). Solution of the $\bar\partial$-equation with uniform estimates on strictly q-convex domains with nonsmooth boundary, *Math. Z.* **202**, pp. 409–430.

[SI1] Siu, Y.T. (1974). The $\bar\partial$ problem with uniform bounds on derivatives, *Math. Ann.* **207**, pp. 163–176.

[SI2] Siu, Y.T. (1996). The Fujita conjecture and the extension theorem of Ohsawa-Takegoshi, *Proc. 3rd Int. RIMSJ., Geometric Complex Analysis*, pp. 577–592.

[STE] Stein, E.M. (1972). Boundary behavior of holomorphic functions of several complex variables, Math. Notes ♯11, Princeton Univ. Press.

[STO] Stout, E.L. (1975). An integral formula for holomorphic functions on strictly pseudoconvex hypersurfaces, *Duke Math. J.* **42**, pp. 347–356.

[TSU] Tsuji, M. (1995). Counterexample of an unbounded domain for Ohsawa's problem, *Complex variables* **27**, pp. 335–338.

[VER] Verdera, J. (1984). L^∞-continuity of Henkin operators solving $\bar\partial$ in certain weakly pseudoconvex domains of $\mathbf{C}^2$, *Proc. Royal Soc. Edinburgh* **99**, pp. 25–33.

Index

C^∞ extension, 196
C^k extensions, 189
H^p extension, 188
L^2 estimate for the $\bar{\partial}$ problem, 91
L^p estimates for the $\bar{\partial}$ problem, 263
$PS(\Omega)$-convex, 33
X intersects $\partial\Omega$ transversally, 158
$\mathcal{O}(\Omega)$-convex, 23

admissible kernel, 217
analytic polyhedron, 43
analytic sheaf, 308
analytic subset, 320
Ascoli-Arzela theorem, 331

Baire's theorem, 14
Banach-Steinhaus theorem, 56
Bell's density lemma, 233
Bergman kernel, 205
Bergman metric, 242
Bergman projection, 228
Bergman space, 202
Berndtsson formula, 265
Berndtsson-Andersson formula, 254
Bessel's inequality, 197
Bochner-Martinelli formula, 125
bounded extension, 181

Carathéodory metric, 291
Cartan theorem A, 321
Cartan theorem B, 321
Cauchy inequality, 13

Cauchy-Fantappiè formula, 132
closed operator, 53
coboundary, 317
cochain, 317
cocycle, 317
coherent sheaf, 316
cohomology group, 317
compact operator, 227, 331
complete orthonormal system, 198
condition $(\mathbf{B}_k)$, 237
condition $(\mathbf{R})$, 230
condition $(\mathbf{R}_k)$, 230
conjugate operator, 331
continuous extension, 181
convex domain of finite type, 288
convex hull, 23

Dini's theorem, 36
domain of convergence, 10
domain of holomorphy, 22

E. M. Stein's counterexample, 150

Fefferman's mapping theorem, 241
first Cousin problem, 326
frame, 43
function element, 307

germ, 307
Gram-Schmidt orthonormalization process, 196
Green's theorem, 48

Hölder estimate for the $\bar{\partial}$ problem, 148
Hölder space, 123
Hahn-Banach theorem, 56
Hartogs theorem, 16
Henkin-Ramirez kernel, 148
holomorphic, 11
holomorphically convex, 23
holomorphically convex hull, 23
homotopy formula, 132

implicit function theorem, 302
inverse mapping theorem, 304
invertible operator, 331

kernel, 60
Kobayashi metric, 292
Koppelman formula, 127
Koppelman-Leray formula, 132

Leray formula, 131
Leray map, 129
Levi polynomial, 143
Levi's problem, 95
Lipschitz space, 123
locally finitely generated, 315

maximum principle, 44
Montel's theorem, 112

Narashimhan's lemma, 136
Noetherian ring, 309
nonsingular mapping, 303
null space, 60

Ohsawa-Takegoshi theorem, 114
Oka's counterexample, 330
Oka's fundamental theorem, 311
open mapping theorem, 46
outward normal vector, 48

Parseval's equality, 199
peak function, 190
plurisubharmonic, 19
Poincaré theorem, 297, 329

Poisson integral, 5
pseudoconvex open set, 32
pullback, 118

regular of order k, 298
relatively compact, 11
Riesz representation theorem, 56
Riesz-Fischer theorem, 199

Schwarz lemma, 45
Schwarz-Pick lemma, 45
second Cousin problem, 328
section, 308
separable, 86
sheaf, 308
sheaf homomorphism, 308
sheaf of Abelian groups, 308
sheaf of ideals, 320
sheaf of modules, 308
sheaf of relations, 316
sheaf of rings, 308
simple admissible kernel, 217
Sobolev space, 81
stalk, 308
Stein cover, 322
Stein manifold, 319
strictly plurisubharmonic, 19
strictly pseudoconvex domain, 33
strictly subharmonic, 4
subharmonic, 3
submanifold in general position, 159
subsheaf, 308
surface element, 48
surface measure, 149

uniqueness theorem, 45
upper semicontinuous, 3

weak convergence, 86
Weierstrass division theorem, 300
Weierstrass polynomial, 299
Weierstrass preparation theorem, 299